Multifunctional Hydrogels

Hydrogels are important polymer-based materials with innate fascinating properties and applications: they are three-dimensional, hydrophilic, polymeric networks that can absorb large amounts of water or aqueous fluids and are biocompatible, mechanically flexible, and soft. The incorporation of functionalities to develop smart and bioactive platforms has led to a myriad of applications. This book offers a comprehensive overview of multifunctional hydrogels, covering fundamentals, properties, and advanced applications in a progressive way. While each chapter can be read stand-alone, together they clearly describe the fundamental concepts of design, synthesis, and fabrication, as well as properties and performances of smart multifunctional hydrogels and their advanced applications in the biomedical, environmental, and robotics fields.

This book:

- Introduces readers to different hydrogel materials and the polymer types used to fabricate them.
- Discusses conducting polymer hydrogels, nanocomposite hydrogels, and self-healing hydrogels.
- Covers synthesis methodologies and fabrication techniques commonly used to confer certain structures and/or architectures.
- Shows how hydrogels can be modified to incorporate new functionalities able to respond to physical and/or chemical changes.
- Examines applications including bioelectronics, sensors and biosensors, tissue engineering, drug delivery, antipathogen applications, cancer theranostics, environmental applications, and soft robotics, with chapters showcasing the main advances achieved up to date in every field.

Multifunctional Hydrogels: From Basic Concepts to Advanced Applications serves as a valuable resource for academic and industry researchers from interdisciplinary fields including materials science, chemistry, chemical engineering, bioengineering, physics, and pharmaceutical engineering.

Multifunctional Hydrogels
From Basic Concepts to Advanced Applications

Edited by
José García-Torres
Carlos Alemán
Ram K. Gupta

CRC Press
Taylor & Francis Group
Boca Raton London New York

CRC Press is an imprint of the
Taylor & Francis Group, an **Informa** business

Designed cover image: ©Shutterstock Images

First edition published 2024
by CRC Press
2385 NW Executive Center Drive, Suite 320, Boca Raton FL 33431

and by CRC Press
4 Park Square, Milton Park, Abingdon, Oxon, OX14 4RN

CRC Press is an imprint of Taylor & Francis Group, LLC

Library of Congress Cataloging-in-Publication Data
Names: García-Torres, José, editor. | Alemán, Carlos, Dr., editor. | Gupta, Ram K., editor.
Title: Multifunctional hydrogels : from basic concepts to advanced applications / edited by José García-Torres, Carlos Alemán, and Ram K. Gupta.
Description: First edition. | Boca Raton : CRC Press, 2024. | Includes bibliographical references and index. |
Identifiers: LCCN 2023049358 (print) | LCCN 2023049359 (ebook) | ISBN 9781032373409 (hbk) | ISBN 9781032375038 (pbk) | ISBN 9781003340485 (ebk)
Subjects: LCSH: Colloids. | Colloids--Analysis. | Colloids--Biotechnology.
Classification: LCC QD549 .M75 2024 (print) | LCC QD549 (ebook) | DDC 660/.294513--dc23/eng/20240129
LC record available at https://lccn.loc.gov/2023049358
LC ebook record available at https://lccn.loc.gov/2023049359

ISBN: 978-1-032-37340-9 (hbk)
ISBN: 978-1-032-37503-8 (pbk)
ISBN: 978-1-003-34048-5 (ebk)

DOI: 10.1201/9781003340485

Typeset in Times
by MPS Limited, Dehradun

Contents

Preface

Hydrogels, a class of soft, water-swollen materials, have been at the forefront of materials science research for decades. Their unique blend of biocompatibility, tunability, and versatility has made them indispensable in a wide range of applications, from biomedicine to environmental remediation. This book, *Multifunctional Hydrogels: From Basic Concepts to Advanced Applications*, represents a comprehensive exploration of the multifaceted world of hydrogels, aiming to provide a definitive resource for researchers, students, and professionals interested in the field.

Hydrogels represent a remarkable type of materials within materials science as they are three-dimensional polymeric networks capable of holding substantial amounts of water while maintaining their structural integrity. Their properties, such as high water content, softness, and resemblance to natural tissues, render them uniquely suited for various applications. This book embarks on a journey through the captivating realm of hydrogels, offering insights into their fundamental principles, synthesis methods, properties, and diverse applications. Our exploration of hydrogels commences with a solid foundation in the fundamental concepts. We delve into the chemistry behind hydrogels, examining the intricacies of polymer structure, followed by their properties (e.g., mechanical, electrical, magnetic), and finally some advanced applications. Thus, readers will gain a deep understanding of main characteristics and how to design hydrogels with tailored properties for specific applications.

The book is composed of 19 chapters in three parts. Part I covers basic concepts of hydrogels, including an introduction to the main characteristics (Chap. 1), the different natural and synthetic hydrogels employed (Chap. 2), and how they can be modified to prepare nanocomposite hydrogels (Chap. 3). Then, it is discussed how they can be synthesized (Chap. 4) and the main fabrication techniques employed (Chap. 5). In Part II, the properties of hydrogels are discussed, covering hydrogels with advanced properties (Chap. 6) as well as chemical (Chap. 7), electrical (Chap. 8), magnetic (Chap. 9), thermal (Chap. 10), and mechanical properties (Chap. 11). Finally, Part III is focused on advanced applications of hydrogels like bioelectronics (Chap. 12), physical and chemical (bio)sensors (Chap. 13), biomedical applications (Chap. 14), drug delivery (Chap. 15), antipathogen applications (Chap. 16), environmental applications (Chap. 17), water cleaning and recovery (Chap. 18), and soft robotics (Chap. 19).

In editing and organizing this book, *Multifunctional Hydrogels: From Basic Concepts to Advanced Applications*, we have made our best efforts to cover the growing field of hydrogels and related technologies. This book will provide basic concepts to more deep insights into the state of the art for new researchers in the field but also for researchers already working in the passionate world of hydrogels. We express our heartfelt gratitude to the numerous scientists and experts who have generously shared their knowledge, insights, and experiences in this book. Their commitment to advancing hydrogel research is evident in the depth and breadth of the content presented here. We would also like to thank Allison Atkins and the rest of the team at CRC/Taylor & Francis for their invaluable help in the publication process.

In conclusion, *Multifunctional Hydrogels: From Basic Concepts to Advanced Applications* is a testament to the extraordinary journey of hydrogels from basic materials to multifunctional wonders with boundless potential. We hope this book serves as both an informative and inspirational resource, fostering further exploration and innovation in the ever-evolving field of hydrogel science and technology.

José García-Torres
Carlos Alemán
Ram K. Gupta

Editor Biographies

Dr. José García-Torres

Dr. José García-Torres is Serra Húnter Associate Professor in the Department of Materials Science and Engineering at Universitat Politècnica de Catalunya (UPC). Dr. García graduated in chemistry (University of Barcelona (UB)) and engineering of materials (UPC) and received a PhD in chemistry at UB in 2010. He has been a postdoctoral researcher at different internationally renowned universities, research institutes, and companies (University of Surrey, Research Institute for Solid State Physics and Optics, Miquel i Costas&Miquel group). His research interests are focused on the development of functional hybrid materials and (bio)inks using different microfabrication techniques (3D printing, inkjet printing) for organic electronics ((bio)sensors, actuators) and tissue engineering (scaffolds, drug delivery). Dr. García is an author of more than 45 peer-reviewed publications in high-impact journals (h-index: 17 (ISI Web of Knowledge)), 5 book chapters, 4 conference proceedings, and he has attended more than 40 conferences and workshops (4 invited conferences). Moreover, Dr. García has participated in more than 24 national and international projects. He has been PI in one European project, three national projects, and two technology transfer projects. His research activity has been recognized along his career: Marie Curie fellowships for European course attendance (2008, 2009), PhD extraordinary mention award (2011), postdoctoral grant from the EPSRC (UK, 2014–2015), Tecniospring-Marie Curie grant, and other conference awards.

Prof. Carlos Alemán

Prof. Carlos Alemán graduated in chemistry from the University of Barcelona (Spain). He received his PhD from Universitat Politècnica de Catalunya (UPC) in 1994, where he was promoted to the position of full professor of physical chemistry. He was postdoctoral researcher at the ETH in Zürich (Switzerland) and visiting professor at the Università di Napoli Federico II (Italy; 6 months), University of Twente (Holland; 1 year), and Universidade Federal do Rio Grande do Sul (Brazil; 3 months). Since 2003, he has been the leader of the IMEM group in the Chemical Engineering Department. His main research interests focus on conducting polymers and biopolymers with biomedical and technological applications. Prof. Alemán has published a book, around 600 scientific articles (around 260 in journals edited by the American Chemical Society and the Royal Chemical Society), and several book chapters. In addition, he is the author of several patents and has acted as organizer of several international congresses and as an editor of books published by a major publishing house. The public funds obtained by Prof. Alemán for his research have been provided by local (Generalitat de Catalunya and Acció), national (MINECO), and international (EU and NIH) bodies, as well as by private companies. h-Index: 48 (ISI Web of Knowledge), 54 (Google Scholar).

Dr. Ram K. Gupta

Dr. Ram Gupta is an associate vice president for research and support and a professor of chemistry at Pittsburg State University. Gupta has been recently named by Stanford University as being among the top 2% of research scientists worldwide. Before joining Pittsburg State University, he worked as an assistant research professor at Missouri State University, Springfield, MO, and then as a senior research scientist at North Carolina A&T State University, Greensboro, NC. Dr. Gupta's research spans a range of subjects critical to current and future societal needs, including semiconducting materials and devices, biopolymers, flame-retardant polymers, green energy production and storage using nanostructured materials and conducting polymers, electrocatalysts, optoelectronics, and photovoltaic devices, organic-inorganic heterojunctions for sensors, nanomagnetism, biocompatible nanofibers for tissue regeneration, scaffold and antibacterial applications, and biodegradable metallic implants. Dr. Gupta has mentored 10 PhD/postdoc scholars, 76 MS students, and 58 undergraduate/high school students. Dr. Gupta has published over 280 peer-reviewed journal articles (9,500+ citations, 55 h-index, 205 i10-index), made over 400 national/international/regional presentations, chaired/organized many sessions at national/international meetings, wrote several book chapters (90+), worked as editor for many books (40+) for American Chemical Society, CRC Press, etc., and has received several millions of dollars for research and educational activities from external agencies. He also serves as an editor, associate editor, guest editor, and editorial board member for various journals.

Contributors

Carlos Alemán
Universitat Politècnica de
  Catalunya-Barcelona Tech
Barcelona, Spain

Nazmi B. Alsaafeen
Khalifa University of Science and
  Technology
Abu Dhabi, UAE

Elaine Armelin
Universitat Politècnica de
  Catalunya-Barcelona Tech
Barcelona, Spain

Cesar A. Barbero
National University of Rio Cuarto
Rio Cuarto, Argentina

Leticia Buendía-González
Universidad Autónoma del Estado de
  México
Toluca, México

Moises Bustamante-Torres
Universidad de Buenos Aires
Buenos Aires, Argentina

Çiğdem Bilici
Istinye University
Istanbul, Turkey

Ashok Bora
Tezpur University
Tezpur, India

Pablo Bosch
National University of Rio Cuarto
Rio Cuarto, Argentina

Brianna
Sunway University
Selangor Darul Ehsan, Malaysia

Martin F. Broglia
National University of Rio Cuarto
Rio Cuarto, Argentina

Emilio Bucio
Instituto de Ciencias Nucleares, UNAM
Ciudad de México, México

Virginia Capella
National University of Rio Cuarto
Rio Cuarto, Argentina

Yadira D. Cerda-Sumbarda
Tecnológico Nacional de México/
  Instituto Tecnológico de Tijuana
Tijuana, México

Xiaobao Chen
Scindy Pharmaceutical Co. Ltd.
Suzhou, China

Estefani Chichande-Proaño
Universidad Central del Ecuador
Quito, Ecuador

Nuraina Anisa Dahlan
Universiti Malaya
Kuala Lumpur, Malaysia

Felipe M. de Souza
Pittsburg State University
Pittsburg, Kansas, USA

Lorena Duarte-Peña
Instituto de Ciencias Nucleares, UNAM
Ciudad de México, México

Y. Aylin Esquivel-Lozano
Instituto de Ciencias Nucleares, UNAM
Ciudad de México, México

Bushara Fatma
Khalifa University of Science and
 Technology
Abu Dhabi, UAE

Guadalupe Gabriel Flores-Rojas
Instituto de Ciencias Nucleares, UNAM
Ciudad de México, México
and
Universidad de Guadalajara
Jalisco, México
and
Instituto de Investigaciones en
 Materiales, UNAM
Toluca, México

José García-Torres
Universitat Politècnica de
 Catalunya-Barcelona Tech
Barcelona, Spain

Lorena Garcia-Uriostegui
Universidad de Guadalajara
Zapopan, México

Garima
CSIR-Central Scientific Instruments
 Organization
Chandigarh, India
and
Academy of Scientific and Innovative
 Research (AcSIR)
Ghaziabad, India

Mirian A. González-Ayón
Tecnológico Nacional de México/
 Instituto Tecnológico de Tijuana
Tijuana, México

Ram K. Gupta
Pittsburg State University
Pittsburg, Kansas, USA

Nicholas G. Hallfors
Khalifa University of Science and
 Technology
Abu Dhabi, UAE

Martin Himly
Paris Lodron University of Salzburg
Salzburg, Austria

Md. Milon Hossain
Cornell University
Ithaca, New York, USA

Fatimah Ibrahim
Universiti Malaya
Kuala Lumpur, Malaysia

Amin Janghorbani
Semnan University
Semnan, Iran

Niranjan Karak
Tezpur University
Tezpur, India

Arezoo Khosravi
Istanbul Okan University
Istanbul, Turkey

Sonia Lanzalaco
Universitat Politècnica de
 Catalunya-Barcelona Tech
Barcelona, Spain

Su Li
Shenzhen Institute of Advanced
 Technology
Shenzhen, China
and
Paris Lodron University of Salzburg
Salzburg, Austria

Yang Li
Shenzhen Institute of Advanced
 Technology
Shenzhen, China

A. C. Liaudat
National University of Rio Cuarto
Rio Cuarto, Argentina

Angel Licea-Claverie
Tecnológico Nacional de México/
   Instituto Tecnológico de Tijuana
Tijuana, México

Yingnan Liu
Shenzhen Institute of Advanced
   Technology
Shenzhen, China
and
Paris Lodron University of Salzburg
Salzburg, Austria

Felipe López-Saucedo
Instituto de Ciencias Nucleares, UNAM
Ciudad de México, México
and
Universidad Autónoma del Estado de
   México
Toluca, México

Ishita Matai
Amity University Punjab
Mohali, India

Eduardo Mendizábal
Universidad de Guadalajara
Jalisco, México

Meng Meng
Nankai University
Tianjin, China

Mojdeh Mirshafiei
University of Tehran
Tehran, Iran

Maria A. Molina
National University of Rio Cuarto
Rio Cuarto, Argentina

Zinnat Morsada
Clarkson University
Potsdam, New York, USA

Ebrahim Mostafavi
Stanford University
Stanford, California, USA

David Naranjo
Universitat Politècnica de
   Catalunya-Barcelona Tech
Barcelona, Spain

Vijay Kumar Pal
Institute of Nano Science and
   Technology
Punjab, India

Sofia Paulo-Mirasol
Universitat Politècnica de
   Catalunya-Barcelona Tech
Barcelona, Spain

Charalampos Pitsalidis
Khalifa University of Science and
   Technology
Abu Dhabi, UAE

Benjamin Punz
Paris Lodron University of Salzburg
Salzburg, Austria

David Romero-Fierro
Instituto de Ciencias Nucleares, UNAM
Ciudad de México, México

Claudia R. Rivarola
National University of Rio Cuarto
Rio Cuarto, Argentina

Nancy Rodríguez
National University of Rio Cuarto
Rio Cuarto, Argentina

Sangita Roy
Institute of Nano Science and Technology
Mohali, India

Abhay Sachdev
CSIR-Central Scientific Instruments
    Organization
Chandigarh, India
and
Academy of Scientific and Innovative
    Research (AcSIR)
Ghaziabad, India

Dimpee Sarmah
Tezpur University
Tezpur, India

Saleheh Shahmoradi
University of Tehran
Tehran, Iran

Serap Sezen
Sabanci University
Istanbul, Turkey

Sin-Yeang Teow
Wenzhou-Kean University
Wenzhou, China

Juan Torras
Universitat Politècnica de
    Catalunya-Barcelona Tech
Barcelona, Spain

Ricardo Vera-Graziano
Instituto de Investigaciones en
    Materiales, UNAM
Ciudad de México, México

Mansi Vij
CSIR-Central Scientific Instruments
    Organization
Chandigarh, India

Shuo Wang
Nankai University
Tianjin, China

Wei Wang
University of Bergen
Bergen, Norway

Heidi Yánez-Vega
Universidad de las Fuerzas Armadas
Sangolquí, Ecuador

Karen Yánez-Vega
Universidad de las Fuerzas Armadas
Sangolquí, Ecuador

Fatemeh Yazdian
University of Tehran
Tehran, Iran

Akhiri Zannat
University of South Asia
Dhaka, Bangladesh

Iman Zare
Sina Medical Biochemistry
    Technologies Co. Ltd.
Shiraz, Iran

Atefeh Zarepour
Saveetha University
Chennai, India

Ali Zarrabi
Istinye University
Istanbul, Turkey

Guofang Zhang
Shenzhen Institute of Advanced
    Technology
Shenzhen, China

Arturo Zizumbo-López
Tecnológico Nacional de México/
    Instituto Tecnológico de Tijuana
Tijuana, México

1 Multifunctional Hydrogels
An Introduction

Felipe M. de Souza and Ram K. Gupta
Pittsburg State University, Pittsburg, Kansas, USA

1.1 INTRODUCTION

Hydrogels consist of materials with a highly porous 3D-networked structure. Such types of arrangement provide these materials an inherently highly active area as well as interstices that allow the adsorption of components such as water, solvents, electrolytes, biological probes, etc. The capability of properly adsorbing certain species in their structure allows hydrogels to be applicable in many sectors, such as energy storage devices like fuel cells, supercapacitors and batteries, sensors, and biomedical components. This broad range of applications has justified the current investment in these materials. Hydrogels have great potential as components for energy storage devices due to their porous semi-solid phase that eases the permeation of electrolytes, which is a core aspect of the improvement in electrochemical properties. Along with that, hydrogels are inherently flexible materials and allow the flow of ionic species through their networked structure. The combination of high surface area and ionic conductivity along with mechanical stability gives an edge on the use of hydrogels compared to traditional electrode materials, which are usually brittle, may display a relatively lower conductivity, and require specific synthetical approaches to attain high surface area. Furthermore, hydrogels present a distinct swelling behavior, as they can adsorb a relatively large quantity of components such as solvents, ions, and biological probes, among other species while maintaining their networked structure despite the variance in volume. Such a factor is often attributed to the presence of hydrophilic groups, such as $-SO_3H$, $-CONH_2$, $-CONH-$, along with others that can adsorb polar solvents and water or retain ionic species. The presence of hydrophilic groups along with the network structure of hydrogels can lead to an intake of water or other polar solvents of more than 20% of its dry weight [1].

Aside from the high adsorption of solvents, another important factor is the reversible stretchability of hydrogels, which allows for cycles of adsorption and desorption of solvents and electrolytes, without noticeable deterioration of mechanical properties. The importance of this factor can be partially associated with electrochemical stability. In the case of supercapacitors, the charging and discharging cycles can eventually cause the electrode material to deteriorate due to the constant insertion and desertion of ions

DOI: 10.1201/9781003340485-1

within the network structure. In the case of batteries, a similar charging and discharging process can also lead to the decomposition of the detachment of electroactive materials on the electrode's surface due to phase transformation of the chemical incorporation of ionic species, which applies a considerably high electrochemical strain on the electrode's components. Hence, the fact that hydrogels present a considerable elasticity and can withstand relatively aggressive mechanical deformations without compromising their structure is a highly desired factor for applications in electrochemical devices. Based on these, the type of network and structure of the hydrogel can greatly influence its water and electrolyte uptake, which can be directly related to its electrochemical performance. Figure 1.1 illustrates some of the main types of hydrogel structures that can be obtained based on their cross-linking.

The versatility of hydrogels extends not only to their field of applications but also to their classification, which can be based on different aspects such as the source from which the starting material or polymer is derived, type of cross-linking, type of response when stimulated, and the electrostatic charge on the structure, which are schematized in Figure 1.2. The hydrogels can be synthetic, natural, or a mix of both. Like polymers, hydrogels can be based on a single polymeric structure, a copolymer, a block copolymer, and so on as most of the materials used for the synthesis of hydrogels are polymers. Some examples of the synthetical ones are poly(hydroxyethyl methacrylate) (PHEMA), poly(hydroxypropyl methacrylate) (PHPMA), and poly(glyceryl methacrylate) (PGMA), among many others. Following that, some of the bio-based materials used for the synthesis of hydrogels are polylactic acid (PLA), and polysaccharides that include cellulose, chitosan, chitin, starch, dextran, pectin, alginates, pullulan, and carrageenan, for instance [2]. Cross-linking can take place through several methods, such as free radical polymerization, UV or γ radiation, interpenetrating networks, and solution casting, among others, which can result in either physical cross-linking, which consists of intermolecular interactions such as hydrogen bonding, van der Waals interactions, or physical entanglement between the polymeric chains. On the other hand, chemical cross-linking takes place through the covalent bonding between reactive pending groups in the polymeric chain, such as -COOH, -NH$_2$, and -OH. Lastly, the dominant charge on the hydrogel's network is also an important factor that can introduce valuable properties such as ionic conductibility, allowing the hydrogel to be used as a solid-state electrolyte [1,3]. Also, the charges on the hydrogel's structure can present antibacterial properties as well as prevent the adhesion of proteins that can lead to infections [4,5]. Thus, throughout these discussions, this chapter provides some of the main concepts related to hydrogels and their applications in the fields of energy production and storage, sensors, and biomedical, along with some future remarks.

1.2 CHEMISTRY AND PROPERTIES OF HYDROGELS

The chemistry and properties of hydrogels are related to their structure and starting materials utilized in their synthesis. One of these factors can be attributed to the presence of hydrophilic groups which, along with the porous and opened networked structure of hydrogels, allow higher water adsorption. Based on that, the mechanical properties of hydrogels drastically change when they achieve their fully swollen

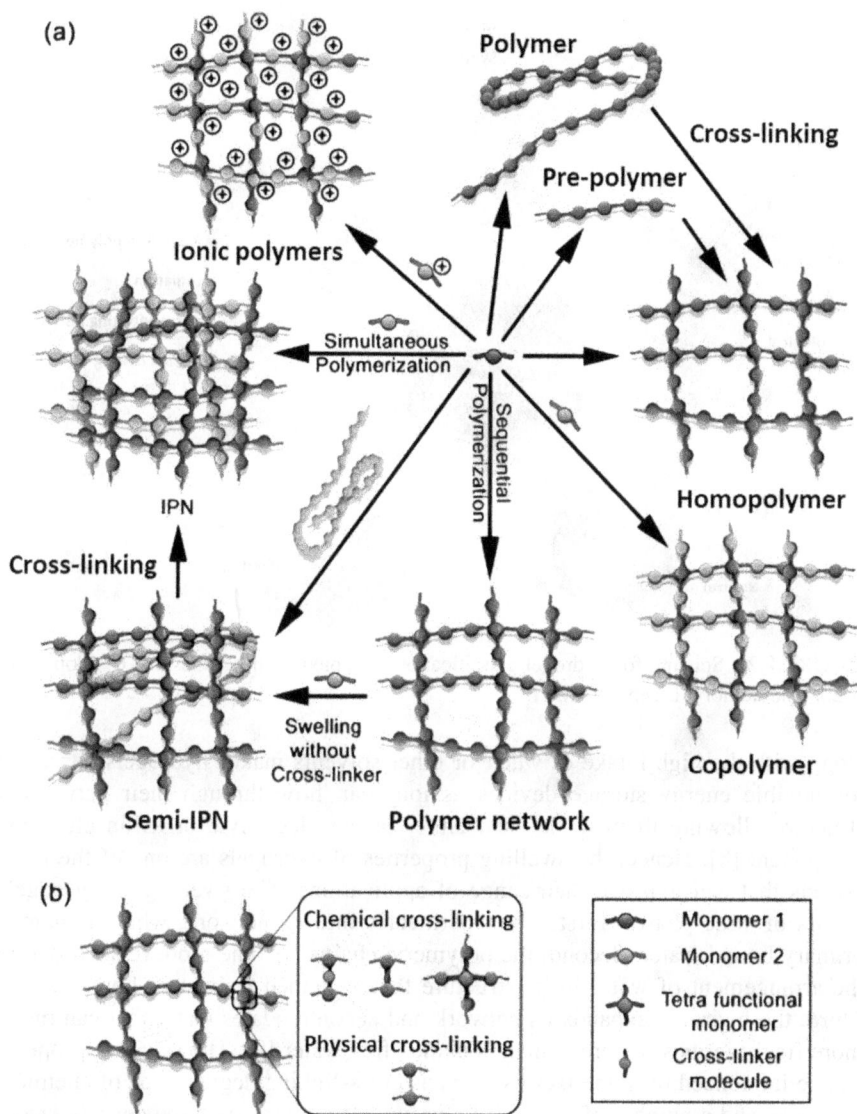

FIGURE 1.1 (a) Schematics for the different types of hydrogel structures. (b) Scheme displaying the chemical and physical cross-linking. Reproduced with permission [1]. Copyright 2020, John Wiley & Sons.

state, as they can have a rubbery, soft, or viscoelastic behavior. Such mechanical behavior is highly desired for biomedical for instance, as their moldability and flexibility allow them to better accommodate the biological system [7]. The hydrophilic nature of hydrogels may increase their interactions with body fluids which may diminish the chances of an allergic reaction and therefore better biocompatibility. For similar reasons, the broad range of mechanical properties

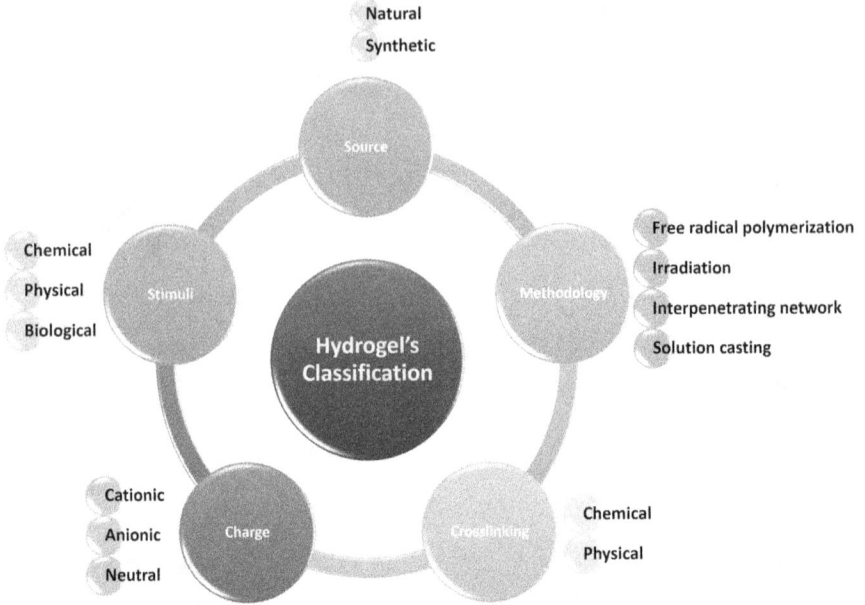

FIGURE 1.2 Scheme for hydrogel classification. Adapted with permission [6]. Copyright 2020, The authors, Licensee MDPI.

along with the high intake of water or other solvents makes hydrogels attractive for flexible energy storage devices as ions can flow through their networked structure allowing them to be used either in the electrolyte or as an electrode component [8]. Hence, the swelling properties of hydrogels are one of the main aspects that can improve their range of applications. The swelling of hydrogels occurs in three phases. First, water permeates into its network, which is named primary bound water. Second, the polymeric chains become more relaxed due to the arrangement of water in its structure that is named secondary bound water. Third, the hydrogel expands its network and accommodates water that can move more freely in its structure, which is named free water [9]. The swelling property can be influenced by some factors, such as cross-linking degree, type of chemical structure, and a number of hydrophilic groups. If a hydrogel is too densely cross-linked, then it may present a lower swelling ratio. Also, a higher number of hydrophilic groups, along with a chemical structure that can properly expose them, promotes a higher swelling ratio in aqueous media. Such an effect can occur because of their response to pH, as the presence of H^+ can ionize the hydrophilic groups. Through that, charges can be formed within the hydrogel's network, which leads to an increase in volume to accommodate the ions as well as repulsive forces due to charges with the same polarity that cause the hydrogel to expand. This phenomenon causes the hydrogel to change from a glassy, or dry state, to a swollen or rubbery state.

Another important property of hydrogels is their response to certain changes in the environment, which can be due to physical, chemical, or biological stimuli.

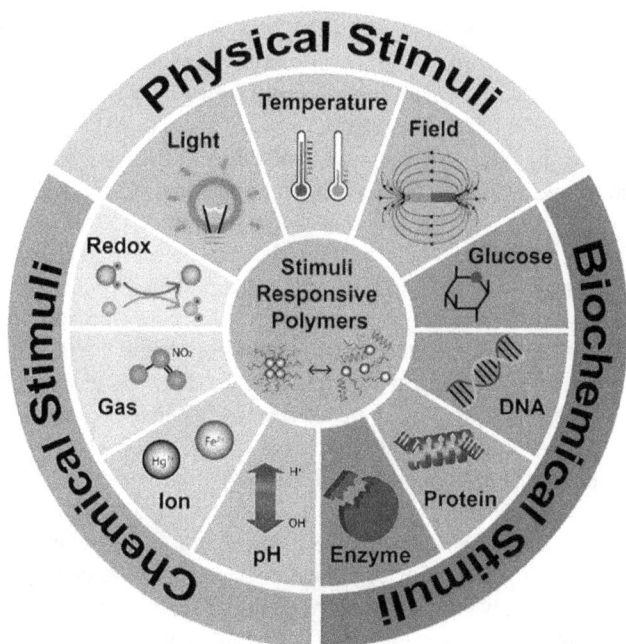

FIGURE 1.3 Different external environments that can promote a response in hydrogels. Reproduced with permission [10]. Copyright 2021, American Chemical Society.

Physical stimuli are a response that occurs due to an applied mechanical force, light, or temperature, for instance. Chemical stimuli can be observed when there is a change in the concentration of chemical species in the media, such as ions or hydronium, which are related to the ionic strength and pH, respectively. Some of the stimuli responses of hydrogels are illustrated in Figure 1.3.

One example of thermo- and pH-sensitive properties of such materials was presented by Zhou et al. [11], who synthesized a chitosan/β-glycerophosphate hydrogel. In their study, it was observed that chitosan solubilizes in acid pH due to the protonation of its amine groups, as this condition can be maintained up to a pH of 6.2. Upon pH values higher than 6.2, a hydrated gel precipitate was formed. Following that, the presence of β-glycerophosphate promoted some important effects, such as adjusting the pH to around 7.0 to 7.4, which is the physiological working range. Also, its solubilization prevents an immediate gelation process which allowed for a tailored gel formation when there was an increase in temperature. Based on these results, the hydrogel could have a transition to a sol state that was influenced by pH, whereas the transition to gel was influenced by temperature.

Hydrogels also display another important set of properties, such as high flexibility, networked, and porous structure, along with self-healing capabilities. By presenting these inherent properties, hydrogels can be used as substrates for the growth of other materials with appreciable electrochemical properties, such as high conductivity and electroactive sites that can be further exposed as it grows on the hydrogel. This concept was explored by Zou et al. [12], who fabricated a polyvinyl alcohol (PVA)-based hydrogel supercapacitor. For the electrode, the PVA hydrogel was sandwiched with polyaniline (PANI) through an *in-situ* polymerization process. Alongside that, 3-aminobenzoic acid (Maba) was co-polymerized with aniline because the presence of carboxy groups from Maba can be ionized by Ca^{2+} cations that were present on the electrolyte. Following that, for the quasi-solid electrolyte, a PVA hydrogel was physically cross-linked with 4-carboxyphenylboronic acid (CPBA) along with complexation with Ca^{2+}. By utilizing the same hydrogel matrix both as an electrode and electrolyte, the supercapacitor could be more flexible, present self-healing properties, and have faster ionic conduction [13,14]. Another important aspect is that utilizing the same material to make the supercapacitor could promote more intimate contact between the electrode/electrolyte interface, reducing some impedance that could arise in the system otherwise. The synthetical diagram for the all-hydrogel supercapacitor is presented in Figure 1.4a. Based on these aspects, the all-hydrogel supercapacitor was assembled by sandwiching two electrodes based on PVA-CPBA-Ca-Maba and PVA-CPBA-Ca in between. Furthermore, the flexible hydrogel-based supercapacitor also presented self-healing properties that were attributed to its network structure that was physically cross-linked, which enabled the reconnection of borate ester bonds along with ionic bonds from the Ca^{2+} species by heating the components at 70°C for 4 h. It was proposed that the relatively high temperature allows the PANI chains to move along the PVA and reestablish the severed sessions. The self-healing process of the PVA-CPBA-Ca-p(Maba-co-aniline) is presented in Figure 1.4b.

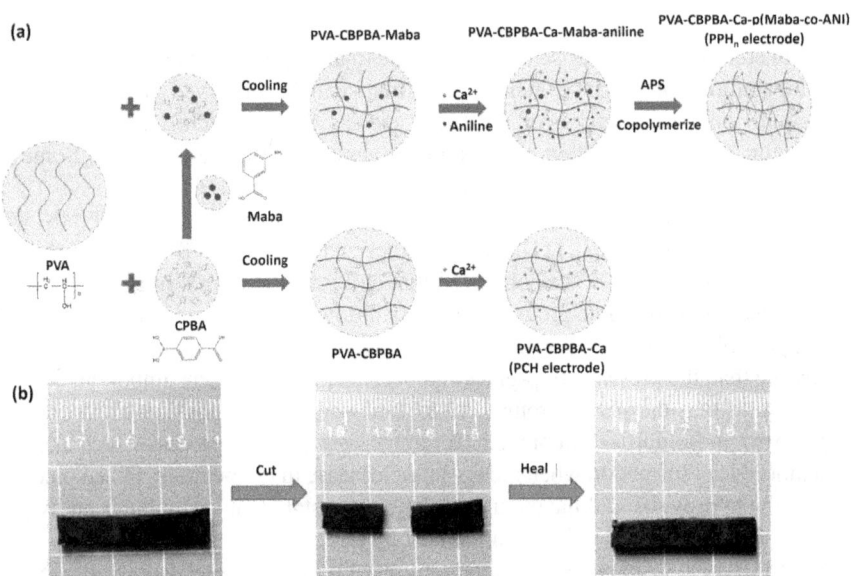

FIGURE 1.4 (a) Synthetic scheme for the fabrication of an all-hydrogel based on PVA-CPBA-Ca-p(Maba-co-aniline) as electrode and PVA-CPBA-Ca as electrolyte. (b) Photocopies displaying the self-healing property of PVA-CPBA-Ca-p(Maba-co-aniline) after being exposed to 70°C for 4 h. Reproduced with permission [12]. Copyright 2021, Elsevier.

1.3 APPLICATIONS OF HYDROGELS

The applications of hydrogels extend to several areas that can go from certain types of foods, such as gelatin, to soft contact lenses and diapers, among others. Yet, these materials are also being applied to energy production, such as fuel cells, and electrocatalysts as well as energy storage such as supercapacitors and batteries, which can be initially attributed to the highly porous network structure that allows it to be embedded with electroactive materials to be used as an electrode or it promotes the flow of ions within its network, which can ensure relatively high conductivity, hence functioning as an electrolyte. Other applications include wearable devices due to their inherent flexibility, sensors due to their response to external stimuli, and water purification systems due to their porous networked structure and swelling behavior. A general schematic displaying the applications of hydrogels is presented in Figure 1.5.

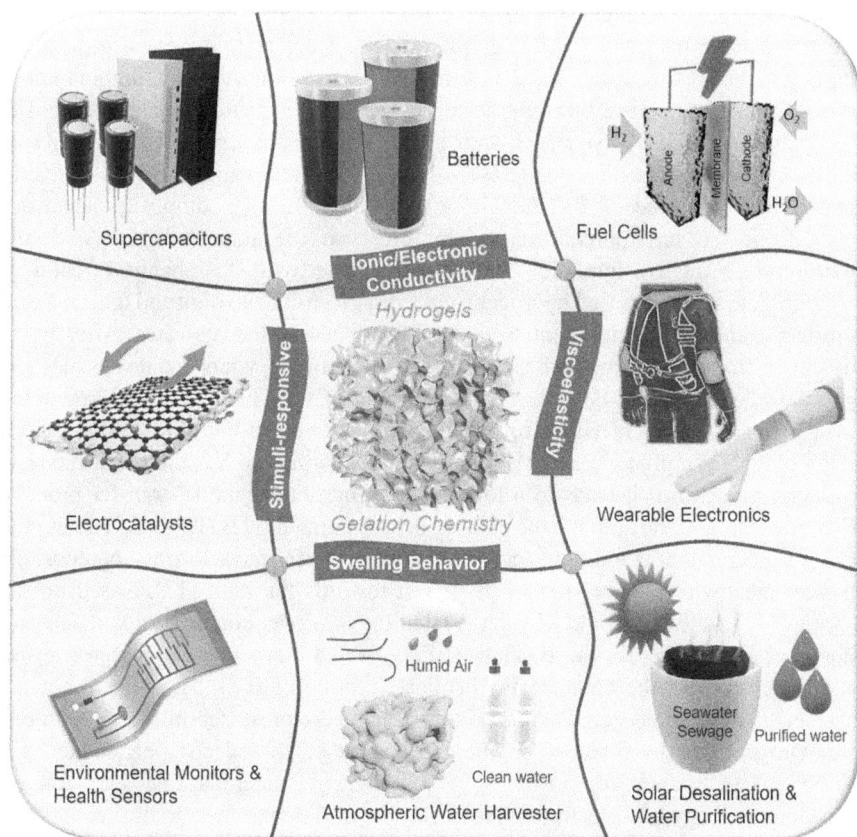

FIGURE 1.5 Schematic displaying the several applications of hydrogels based on their inherent properties, such as conductivity (electronic and ionic), environmental stimuli, flexibility, and swelling behavior. Reproduced with permission [15]. Copyright 2020, American Chemical Society.

1.3.1 HYDROGELS FOR ENERGY PRODUCTION

Energy production is one of the most regarded topics due to its pivotal importance for the working of society as it is a resource so intrinsic that became nearly impossible to leave without it. Solar energy is an attractive aspect as it is a highly sustainable, renewable, and zero-carbon emission process. Based on that, Yang *et al.* [16] proposed the fabrication of a hydrogel composite by using $BaTiO_3@BiVO_4$, a ferroelectric-semiconductor that can improve the charge separation as well as the transfer process during the oxidation of H_2O. This process can be attributed to some of the properties of $BaTiO_3$ that can induce the formation of an electric field at the end of the positive polarization. Through that, the photovoltage and the acceleration of holes (h^+) can be increased, facilitating its transfer to the surface of $BiVO_4$. The hydrogel serves as a substrate that can adsorb the moisture from the atmosphere's humidity, which allows the system to be constantly supplied to carry on the H_2O oxidation process. Highly hygroscopic hydrogels, one based on Zn and the other Co, played the role of carrying H_2O to the photoanode, which could enhance the mobility of carriers. The photoanode-hydrogel system, along with the solar cells in series, provided a photocurrent of 0.4 mA/cm^2 with a 12% humidity reduction and at an illumination of 10 mW/cm^2. The application of the ferroelectric-semiconductor hydrogel composite is displayed in Figure 1.6a, which shows a painting board covered with $BaTiO_3@BiVO_4$ over a fluorine tin oxide (FTO) substrate. Then, Co and Zn hydrogels were applied over the surface of $BaTiO_3@BiVO_4$. Through that, the composite can adsorb water from the environment under the presence of ambient light. Also, the environment's relative humidity (RH) was associated with the change in color of the hydrogel as the Co-based hydrogel went from green-blue to pink and the Zn-based hydrogel went from transparent to opaque upon adsorption of water. After water adsorption from the hydrogel, the $BiVO_4$ absorbs a photon, which excites its electron from the valence band (VB) to the conducting band (CB), which leads to a drop of the quasi-Fermi energy of h^+ ($E_{f,p}$) from the electrons (e^-) quasi-Fermi level ($E_{f,n}$). Such an effect creates a photovoltage (V_{ph}) at which the larger the V_{ph} the better the H_2O oxidation process, as it leads to a lower energy barrier for the h^+ transfer process. Also, according to the authors, the electric field generated by $BaTiO_3$ (E_{in}) due to the positive polarization could have facilitated the h^+ transfer process to the composite's surface; meanwhile, the e^- would be driven towards the bulk [17]. Based on the mechanistic process, it could be inferred that the hydrogel could provide a constant supply of H_2O, whereas the $BaTiO_3@BiVO_4$ could carry out the charge transfer process to promote the photoelectrochemical splitting of H_2O. The scheme for the mechanism and the energy diagram of this process are presented in Figure 1.6b-c.

Hydrogels can be included in the fabrication of triboelectric nanogenerators (TENGs), which are composed of a junction of materials that can convert mechanical energy into electricity. However, the initial hindrance to this technology is that traditional supercapacitors and batteries tend to be rigid. Yet, the introduction of hydrogels can lead to advantageous processes as they can be highly flexible and stretchable, conducting, and self-healing. Through that, hydrogels may provide an alternative for the development of triboelectric devices [18]. Despite that, other components are required to further optimize their properties to make them suitable

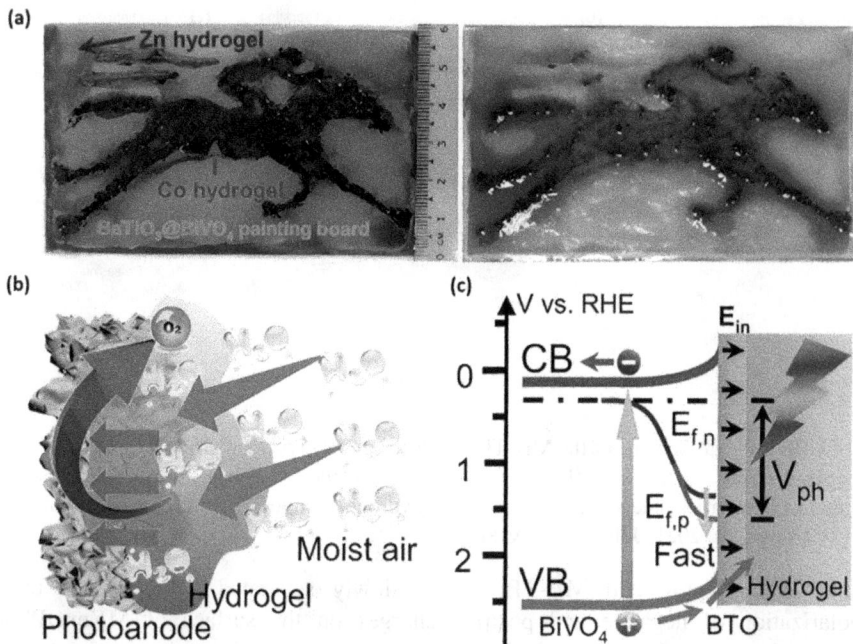

FIGURE 1.6 (a) Photo of the ferroelectric-semiconductor hydrogel composite at which $BaTiO_3@BiVO_4$ was applied over the FTO substrate, whereas the Co and Zn hydrogels were placed. (b) Mechanism of H_2O adsorption by the hydrogel as the H_2O is transferred to the photoanode to undergo the splitting process, realizing O_2. (c) The energy diagram displays the enhancement of the $e^- - h^+$ charge separation and transfer. The occurrence of the photochemical process was attributed to an electric field generated by $BaTiO_3$, along with the upward bending of $BiVO_4$ to the generation and transfer of h^+ to the composite's surface where the H_2O oxidation process occurs. Reproduced with permission [16]. Copyright 2019, Elsevier.

for flexible energy generation devices. Based on these aspects, MXenes can serve as an addition to the hydrogels, as these 2D transition metal carbide-based nanomaterials present relatively higher conductivity, high surface area, hydrophilic nature, and satisfactory mechanical properties. Luo *et al.* [19] fabricated a composite composed of PVA hydrogel and MX nanosheets (MH-TENG). Furthermore, a Si rubber (Ecoflex) was used as a triboelectric layer to avoid the loss of water from the hydrogel. The tests revealed that the MH-TENG could harvest vibrational energy, showing potential uses as a wearable energy harvesting device as well as a wearable sensor (Figure 1.7a). For that, the MXene/PVA hydrogel functioned as the electrode. Also, in the Si rubber, ion transport could occur for electrostatic screening of triboelectric charges. This process occurred in several steps, as shown in Figure 1.7b. Step i: first, a Kapton film is placed in contact with the Ecoflex, which is sandwiched with the MH-TENG through mechanical force. This induces the formation of the same amount of positive and negative charges on the surface of Ecoflex and Kapton, respectively, which is known as the triboelectric effect. Step ii:

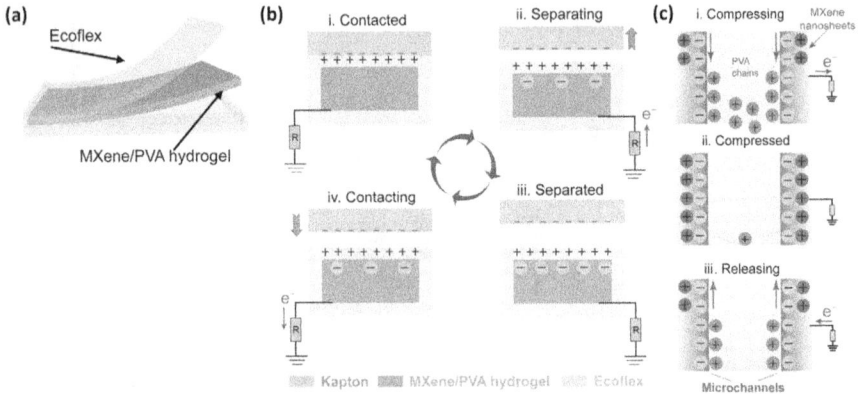

FIGURE 1.7 (a) Scheme of the MH-TENG electrode. (b) Working principle scheme for the single-electrode system of MH-TENG. (c) A triboelectric mechanism consisting of the presence of microchannels of the MXene/PVA hydrogel composite. Adapted with permission [19]. Copyright, 2021, John Wiley & Sons.

after that, Kapton and MH-TENG are slowly separated, which leads to a polarization of negative and positive charges on the surface of MXene/PVA hydrogel and Ecoflex, respectively. Because of that, e^- from the connected wire flows toward the hydrogel. Step iii: once the Kapton is far enough, the system stops generating charges because of the complete screening of the positive and negative charges from the Ecoflex and MXene/PVA hydrogel, respectively. Step iv: upon approaching Kapton and Ecoflex, a decrease of potential difference occurs between the two, which is because of an increment in the electrostatic shielding. That effect leads to the repulsion of the negative ions from the upper surface of the MXene/PVA hydrogel, leading to an excess of e^- flow out of the system. Upon contacting Kapton and Ecoflex, the system returns to step i. Furthermore, it is also worth describing the mechanism that takes place on the MXene/PVA hydrogel, known as streaming vibrational potential (SVP), which is described in Figure 1.7c. In step i there is a formation of hydrogen bonds between the PVA and the MXene nanosheets. This interaction leads to the formation of microchannels in the MXene, which can be filled with PVA chains and water. Upon the interaction between water and MXenes, polarization happens in a way that positive and negative charges are generated in the former and latter, respectively. This process leads to an electric double layer, which can generate a non-Faradaic current. When the MH-TENG is pressurized, water can flow out of the positive ions while maintaining the negative charges on the MXene's surface. That process leads e^- from the external circuit to be repelled out of the system, whereas positive charges are induced onto the MXene. Step ii, when maximum compression is applied, most of the positive ions present in the microchannel are expelled to the outside microchannels, which causes the MXene's positive and negative charges to be in equilibrium; hence, no external current passes through the system. Step iii, upon release of pressure, the water can flow back, which leads to the buildup of the electric double layer and therefore a reversed external current.

1.3.2 HYDROGELS FOR ENERGY STORAGE

There has been a plethora of materials that have been employed in the fabrication of energy storage devices, such as transition metal derivatives that include oxides, sulfides, and bimetallics, along with conducting polymers, metal-organic frameworks, and covalent organic frameworks, among many others. Yet, hydrogels can also be incorporated into supercapacitors and batteries, given their highly porous networked structure, conductivity, and stimuli response to external gradients, which may add a novel factor to energy storage devices derived from them. Some electrochemical devices may undergo an increase in temperature that can lead to fires or explosions due to overcharging, aggressive use, or inappropriate environments [20,21]. Because of that, introducing active ways to prevent thermal runaway is an important factor to avoid failure of the device, along with longer cycle life and electrochemical stability. Some examples of polymers that can perform phase separation or sol-gel transitions at high temperatures have been used as electrolyte matrices in Li-ion batteries [22,23]. This feature provides a safety step in case of overheating, as the reversible sol-gel transition can protect the device by temporarily stopping the charging or discharging process until the temperature is back to the safe working range. One of the polymers that have been used for this end is poly(N-isopropylacrylamide) (PNIPAM) [21,23]. Yet, a broader transition temperature range along with more feasible synthetical approaches are desired. Shi *et al.* [24] utilized a block polymer composed of poly (propylene oxide) in between poly(ethylene oxide) (PEO-PPO-PEO), which was employed as a smart electrolyte that could shift to the gel state based on the temperature. Based on that, it was observed that at low temperatures the polymer was in the solution state, which allowed the ions to move freely within its structure. Yet, with the increase in temperature, the PEO-PPO-PEO transitioned to the gel state, which hindered the ionic flow, causing the device to cease working. This type of feature provides some advantages, such as a reversible self-protection mechanism. Also, the trigger temperature for the gel to the formed as well as the capacity loss can be tuned based on the molecular weight. Lastly, the polymers employed to make the electrolyte matrix are commonly used, along with their proper interaction with different electrode materials may present some potential for larger-scale applications. The schematics for the representation of ionic motion and the physical aspect of the PEO-PPO-PEO in its solution and gel state are presented in Figure 1.8.

The versatility of hydrogels can go beyond the types of polymers that can be employed to synthesize them. Aside from it, hydrogels can be mixed with several nanomaterials to enhance their properties and perform the transport of different ionic species. Also, hydrogels can function as a core component for the development of flexible energy storage devices, which have attracted some attention from the scientific community due to the novel products that can be obtained, such as smart watches, e-skins, portable devices, and so on [25–27]. Chen *et al.* [28] studied the effect of different electrolytes on a composite electrolyte based on graphene thin films and PVA gels as the schematics for the device are presented in Figure 1.9a. Several electrolytes, such as NaCl, KCl, NaOH, KOH, H_2SO_4, and H_3PO_4, were studied. After performing the cyclic voltammetry (CV), it was observed that H_3PO_4 had the best overall performance by presenting a typical

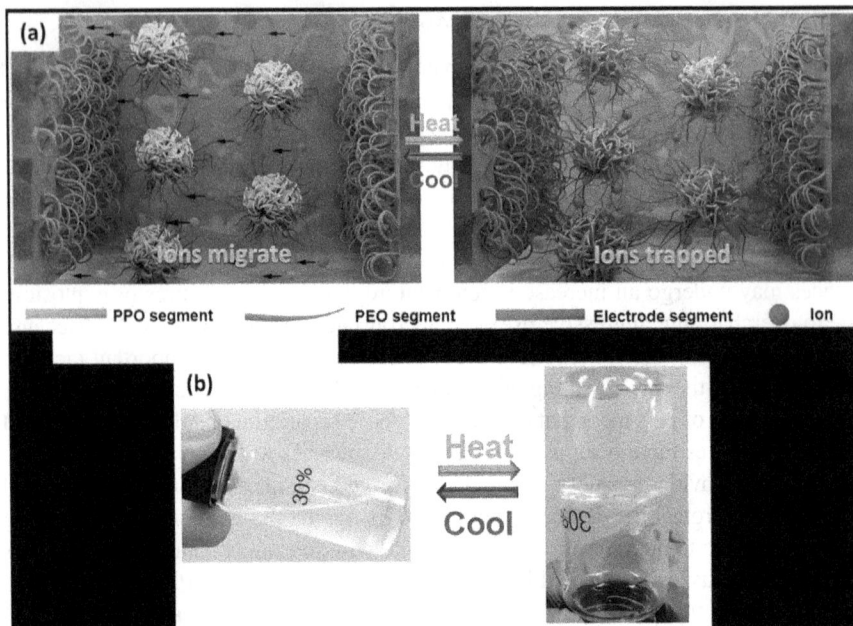

FIGURE 1.8 (a) Illustration of the hydrogel thermal response upon an increase in temperature that leads to the formation of a gel system that hinders the movement of the ions in the electrolyte preventing the storage device from overheating. (b) PEO-PPO-PEO in the solution state at cooler temperatures and in the gel state at higher temperatures. Adapted with permission [24]. Copyright 2016, John Wiley & Sons.

electric double-layer capacitance mechanism based on the squared shape of the CV plot. The authors proposed that the higher capacitance of H_3PO_4 was attributed to the higher ionic strength of the acid when compared to the other electrolytes. Also, the H^+ could diffuse more easily through the PVA-graphene structure when compared to relatively larger Na^+ and K^+. Alongside that, the lower amount of H^+ derived from the H_2SO_4 when compared with H_3PO_4 for the same concentration led to a decrease in the capacitance of H_2SO_4. In addition, the presence of flexible PVA gels allowed the device to perform at several bending angles with virtually no decrease in performance, along with satisfactory electrochemical stability after 5,000 CV, which is demonstrated in Figure 1.9b-c.

1.3.3 HYDROGELS FOR SENSORS

The external stimuli response of hydrogels is one of their unique properties that allow their use in the fabrication of varied sensors, such as for the identification of humidity, pressure, temperature, ionic strength, and biomolecules, among many others. Several synthetic polymers can be used in the development of sensors, such as PVA, poly(ethylene glycol) (PEG), poly(vinylpyrrolidone) (PVP), poly(vinyl imidazole) (PVI), and poly(acrylamide) (PAM), among others [29]. Based on that,

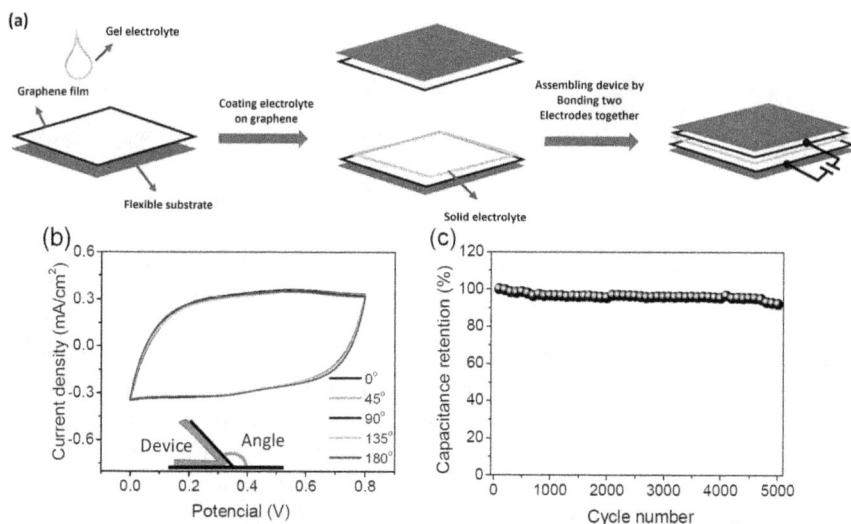

FIGURE 1.9 (a) Scheme for the fabrication of the composite PVA-graphene hydrogel-based electrode. (b) CV plot for the PVA-graphene hydrogel with H_3PO_4 as an electrolyte at different bending degrees. (c) Electrochemical stability test performed up to 5,000 cycles. Adapted with permission [28]. Copyright 2014, Royal Society of Chemistry.

it is possible that these hydrogels can convert some type of mechanical deformation, such as compression or strain, into an appreciable electrical signal, making them applicable as sensors. However, one of the recurrent challenges lies in providing an external energy source that allows the generation of detectable and stable signals [30–32]. Under this line, it is deemed necessary to develop hydrogel-based sensors that can function without the need for a power supply, as this can expand their applications as sensors. In a previous study, hydrogel-based sensors for the generation of electricity through moisture were developed [33].

In another work, a TENG capable of converting applied mechanical energy into electricity was fabricated [34]. In this case, the device was self-powered and could effectively perform as a strain sensor. Another work based on the use of hydrogels for the development of sensors was performed by Wang et al. [35], who used a PVA hydrogel with malic acid (MA). It is worth noting that MA has a relatively high acidity, which leads to an increase in output voltage. Alongside that, when compared to other organic acids, such as acrylic acid, citric acid, and tannic acid, for instance, MA displays the highest redox potential, ionizability, and conductivity, making it a feasible candidate as an electrolyte. Also, Ca^{2+}, in the form of $CaCl_2$, was introduced to improve conductivity, enhance the self-powering properties, and prevent the hydrogel from drying and freezing. Through that, the MA–CA–PVA could deliver a voltage of around 0.55 V along with a conductivity of around 5.3 S/m. Also, the hydrogel could provide strain-related signals based on the changes in the output current without an external supply for power. In addition, the hydrogel could monitor human motion at temperatures as low as −20°C. Based on that, the scheme for the fabrication of the MA–Ca–PVA hydrogel is presented in Figure 1.10.

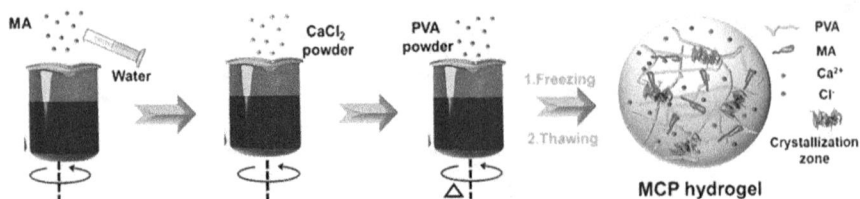

FIGURE 1.10 Scheme for the synthesis process of MA–CA–PVA (MCP) hydrogel. Adapted with permission [35]. Copyright 2020, Royal Society of Chemistry.

There is a need for the development of a fast responsive sensor that can be obtained on a large scale. One of the fields that has a higher demand for this technology is biomedical, for accurate and fast diagnosis, for instance. Based on these aspects, Choe *et al.* [36] fabricated a PNIPAM microgel that was decorated with Au nanoparticles (AuNP) over a flexible PAA hydrogel matrix. The composite hydrogel presented a thermosensing color gradient that presented a resolution of temperature sensing of around 0.2°C. The device was fabricated using the composite hydrogel that functioned as a color-based thermometer. The color change properties were attributed to the localized surface plasmon resonance (LSPR), which is the phenomenon by which a collection of oscillatory conducting electrons become excited by the incidence of light in a metallic nanoparticle [37,38]. The LSPR is the main principle of operation for the PNIPAM AuNP-hydrogel, as at around 24°C, the microgel has a loose structure, whereas at temperatures of around 50°C, the structure shrinks. Through that, the system presents an uncoupled/couple plasmon or swollen/shrink mode around 24 and 50°C, respectively (Figure 1.11a). The scanning electron microscopy (SEM) of the plasmonic microgel at 24 and 50°C shows the swelling/shrinking effect in the function of temperature that leads to a variety of colors, which is shown in Figure 1.11b. Based on this property, the thermo-sensitive hydrogel can be incorporated into an array of sensors that can provide a broader range of temperatures, leading to a more accurate measurement of the temperature, as shown in Figure 1.11c. Such a type of device can find several fields of applications, aside from biomedical, yet it shows a convenient and effective way to analyze the temperature based on a thermo response of a hydrogel decorated with nanoparticles.

1.3.4 Hydrogels for Biomedicals

Hydrogels present a vast range of applications within the biomedical field, as they can function as drug delivery systems, scaffolds for tissue regeneration and engineering, antibacterial agents, and biosensors for the identification of biomolecules, along with presenting good biodegradability and biocompatibility. The combination of such properties makes it an attractive component for the design of medical products [39–41]. Cui *et al.* [42] fabricated a hydrogel reinforced with tunicate cellulose nanocrystals (TCNCs) that presented a tendril shape that could potentially function as an artificial muscle. The hydrogel was obtained through the polymerization of acrylic acid, acrylamide, and adamantly acrylamide in the

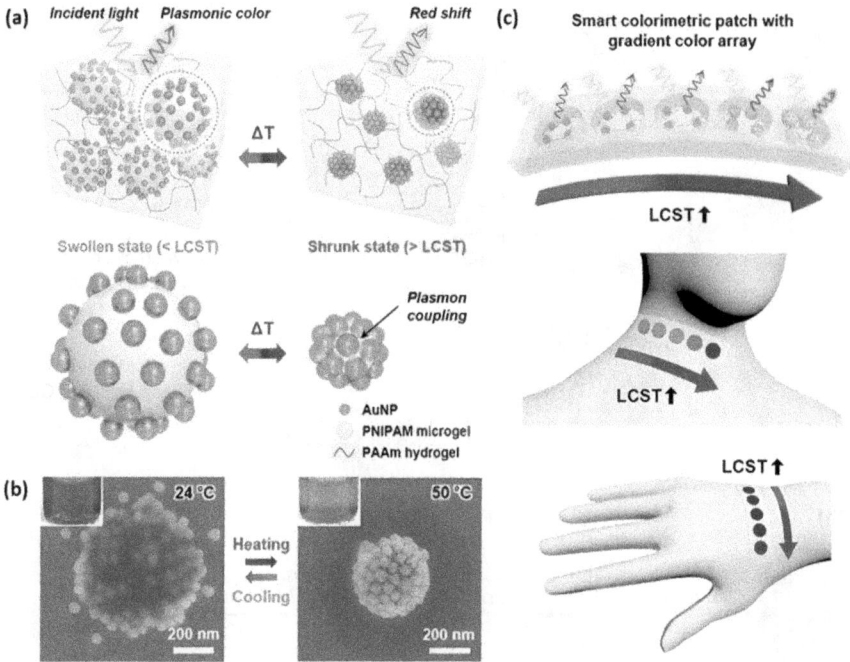

FIGURE 1.11 (a) Scheme of the temperature response of the PNIPAM microgel decorated with AuNP at a PAAM hydrogel substrate that is based on the LCST phenomenon. (b) SEM images showing the swollen and shrunk state of the plasmonic microgel and the inlet photocopy displaying the color based on the temperature of 24 and 50°C, respectively. (c) Schematic for the sensor array that can be placed on human skin for temperature measurement. Reproduced with permission [36]. Copyright 2018, The authors, Licensee NPG Asia Materials.

presence of TCNC modified with β-cyclodextrin (β-CD-TCNCs). Furthermore, the tendril shape was obtained by submersing the hydrogel into a $FeCl_3$ solution that would promote Fe^{3+}/-COO$^-$ ionic coordination bonds in the structure. During the shaping process, the chirality of twisting and coiling were controlled which yield artificial muscles with both homo- and heterochiral. Because of that, the hydrogels could respond to the presence of solvents by contracting or expanding, allowing the hydrogel to function as a muscle. The full schematic for the synthetic process of the hydrogel artificial muscle is provided in Figure 1.12.

Tissue engineering technologies have the main goal of promoting the growth of healthy tissue. Yet, such a process is inherently challenging, as there are several factors that one must consider before introducing certain types of material within a physiological system. Some of these aspects include introducing a material that can imitate the natural extracellular matrix (ECM), to serve as a substrate at which the cells can grow without triggering an inflammatory reaction. Also, it is desired that the material used as a scaffold for the growth of cells or as tissue recovery can slowly degrade as the healthy tissue is regenerated. Also, during its *in-vivo* decomposition

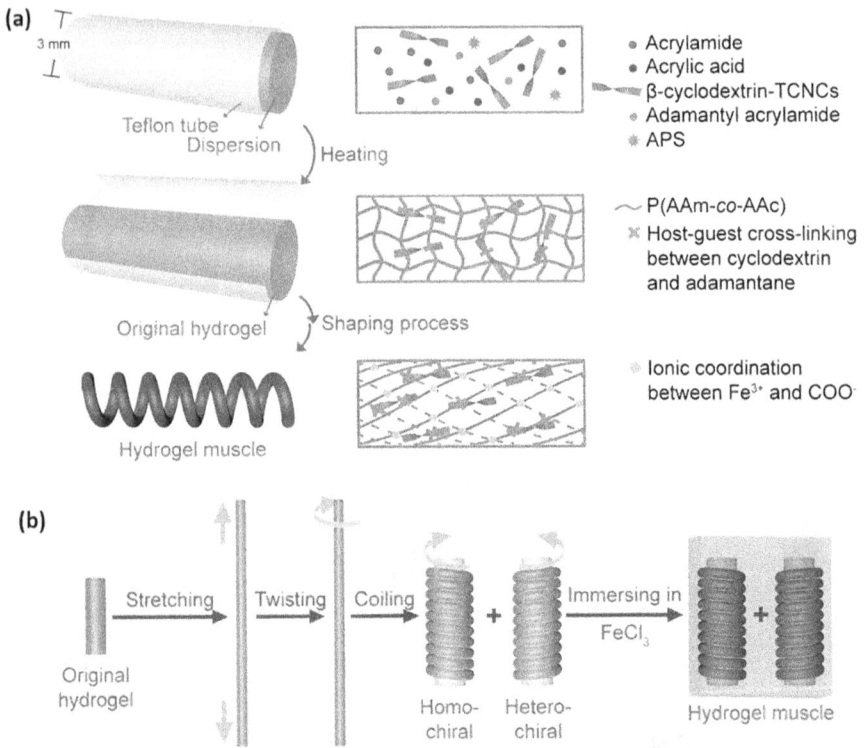

FIGURE 1.12 (a) Schematics for the synthesis of hydrogel-based artificial muscle. (b) The approach adopted to shape the hydrogel artificial muscle to obtain homo- and heterochiral materials. Reproduced with permission [42]. Copyright, 2021. American Chemical Society.

process, there should be the formation of non-toxic compounds of fragments to provide a safer excretion from the body [43,44]. Another important aspect of tissue recovery lies in the controlled dosage of growth factors or drugs that are associated with delivery system technology. Based on that, the material must release the drug in concentrations that are within the therapeutic range to prevent the drug from becoming poisonous. One reported example is the case of nerve growth factor, which has been shown to have an optimal dosage of 800 pg/μL in rats to promote the growth of neurites. However, at higher concentrations, it hinders its growth [45,46]. Hydrogels are feasible components for the development of scaffolds, tissue recovery, and drug delivery systems due to their porous structure that can enable the growth of cells and tissue recovery. Also, the high porosity enables the adsorption of drugs that can be released in a target region in the body. Alongside that, the versatile chemical nature of hydrogels permits the use of materials that can be biocompatible as well as biodegradable. Despite that, there are several that must be considered for the development of an ideal tissue engineering material. In this sense, mimicking the cellular structure includes presenting similar mechanical, structural, and chemical properties of the cell. Also, if the hydrogel carries a drug for cell growth or treatment, it should present a controlled release.

Yan *et al.* [47] fabricated a β-hairpin peptide hydrogel that could be potentially employed as an injectable solid. The main feature of this material was based on its reversible mechanical properties, as under an applied shear, the hydrogel network was disrupted into domains that allowed it to flow. However, once the shear force was removed, these domains could promptly recover to their initial state. Such a process occurred due to the physically cross-linked structure of fibrillar branching and entanglements [48,49]. Hydrogels with such behavior can potentially find applications in the biomedical field. Yet, several factors must be considered, such as biocompatibility, biodegradability, and capability of the hydrogel to either host cells for their growth if used for tissue engineering or recovery. Also, if the aim is focused on drug delivery systems, the hydrogel should be able to adsorb the drug and release it in the targeted area.

1.4 CONCLUSION

Hydrogels have been growing within research due to the vast number of commonly known compounds that can be used for their synthesis, which expands their classification in terms of materials used, types of cross-linking, type of bonding, type of stimulus, and surface charge, among others. Aside from the chemical and synthetical versatility, their most attractive point is their properties that arise from their networked structure that promotes a swelling behavior. Along with that, hydrogels present a distinct property of sense that can take place through different stimuli such as physical, chemical, or biological. The combination of all these factors made hydrogels attractive to the scientific community in several areas of research, such as energy generation and storage, sensors, biosensors, and biomedical to name a few. Hydrogels are applicable in energy generation, such as fuel cells due to their high porosity, which allows the diffusion of water to properly perform its redox process. Also, by presenting a chemically stable and open morphology, electroactive materials can be easily grown in their structure to obtain a composite. Similarly, hydrogels can be used in energy storage devices, such as supercapacitors and batteries, as their porous structure enables the permeation of electrolytes to promote the charging process in supercapacitors. Also, in the case of batteries, hydrogels can be particularly useful because of their swelling properties, as they can withstand chemical as well as physical strain after the incorporation of metal atoms, which could lead to an improvement in electrochemical stability. Yet, further research is required to optimize this process as there is a relatively smaller number of studies in this specific area. Hydrogels can respond to different external factors such as temperature, pH, mechanical forces, specific chemical compounds, and so on. This unique property is highly desired for the application of sensors. Lastly, hydrogels have been widely used in the biomedical field as tissue engineering and drug delivery systems. Their networked structure functions as nucleation sites for the growth of cells, along with the possibility of carrying growth factors or drugs to expedite this process. Also, hydrogels can be obtained from biocompatible and biodegradable materials, which allows for them to be inserted into a biological system and released without harming the body. Despite the vast realm of applications of hydrogels, there is still the need to further tailor their properties to optimize their uses.

REFERENCES

[1] U.S.K. Madduma-Bandarage, S.V. Madihally, Synthetic hydrogels: Synthesis, novel trends, and applications, J. Appl. Polym. Sci. 138 (2021) e50376/1–e50376/23.

[2] I. Gholamali, Stimuli-responsive polysaccharide hydrogels for biomedical applications: A review, Regen. Eng. Transl. Med. 7 (2021) 91–114.

[3] H. Dechiraju, M. Jia, L. Luo, M. Rolandi, Ion-conducting hydrogels and their applications in bioelectronics, Adv. Sustain. Syst. 6 (2022) 2100173.

[4] D.A. Salick, J.K. Kretsinger, D.J. Pochan, J.P. Schneider, Inherent antibacterial activity of a peptide-based β-hairpin hydrogel, J. Am. Chem. Soc. 129 (2007) 14793–14799.

[5] D. Jiang, Z. Liu, X. He, J. Han, X. Wu, Polyacrylamide strengthened mixed-charge hydrogels and their applications in resistance to protein adsorption and algae attachment, RSC Adv. 6 (2016) 47349–47356.

[6] S. Bashir, M. Hina, J. Iqbal, A.H. Rajpar, M.A. Mujtaba, N.A. Alghamdi, S. Wageh, K. Ramesh, S. Ramesh, Fundamental concepts of hydrogels: Synthesis, properties, and their applications, Polymers (Basel). 12 (2020) 1–60.

[7] J. Saroia, W. Yanen, Q. Wei, K. Zhang, T. Lu, B. Zhang, A review on biocompatibility nature of hydrogels with 3D printing techniques, tissue engineering application and its future prospective, Bio-Design Manuf. 1 (2018) 265–279.

[8] S. Sardana, A. Gupta, K. Singh, A.S. Maan, A. Ohlan, Conducting polymer hydrogel based electrode materials for supercapacitor applications, J. Energy Storage. 45 (2022) 103510.

[9] D. Buenger, F. Topuz, J. Groll, Hydrogels in sensing applications, Prog. Polym. Sci. 37 (2012) 1678–1719.

[10] X. Sun, S. Agate, K.S. Salem, L. Lucia, L. Pal, Hydrogel-based sensor networks: Compositions, properties, and applications—A review, ACS Appl. Bio Mater. 4 (2021) 140–162.

[11] H.Y. Zhou, L.J. Jiang, P.P. Cao, J.B. Li, X.G. Chen, Glycerophosphate-based chitosan thermosensitive hydrogels and their biomedical applications, Carbohydr. Polym. 117 (2015) 524–536.

[12] Y. Zou, C. Chen, Y. Sun, S. Gan, L. Dong, J. Zhao, J. Rong, Flexible, all-hydrogel supercapacitor with self-healing ability, Chem. Eng. J. 418 (2021) 128616.

[13] P. Li, Z. Jin, L. Peng, F. Zhao, D. Xiao, Y. Jin, G. Yu, Stretchable all-gel-state fiber-shaped supercapacitors enabled by macromolecularly interconnected 3D graphene/nanostructured conductive polymer hydrogels, Adv. Mater. 30 (2018) 1800124.

[14] L. Pan, G. Yu, D. Zhai, H.R. Lee, W. Zhao, N. Liu, H. Wang, B.C.-K. Tee, Y. Shi, Y. Cui, Z. Bao, Hierarchical nanostructured conducting polymer hydrogel with high electrochemical activity, Proc. Natl. Acad. Sci. 109 (2012) 9287–9292.

[15] Y. Guo, J. Bae, Z. Fang, P. Li, F. Zhao, G. Yu, Hydrogels and hydrogel-derived materials for energy and water sustainability, Chem. Rev. 120 (2020) 7642–7707.

[16] L. Yang, D.K. Nandakumar, L. Miao, L. Suresh, D. Zhang, T. Xiong, J.V. Vaghasiya, K.C. Kwon, S. Ching Tan, Energy harvesting from atmospheric humidity by a hydrogel-integrated ferroelectric-semiconductor system, Joule. 4 (2020) 176–188.

[17] S. Wang, X. Liu, L. Li, C. Ji, Z. Sun, Z. Wu, M. Hong, J. Luo, An unprecedented biaxial trilayered hybrid perovskite ferroelectric with directionally tunable photovoltaic effects, J. Am. Chem. Soc. 141 (2019) 7693–7697.

[18] D. Bao, Z. Wen, J. Shi, L. Xie, H. Jiang, J. Jiang, Y. Yang, W. Liao, X. Sun, An anti-freezing hydrogel based stretchable triboelectric nanogenerator for bio-mechanical energy harvesting at sub-zero temperature, J. Mater. Chem. A. 8 (2020) 13787–13794.

[19] X. Luo, L. Zhu, Y.-C. Wang, J. Li, J. Nie, Z.L. Wang, A flexible multifunctional triboelectric nanogenerator based on MXene/PVA hydrogel, Adv. Funct. Mater. 31 (2021) 2104928.

[20] Y. Tang, Y. Zhang, J. Deng, J. Wei, H. Le Tam, B.K. Chandran, Z. Dong, Z. Chen, X. Chen, Mechanical force-driven growth of elongated bending TiO_2-based nanotubular materials for ultrafast rechargeable lithium ion batteries, Adv. Mater. 26 (2014) 6111–6118.

[21] H. Yang, Z. Liu, B.K. Chandran, J. Deng, J. Yu, D. Qi, W. Li, Y. Tang, C. Zhang, X. Chen, Self-protection of electrochemical storage devices via a thermal reversible Sol–Gel transition, Adv. Mater. 27 (2015) 5593–5598.

[22] J.C. Kelly, N.L. Degrood, M.E. Roberts, Li-ion battery shut-off at high temperature caused by polymer phase separation in responsive electrolytes, Chem. Commun. 51 (2015) 5448–5451.

[23] J.C. Kelly, M. Pepin, D.L. Huber, B.C. Bunker, M.E. Roberts, Reversible control of electrochemical properties using thermally-responsive polymer electrolytes, Adv. Mater. 24 (2012) 886–889.

[24] Y. Shi, H. Ha, A. Al-Sudani, C.J. Ellison, G. Yu, Thermoplastic elastomer-enabled smart electrolyte for thermoresponsive self-protection of electrochemical energy storage devices, Adv. Mater. 28 (2016) 7921–7928.

[25] G. Xiong, C. Meng, R.G. Reifenberger, P.P. Irazoqui, T.S. Fisher, Graphitic petal electrodes for all-solid-state flexible supercapacitors, Adv. Energy Mater. 4 (2014) 1300515.

[26] X. Zang, Q. Chen, P. Li, Y. He, X. Li, M. Zhu, X. Li, K. Wang, M. Zhong, D. Wu, H. Zhu, Highly flexible and adaptable, all-solid-state supercapacitors based on graphene woven-fabric film electrodes, Small. 10 (2014) 2583–2588.

[27] X. Li, T. Zhao, Q. Chen, P. Li, K. Wang, M. Zhong, J. Wei, D. Wu, B. Wei, H. Zhu, Flexible all solid-state supercapacitors based on chemical vapor deposition derived graphene fibers, Phys. Chem. Chem. Phys. 15 (2013) 17752–17757.

[28] Q. Chen, X. Li, X. Zang, Y. Cao, Y. He, P. Li, K. Wang, J. Wei, D. Wu, H. Zhu, Effect of different gel electrolytes on graphene-based solid-state supercapacitors, RSC Adv. 4 (2014) 36253–36256.

[29] Dhanjai, A. Sinha, P.K. Kalambate, S.M. Mugo, P. Kamau, J. Chen, R. Jain, Polymer hydrogel interfaces in electrochemical sensing strategies: A review, TrAC Trends Anal. Chem. 118 (2019) 488–501.

[30] B. Ying, Q. Wu, J. Li, X. Liu, An ambient-stable and stretchable ionic skin with multimodal sensation, Mater. Horizons. 7 (2020) 477–488.

[31] R. Fu, L. Tu, Y. Zhou, L. Fan, F. Zhang, Z. Wang, J. Xing, D. Chen, C. Deng, G. Tan, P. Yu, L. Zhou, C. Ning, A tough and self-powered hydrogel for artificial skin, Chem. Mater. 31 (2019) 9850–9860.

[32] X. Pan, Q. Wang, R. Guo, S. Cao, H. Wu, X. Ouyang, F. Huang, H. Gao, L. Huang, F. Zhang, L. Chen, Y. Ni, K. Liu, An adaptive ionic skin with multiple stimulus responses and moist-electric generation ability, J. Mater. Chem. A. 8 (2020) 17498–17506.

[33] Z. Lei, P. Wu, A highly transparent and ultra-stretchable conductor with stable conductivity during large deformation, Nat. Commun. 10 (2019) 3429.

[34] X. Pu, M. Liu, X. Chen, J. Sun, C. Du, Y. Zhang, J. Zhai, W. Hu, Z.L. Wang, Ultrastretchable, transparent triboelectric nanogenerator as electronic skin for biomechanical energy harvesting and tactile sensing, Sci. Adv. 3 (2023) e1700015.

[35] Q. Wang, X. Pan, H. Zhang, S. Cao, X. Ma, L. Huang, L. Chen, Y. Ni, Fruit-battery-inspired self-powered stretchable hydrogel-based ionic skin that works effectively in extreme environments, J. Mater. Chem. A. 9 (2021) 3968–3975.

[36] A. Choe, J. Yeom, R. Shanker, M.P. Kim, S. Kang, H. Ko, Stretchable and wearable colorimetric patches based on thermoresponsive plasmonic microgels embedded in a hydrogel film, NPG Asia Mater. 10 (2018) 912–922.

[37] S.K. Ghosh, T. Pal, Interparticle coupling effect on the surface plasmon resonance of gold nanoparticles: From theory to applications, Chem. Rev. 107 (2007) 4797–4862.

[38] K.A. Willets, R.P. Van Duyne, Localized surface plasmon resonance spectroscopy and sensing, Annu. Rev. Phys. Chem. 58 (2007) 267–297.

[39] M.C. Catoira, L. Fusaro, D. Di Francesco, M. Ramella, F. Boccafoschi, Overview of natural hydrogels for regenerative medicine applications, J. Mater. Sci. Mater. Med. 30 (2019) 115.

[40] Q. Chai, Y. Jiao, X. Yu, Hydrogels for biomedical applications: Their characteristics and the mechanisms behind them, Gels. 3 (2017) 6/1–6/15.

[41] D. Seliktar, Designing cell-compatible hydrogels for biomedical applications, Science. 336 (2012) 1124–1128.

[42] Y. Cui, D. Li, C. Gong, C. Chang, Bioinspired shape memory hydrogel artificial muscles driven by solvents, ACS Nano. 15 (2021) 13712–13720.

[43] F.M. Watt, W.T.S. Huck, Role of the extracellular matrix in regulating stem cell fate, Nat. Rev. Mol. Cell Biol. 14 (2013) 467–473.

[44] C. Frantz, K.M. Stewart, V.M. Weaver, The extracellular matrix at a glance, J. Cell Sci. 123 (2010) 4195–4200.

[45] S.W.P. Kemp, A.A. Webb, S. Dhaliwal, S. Syed, S.K. Walsh, R. Midha, Dose and duration of nerve growth factor (NGF) administration determine the extent of behavioral recovery following peripheral nerve injury in the rat, Exp. Neurol. 229 (2011) 460–470.

[46] K.F. Bruggeman, R.J. Williams, D.R. Nisbet, Dynamic and responsive growth factor delivery from electrospun and hydrogel tissue engineering materials, Adv. Healthc. Mater. 7 (2018) 1700836.

[47] C. Yan, A. Altunbas, T. Yucel, R.P. Nagarkar, J.P. Schneider, D.J. Pochan, Injectable solid hydrogel: Mechanism of shear-thinning and immediate recovery of injectable β-hairpin peptide hydrogels, Soft Matter. 6 (2010) 5143–5156.

[48] T. Yucel, C.M. Micklitsch, J.P. Schneider, D.J. Pochan, Direct observation of early-time hydrogelation in β-hairpin peptide self-assembly, Macromolecules. 41 (2008) 5763–5772.

[49] B. Ozbas, J. Kretsinger, K. Rajagopal, J.P. Schneider, D.J. Pochan, Salt-triggered peptide folding and consequent self-assembly into hydrogels with tunable modulus, Macromolecules. 37 (2004) 7331–7337.

2 Hydrogels Based on Natural and/or Synthetic Polymers

Felipe López-Saucedo

Instituto de Ciencias Nucleares, UNAM Ciudad de México, México and Universidad Autónoma del Estado de México, Toluca, México

Guadalupe Gabriel Flores-Rojas

Instituto de Ciencias Nucleares, UNAM Ciudad de México, México and Universidad de Guadalajara Jalisco, México and Instituto de Investigaciones en Materiales, UNAM, Toluca, México

Ricardo Vera-Graziano

Instituto de Investigaciones en Materiales, UNAM, Ciudad de México, México

Lorena Garcia-Uriostegui

Universidad de Guadalajara, Zapopan, México

Eduardo Mendizabal

Universidad de Guadalajara, Jalisco, México

Leticia Buendía-González

Universidad Autónoma del Estado de México, Toluca, México

Emilio Bucio

Instituto de Ciencias Nucleares, UNAM, Ciudad de México, México

2.1 INTRODUCTION

Hydrogels are mixtures of cross-linked polymeric chains with a three-dimensional structure, which are capable of absorbing and retaining a considerable amount of water inside due to their hydrophilic nature. Hydrogels can contain polar groups such as

DOI: 10.1201/9781003340485-2

-COOH, -OH, -NH$_2$, -CONH, -CONH$_2$, and -SO$_3$H. These functional groups provide the polymeric network with the characteristic appearance of hydrogels on a microscopic and macroscopic scale [1,2]. Depending on the source of origin, the polymer network can be synthesized by various standardized methods that are generally reproducible without difficulty. Cross-linking is a key step that can be accomplished by strategies involving covalent bonding or by non-covalent interactions, such as ionic bonding or hydrogen bonding [3,4]. Whether by the selected method of synthesis or polymers (hence, the functional groups present in the hydrogel) will result in the characteristics of the hydrogel, in other words, water retention capacity; the percentages of cross-linking; and usefulness at a determined pH [5–7], temperature [8], or ionic strength [9].

There are several types of hydrogels and are classified according to the origin of components, which means natural or synthetic; another way to classify them is by the synthesis method, which can involve physical or chemical cross-linking. The characterization of hydrogels is performed through various physical parameters, such as pore size, elastic modulus, degree of swelling, degradation rate, and biocompatibility degree [10]. Currently, research has focused on hydrogels prepared from biological materials of protein origin, such as collagen or elastin, and polysaccharide polymers such as glycosaminoglycans, chitosan, and alginate, among others. Synthetic polymer-based hydrogels are obtained from polymers such as poly(acrylic acid) (PAAc) [11], poly(N-isopropylacrylamide) (PNIPAm) [12], poly(ethylene glycol) (PEG) [13], poly(N-vinyl caprolactam) (PNVCL) [14], or poly(2-hydroxyethylmethacrylate) (HEMA) [15], which are obtained using techniques of addition polymerizations, and may have stimuli-responsive properties such as pH, temperature, magnetic, and/or ionic strength [16].

The characteristics of polymers present in a hydrogel should include high biocompatibility, biodegradability, and low cytotoxicity. Fortunately, these positive properties are shared by many natural and synthetic hydrogels. Certain characteristics, such as the chemical reactivity of the hydrogel, can be manipulated by adding, removing, or blocking specific functional groups. The chemical formula, in turn, affects the physical rheology in a solution; so, it is important to control the swelling degree, network size, or viscosity, to control the capacity of the hydrogel to diffuse and trap molecules [17], and/or cells within its network [11]. These properties allow hydrogels to be applied in multiple industrial, pharmaceutical, and biomedical fields [18], due to the great similarity and acceptance of hydrogels with soft tissue [19]. All hydrogels, both natural and synthetic, have qualities, limitations, and disadvantages; for example, some hydrogels that contain only natural polymers in their structure have limitations, such as poor mechanical properties or rapid degradation [19]. For this reason, natural polymers are often combined with synthetic polymers to create composite hydrogels. In general, composite polymers contain in their structure at least one polymer matrix that is reinforced by fillers, which produces an improvement in overall performance; therefore, their possible applications are increased [20], since synthetic polymer hydrogels can be designed to obtain customized mechanical properties, which otherwise, by themselves, do not usually have an inherent bioactivity. Therefore, adapting the hydrogels based on a potential application is the most useful tool in the engineering of hydrogels [21].

Hydrogels are widely studied in most of the materials science disciplines due to the number of possibilities for synthesis and their possible uses and applications

[18]; it is common to find hydrogel materials for heavy-duty use but also in biomedical microdevices [22]. Many hydrogel systems have already been success-fully tested in biomedical applications [23], including drug delivery systems [10], three-dimensional (3D) cell cultures [24], tissue implants [25], tissue regeneration [26], contact lenses [27], and other personal hygiene products [28]. Each type of hydrogel can be adapted according to the application for which it is designed. This also includes choosing of a specific method and technique to achieve the product of interest with the chemical and physical properties required.

Although it has already been said that hydrogels may have their special characteristics depending on the final use, certain general parameters are expected to be mostly covered, such as desirable characteristics all around their high water absorption or a well-known water retention capacity under a controlled time [29].

2.2 NATURAL HYDROGELS

Natural hydrogels are made up of polymers extracted from plants, animals, or microorganisms [30]. The most used natural polymers for the synthesis of hydrogels are macromolecules such as carbohydrates and proteins, which are essential components of organisms, and therefore their properties can be transferred to the biomedical device with a specific physicochemical characteristic or resistance [31]. In addition, natural polymers exhibit high biocompatibility, biodegradability, accessibility, stability, lack of toxicity, and low cost [32]. The inherent advantages of these materials are evident once they are used and the life cycle is complete since the residues are reincorporated into the environment in a more harmonious way to be used later as all organic matter [33]. There are several types of natural polymers, of which the most representative are proteins, polysaccharides, polynucleotides, polyisoprenes, and lignin. Protein and polysaccharide polymers are inexpensive compared to other natural polymers and are highly available, biodegradable, and biocompatible [33]. Therefore, these polymers are recommended for applications in tissue scaffolds [34] and regenerative medicine because hydrogels can easily reach this functionality, given that they can replicate an extracellular matrix, which is a common property among natural raw materials available in the market [35].

2.2.1 CELLULOSE AND STARCH

Cellulose is the most abundant polysaccharide in plants, consisting of a linear chain of β (1→4) linked D-glucose units. The first use of cellulose as a material is in the paper industry, where the main sources are bamboo, cotton, linen, and practically every fiber from plants. Cellulose can interact well with water but it is insoluble in it, which is a limitation [36]. Nonetheless, cellulose is an interesting biopolymer because it is biodegradable and biocompatible. For these reasons, it is a sturdy candidate as a biomaterial for medical devices or sanitary disposables. The potential of cellulose-based hydrogels is big, despite the problems related to chemical modifica-tions needed before the formation of the hydrogel, which is essential because cellulose is a stable macromolecule. Therefore, cellulose derivatives are prepared as precursors of hydrogels. The list includes, for example, methylcellulose, hydroxyethyl cellulose, hydroxypropyl cellulose, and carboxymethyl cellulose sodium [37]. These previous modifications aim to provide the hydrophilic groups to make the reaction easier with the cross-linking agents that improve the hydrophilicity eventually. Methods employed are also vast and comprise of gamma radiation, chemical cross-linking, photocross-linking, freeze-thaw, etc. [38]. These changes enhance the absorption of water and swelling without dissolving the material. Using cellulose as a hydrogel matrix is only the first example of a natural matrix because other natural polymers attain superior performance and other capabilities, as we see with starch below.

Starch and cellulose are the most known carbohydrates. Starch is present in plants and is one of the main sources of food for human consumption, so our body can easily recognize it. Starch is found abundantly in plants in green leaf chloroplasts and in seed amyloplasts [39]. Starch has in its structure two types of glucose units. These are the parts of the polymer chain and are covalently bound through glycosidic bonds, mainly with linear and helical amylose (containing D-glucose monomers bound to α-1,4) in a ratio ranging 20–25% and by amylopectin (which contains α-D-(1,6) bonds, located each 25–30 linear units of glucose) in a percentage of 75 to 80% (Figure 2.1). The exact

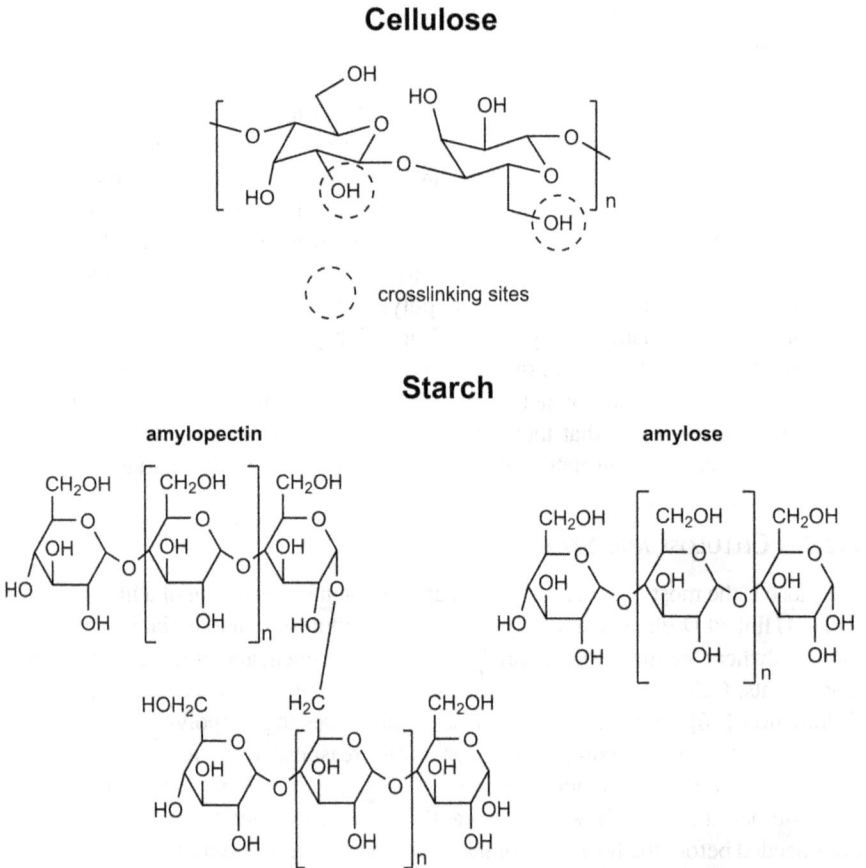

Cellulose

crosslinking sites

Starch

amylopectin   amylose

FIGURE 2.1   Structure of polysaccharides cellulose (up) and starch (down).

composition varies depending on the starch source, but the microstructure of the starch remains identical, regardless of the source. These two types of starch components also have different physical properties because amylose is a semicrystalline biopolymer soluble in hot water, while amylopectin is crystalline and almost insoluble in hot water [40].

Assets of starch-based hydrogels lie in their hydrophilicity; this property is translated as high swelling and hydration capacity since these hydrogels are characterized by absorbing a large volume of water. In fact, starch chains can retain water through two mechanisms, besides direct hydrogen bonding with hydrophilic groups; starch hydrogels can also retain water through hydrophobic groups and cross-linking chains, which increases the total retention capacity [41].

Starch-based hydrogels are suitable for the load and controlled release of many drugs. For example, quercetin, a low molecular weight polycyclic molecule, mainly contains aromatic -OH groups, that form hydrogen bonds with the starch in an aqueous medium [42].

2.2.2 ALGINATES

Alginate is a highly hydrophilic linear chain polysaccharide that contains two types of monosaccharide units in the backbone. These are 1,4-α-L-guluronic acid and 1,4-β-D-mannuronic acid, which allows the mannuronic and guluronic blocks (Figure 2.2). Alginate hydrogels can absorb huge amounts of water up to 20 times its weight.

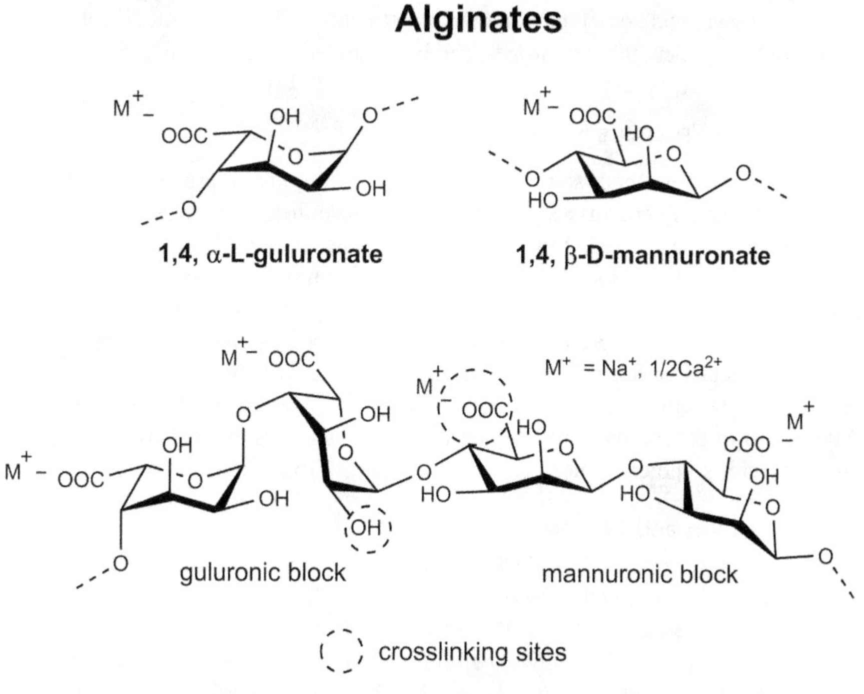

FIGURE 2.2 Structure of alginates consisting of guluronic and mannuronic blocks.

Linear alginate chains contain multiple carboxyl groups able to bind cations such as Ca^{2+} and Ba^{2+} to promote the formation of cross-linked structures [43,44]. The main source of alginates is marine algae of the phylogenetic class *Phaeophyceae* [45], although it is also possible to find alginates in *Azotobacter* [46] and *Pseudomonas* [47], in which biosynthesis is being increasingly studied in recent years as an alternative to marine algae sources. Nevertheless, the extraction of alginates from algae is still the most important [48] and the extraction methods are continuously improved [49]. The alginate has a structure with carboxylate groups, so it can be modified to obtain derivatives with specific properties to retain water inside the hydrogel structure that is necessary for biocompatible hydrogels.

The formation of hydrogels with alginates can be carried out by cross-linking different methods, considering the reactivity of the carboxylates of the mannuronic and guluronic units [50]; for example, through oxidation reactions [51] and sulfonations [52], but the most common are esterification [53], amidation [54], and reductive amination reactions of oxidized alginates with $NaIO_4$ [55] of alginates. The last three cross-linking reactions are especially interesting for alginate hydrogels since esters, amides, and amines are quite stable and have potentially biocompatible groups, as these hydrogels are viable for biomedical applications [18] due to their biocompatibility, availability, low toxicity, a good rate of flexible gelation, and low cost. Therefore, it is a standard tool for further objectives regarding cell cultures for wound healing [56], encapsulation of therapeutic agents [57], and drug delivery [58]. However, alginate hydrogels have some drawbacks, such as slow degradation and poor mechanical stability [59], which have limited their use for certain biomedical applications. But when combined with other biopolymers or structural chain size modifications are carried out [59], then, these hydrogels can be improved in performance aspects.

2.2.3 Non-Vegetable Sources

Differently from vegetal sources, other sources require a previous purification process that usually comprises several steps because these polymers, obtained from animals, must be extracted, and separated from other tissue. The most used proteins to create hydrogels consist of collagen, gelatin, fibrin, silk fibroin, or chitosan. The main difference between cellulose and other natural polymers is the presence of amino acid residues, where the N heteroatom plays a main role in the reactivity of hydrogels because it takes an active part in the hydrogel network and the formation of hydrogen bonding with water molecules. For this reason, protein hydrogels are different from others that only contain carbohydrates or polysaccharides with -OH groups. Here, some examples are mentioned in the following sections.

2.2.3.1 Chitin and Chitosan

The chemical structure of chitin is like cellulose in the way that both polymers have chains linked with ether bonds of the type $R-CH_2-O-CH_2-R$ (Figure 2.3). The difference between chitin and cellulose is that in chitin there are *N*-methylamide groups on the ring [60]. Unlike cellulose, which is abundant in organisms of the plant kingdom, chitin is a polysaccharide that is mainly found in certain species of animal origin, such as arthropods, and in fungi organisms in general, such as crab

Chitin

Chitosan

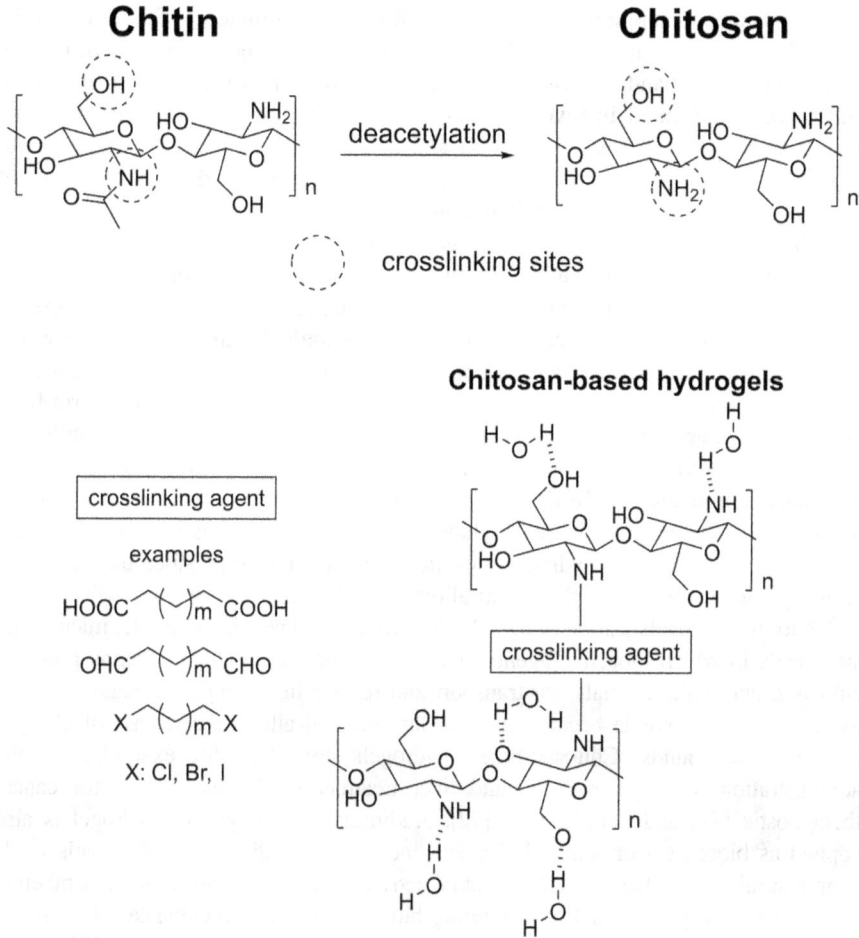

crosslinking sites

crosslinking agent

examples

HOOC $\overset{}{\underset{m}{\wedge}}$ COOH

OHC $\overset{}{\underset{m}{\wedge}}$ CHO

X $\overset{}{\underset{m}{\wedge}}$ X

X: Cl, Br, I

Chitosan-based hydrogels

crosslinking agent

FIGURE 2.3 Natural polymers chitin and chitosan (up), of which hydrogels are formed using a cross-linking agent.

peritrophic membranes, lobsters, and cocoons of insects [61]. In arthropods, chitin is mainly distributed in the exoskeleton, while in fungi, chitin is found in the cell walls [62]. Chitin polymer contains hydroxyl groups that give it a certain hydrophilic behavior and the possibility to obtain chitin-based hydrogels through physical cross-linking; for example, with N-N-dimethylacetamide and lithium chloride (LiCl) aqueous solution by water-vapor induced phase inversion [63]. Nevertheless, the secondary amide can be reduced to the corresponding amine, making it the crucial point in the formation of hydrogels and conferring new physicochemical properties [64]. Deacetylation of chitin is the key before its use as a hydrogel, either in an aqueous alkaline medium or by enzymatic action and this causes the reduction of the amide group to obtain the amine group -NH_2 in a yield that depends on the method employed, giving the formation of chitosan, which is an important polymer for hydrogels. The last words about chitin are its use as an active ingredient in food supplements [65]. Thus, it is not uncommon to find chitin or its derivatives in the design of new drug transporters due to their innocuousness [66].

On the other hand, chitosan (the deacetylated form of chitin) is an amino polysaccharide formed by two fundamental units: β-(1–4) D-glucosamine (which corresponds to the deacetylated units) and N-acetyl-D-glucosamine, which corresponds to the acetylated units (Figure 2.3) [60,67]. There are different methods of obtaining chitosan from chitin [68], but one of the easiest ways to obtain it consists in the elimination of basic salts of the organic source where the chitin was extracted using an acid medium (e.g., HCl 10%), followed by a treatment in an aqueous alkaline medium (e.g., 10% NaOH under heating and constant stirring) to eliminate proteins, excess acids, and other organic components commonly found in chitin; lastly, the deacetylation is carried out in a more concentrated alkaline medium and with a second heating to a base concentration of around 50% to obtain the deacetylated form [69]. Hence, this three-step route, demineralization, deproteinization, and deacetylation, allows chitosan to be easily obtained. Finally, it must be considered that both 100% acetylated chitin and 100% deacetylated chitosan are idealized structures, but in practice, they coexist with deacetylation/acetylation percentages. Therefore, the results when preparing a hydrogel with this polymer may vary since the degree of amine groups available are those that allow cross-linking.

Chitosan hydrogels can be designed with different shapes, for example, micro/nano size beads in which bioactive agents or drugs can be incorporated, such as proteins, enzymes, and antimicrobials, for transport and release in a controlled manner in acid medium, this is possible because the swollen material allows the release of charged bioactive compounds. Chitosan-based hydrogels are ideal, for example, for the administration of drugs in colon and liver treatments [70], as well as for cancer theragnostic [71] and radiolabeling [72]. Additionally, this type of hydrogel is also applied as biosensors or scaffolds for enzyme immobilization [73]. Hydrogels made from natural sources have no antimicrobial activity except chitosan, which has inherent antimicrobial properties, and incorporating bioactive agents can enhance it [74–76].

Chitosan is not soluble in water, but it is quite soluble in acetic acid [77], so the chitosan/acetic acid mixture is usually used prior to the formation of the hydrogel. But once the chitosan is cross-linked, the network has more affinity for water, and more of these molecules are retained in the network, reaching very high swellings [78]; thus, the cross-linked chitosan (after washed with water) can function as a hydrogel, because an acid pH medium is no longer required to obtain a chitosan hydrogel-based material, which is not only biodegradable but also absorbable [79].

2.2.3.2 Collagen and Gelatin

Collagen is a fibrous protein of animal origin generally produced by fibroblast cells and forms the main component of the extracellular matrix and is formulated of three polypeptide chains folded into a triple helix structure [80]. There are different types of collagens, depending on the source; the most common being type I collagen present in the skin, cartilage, and ligaments, obtained through enzymatic and acid processes [81]. The collagen peptide sequence confers low immunogenicity, high hydrophilicity, and stability, allowing its application in tissue engineering due to the cell adhesion it presents. Other applications in the medical field include the processes of general surgery, orthopedics, cardiovascular, dermatology, otorhinolaryngology, urology, dentistry, ophthalmology, plastic, and reconstructive surgery, due to its biodegradability, biocompatibility, availability, and versatility properties. However, collagen shows certain drawbacks, such as unsuitable mechanical properties for hydrogels [81], difficulty in handling, and the possibility of inducing immune reactions, as well as less control over biodegradability and water resistance [82].

The modification of the triple helix structure of collagen through denaturation or physicochemical degradation forms a derivative known as gelatin, which is a high molecular weight polypeptide constituted of glycine, proline, and 4-hydroxy proline residues; its appearance is colorless and translucent, but also edible. For that reason, it can be orally ingested and 100% digested by the organism. Gelatin has a thermosensitive nature and can undergo a sol-gel transition at physiological temperature (37°C), as this is a reversible state if the temperature exceeds 90°C [83]. The cross-linking is a crucial step, either for enzymatic or chemical methods, where the enzymatic cross-linking may be carried out using enzymes such as tyrosinase and transglutaminase [84] or chemically using aldehydes, which is the most common [85], but no matter the method, in any case, it may affect the biocompatibility of the hydrogel.

Gelatin's mechanical properties require slight modifications, using physical, chemical, or enzymatic methods, to improve the material and to maintain a porous structure and biocompatibility [86]. Gelatin is perhaps an ideal candidate in the design of cell scaffolds and other applications such as stabilizers for emulsifiers, foams, colloids, or biodegradable disposables, among other biomimetic materials [87]. Therefore, this hydrogel-based material is perfect for oral administration and a first-pass effect medication needs to be diminished as much as possible.

2.2.3.3 Silk Fibroin

Silk fibroin is a protein produced by the *Bombyx mori* worm, its primary structure consists of a mixture of glycine and alanine amino acids [88]. What is most outstanding about this fiber is that a living organism produces it, and the specimen is not sacrificed for its production; vice versa, the culture of the silkworm is necessary. This fibroin has excellent elasticity, biocompatibility, high toughness, sturdy, and flexibility, as its mechanical properties are superior compared to other synthetic and natural polymers. It is mainly structured of both amorphous and β-sheet crystallite domains [89]. The basic structure contains polypeptide chains; the most abundant amino acids are glycine (Gly) (42.9%), alanine (Ala) (30.0%), and serine (Ser) (12.2%), and the other 13 amino acid residues constitute around 14.9% (Figure 2.4) [90].

Silk protein-based hydrogels may be prepared using other natural and synthetic polymers. There are both physical and chemical hydrogels with special characteristics conferred by the cross-linking method [91]. Physical cross-linking involves changes in the conformational state of silk from random coil to β sheet conformation. The physical methods of cross-linking include self-assembly, ultrasonication, heating, and the use of surfactants. In chemical cross-linking, the silk chains are bonded through covalent bonds, in cross-linking sites, which allow the final stable network structure. Preparation methods include ultraviolet (UV), gamma irradiation, and enzymatic and chemical initiator cross-linking.

Silk fibroin

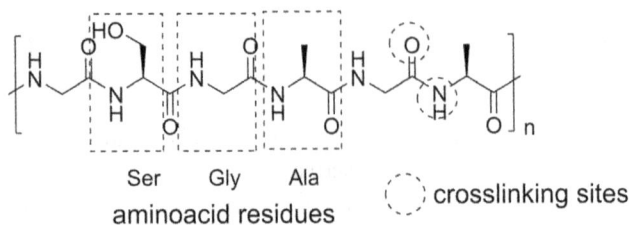

Ser   Gly   Ala
aminoacid residues   ⟨ ⟩ crosslinking sites

FIGURE 2.4 Silk fibroin is a natural polymer that mostly consists of serine, glycine, and alanine residues.

A disadvantage of this natural polymer is that it requires organic solvents such as *N*-methylmorpholine *N*-oxide, since saltwater systems are not strong enough or suitable to dissolve the macromolecule completely [90]. Nonetheless, silk fibroin gels are very interesting due to their improved mechanical properties [92], for that reason, it is possible to find potential applications as scaffolds in tissue engineering, wound dressings, and articular cartilage repair [93].

2.2.3.4   Fibrin

Fibrin is a biopolymer formed by the enzymatic polymerization of fibrinogen in the presence of thrombin with a high water content. This biopolymer is derived from mammals and plays an important role in blood coagulation mechanisms. Some of the drawbacks of fibrin are its low mechanical resistance and rapid degradation that affect its stability. In this sense, agents that inhibit the enzymatic action responsible for degradation are added, controlling to a certain extent its rapid degradation [94]. Their excellent biocompatibility and biodegradability make them suitable for use in the tissue engineering of cartilage [95], bone [96], and adipocytes, as well as a nerve regeneration [97] substrate for cell adhesion and guided cell migration. Other uses include sealant, adhesive in surgery, and for fixation of skin grafting [95].

2.2.3.5   Hyaluronic Acid

Hyaluronic acid (HA) is a macromolecule that has gained relevance in recent years, despite its discovery in the third decade of the last century [98]. The gained relevance of HA is because of its anti-aging properties that are exploited in the cosmetic and reconstructive medicine sector. HA is a natural polysaccharide, specifically a glycos-aminoglycan conformed of D-glucuronic acid and N-acetyl-D-glucosamine units linked by a β (1→3) bond [99]. This macromolecule is found in humans, especially in the skin, but also in skeletal tissue, heart valves, and lungs [100]. Moreover, the presence of HA in the skin is relevant in anti-aging because HA has a unique capacity in retaining water. For production, HA is extracted from natural sources, including rooster combs, shark fins, and umbilical cords, although HA is also obtained by biosynthesis methods [100].

The mechanical properties and hydrophilic groups of HA increase the possibili-ties of cross-linking to form hydrogels, becoming more attractive in the research on HA-based hydrogels. Cross-linking of HA occurs mainly in the carboxylate or acetamide group through condensation reactions [101], but also vinyl cross-linking agents are active in addition reactions to form the HA network. For example, when using divinyl sulfone, the cross-linking is simple at room temperature and relatively safe because solvent-free conditions are enough to produce hydrogels with high cell viability [102]. In general, HA hydrogels are biocompatible, even in hydrogels mixed with proteins such as collagen [103]. HA are absorbable and fast-delivery drug vehicles because of the polymer network architecture, i.e., mesh size, molecular weight, and interactions between drug-polymer that facilitate the diffusion during the release of some drugs. This behavior of HA-based hydrogels was already reported in sodium salicylate fast diffusion [104]. HA hydrogels are not only limited to anti-inflammatory drug delivery systems, but recently HA has begun to be studied as a carrier in delivery systems of anticancer drugs, such as doxorubicin (DOX) [105] and quercetin [106], with the firm aim of improving the cancer cell-targeted chemotherapy [107].

In conclusion, the uses of HA hydrogels have begun to expand beyond cosmetic medicine to become components of oncological materials. This is plausible because HA can be totally absorbed and even degrade itself quickly, so there are no problems associated with bioaccumulation.

2.3 SYNTHETIC HYDROGELS

Hydrogels based on synthetic polymers are a reliable tool of bioengineering. Unlike natural polymers, for synthetic hydrogels, it is possible to control the reaction conditions to design the size of the main polymer chain (backbone), as well as the option to choose the best cross-linking agents to add hydrogels with a fine (and final) customization [108]. The synthesis allows a total control from the beginning and adding advantages such as better durability, a well-known water permeability, precise drug loading capacity, and controlled drug release (for stimuli-responsive polymer materials) [12], besides the versatility of adding a metal [109] or immobilizing an enzyme [110] in a better-ordered structure. Depending on the final objectives for which a hydrogel is designed, which usually includes the possibility of modulating the loading capacity or release rate of a drug, this can be achieved simply with the incorporation of a smart polymer to the hydrogel, usually with response to light, pH, temperature, field magnetic, or ionic strength [16,111]. Either with a physical or chemical cross-linking, or with a gelling agent, the stimulus-response is compatible with other positive properties pursued in biomedical devices [112], either for drug carriers [113] or self-healing systems [114].

Among the most important synthetic hydrogels are vinyl and allyl polymers; these polymers are derivatives from the corresponding vinyl and allyl monomers, and are characteristic due to the terminal alkene group, named vinyl ($R-CH=CH_2$) [115] or allyl ($R-CH_2-CH=CH_2$) (Figure 2.5). Subgroups of relevance are acrylates and methacrylates and are mentioned below with some examples. All these groups of molecules have a similar reactivity that allows addition reactions through various mechanisms, whether radical or ionic, to form the corresponding polymer. The synthesis of these polymers is quite versatile since it can also be obtained through

Vinyl polymer

polymerization

Allyl polymer

polymerization

examples
R = COOH; COOCH$_3$; NH$_2$; OH

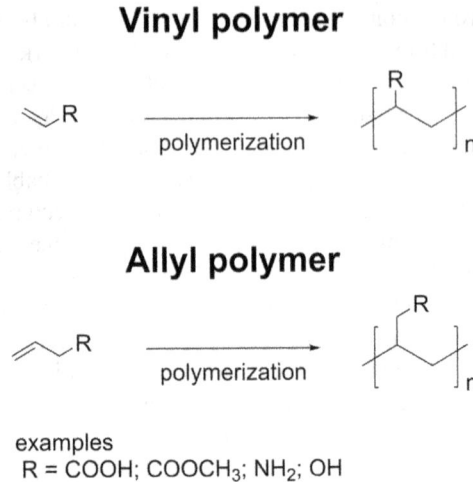

FIGURE 2.5   General structure of vinyl and allyl polymers, the backbone of some synthetic hydrogels.

chemical and physical activation methods, in an aqueous reaction medium or with organic solvents, and even under solvent-free conditions [116]. Another type of synthetic polymer used for hydrogels is based on polyesters and polyamides; the unique problem is their limited swelling capacity, and for that reason their use as hydrogels requires copolymerization or branched structures. These details are mentioned in the next section.

Hydrogels with synthetic components, at first instance, could suppose a certain risk, due to the concern caused by possible problems such as incompatibility, bioaccumulation, or allergy, among other toxicity problems. These problems usually arise when hydrogels are implanted, ingested, injected, or get into contact with the tissue where they are destined [30]. But these fears have no place in modern medicine, and part of basic research contemplates a series of "protocols to follow," including the testing of viability before considering prospects as biomaterials. In this way, those unsafe mixtures for human health and sources of environmental pollution are always rejected [117]. Hence, the implementation of protocols for synthetic biomaterials is obligated to prevent side effects, at least, in the short and medium term [118].

2.3.1   POLYESTERS AND POLYAMIDES

Polyester is a category of polymers that contains the functional group -RCOO- in its main chain, while in its counterpart, polyamides contain the -RCON- group, both structures sharing the carbonyl -RCO group (Figure 2.6). Polyesters and polyamides are inspired in structures found in nature, with synthetics as a better molecular weight control, superior mechanical properties, and higher purity [119]. Polyesters and polyamides can be thermoplastic or thermoset. Synthesis of these polymers includes various methods; among the most used are polycondensation of a diacid

Polyesters and polyamides

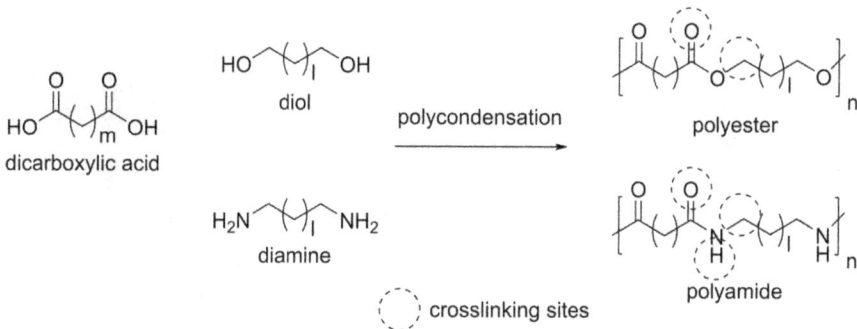

FIGURE 2.6 Polycondensation of dicarboxylic acids with diol and diamines, structures have potential cross-linking sites for hydrogel formation.

and a diol, ring-opening polymerization of cyclic lactones, and alcoholic transesterification [120]. Also, the polyesters and polyamides allow multicomponent condensations or sidechain copolymerizations, which can be directed to adding hydrophilic groups to enhance the water retention capacity, which is necessary in a hydrogel.

An example of structural modification of a polyester-based hydrogel is the "hyperbranching" that can be achieved using small polyol molecules and a carboxylic acid in several condensation steps to control the size of the "dendritic core" in a divergent method. Then, in the next step, vinyl end groups are added by condensation of acryloyl chloride to provide the reactivity in the outside of the macromolecule. This technique guarantees adequate functional groups for polymerization via an addition mechanism, using a photoinitiator that produces the hydrogel [121]. Hyperbranched polyester hydrogels are suitable to encapsulate bioactive agents through non-covalent interactions due to the bulky structure and the cavities inside, which are also adequate for cell adhesion in a hydrated medium.

Copolymerization and ring-opening are other ways to enhance the hydrophilicity [122]; for example, polymers such as poly(ε-caprolactone) (PCL) and poly(ethylene oxide terephthalate)-poly(butylene terephthalate), abbreviated PEOT/PBT, produce hydrogel scaffolds via the electrospinning method [123]. This structure is biocompatible and can mimic an extracellular matrix for in vitro cell culture.

2.3.2 POLYACRYLATES AND POLYMETHACRYLATES

Acrylate and methacrylate polymers are formed from addition reactions. These are usually stable polymers with a high degree of biocompatibility and low toxicity (depending on the use) [124]. This family of molecules is quite large and is identified for the carbonyl group (COOR) of acids, esters, and amides. The reactivity of carbonyl polar group is characteristic because it is polarizable, has a certain degree of hydrophilicity, and has the capability to form multiple hydrogen bonds with water, which is necessary in the network of hydrogels [124]. These are

PAAc/PEG hydrogel

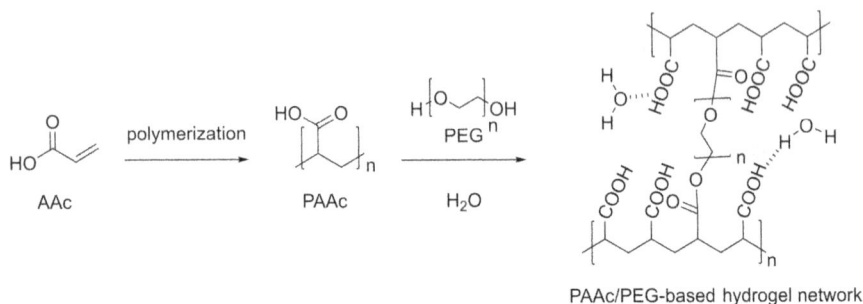

PAAc/PEG-based hydrogel network

FIGURE 2.7   Polymerization of AAc and hydrogel formation with PEG via polycondensation.

the main reasons why the polymeric derivatives of acrylates and methacrylates are of special interest; some representative examples are mentioned as follows.

Acrylic acid (AAc) is the flagship of acrylic hydrogels because this molecule has the carboxylic group (to form hydrogen bonds) and a lineal structure ideal for additional polymerization. For example, the network formation of PAAc hydrogels is easily prepared via a free radical mechanism, where reaction conditions such as monomer concentration, pH, and ionic strength alter the final structure of polymer. For example, the cross-linking of PAAc with PEG produces PAAc/PEG (Figure 2.7), a hydrogel with diester bonds that shows a long-term biocompatibility in vivo [125].

2.3.3   POLYMER AMINES

Amines have multi-reactivity features because, due to their alkaline pH easily, they react with organic and inorganic acids. Besides, primary and secondary amines are nucleophiles [126] and also are well known as ligands of several metal centers. Polymers containing amine groups share reactivity with small monomers; the polarizable C-N and N-H bonds are appropriate to synthesize hydrogels because primary ($R-NH_2$), secondary ($R-NR_2H$), and tertiary ($R-NR_2R_3$) amines form covalent and non-covalent bonds for gelation in an aqueous medium. Usually polymer amine hydrogels are pH-stimulus-responsive, a property useful in cases when the swelling degree must be controlled at a specific pH. This provides a better control on the functionality of the material, being this relevant, for example in drug delivery systems [127] or for an enhanced cell adhesion [128].

Permanent hydrogels with polymeric amines can be synthesized through a condensation reaction of a polymeric amine with a dialdehyde to form the corresponding secondary amine (via imine reduction) and yielding a cross-linked structure. In a specific case, polyallylamine is cross-linked in an acidic medium (e.g., HCl_{aq}) with polyprotic acids, such as phosphoric acid (H_3PO_4), where once the acid-base reaction takes place, strong ionic interactions are formed between the $R-NH_3^+$ residues of the amine with anionic species of phosphates (PO_4^{3-}, HPO_4^{2-}, $H_2PO_4^-$) (Figure 2.8) [129]. This type of hydrogel can work for

Polyallylamine/phosphate hydrogel

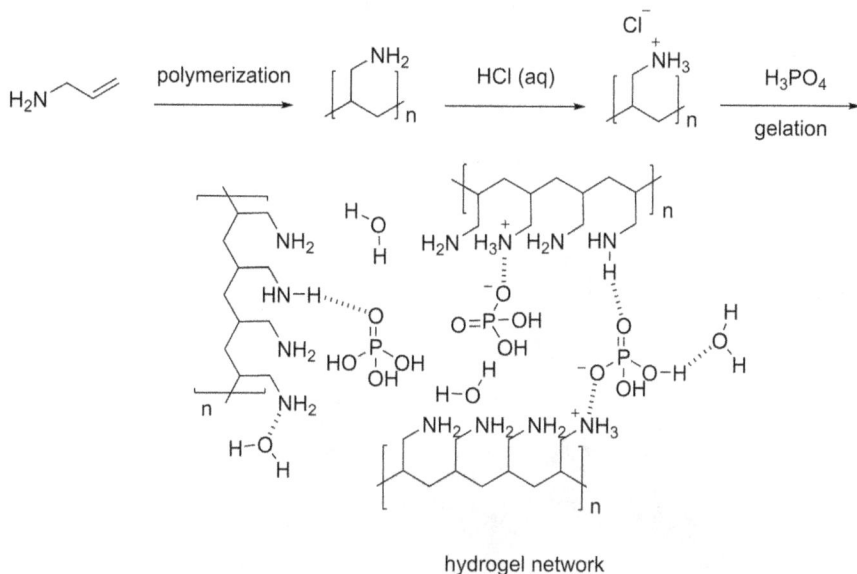

FIGURE 2.8 Polymerization of polyallylamine and interactions present in a hydrogel with phosphoric acid.

controlled release of drugs at an internal physiological pH, which is more acid than in other parts of an organism. Several examples with different types of cross-linking agents can be found, with small molecules such as aldonic acid (a sugar acid) [130], 2,3-dimercaptosuccinic acid [131], and polysaccharides in general, which are modified with pendant groups to favor the gelation reaction with the amines [132].

2.3.4 POLYVINYL ALCOHOL HYDROGELS

Polyvinyl alcohol (PVA) hydrogels have three-dimensional network structures and can absorb a high ratio of water, while biocompatibility is kept. PVA hydrogels are broadly explored due to their stability, reactivity, and resistance to environmental conditions. PVA synthetic polymers have been studied since the 1970s for biomedical applications, which means these are among the first to be developed [133].

Polyvinyl alcohol is a polymer backbone with a double purpose because on the one hand -OH pendant groups form hydrogen bonds with water and on another hand, polyols serve as a site of cross-linking, commonly achieved with dicarbonyl molecules, such as dialdehydes or dicarboxylic acids (Figure 2.9). For example, using citric acid as the cross-linking agent provides excellent water-resistance properties [134]. Also, it is common to find gelation of PVA with aldehydes compounds to form acetals, which are stable compounds. For this reason, chemical cross-linking of permanent hydrogels is widely reported. A PVA-based hydrogel

Polyvinyl alcohol hydrogels

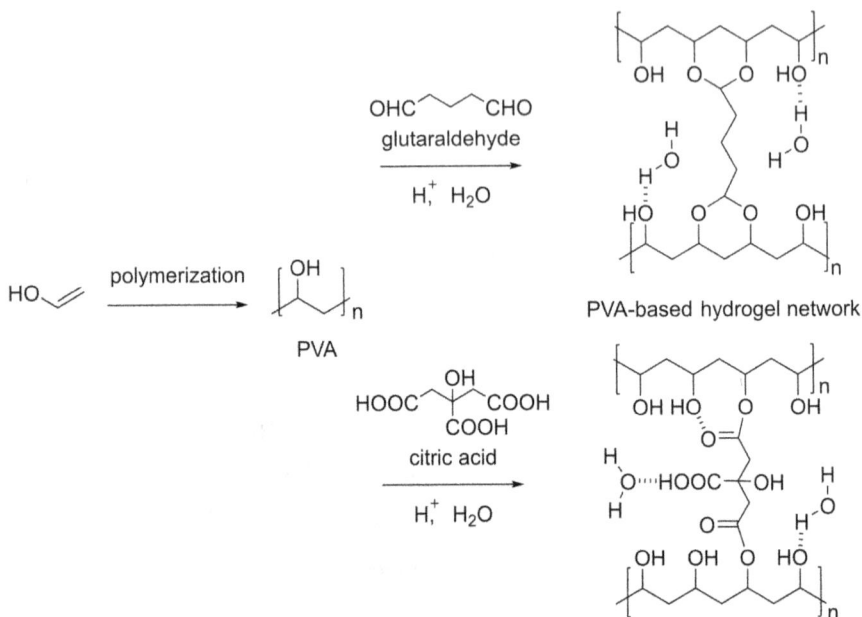

FIGURE 2.9 Cross-linking of PVA using glutaraldehyde (up) and citric acid (down).

prepared with glutaraldehyde is an example of cross-linking forming stable acetals through acid condensation of carbonyl that produces a pH-sensitive hydrogel network [135].

As another example, PVA hydrogels are obtained through a physical cross-linking method. A PVA hydrogel was prepared using a reactant only DMSO (dimethyl sulfoxide) and DMF (dimethyl formamide), which are regular solvents; so with this simple method, a hydrogel structure is obtained exclusively constructed by hydrogen bonds, where hydrogen bonding strength is corroborated [136] with the network of a hydrogel and is exclusively conformed by molecules (no ionic species). Nonetheless, this PVA-based hydrogel has a high water content and reswelling rate, which is able to work in a wide range of temperatures (25–65°C) [137].

2.4 CONCLUSIONS

As we have checked in this brief chapter, there is a wide variety of hydrogels with different attributes, some more favorable than others. On the one hand, most natural hydrogels are biocompatible and even absorbable or biodegradable; on the other hand, synthetic hydrogels can be customized for better performance, for example, adding or blocking a pendant group to control the capacity of water retention under specific reaction conditions. For that reason, the best way to optimize the hydrogel performance would be a combination of natural and synthetic polymer backbones in the same network during the cross-linking or gelation. The final remarks are no more but highlight the versatility that provides the synthesis or modification of hydrogels as well as their direct applications in vast areas of advanced medicine, such as nuclear magnetic resonance imaging, chemotherapy, radiotherapy, gene carriers, drug delivery, and cell culture. Therefore, regardless of their origin (natural or synthetic), hydrogels are well-structured materials obtained in high yields without any problem by standard methods, which is nothing but positive points for hydrogels.

ACKNOWLEDGMENTS

This work was supported by the Dirección General de Asuntos del Personal Académico (DGAPA), Universidad Nacional Autónoma de México under Grant IN 204223. University of Guadalajara under PROSNI 2021. Support Program for Technological Research and Innovation Projects under PAPIIT IG100220 and National Council of Science and Technology under CONAHCyT CF-19 No 140617. Call for basic scientific research CONACYT 2017–2018 del "Fondo Sectorial de Investigación para la educación CB2017-2018" (A1-S-29789). Universidad Autónoma del Estado de México grant number 6187/2020CIF. FLS (CVU 409872) and GGFR (CVU 407270) thanks to CONAHCyT for the grant awarded during postdoctoral stay.

REFERENCES

[1] Kondiah PJ, Choonara YE, Kondiah PPD, Marimuthu T, Kumar P, Toit LC, Pillay V A review of injectable polymeric hydrogel systems for application in bone tissue engineering. Molecules. 2016;21(11):1580.

[2] Maitra J, Shukla VK Cross-linking in hydrogels - a review. Am J Polym Sci. 2014;4(2):25–31.

[3] Gabriele F, Donnadio A, Casciola M, Germani R, Spreti N Ionic and covalent cross-linking in chitosan-succinic acid membranes: Effect on physicochemical properties. Carbohydr Polym. 2021 Jan;251:117106.

[4] Parhi R Cross-linked hydrogel for pharmaceutical applications: A review. Adv Pharm Bull. 2017 Dec 31;7(4):515–530.

[5] Rizwan M, Yahya R, Hassan A, Yar M, Azzahari AD, Selvanathan V, Sonsudin F, Abouloula CN pH sensitive hydrogels in drug delivery: Brief history, properties, swelling, and release mechanism, material selection and applications. Polymers (Basel). 2017;9(137):1–137.

[6] Hendi A, Umair Hassan M, Elsherif M, Alqattan B, Park S, Yetisen AK, Butt H Healthcare applications of pH-sensitive hydrogel-based devices: A review. Int J Nanomedicine. 2020 Jun;15:3887–3901.

[7] Kocak G, Tuncer C, Bütün V pH-Responsive polymers. Polym Chem. 2017;8(1):144–176.

[8] Zarrintaj P, Jouyandeh M, Ganjali MR, Hadavand BS, Mozafari M, Sheiko SS, Vatankhah-Varnoosfaderani M, Gutiérrez TJ, Saeb MR Thermo-sensitive polymers in medicine: A review. Eur Polym J. 2019;117:402–423.

[9] Shi Z, Gao X, Ullah MW, Li S, Wang Q, Yang G Electroconductive natural polymer-based hydrogels. Biomaterials. 2016;111:40–54.

[10] Qiu Y, Park K Environment-sensitive hydrogels for drug delivery. Adv Drug Deliv Rev. 2001;53(3):321–339.

[11] Elliott JE, Macdonald M, Nie J, Bowman CN Structure and swelling of poly(acrylic acid) hydrogels: Effect of pH, ionic strength, and dilution on the cross-linked polymer structure. Polymer (Guildf). 2004 Mar;45(5):1503–1510.

[12] López-Barriguete JE, Isoshima T, Bucio E Development and characterization of thermal responsive hydrogel films for biomedical sensor application. Mater Res Express. 2018;5(4):45703.

[13] Zhang S, Xu K, Ge L, Darabi MA, Xie F, Derakhshanfar S, Liu Y, Xing MMQ, Wei H A novel nano-silver coated and hydrogel-impregnated polyurethane nanofibrous mesh for ventral hernia repair. RSC Adv. 2016;6(93):90571–90578.

[14] Rao MK, Rao SK, Ha C-S Stimuli responsive poly(vinyl caprolactam) gels for biomedical applications. Gels. 2016;2(1):6–24.

[15] Siddiqui MN, Redhwi HH, Tsagkalias I, Softas C, Ioannidou MD, Achilias DS Synthesis and characterization of poly(2-hydroxyethyl methacrylate)/silver hydrogel nanocomposites prepared via in situ radical polymerization. Thermochim Acta. 2016;643:53–64.

[16] Aguilar MR, Elvira C, Gallardo A, Vázquez B, Román JS Smart polymers and their applications as biomedicals. In: Ashammakhi N, Reis R, Chiellini E, editors. Topics in Tissue Engineering. 2007. pp 225–248, Oulu, Finland.

[17] Wu F, Pang Y, Liu J Swelling-strengthening hydrogels by embedding with deformable nanobarriers. Nat Commun. 2020 Sep 9;11(1):4502.

[18] Cabral J, Moratti SC Hydrogels for biomedical applications. Future Med Chem. 2011 Nov;3(15):1877–1888.

[19] Pertici V, Pin-Barre C, Rivera C, Pellegrino C, Laurin J, Gigmes D, Trimaille T Degradable and injectable hydrogel for drug delivery in soft tissues. Biomacromolecules. 2019;20(1):149–163.

[20] Wang G, Yu D, Kelkar AD, Zhang L Electrospun nanofiber: Emerging reinforcing filler in polymer matrix composite materials. Prog Polym Sci. 2017;75:73–107.

[21] Nasef MM, Gupta B, Shameli K, Verma C, Ali RR, Ting TM Engineered bioactive polymeric surfaces by radiation induced graft copolymerization: Strategies and applications. Polymers (Basel). 2021 Sep 15;13(18):3102.

[22] Wechsler ME, Stephenson RE, Murphy AC, Oldenkamp HF, Singh A, Peppas NA Engineered microscale hydrogels for drug delivery, cell therapy, and sequencing. Biomed Microdevices. 2019 Jun 23;21(2):31.

[23] Hoffman AS Hydrogels for biomedical applications. Adv Drug Deliv Rev. 2012 Dec;64:18–23.

[24] Tibbitt MW, Anseth KS Hydrogels as extracellular matrix mimics for 3D cell culture. Biotechnol Bioeng. 2009 Jul 1;103(4):655–663.

[25] Mateescu M, Baixe S, Garnier T, Jierry L, Ball V, Haikel Y, Metz-Boutigue MH, Nardin M, Schaaf P, Etienne O, Lavalle P Antibacterial peptide-based gel for prevention of medical implanted-device infection.PLoS One. 2015 Dec 14;10(12):e0145143–e0145155.

[26] Mantha S, Pillai S, Khayambashi P, Upadhyay A, Zhang Y, Tao O, Pham HM, Tran SD Smart hydrogels in tissue engineering and regenerative medicine. Materials (Basel). 2019 Oct 12;12(20):3323.

[27] Gallagher AG, Alorabi JA, Wellings DA, Lace R, Horsburgh MJ, Williams RL A novel peptide hydrogel for an antimicrobial bandage contact lens. Adv Healthc Mater. 2016;5(16):2013–2018.

[28] Haque O, Mondal IH Cellulose-based hydrogel for personal hygiene applications. In: Mondal M, editor. Cellulose-Based Superabsorbent Hydrogels Polymers and Polymeric Composites: A Reference Series. Cham: Springer; 2018. pp. 1–21.

[29] Lv Q, Wu M, Shen Y Enhanced swelling ratio and water retention capacity for novel super-absorbent hydrogel. Colloids Surfaces A Physicochem Eng Asp. 2019 Dec;583:123972.

[30] Gyles DA, Castro LD, Silva JOC, Ribeiro-Costa RM A review of the designs and prominent biomedical advances of natural and synthetic hydrogel formulations. Eur Polym J. 2017 Mar;88:373–392.

[31] Catoira MC, Fusaro L, Di Francesco D, Ramella M, Boccafoschi F Overview of natural hydrogels for regenerative medicine applications. J Mater Sci Mater Med. 2019 Oct 10;30(10):115.

[32] Khan MJ, Svedberg A, Singh AA, Ansari MS, Karim Z Use of nanostructured polymer in the delivery of drugs for cancer therapy. In: Swain SK, Jawaid M, editors. Nanostructured Polymer Composites for Biomedical Applications. Amsterdam: Elsevier; 2019. pp. 261–276.

[33] Cui X, Lee JJL, Chen WN Eco-friendly and biodegradable cellulose hydrogels produced from low cost okara: Towards non-toxic flexible electronics. Sci Rep. 2019 Dec 3;9(1):18166.

[34] Gupta K, Patel R, Dias M, Ishaque H, White K, Olabisi R Development of an electroactive hydrogel as a scaffold for excitable tissues. Int J Biomater. 2021 Jan 30;2021:1–9.

[35] Caló E, Khutoryanskiy VV Biomedical applications of hydrogels: A review of patents and commercial products. Eur Polym J. 2015 Apr;65:252–267.

[36] Etale A, Onyianta AJ, Turner SR, Eichhorn SJ Cellulose: A review of water interactions, applications in composites, and water treatment. Chem Rev. 2023 Mar 8;123(5):2016–2048.

[37] Nasution H, Harahap H, Dalimunthe NF, Ginting MHS, Jaafar M, Tan OOH, Aruan HK, Herfananda AL Hydrogel and effects of cross-linking agent on cellulose-based hydrogels: A review. Gels. 2022 Sep 7;8(9):568.

[38] Zainal SH, Mohd NH, Suhaili N, Anuar FH, Lazim AM, Othaman R Preparation of cellulose-based hydrogel: A review. J Mater Res Technol. 2021 Jan;10:935–952.

[39] Yahia EM, Carrillo-López A, Bello-Perez LA Carbohydrates. In: Yahia EM, editor. Postharvest Physiology and Biochemistry of Fruits and Vegetables. Duxford, United Kingdom: Woodhead Publishing; 2019. pp. 175–205.

[40] Martens BMJ, Gerrits WJJ, Bruininx EMAM, Schols HA Amylopectin structure and crystallinity explains variation in digestion kinetics of starches across botanic sources in an in vitro pig model. J Anim Sci Biotechnol. 2018 Dec 29;9(1):91.

[41] Ismail H, Irani M, Ahmad Z Starch-based hydrogels: Present status and applications. Int J Polym Mater. 2013 Mar;62(7):411–420.

[42] Moghadam M, Seyed Dorraji MS, Dodangeh F, Ashjari HR, Mousavi SN, Rasoulifard MH Design of a new light curable starch-based hydrogel drug delivery system to improve the release rate of quercetin as a poorly water-soluble drug. Eur J Pharm Sci. 2022 Jul;174:106191.

[43] Hamai R, Anada T, Suzuki O Novel scaffold composites containing octacalcium phosphate and their role in bone repair. In: Octacalcium Phosphate Biomaterials. 1st ed. Amsterdam: Elsevier; 2020. pp. 121–145.

[44] Ågren MS Wound Healing Biomaterials. Cambridge: Elsevier; 2016. p. 519.

[45] Wouthuyzen S, Herandarudewi SMC, Komatsu T Stock assessment of brown seaweeds (Phaeophyceae) along the Bitung-Bentena coast, North Sulawesi province, Indonesia for alginate product using satellite remote sensing. Procedia Environ Sci. 2016;33:553–561.

[46] Clementi F Alginate production by Azotobacter vinelandii. Crit Rev Biotechnol. 1997 Jan 27;17(4):327–361.

[47] Valentine ME, Kirby BD, Withers TR, Johnson SL, Long TE, Hao Y, Lam JS, Niles RM, Yu HD Generation of a highly attenuated strain of Pseudomonas aeruginosa for commercial production of alginate. Microb Biotechnol. 2020 Jan;13(1):162–175.

[48] Sellimi S, Younes I, Ayed H Ben, Maalej H, Montero V, Rinaudo M, Dahia M, Mechichi T, Hajji M, Nasri M Structural, physicochemical and antioxidant properties of sodium alginate isolated from a Tunisian brown seaweed. Int J Biol Macromol. 2015 Jan;72:1358–1367.

[49] Trica, D, Gros, U, Dobre, D, Michaud, O Extraction and characterization of alginate from an edible brown seaweed (Cystoseira barbata) harvested in the Romanian Black Sea. Mar Drugs. 2019 Jul 8;17(7):405.

[50] Yang J-S, Xie Y-J, He W Research progress on chemical modification of alginate: A review. Carbohydr Polym. 2011 Feb;84(1):33–39.

[51] Liu H, Zhou H, Lan H, Liu T, Liu X, Yu H 3D printing of artificial blood vessel: Study on multi-parameter optimization design for vascular molding effect in alginate and gelatin. Micromachines. 2017 Jul 31;8(8):237.

[52] Yu M, Ji Y, Qi Z, Cui D, Xin G, Wang B, Cao Y, Wang D Anti-tumor activity of sulfated polysaccharides from Sargassum fusiforme. Saudi Pharm J. 2017 May;25(4):464–468.

[53] Chen X, Zhu Q, Liu C, Li D, Yan H, Lin Q Esterification of alginate with alkyl bromides of different carbon chain lengths via the bimolecular nucleophilic substitution reaction: Synthesis, characterization, and controlled release performance. Polymers (Basel). 2021 Sep 30;13(19):3351.

[54] Heo EY, Ko NR, Bae MS, Lee SJ, Choi B-J, Kim JH, Kim HK, Park SA, Kwon IK Novel 3D printed alginate–BFP1 hybrid scaffolds for enhanced bone regeneration. J Ind Eng Chem. 2017 Jan;45:61–67.

[55] Kang H-A, Jeon G-J, Lee M-Y, Yang J-W Effectiveness test of alginate-derived polymeric surfactants. J Chem Technol Biotechnol. 2002 Feb;77(2):205–210.

[56] Aderibigbe B, Buyana B Alginate in wound dressings. Pharmaceutics. 2018 Apr 2;10(2):42.

[57] Dhamecha D, Movsas R, Sano U, Menon JU Applications of alginate microspheres in therapeutics delivery and cell culture: Past, present and future. Int J Pharm. 2019 Oct;569:118627.

[58] Ray P, Maity M, Barik H, Sahoo GS, Hasnain MS, Hoda MN, Nayak AK Alginate-based hydrogels for drug delivery applications. In: Nayak AK, Hasnain S, editors. Alginates in Drug Delivery. Cambridge: Academic Press; 2020. pp. 41–70.

[59] Kong HJ, Kaigler D, Kim K, Mooney DJ Controlling rigidity and degradation of alginate hydrogels via molecular weight distribution. Biomacromolecules. 2004 Sep 1;5(5):1720–1727.

[60] Ravi Kumar MN A review of chitin and chitosan applications. React Funct Polym. 2000 Nov;46(1):1–27.

[61] Moeini A, Pedram P, Makvandi P, Malinconico M, Gomez d'Ayala G. Wound healing and antimicrobial effect of active secondary metabolites in chitosan-based wound dressings: A review. Carbohydr Polym. 2020 Apr;233:115839.

[62] Brown HE, Esher SK, Alspaugh JA Chitin: A "hidden figure" in the fungal cell wall. Curr Top Microbiol Immunol. 2020;425:83–111.

[63] Nguyen KD, Kobayashi T Chitin hydrogels prepared at various lithium chloride/ N,N-dimethylacetamide solutions by water vapor-induced phase inversion. J Chem. 2020 Dec 5;2020:1–16.

[64] Muzzarelli RAA, Muzzarelli C Chitin and chitosan hydrogels. In: Phillips GO, Williams PA, editors. Handbook of Hydrocolloids. Cambridge: Woodhead Publishing; 2009. pp. 849–888.

[65] Lv J, Lv X, Ma M, Oh D-H, Jiang Z, Fu X Chitin and chitin-based biomaterials: A review of advances in processing and food applications. Carbohydr Polym. 2023 Jan;299:120142.

[66] Miyazaki S, Ishii K, Nadai T The use of chitin and chitosan as drug carriers. Chem Pharm Bull. 1981;29(10):3067–3069.

[67] Pillai CKS, Paul W, Sharma CP Chitin and chitosan polymers: Chemistry, solubility and fiber formation. Prog Polym Sci. 2009;34(7):641–678.

[68] Kou S (Gabriel), Peters LM, Mucalo MR Chitosan: A review of sources and preparation methods. Int J Biol Macromol. 2021 Feb;169:85–94.

[69] No HK, Meyers SP Preparation and characterization of chitin and chitosan—A review. J Aquat Food Prod Technol. 1995 Oct 3;4(2):27–52.

[70] Kulkarni N, Jain P, Shindikar A, Suryawanshi P, Thorat N Advances in the colon-targeted chitosan based multiunit drug delivery systems for the treatment of inflammatory bowel disease. Carbohydr Polym. 2022 Jul;288:119351.

[71] Li X, Wang Y, Feng C, Chen H, Gao Y Chemical modification of chitosan for developing cancer nanotheranostics. Biomacromolecules. 2022 Jun 13;23(6):2197–2218.

[72] Ekinci M, Koksal-Karayildirim C, Ilem-Ozdemir D Radiolabeled methotrexate loaded chitosan nanoparticles as imaging probe for breast cancer: Biodistribution in tumor-bearing mice. J Drug Deliv Sci Technol. 2023 Dec;80:104146.

[73] Smith SK, Lugo-Morales LZ, Tang C, Gosrani SP, Lee CA, Roberts JG, Morton SW, McCarty GS, Khan SA, Sombers LA Quantitative comparison of enzyme immobilization strategies for glucose biosensing in real-time using fast-scan cyclic voltammetry coupled with carbon-fiber microelectrodes. ChemPhysChem. 2018 May 22;19(10):1197–1204.

[74] Tsai G-J, Su W-H Antibacterial activity of shrimp chitosan against Escherichia coli. J Food Prot. 1999 Mar;62(3):239–243.

[75] Kumar S, Mukherjee A, Dutta J Chitosan based nanocomposite films and coatings: Emerging antimicrobial food packaging alternatives. Trends Food Sci Technol. 2020 Mar;97:196–209.

[76] Kaczmarek MB, Struszczyk-Swita K, Li X, Szczęsna-Antczak M, Daroch M Enzymatic modifications of chitin, chitosan, and chitooligosaccharides. Front Bioeng Biotechnol. 2019 Sep 27;7:243.

[77] Lu S, Song X, Cao D, Chen Y, Yao K Preparation of water-soluble chitosan. J Appl Polym Sci. 2004 Mar 15;91(6):3497–3503.

[78] Michalik R, Wandzik I A mini-review on chitosan-based hydrogels with potential for sustainable agricultural applications. Polymers (Basel). 2020 Oct 21;12(10):2425.

[79] Sanchez-Salvador JL, Balea A, Monte MC, Negro C, Blanco A Chitosan grafted/cross-linked with biodegradable polymers: A review. Int J Biol Macromol. 2021 May;178:325–343.

[80] Antoine EE, Vlachos PP, Rylander MN Review of collagen I hydrogels for bioengineered tissue microenvironments: Characterization of mechanics, structure, and transport. Tissue Eng Part B Rev. 2014 Dec;20(6):683–696.

[81] Sarrigiannidis SO, Rey JM, Dobre O, González-García C, Dalby MJ, Salmeron-Sanchez M A tough act to follow: collagen hydrogel modifications to improve mechanical and growth factor loading capabilities. Mater Today Bio. 2021 Mar;10:100098.

[82] Sommer I, Kunz PM Improving the water resistance of biodegradable collagen films. J Appl Polym Sci. 2012 Sep 25;125(S2):E27–E41.

[83] Suwa Y, Nam K, Ozeki K, Kimura T, Kishida A, Masuzawa T Thermal denaturation behavior of collagen fibrils in wet and dry environment. J Biomed Mater Res Part B Appl Biomater. 2016 Apr;104(3):538–545.

[84] Chen R-N, Ho H-O, Sheu M-T Characterization of collagen matrices cross-linked using microbial transglutaminase. Biomaterials. 2005 Jul;26(20):4229–4235.

[85] Abou-Yousef H, Dacrory S, Hasanin M, Saber E, Kamel S Biocompatible hydrogel based on aldehyde-functionalized cellulose and chitosan for potential control drug release. Sustain Chem Pharm. 2021 Jun;21:100419.

[86] Duconseille A, Astruc T, Quintana N, Meersman F, Sante-Lhoutellier V Gelatin structure and composition linked to hard capsule dissolution: A review. Food Hydrocoll. 2015 Jan;43:360–376.

[87] Fang X, Xie J, Zhong L, Li J, Rong D, Li X, Ouyang J Biomimetic gelatin methacrylamide hydrogel scaffolds for bone tissue engineering. J Mater Chem B. 2016 Aug 1;4(6):1070–1080.

[88] Zuluaga-Vélez A, Quintero-Martinez A, Orozco LM, Sepúlveda-Arias JC Silk fibroin nanocomposites as tissue engineering scaffolds – A systematic review. Biomed Pharmacother. 2021 Sep;141:111924.

[89] Cheng Y, Koh L-D, Li D, Ji B, Han M-Y, Zhang Y-W On the strength of β-sheet crystallites of Bombyx mori silk fibroin. J R Soc Interface. 2014 Jul 6;11(96):20140305.

[90] Sashina ES, Bochek AM, Novoselov NP, Kirichenko DA Structure and solubility of natural silk fibroin. Russ J Appl Chem. 2006 Jun;79(6):869–876.

[91] Zheng H, Zuo B Functional silk fibroin hydrogels: preparation, properties and applications. J Mater Chem B. 2021;9(5):1238–1258.

[92] Kim U-J, Park J, Li C, Jin H-J, Valluzzi R, Kaplan DL Structure and properties of silk hydrogels. Biomacromolecules. 2004 May 1;5(3):786–792.

[93] Ki CS, Park YH, Jin H-J Silk protein as a fascinating biomedical polymer: Structural fundamentals and applications. Macromol Res. 2009 Dec;17(12):935–942.

[94] Janmey PA, Winer JP, Weisel JW Fibrin gels and their clinical and bioengineering applications. J R Soc Interface. 2009 Jan 6;6(30):1–10.

[95] Noori A, Ashrafi SJ, Vaez-Ghaemi R, Hatamian-Zaremi A, Webster TJ A review of fibrin and fibrin composites for bone tissue engineering. Int J Nanomedicine. 2017 Jul;12:4937–4961.

[96] Pathmanapan S, Periyathambi P, Anandasadagopan SK Fibrin hydrogel incorporated with graphene oxide functionalized nanocomposite scaffolds for bone repair — In vitro and in vivo study. Nanomedicine Nanotechnology, Biol Med. 2020 Oct;29:102251.

[97] Yu Z, Li H, Xia P, Kong W, Chang Y, Fu C, Wang K, Yang X, Qi Z Application of fibrin-based hydrogels for nerve protection and regeneration after spinal cord injury. J Biol Eng. 2020 Dec 3;14(1):22.

[98] Pérez LA, Hernández R, Alonso JM, Pérez-González R, Sáez-Martínez V Hyaluronic acid hydrogels cross-linked in physiological conditions: Synthesis and biomedical applications. Biomedicines. 2021 Aug 30;9(9):1113.

[99] Nobile V, Buonocore D, Michelotti A, Marzatico F Anti-aging and filling efficacy of six types hyaluronic acid based dermo-cosmetic treatment: Double blind, randomized clinical trial of efficacy and safety. J Cosmet Dermatol. 2014 Dec 17;13(4):277–287.

[100] Papakonstantinou E, Roth M, Karakiulakis G Hyaluronic acid: A key molecule in skin aging. Dermatoendocrinol. 2012 Jul 27;4(3):253–258.

[101] Xu X, Jha AK, Harrington DA, Farach-Carson MC, Jia X Hyaluronic acid-based hydrogels: from a natural polysaccharide to complex networks. Soft Matter. 2012;8(12):3280.

[102] Borzacchiello A, Russo L, Malle BM, Schwach-Abdellaoui K, Ambrosio L Hyaluronic acid based hydrogels for regenerative medicine applications. Biomed Res Int. 2015;2015:1–12.

[103] Xu Q, Torres JE, Hakim M, Babiak PM, Pal P, Battistoni CM, Nguyen M, Panitch A, Solorio L, Liu JC Collagen- and hyaluronic acid-based hydrogels and their biomedical applications. Mater Sci Eng R Reports. 2021 Oct;146:100641.

[104] Vanoli V, Delleani S, Casalegno M, Pizzetti F, Makvandi P, Haugen H, Mele A, Rossi F, Castiglione F Hyaluronic acid-based hydrogels: Drug diffusion investigated by HR-MAS NMR and release kinetics. Carbohydr Polym. 2023 Feb;301:120309.

[105] Kim J, Moon M, Kim D, Heo S, Jeong Y Hyaluronic acid-based nanomaterials for cancer therapy. Polymers (Basel). 2018 Oct 12;10(10):1133.

[106] Jia Y, Chen S, Wang C, Sun T, Yang L Hyaluronic acid-based nano drug delivery systems for breast cancer treatment: Recent advances. Front Bioeng Biotechnol. 2022 Aug 24;10:990145.

[107] Dosio F, Arpicco S, Stella B, Fattal E Hyaluronic acid for anticancer drug and nucleic acid delivery. Adv Drug Deliv Rev. 2016 Feb;97:204–236.

[108] Madduma-Bandarage USK, Madihally SV Synthetic hydrogels: Synthesis, novel trends, and applications. J Appl Polym Sci. 2021 May 15;138(19):50376.

[109] López-Barriguete JE, Flores-Rojas GG, López-Saucedo F, Isoshima T, Bucio E Improving thermo-responsive hydrogel films by gamma rays and loading of Cu and Ag nanoparticles. J Appl Polym Sci. 2021 Feb 15;138(7):49841.

[110] Meyer J, Meyer L, Kara S Enzyme immobilization in hydrogels: A perfect liaison for efficient and sustainable biocatalysis. Eng Life Sci. 2022 Mar 21;22(3–4):165–177.

[111] Kabanov VY Radiation chemistry of smart polymers (A Review). High Energy Chem. 2000 Jul 11;34(4):203–211.

[112] Bajpai AK, Bajpai J, Kumar R, Agrawal P, Tiwari A Smart Biomaterial Devices: Polymers in Biomedical Sciences. I. Boca Raton, FL: CRC Press; 2017. p. 448.

[113] Liu G, Lovell JF, Zhang L, Zhang Y Stimulus-responsive nanomedicines for disease diagnosis and treatment. Int J Mol Sci. 2020 Sep 2;21(17):6380.

[114] Hasanzadeh M, Shahidi M, Kazemipour M Application of EIS and EN techniques to investigate the self-healing ability of coatings based on microcapsules filled with linseed oil and CeO_2 nanoparticles. Prog Org Coatings. 2015;80:106–119.

[115] Lu Y, Larock RC Novel polymeric materials from vegetable oils and vinyl monomers: Preparation, properties, and applications. ChemSusChem. 2009 Feb 23;2(2):136–147.

[116] Satoh K Controlled/living polymerization of renewable vinyl monomers into bio-based polymers. Polym J. 2015 Aug 13;47(8):527–536.

[117] Nikolić LB, Zdravković AS, Nikolić VD, Ilić-Stojanović SS Synthetic hydrogels and their impact on health and environment. In: Mondal M, editor. Cellulose-Based Superabsorbent Hydrogels Polymers and Polymeric Composites: A Reference Series. Cham: Springer; 2018. pp. 1–29.

[118] Ahmad Z, Salman S, Khan SA, Amin A, Rahman ZU, Al-Ghamdi YO, Akhtar K, Bakhsh EM, Khan SB Versatility of hydrogels: From synthetic strategies, classification, and properties to biomedical applications. Gels 2022 Mar 7;8(3):167.

[119] Boddu SHS, Bhagav P, Karla PK, Jacob S, Adatiya MD, Dhameliya TM, Ranch KM, Tiwari AK Polyamide/poly(amino acid) polymers for drug delivery. J Funct Biomater. 2021 Oct 8;12(4):58.

[120] Mighani H Synthesis of thermally stable polyesters. In: Saleh HM, editor. Polyester. London: IntechOpen; 2012.

[121] Zhang H, Patel A, Gaharwar AK, Mihaila SM, Iviglia G, Mukundan S, Bae H, Yang H, Khademhosseini A Hyperbranched polyester hydrogels with controlled drug release and cell adhesion properties. Biomacromolecules. 2013 May 13;14(5):1299–1310.

[122] Lina G, Chun F, Dong Y, Yaogong L, Jianhua H, Guolin L, Xiaoyu H PPEGMEA-g-PDEAEMA: Double hydrophilic double-grafted copolymer stimuli-responsive to both pH and salinity. J Polym Sci Part A Polym Chem. 2009;47(12):3142–3153.

[123] Gonçalves de Pinho AR, Odila I, Leferink A, van Blitterswijk C, Camarero-Espinosa S, Moroni L Hybrid polyester-hydrogel electrospun scaffolds for tissue engineering applications. Front Bioeng Biotechnol. 2019 Sep 25;7:231.

[124] Serrano-Aroca Á, Deb S Acrylic-based materials for biomedical and bioengineering applications. In: Serrano-Aroca A, Deb S, editors. Acrylate Polymers for Advanced Applications. London, United Kingdom: IntechOpen; 2020.

[125] Luo Zheng L, Vanchinathan V, Dalal R, Noolandi J, Waters DJ, Hartmann L, Cochran JR, Frank CW, Yu CQ, Ta CN Biocompatibility of poly(ethylene glycol) and poly(acrylic acid) interpenetrating network hydrogel by intrastromal implantation in rabbit cornea. J Biomed Mater Res Part A. 2015 Oct;103(10):3157–3165.

[126] Kanzian T, Nigst TA, Maier A, Pichl S, Mayr H Nucleophilic reactivities of primary and secondary amines in acetonitrile. European J Org Chem. 2009 Dec;2009(36):6379–6385.

[127] Yoshida T, Lai TC, Kwon GS, Sako K pH- and ion-sensitive polymers for drug delivery. Expert Opin Drug Deliv. 2013;10(11):1497–1513.

[128] Han J-H, Kim J, Acter S, Kim Y, Lee H-N, Chang H-K, Suh K-D, Kim JW Uniform hollow-structured poly(vinyl amine) hydrogel microparticles with controlled mesh property and enhanced cell adhesion. Polymer (Guildf). 2014 Mar;55(5):1143–1149.

[129] Elsiddig R, O'Reilly NJ, Hudson SP, Owens E, Hughes H, O'Grady D, McLoughlin P The influence of poly(allylamine hydrochloride) hydrogel cross-linking density on its thermal and phosphate binding properties. Int J Pharm. 2022 Jun;621:121806.

[130] Andrews MA, Figuly GD, Chapman JS, Hunt TW, Glunt CD, Rivenbark JA, Chenault HK Antimicrobial hydrogels formed by cross-linking polyallylamine with aldaric acid derivatives. J Appl Polym Sci. 2011 Mar 15;119(6):3244–3252.

[131] Mohammadi Z, Shangbin S, Berkland C, Liang J Chelator-mimetic multi-functionalized hydrogel: Highly efficient and reusable sorbent for Cd, Pb, and As removal from waste water. Chem Eng J. 2017 Jan;307:496–502.

[132] O'Connor NA, Abugharbieh A, Yasmeen F, Buabeng E, Mathew S, Samaroo D, Cheng H-P The cross-linking of polysaccharides with polyamines and dextran–polyallylamine antibacterial hydrogels. Int J Biol Macromol. 2015 Jan;72:88–93.

[133] Wang M, Bai J, Shao K, Tang W, Zhao X, Lin D, Huang S, Chen C, Ding Z, Ye J Poly(vinyl alcohol) hydrogels: The old and new functional materials. Int J Polym Sci. 2021 Nov 30;2021:1–16.

[134] Shijie Xu, Zhang P, Ma W, Yang H, Cao Z, Gong F, Zhong J High water resistance polyvinyl alcohol hydrogel film prepared by melting process combining with citric acid cross-linking. Polym Sci Ser B. 2022 Apr 21;64(2):198–208.

[135] Mansur HS, Sadahira CM, Souza AN, Mansur AAP FTIR spectroscopy characterization of poly (vinyl alcohol) hydrogel with different hydrolysis degree and chemically cross-linked with glutaraldehyde. Mater Sci Eng C. 2008 May;28(4):539–548.

[136] Wendler K, Thar J, Zahn S, Kirchner B Estimating the hydrogen bond energy. J Phys Chem A. 2010 Sep 9;114(35):9529–9536.

[137] Ma S, Wang S, Li Q, Leng Y, Wang L, Hu G-H A novel method for preparing poly (vinyl alcohol) hydrogels: Preparation, characterization, and application. Ind Eng Chem Res. 2017 Jul 19;56(28):7971–7976.

3 Nanocomposite Hydrogels

Garima and Abhay Sachdev
CSIR-Central Scientific Instruments Organization
Chandigarh, India and Academy of Scientific and Innovative
Research (AcSIR), Ghaziabad, India

Mansi Vij
CSIR-Central Scientific Instruments Organization,
Chandigarh, India

Ishita Matai
Amity University Punjab, Mohali, India

3.1 INTRODUCTION

Hydrogels, with a three-dimensional (3D) hydrophilic polymeric structure, popularly known as "aqua gels", are cross-linked together via covalent interactions, hydrogen bonding, ionic forces, hydrophobic interactions, and physical entanglements of individual polymer chains. The hydrogel structure comprises abundant functional groups (-OH, -COO, -NH$_2$, and -SO$_3$H), which enable them to entrap and stabilize water as well as solute molecules [1]. Interestingly, the properties of hydrogels are similar to living tissues in terms of high oxygen permeability, water absorbing capacity, excellent porosity, ability to incorporate bioactive substances, and viscoelastic properties that are responsive to an external environment [2], while with synthesis, the porous microstructure of the hydrogels can be tailored to achieve the desired physico-chemical properties at the microscopic scale. External influences such as pH, temperature, ionic strength, electric field, magnetic field, and light have been ascertained to modulate the properties of hydrogels [3,4]. Remotely controlled stimuli-responsiveness has enabled the widespread application of hydrogels in diverse fields such as sensors, delivery systems, and prompt tissue healing agents. Moreover, the degree of swelling (water-incorporation capacity) and porosity can be intrinsically varied via optimizing the nature and density of cross-linking, and polymer composition, to ultimately impact the overall gel macromolecular structure. Despite such significant characteristics, hydrogels suffer from several drawbacks, which include poor mechanical strength and limited functionality [5]. With the advent of nanotechnology, nanocomposite hydrogels have been designed and developed as improved versions of hydrogels with advanced physical and chemical characteristics. The choice of nanoentities for incorporation within the hydrogel molecular framework spans from carbon-based nanostructures (carbon nanotubes, graphene, fullerene, nanodiamonds, etc.), polymeric nanoparticles (dendrimers/hyper branched polymers, liposomes, polymeric micelles, nanogels, core-shell polymeric particles), metal/metal oxide nanoparticles (gold, silver, iron oxide, titania, alumina, and zirconia), to inorganic nanoparticles (nano-hydroxyapatite, nanoclays, bioactive glasses, silica, calcium phosphate, ceramic, and wollastonite) via *in situ* polymerization or pre-mixing of pre-synthesized nanoparticles. The resulting nanostructure-reinforced hydrogels have been referred to as nanocomposite hydrogels. The nanostructure inclusion within the hydrogel matrix potentially leads to the creation of structurally diverse yet stable materials, thanks to multiple nanoparticles-polymeric chain interactions such as electrostatic interactions, hydrogen bonding, and van der Waals forces [6,7]. Further, the porous microstructure of a hydrogel can serve as a suitable matrix to enable host-guest assembly for the accommodation of different kinds, classes, and dimensions of nanostructures. Such nanocomposite hydrogel matrix possesses exceptional chemical, biological, electrical, and physical properties that potentially transcend the limits of conventional chemically cross-linked hydrogels. Moreover, such nanocomposite hydrogels hold tremendous and diverse applications including soft robotics, environment friendly materials, energy devices, bioelectronics, biosensors, and biomedical applications [8]. In this chapter, diverse strategies for the synthesis of various nanocomposite hydrogels and their unique properties are discussed. Subsequently, an overview on types of nanostructures as fillers that can be incorporated within the hydrogel framework in light of their respective properties for intended applications has been described.

3.2   SYNTHESIS OF NANOCOMPOSITE HYDROGELS

The synthesis route that is adopted for the formation of nanocomposite hydrogels is critical in determining its structural foundation, complexity, and hence functionality. Indeed, the varied synthetic schemes followed for nanocomposite hydrogel development have enabled a better understanding of the reaction and gelation mechanisms, and understanding the effects following incorporation of different nanomaterials in the polymeric hydrogel base mixture. As a result, a wide range of nanocomposite hydrogels have been synthesized using either conventional or new hydrogel synthetic methods. The formation of nanocomposite hydrogels can be broadly classified on the basis of methodology and reaction mechanism, which has been discussed in the subsequent section.

a. In situ hydrogel formation in the presence of nanoparticles

This synthetic strategy is one of the most commonly employed methods for preparing the nanocomposite hydrogels. In this process, a suspension of nanoparticles is uniformly dispersed and added to the synthetic or natural monomer solution, followed by hydrogelation (Figure 3.1(A)). Usually the nanoparticles do not interfere much in the polymerization process, leading to the formation of nanocomposite hydrogel. A classic example of formation of a nanocomposite hydrogel was reported by Willner and group, in which gold nanoparticles (AuNPs) were immobilized in polyacrylamide gel by

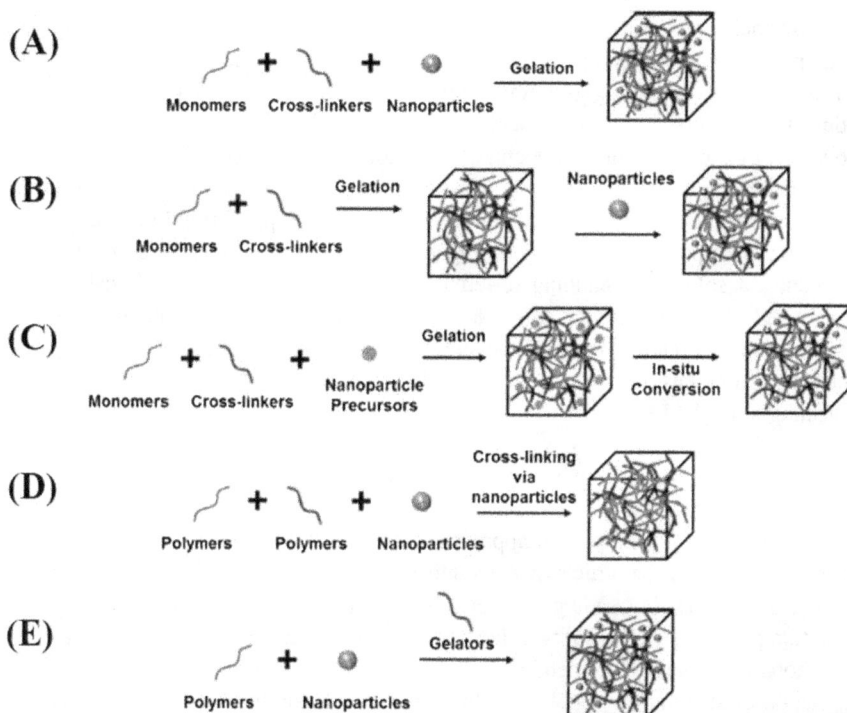

FIGURE 3.1 Approaches employed for the formation of nanocomposite hydrogel: A) hydrogel formation in a nanoparticle suspension; B) incorporation of nanoparticles into hydrogel matrix after gelation; C) reactive nanoparticle formation within a preformed hydrogel; D) cross-linking of hydrogel using nanoparticles to form nanocomposite hydrogels; E) gel formation using nanoparticles, polymers, and distinct cross-linking agent. Reproduced with permission from [11], Copyright (2015), John Wiley and Sons.

placing the dehydrated gel in AuNPs suspension and further allowed to swell, leading to uniform distribution of AuNPs in the gel matrix [9]. The advantages of the method include even distribution of nanoparticles in the hydrogel network without any agglomeration. However, the weak nanoparticles-hydrogel matrix interactions and low cross-link density (if any) may lead to leaching of nanoparticles from the hydrogel network under swelling conditions [10,11].

b. **Physical mixing of nanoparticles with hydrogels**

Physical mixing of nanoparticles with a hydrogel pre-mix is a facile strategy to gather them in a nanocomposite hydrogel form. The pre-synthesized metal/metal oxide, or any other non-metallic/ceramic-nanoparticles, can be introduced into the pre-formed hydrogels prior to the addition of a cross-linker (Figure 3.1(B)). Electro-polymerization of hydrogels with Au-NPs is the best illustration of the aforementioned technique. The addition of Au-NPs during the polymerization process leads to aggregation of nanoparticles under the influence of an electric field. To circumvent the drawbacks, the nanoparticles were incorporated into the hydrogel matrix via a breathing cycle mechanism [12]. In this process, the nanoparticles were adsorbed on the porous network of hydrogels through physical entanglement and hydrogen bonding induced by a chelating agent without any undesirable release of nanoparticles. However, the above-mentioned technique has some discrepancies. including repetition of the breathing cycle to obtain the desired nanoparticle density and time-consuming process. Alternatively, nanoparticles were embedded into pre-formed colloidal hydrogels through repeated heating, centrifugation, and re-dispersion cycle. In a typical procedure, a solution containing different ratios of micro-gel and colloidal Au was centrifuged. Au-NP solution was added to the micro-gel pellet obtained and re-dispersed throughout the micro-gel pellet by repeated heating, agitating, and sonication. The suspension was then re-centrifuged and the protocol was repeated before the final annealing step [10,13].

c. In situ conversion of nanoparticles within hydrogels

The Langer group developed an approach in which a pre-formed hydrogel acted as a micro reactor and nanoparticles were synthesized through an *in situ* reduction method (Figure 3.1(C)) [14]. Loading of metal precursors into a gel and their subsequent reduction produced well-dispersed nanoparticles inside the matrix. The *in situ* synthesis of nanoparticles gained tremendous attention over the *ex situ* method, owing to the regulation of size and morphology of nanoparticles in the former case. Moreover, the method offers other advantages, like uniform distribution of nanoparticles and inhibition of agglomeration. Besides, the size of nanoparticles does not vary with the concentration of a precursor. In a typical procedure, a chitosan-based cryogel was prepared by the photo-polymerization method. A lyophilized polymeric cryogel was immersed in a silver nitrate ($AgNO_3$) solution until the swelling equilibrium was attained. To the above recipe, sodium borohydride ($NaBH_4$), a reducing agent, was added, leading to immediate *in situ* formation of AgNPs in the hydrogel networks. The Ag^+ ions interact with the functional groups of hydrogels through an electrostatic or dipole ion interaction and Ag^+ ions, leading to formation of metal/metal oxide nanoparticles by means of self-assembly, electrochemical deposition, co-precipitation, etc. During the absorption process, the ion exchange phenomenon occurs between metal ions present in the solution and the functional group present within the hydrogel, resulting in oxidation of metal ions into metal oxide nanoparticles. The reducing agent plays a crucial role in the formation of metal/metal oxide nanoparticles. The reducing agent, such as ammonium hydroxide (NH_4OH), causes formation of metal oxide nanoparticles, while $NaBH_4$ and citrate lead to the formation of metal nanoparticles in addition to a contribution towards different particle size [11,15].

d. Cross-linking of hydrogels using nanoparticles

Nanoparticles with surface functional groups have been employed as cross-linking agents for the development of nanocomposite hydrogels during a gelation process (Figure 3.1(D)). In this strategy, nanoparticles serve as a connector between the hydrogel polymeric chain surface via covalent or physical interactions. For instance, the hydrophilic nature of clays makes them an ideal candidate to be used as multifunctional cross-linkers. In a typical process, radicals transfer to the surface of nanoparticles during the radical polymerization process, causing grafting at the surface of nanoparticles. However, clays, popularly known as "radical killers", are known to prevent the radical polymerization and inhibit the gelation process. Moreover, the excellent mechanical characteristics of the nanocomposite (NC) hydrogels can be indebted to the multiple non-covalent effects between clay nanosheets (Clay-NS) and the polymeric chains [16]. Apart from clays, silica nanoparticles, and functionalized (vinyl, carboxylic, and thiol groups), Au-NPs have also been explored as multifunctional cross-linkers that can potentially adhere to different domains of hydrogels. The incorporation of nanoparticles into the hydrogel matrix resulted in increased interfacial binding between the network and nanoparticles, leading to increased stiffness as well as excellent energy dissipation capability [11]. The key factors that predominantly enable rapid and robust adhesion of hydrogels include the following: i) inherent capability of nanoparticles to be adsorbed onto the polymeric gels, ii) nanoparticles may act as connectors between polymer chains, and iii) the ability of the polymer chains to reorganize and dissipate energy under stress after adsorption of nanoparticles. Because of its high energy dissipation, the nanocomposite hydrogel exhibits higher toughness and stretchability than conventional covalently cross-linked hydrogels, offering a huge potential for surgical and tissue engineering applications [17,18].

e. **Hydrogel formation using nanoparticles, polymers, and distinct cross-linking agents**

Das and co-workers produced a well-connected three-dimensional micro-fibrillar hydrogel network of Ag-NPs grown on the conducting polymer network of folic acid and polyaniline (PANI) hydrogels (Figure 3.1(E)) [19]. The hydrogel was formed by mixing a folic acid, aniline monomer, ammonium persulfate, and silver nitrate solution, in which folic acid acted as a supramolecular cross-linker for PANI hydrogel. Following, aniline was polymerized and cross-linked, yielding a deep greenish hydrogel. In light of this, *in situ* generated Ag-NPs into the PANI hydrogel demonstrated excellent conductivity, suggesting potential application for energy storage. The hydrogel framework had several distinguishing characteristics, including porous structure, high mechanically strength, and increased electrical conductivity. In addition, graphite and silica nanoparticles (Si-NPs) have been used for the formation of conductive nanocomposite hydrogels and their potential application in the field of rechargeable lithium ion batteries has been recently explored. A slurry of silica nanoparticles was deposited on the working area of electrode using binders. The binders helped to boost the efficiency of batteries through interactions (either hydrogen bonding or electrovalent bonding) between the binders and Si-NPs. During the polymerization process, different binders, including PVDF (polyvinylidene fluoride), PAA (polyacrylic acid), CMC (carboxymethyl cellulose), alginate, and phytic acid, were utilized and resulted in the formation of conductive nanocomposite hydrogels. The polymeric binders may serve as dual-purpose materials that cause the enhancement of binding and conductivity of the electrodes [13,20].

3.2.1 CARBON NANOSTRUCTURE-BASED NANOCOMPOSITE HYDROGELS

Carbon-based nanostructures, such as carbon nanotubes (CNTs), graphene, buckminsterfullerene (C60), nanodiamonds, and carbon dots, have demonstrated potential applications in various fields [21]. In particular, graphene and CNTs evince excellent mechanical strength, electrical conductivity, and optical properties compared to other carbon-based nanostructures. However, CNT does not possess adhesive properties, which can limit their practical applicability [22]. To address the aforementioned issue, CNTs were modified with a variety of functional groups, including tannic acid, urea, and dopamine. A glycerol–water (GW) hydrogel with dopamine (PDA)-decorated CNTs as conducting nanofillers demonstrated long-term stability, excellent conductivity, and strong adhesion under extreme temperature conditions. The synthesis of nanocomposite hydrogels was based on a UV-initiated polymerization process. To overcome the poor water dispersion ability of CNTs, binary solvents composed of glycerol and water were utilized. Furthermore, the water-locking effect of glycerol in the hydrogel results in superior antifreezing/antiheating properties of GW-hydrogels when compared to conventional hydrogels. Moreover, the presence of non-covalent interactions within the polymer chains of hydrogels as a result of binary solvent augments the mechanical strength of GW-hydrogels. The CNTs imparted the conductive characteristics to the hydrogel, while polydopamine facilitated the uniform distribution of CNTs in the polymer network, causing strong interactions between the CNTs and polymer chains. Thus, even after large deformation, the GW-hydrogels demonstrated high flexibility, recoverability, and adhesion [23].

Graphene oxide (GO) is a two-dimensional carbon-based nanomaterial with larger specific surface area, high mechanical strength, and diverse functionality due to the presence of abundant oxygen-containing moieties [24]. GO is a popular 2D nanomaterial for the formation of nanocomposite hydrogels, due to its unique characteristics. Zhong *et. al.* created graphene oxide (GO)–poly(acrylic acid) (PAA) nanocomposite hydrogels with self-healable and toughness properties. GO was synthesized using a modified Hummer's method, and a GO-PAA nanocomposite hydrogel was prepared using an *in situ* free-radical polymerization process that included a GO nanosheet as a reinforcer and the cross-linker sites, acrylic acid as the monomer, Fe^{3+} ions as the cross-linker, and ammonium persulfate (APS) as the initiator. The addition of a metal ion, namely Fe^{3+}, enabled the strong, dynamic, and reversible ionic bonding among carboxyl functionalized PAA chains rather than weak hydrogen bonding without the use of any chemical cross-linkers. Furthermore, Fe^{3+} ions facilitated dual cross-linking in GO-PAA nanocomposite hydrogels by causing cross-linking between GO nanosheets and PAA chains via coordination assembly. The superior mechanical properties and thermal stability of the GO-PAA nanocomposite hydrogels could be attributed to the dual cross-linking effects and remarkable synergistic effect among the Fe^{3+} ions, PAA chains, and GO nanosheets that are created in the GO-PAA nanocomposite hydrogels, imparting both toughness and flexibility [25]. Similarly, a hybrid hydrogel comprising of reduced graphene oxide–cerium oxide (rGO–CeO_2) nanocomposite and cytochrome c (a redox protein) was readily prepared via covalent cross-linking and *in situ* self-assembly. In this process, the 3D continuous network of the hydrogel provides entraps cytochrome c, besides attaining increased electrical conductivity due to the dispersion of rGO–CeO_2 [26].

3.2.2 POLYMERIC NANOPARTICLE-BASED NANOCOMPOSITE HYDROGELS

Polymeric nanoparticles are largely composed of micelles, dendrimers, liposomes, nanogels, hyper branched polymers, and core-shell nanoparticles [18]. The ability to encapsulate and ferry hydrophobic and hydrophilic substances including drugs, proteins, enzymes, inorganic nanoparticles, and other active agents, makes them much suitable for medical and pharmaceutical applications. Hence, reinforcement of such nanoparticles into the hydrogel matrix could impart multifunctional character-istics to the underlying composite hydrogel. Dendrimers and other hyper branched polymeric nanoparticles with a highly branched architecture and a large number of peripheral functional groups offer high reactivity and loading efficiency towards a multitude of bioactive agents. Such polymeric nanoparticles can tenaciously interact with the hydrogel matrix via covalent interactions, physical interaction/self-assembly, and inverse nanoprecipitation to form a stable nanocomposite hydrogel with uniform size and distribution. For example, polymeric nanoparticles with different shapes (spheres, cylinders, and platelets) were assembled on the surface of calcium-alginate hydrogels through a crystallization-driven self-assembly technique [27]. It was found that the size, morphology, and surface charge of nanoparticles play an important role for the resultant properties of a composite hydrogel. The experimental results demonstrated that platelet-shaped polymeric nanoparticles exhibited superior uniaxial tensile adhesive strength compared to spherical- and cylinder-shaped nanoparticles. The large surface area of platelets with high binding affinity and ability to interact with a polymeric network chain of hydrogel resulted in enhanced mechanical and adhesive properties [22,28]. Wang and co-workers utilized the aza-Michael addition reaction for the synthesis of polyamidoamine (PAMAM) dendrimer hydrogels (DH) with tunable properties. PAMAM dendrimer G5 was chosen as the underlying core and functionalized with acetic anhydride to varying degrees of acetylation (Figure 3.2). PAMAM dendrimers have a high concentration of primary amines on the surface and secondary amines in the core. However, secondary amines are found to be more reactive than the primary amines during the reaction. Because of steric hindrance, secondary amines are less available for the reaction. As a result, the reaction lies heavily on the primary amines present on the dendrimer surface. The nucleophilic amines on the dendrimer surface reacted with α, β unsaturated ester in acrylate groups of polyethylene glycol diacrylate (PEG-DA) via the aza-Michael addition reaction to form a cross-linked network of dendrimer hydrogels without employing any catalyst. The degree of acetylation has been found to affect the swelling and disintegrating properties of dendrimers hydrogels. Furthermore, the DHs were discovered to be highly cytocompatible and promote cell adhesion and proliferation [29]. Matai and co-workers demonstrated the improved anticancer potential of conventional chemotherapeutic drugs such as 5-Fluorouracil, Epirubicin, and Tamoxifen when complexed with PAMAM G5 dendrimers. Such dendritic nanoentities could facilitate drug release over an extended period of time in a tumor microenvironment pH [30–32]. These hyper branched PAMAM G5 dendrimers could also form nanogels with natural polymers such as sodium alginate after ionic gelation with calcium chloride, thereby improving the mechanical properties and impart pH-dependent drug release characteristics [33].

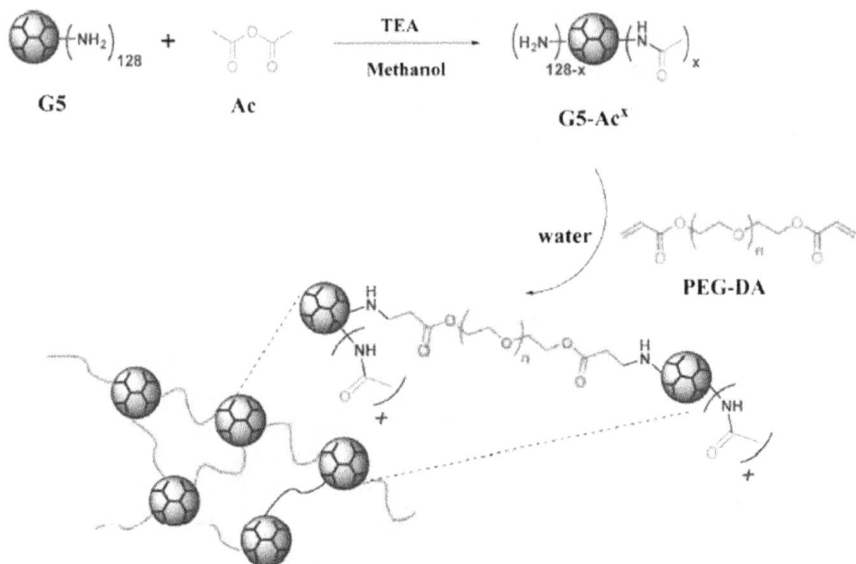

FIGURE 3.2 Schematic representation of the synthesis of acetylated G5 (G5-Acx) and PEG-DA using aza-Michael addition reaction. Reproduced with permission from [29], Copyright (2017), American Chemical Society.

3.2.3 METAL NANOPARTICLE-BASED NANOCOMPOSITE HYDROGELS

Metallic nanoparticles (NPs) are crystalline materials that are typically 1–100 nm in each of the three spatial dimensions with a wide range of intriguing properties and have been explored for nanocomposite hydrogels. The unique combination of colloidal metallic nanoparticles derived of noble metals such as gold nanoparticles (Au NPs), silver nanoparticles (Ag NPs), and platinum nanoparticles (Pt NPs) as well as other metals including cobalt (Co) and nickel (Ni) and metal oxide nanoparticles including iron oxide (Fe_3O_4, Fe_2O_3), titania (TiO_2), alumina (Al_2O_3), and zirconia (ZrO_2) in the polymeric network of hydrogels can endow structural and functional diversity. Pores in the hydrogel network (smaller/larger) can act as a reservoir and nanoreactor templates to promote nucleation and growth of metal/metal oxide nanoparticles without aggregation concerns.

On the other hand, the interactions between polymer chains and metal/metal oxide nanoparticles are known to be weak in nature. Therefore, functionalization of nanoparticles is an excellent alternative for efficient interaction between the polymer and the nanoparticles. Furthermore, the loading of metal or metal oxide nanoparticles into hydrogel networks can provide an environment for stabilization and prevention of oxidation of nanoparticles. Moreover, the inclusion of bimetallic nanoparticles instead of single-material nanoparticles is known to improve the catalytic activity, selectivity, and stability of the nanocomposite hydrogel system. Therefore, the incorporation of nanoparticles into hydrogels leads to generating new nanocomposite materials with novel collective physiochemical properties that are different from individual components or even from the bulk material [34–36].

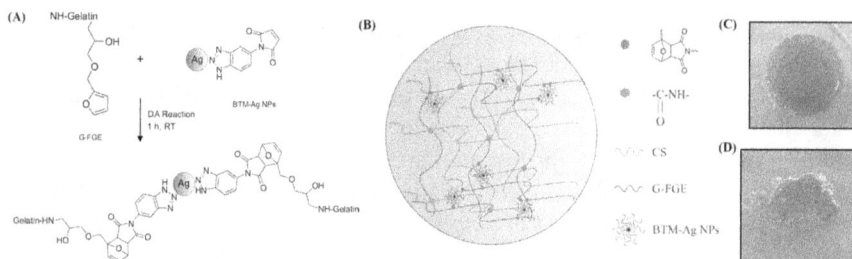

FIGURE 3.3 A) Cross-linking of furan-modified gelatin (G-FGE) with benzotriazole maleimide-coated silver nanoparticles (BTM-Ag NPs) through Diels–Alder Cycloaddition reaction. B) Schematic representation of G-CS-BTM-Ag hydrogel formation. C) Pictorial representation of G-CS-BTM-Ag hydrogel, and (D) G-CS hydrogel. Reproduced with permission from [37], Copyright (2015), American Chemical Society.

Garcia-Astrain and his co-workers proposed a biocompatible hydrogel based on Ag-NPs with remarkable antibacterial activity. Briefly, benzotriazole maleimide surface-modified Ag-NPs were prepared using NaBH$_4$ as a reducing agent, and the surface-modified Ag-NPs were used as multifunctional cross-linkers for the formation of nanocomposite hydrogels. Furthermore, carbodiimide chemistry-based furan-modified gelatin and chondroitin sulfate (CS) were coupled via amide coupling between free amino acids of G-FGE and the carboxylic group of CS to form a stable hydrogel. The furfural-modified gelatin was then covalently cross-linked with surface-modified AgNPs using Diels–Alder Cycloaddition (Figure 3.3). It was observed that functionalized Ag-NPs were uniformly dispersed into the hydrogel matrix with no visible particle agglomeration that could impair the properties and stability of the formed gel. Furthermore, the cytotoxicity and cell viability studies suggested that the bionanocomposite hydrogel could be potentially used in biomedical applications such as for controlled therapeutic delivery or tissue engineering [37]. In another study, zinc oxide nanoparticles (ZnO) and sodium alginate (SA) based bilayered hydrogel matrix prepared via casting/solvent evaporation technique were explored as wound dressing materials. The bilayered hydrogel film had a dense outer layer and porous sublayer, which improved the wound dressing characteristics. Firstly, the pre-formed ZnO nanoparticles were ultrasonically dispersed in distilled water. Following, SA and glycerol as a plasticizer were dissolved into the above suspension. The bilayered hydrogel films were then immersed in CaCl$_2$ to cross-link. ZnO/SA bilayered hydrogel films had a porous structure with high transparency, good hydrophilicity, swelling abilities, and healing properties. Moreover, the ZnO/SA bilayered hydrogel films exhibit good antibacterial properties with no significant cytotoxicity [38].

Similarly, researchers have synthesized thermosensitive, electro-conductive injectable hydrogels based on chitosan (CS) and gold nanoparticles (GNPs) that mimic the electromechanical properties of the myocardium for cardiac tissue engineering applications. CS was used to create a thermosensitive conductive hydrogel with a network of interconnected pores. GNPs, on the other hand, as biocompatible nanostructures, were found to improve intercellular electrical communications. As a result, incorporation of GNPs into the CS matrix could provide electrical cues. Chitosan-stabilized gold nanoparticles (CS-GNP) were prepared by dissolving a tetrachloroauric acid and chitosan solution in distilled water in a molar ratio of 1:1, followed by the addition of sodium citrate as a reducing agent. Thereafter, for preparation of a thermoresponsive hydrogel, CS-GNP powders were dissolved in acetic acid, followed by the dropwise addition of a β-glycerophosphate disodium salt solution (β-GP). The electrostatic interactions between chitosan and β-GP, as well as hydrophobic interactions and hydrogen bonding between chitosan chains caused thermoresponsive gelation in CS. The gel was created by incubating the resulting solution at 37°C. The hydrogel displayed a fast gelation time, stability, and electrical conductivity. The experimental results confirmed that GNPs, as electrical stimuli, provided a pathway to direct MSCs differentiation towards cardio myocyte lineages, while having no adverse effect on the physico-mechanical properties of the hydrogel [39].

The hydrogels based on natural polymers have inherent biocompatibility and biodegradability, which enhances their biomedical application prospects. Compared with synthetic polymer hydrogels, the mechanical stability of natural hydrogels is quite low. Therefore, a variety of synthetic methods have been explored for enhancing the stability and mechanical strength of natural hydrogels. Most of the natural polymer hydrogels have been reinforced with metal nanoparticles and prepared through blending or *in situ* method, wherein the functional groups of natural polymers play a key role in synthesis as well as stabilization of nanoparticles inside the gel network. Magnetic nanocomposite hydrogels were prepared by blending the cellulose/β-cyclodextrin solution with the ultrasonically dispersed, pre-synthesized magnetic Fe_3O_4 nanoparticles. The dropwise addition of the resulting mixture into $CaCl_2$ solution was done to cross-link the hydrogels as well as encapsulate the magnetic nanoparticles. The cellulose chain backbone stabilized the Fe_3O_4 nanoparticles through electrostatic adsorption and hydrogen bonds interactions and the resulting 3D network demonstrated unique magnetic stimulated drug release [40]. The *in situ* method is another interesting approach for natural nanocomposite hydrogel preparation, wherein metal salts are reduced to form nanoparticles inside the polymer hydrogel network. In one such approach, the chelating effect between iron ions and amino groups of chitosan was used to prepare magnetic chitosan hydrogels. The chitosan-iron complex served as nucleation sites for *in situ* precipitation of Fe_3O_4 nanoparticles inside the gel network [41], an interesting green approach to synthesize a bacterial cellulose nanocomposite hydrogel containing silver nanoparticles. Bacterial cellulose was extracted from gram negative bacterium *Acetobacter xylinum* using green tea. Additionally, the green tea, which contains various functional biomolecules groups, acted as a reducing and capping agent for the synthesis of silver nanoparticles. The silver nanoparticles were *in situ* synthesized by the addition of an $AgNO_3$ salt solution with a concentration of 0.005 g along with bacterial cellulose growth in the presence of green tea. The hydroxyl groups of bacterial cellulose further acted as an active site for the nucleation and growth of silver nanoparticles [42]. Similarly, gold nanoparticles have been synthesized in a polysaccharide-based hydrogel using a simple and environment friendly approach. Aliquot of $HAuCl_4$ solution was introduced into the binary blend of carrageenan (κCG) and locust bean gum (LBG). The *in situ* synthesis of Au-NPs was facilitated by the hydroxyl groups in the polysaccharide chains, while the hemiacetal groups in κCG and LBG contributed to the oxidation of metal salts and the stabilization of metal nanoparticles [43].

3.2.4 INORGANIC NANOPARTICLE-BASED NANOCOMPOSITE HYDROGELS

Advanced nanocomposite hydrogels based on inorganic nanoparticles were prepared using a combination of inorganic nanoparticles with the natural or synthetic polymer hydrogels. These nanoparticles are identified as bioactive and resorbable in nature and exhibit good mechanical properties, stretching, and self-healing properties due to surface structure and charging nature of inorganic nanoparticles. As a result, inclusion of inorganic nanoparticles within polymeric hydrogels is anticipated to introduce bioactive characteristics to the network, increased mechanical strength, and adhesion features of hydrogels. Furthermore, nanocomposite hydrogel showed consistent and long-lasting adhesion capacity [1,7,17]. As a matter of fact, inorganic nanoparticle-based nanocomposite hydrogels such as hydroxyapatite nanoparticles (nHAP) have been found to be suitable candidates for designing orthopedic biomaterials due to their biocompatible and osteoconductive nature. Gaharwar *et al.* established highly extensible, tough, and elastomeric nanocomposite hydrogels based on poly(ethylene glycol) and hydroxy-apatite nanoparticles. The nanocomposite hydrogels were prepared at room temperature. Firstly, the nHAP was uniformly dispersed in deionized water through a vortex and ultrasonication method. Afterwards, the PEG diacrylate was added to the nHAp solution and allowed to mix for around 20 min. Following, the composite solution was injected into a mold, followed by photo-polymerization for 10 min using a high-intensity UV lamp. The nanocomposite hydrogels possessed unique mechanical properties with regard to elongation, compressibility, and toughness compared to PEG hydrogels because of a unique combination of covalent polymer cross-linking (which leads to elastic properties) with physical polymer nanoparticle interactions (which induce viscoelasticity). Also, the combination of PEG and nHAp nanoparticles significantly improved the physical and chemical hydrogel properties as well as biological characteristics such as mammalian cell adhesion [44]. Sujan *et al.* explored the bi-functional silica nanoparticles for the simultaneous enhancement of mechanical strength and swelling properties of hydrogels. Following, silica nanoparticles were synthesized using the modified Stober's method and then bi-functionalized with APTES (3-aminopropyl triethoxysilane) and methacrylic anhydride to obtain amine and acrylic acid bi-functionalized silica nanoparticles (BF-SiNPs). Polyacrylic acid (PAc) and polyacrylamide (PAm) have been extensively researched for the formation of stimuli-responsive hydrogels. Both PAc-Si and PAm-Si hydrogels were prepared by free-radical polymerization. For the preparation of PAc-Si, acrylic acid as a monomer, different amounts of BF-Si NPs as a cross-linker, and potassium persulfate (KPS) as an initiator, were used. The PAm-Si hydrogel was also prepared following the same approach, with the exception of a difference in polymerization time and temperature. The bi-functionalization on silica nanoparticles aids in the formation of covalent and non-covalent bonding (mechanical entanglement, hydrogen bonding, and hydrophobic bonding) with the functional groups present on the hydrocarbon backbone of the polymer chain. In comparison to the conventional MBA cross-linker, the BF-Si NPs added structural integrity to the hydrogel as well as improved mechanical strength and an efficient energy dissipation mechanism during deformation [45].

3.3 NATURAL POLYMER-BASED NANOCOMPOSITE HYDROGELS

Hydrogels prepared from natural hydrophilic polymers have gained wide attention due to their inherent biocompatibility, ease of preparation, and abundant chemical functionality (hydroxyl, amine, and amide). Under specific conditions, natural polymers like polysaccharides, DNA, and proteins can be assembled to form a 3D cross-linked network to absorb water, hence forming the hydrogel. In the recent past, upon incorporation of functional nanomaterials either by blending or as a multifunctional cross-linking agent within the hydrogels has endowed them with exceptional physico-chemical properties, forming a nanocomposite hydrogel. Thus, these natural nanocomposite hydrogels have been considered good candidates for some potential use as drug delivery carriers, electro-stimulated systems, and biosensors, as well as tissue engineering. Considering the importance of such hydrogels, the current state of synthesis strategies for these new types of biomaterials as well as their potential properties have been illustrated.

3.3.1 POLYSACCHARIDE-BASED NANOCOMPOSITE HYDROGELS

Hydrogels based on polysaccharides like sodium alginate, cellulose, chitosan, and tragacanth gum are widely used, but these hydrogels often have poor mechanical strength and stability under physiological conditions. To address this issue, a variety of nanomaterials have been incorporated/conjugated in hydrogels to enhance their mechanical performance without altering their functional properties. Sodium alginate, consisting of mannuronic acid (M unit) and guluronic acid (G unit), has been frequently used to form hydrogels by divalent ions mediated ionic cross-linking. Sodium alginate hydrogel reinforced with pre-electrospun SiO_2 nanofibers was prepared by homogenization of an aqueous sodium alginate solution and SiO_2 nanofibers. Such a dispersion was then lyophilized to make it suitable for immersion into an aqueous solution of Al^{3+} for performing ionic cross-linking to generate a highly ordered nanofiber structure inside the composite hydrogels [46]. Chitosan (CS) is a natural linear cationic polysaccharide composed of varying amounts of glucosamine and N-acetyl-glucosamine residues. For instance, pH-sensitive core-shell ZnO/carboxymethyl cellulose (CMC)/CS nanocomposite hydrogel beads were prepared by self-assembly of ZnO nanoparticles inside CS-coated CMC beads for improved drug release applications [47]. Similarly, CS was used as a base polymer to create GO-Ag nanocomposite hybrid hydrogel with high mechanical strength and antibacterial activity. GO–Ag nanohybrid with antibacterial properties was synthesized via reduction of $AgNO_3$ by $NaBH_4$ in the GO suspension, followed by uniform dispersion of as-formed GO-Ag nanohybrid particles in the polyanion cross-linked chitosan hydrogel matrix. The resulting nanocomposite hydrogel showed higher swelling efficiency at a lower pH due to conversion of amine functionalized chitosan into ammonium ion, leading to higher electrostatic repulsion between polymeric chains [48]. Tragacanth gum (TG), an anionic, highly branched, and heterogeneous polysaccharide, has been explored for synthesizing nanocomposite hydrogels, owing to its diverse functional groups, such as carboxyl and hydroxyl, which facilitate the interaction of TG with different cross-linking agents and nanoparticles. A hybrid hydrogel of TG and sodium alginate was formulated by a simple ionic cross-linking of hydroxyl and carboxyl groups by Ca^{2+}, with *in situ* incorporation of silver nanotriangles (AgNTs). The hydrophilic nature of AgNTs played a significant role for their incorporation inside hydrogel networks due to multiple electrostatic interactions and hydrogen bonding between surface functional groups of hydrogels [49]. In another study, Ag-NP loaded TG hydrogel was generated via a conventional redox polymerization technique. A leaf extract, *Terminalia chebula*, was utilized as a stabilizing and reducing agent for *in situ* Ag-NP synthesis within the hydrogel network [50].

3.3.2 DNA-BASED NANOCOMPOSITE HYDROGELS

Deoxyribonucleic acid, also known as DNA, is a molecule that encodes genetic material and serves as the guide for the growth and functioning of prokaryotes and eukaryotes. The DNA polymeric chain is composed of four deoxyribonucleotide monomers (abbreviated as A, T, C, and G), which can spontaneously undergo self-assembly to form novel materials with specific functions [51]. DNA hydrogels can be rationally synthesized as long polymer chains through physical entanglement or chemical cross-linking. Such hydrogels are endowed with unique properties like pH/temperature responsiveness, enzyme specificity, and biomolecule recognition [52]. DNA-based hydrogels have been mainly synthesized by DNA self-assembly and hybridization methods. In the DNA self-assembly, the branched DNA blocks with sticky ends can be assembled to form 3D networks by ligating the complementary palindromic ends via ligase catalyzed reaction. On the other hand, hybrid DNA hydrogels involve the coupling and cross-linking natural/ artificial polymers/biomolecules or nanoparticles to DNA scaffolds [53]. Notably, functional nanoparticles can be introduced inside DNA hydrogels to create smart nanocomposite hydrogels for diverse applications. Molybdenum sulfate (MoS_2) quantum dots as pseudo-cross-linkers in DNA hydrogels produced materials with synergistic properties of fluorescence and mechanical robustness [54]. Similarly, a pH-controllable sol-gel switchable DNA-carbon nanotube hybrid hydrogel was designed, wherein a carbon nanotube acted as a cross-linker for DNA strands [55]. Mesoporous silica nanoparticle-DNA hydrogels as smart multicomponent carriers were designed, wherein the enzyme could be encapsulated inside the gel phase and substrate inside the silica. This system mediated enzyme-catalyzed reactions based on an external stimulus [56].

3.3.3 PROTEIN NANOCOMPOSITE HYDROGELS

Natural biomolecules, or proteins, are composed of polypeptide amino acid chains that serve as building blocks for all living beings. In the recent past, natural protein hydrogels have drawn considerable interest owing to their biocompatibility and biodegradability. Collagen is the most abundant fibrous protein that can be chemically cross-linked to form a hydrogel. Antibiotics (rifamycin and gentamicin) were encapsulated inside silica nanoparticles, simply mixed with collagen sol, and kept for gelation to synthesize drug-loaded gels [57]. The prepared hydrogels had an intact fibrillar structure and demonstrated sustained drug release for long-term antibacterial activity. Gelatin is a biopolymer derived from hydrolytic degradation of collagen that has been suitably cross-linked to form hydrogels. Piao *et al.* created a hydrogel composed of rGO and gelatin that displayed a high storage modulus and swelling ratio. The hydrogel was made by combining an aqueous GO suspension with a gelatin solution at a specified weight ratio and then subjecting the mixture to mild heating at 95°C for 24 h. Unlike other methods, no additional chemical cross-linker was employed during the process. The gelatin played a dual role in cross-linking the polymer chains as well as acted as a reducing agent in forming rGO from GO in the nanocomposite hydrogels [58]. Peptide-based hydrogels are prepared by mixing short amino acid sequences with suitable gelator molecules to form a self-assembled fibrillar network [59]. The addition of non-gelling agents like nanoparticles in such hydrogels has resulted in improved gelation and introduction of specific properties, thus forming peptide hybrid hydrogels. Photochemical cross-linking of triblock polypeptide $PC_{10}A(RGD)$ in the presence of photoinitiator I-2959 under 365 nm UV light irradiation served as a template for synthesis of Au-NPs to form hybrid nanogels with photoacoustic effects [57]. Nonetheless, nanoparticles have been synthesized inside a peptide-hydrogel matrix. Self-assembled Naphthalene (Nap)-protected tripeptide nanofibrous framework served as absorption site for Ag^+ ions. Further, the Ag^+ conjugated framework was subjected to *in situ* reduction to synthesize stabilized AgNPs [60].

3.4   PROPERTIES OF NANOCOMPOSITE HYDROGELS

3.4.1   STIMULI RESPONSE

Nanocomposite hydrogels are highly responsive to external environmental stimuli alterations such as in temperature, solvent type, pH, NIR light, electrical field, magnetic field, sound waves, etc. Heat is the most common stimulus used to achieve the desired function. Deng *et al.* reported a temperature- and light-sensitive nanocomposite hydrogel with conducting and self-healing properties. The nanocomposite hydrogels were composed of nanoclay (laponite) as a physical cross-linker, MWNTs as a conductive filler, and N-isopropylacrylamide (NIPAM). The incidence of NIR light with an intensity of 0.50 W/cm^2 caused the least amount of temperature increase. However, when the intensity of NIR light was greater than 0.75 W/cm^2, the PNIPAM/L/CNT hydrogels demonstrated a remarkable temperature change within several minutes, demonstrating high-efficiency NIR light transformation into heat. However, the PNIPAM/L/CNT hydrogels showed a notable temperature shift within a few minutes when the NIR light intensity was above 0.75 W/cm^2, indicating high-efficiency conversion of NIR light into heat. The temperature change caused by a further increase in NIR light intensity up to

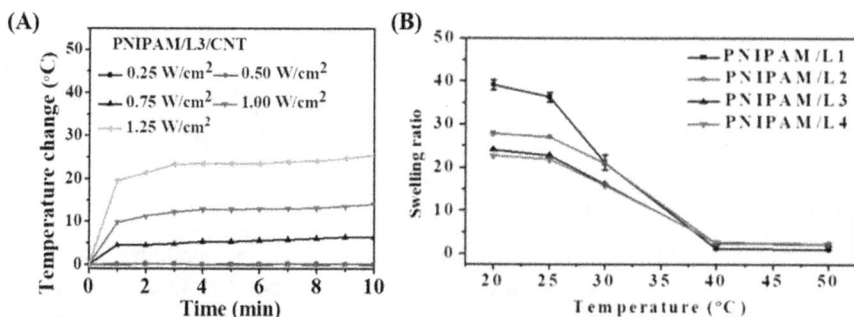

FIGURE 3.4 A) Change in temperature of PNIPAM/L3/CNT hydrogels under NIR light exposure. B) Swelling rate of hydrogels at different temperatures. Reproduced with permission from [61], Copyright (2019), American Chemical Society.

1.25 W/cm^2 during 2 minutes of exposure was around 20–25°C, indicating that the clay content played no effect on the photothermal property of the hydrogels (Figure 3.4(A)). Interestingly, the PNIPAM hydrogel showed a clear volume transition at an acceptable rate of swelling and deswelling in response to temperature fluctuations (Figure 3.4(B)). It has been noted that the cross-linking density of the hydrogel causes the swelling rate (SR) to decrease as clay content increases. Because of the hydrophobic nature of CNTs, further CNT addition to PNIPAM/L hydrogel resulted in reduced SR and increased weight in the dry state. The nanocomposite hydrogels clearly displayed a significant volume transition, and the deswelling process resulted in a fall in SR from 25.4 to 1.9 [61].

Bardajee *et al.* fabricated a nanocomposite hydrogel highly responsive to changes in pH, solvent, temperature, pressure, and magnetic field using poly(N-isopropylacrylamide) (PNIPAM), poly(acrylic acid) (PAA), and Fe_3O_4 nanoparticles. Within a 150-minute interval, the soft nanocomposite hydrogel demonstrated rapid SR up to 660 g/g in the absence of magnetic field and pressure. However, as pressure was increased up to 0.3 psi, the SR decreased due to decreased pore dimensions and pore deformations, resulting in less space for water uptake. In the presence of a magnetic field, a similar response was observed due to the collection of the Fe_3O_4 NPs in the vicinity of magnet field lines, which reduced the distance between the Fe_3O_4 NPs and produced close packed shells of Fe_3O_4 NPs with a reduction in the pores size of the hydrogel matrix. The swelling ratio (SR) of nanocomposite hydrogels (NCHs) as discovered to be less affected by temperature increase. The NCHs hydrogel exhibited significant SR with temperature increases up to 37°C; further temperature increases could cause breakage of hydrogen bonds between water molecules and the hydrogel structure, resulting in hydrogel shrinkage and network collapse. The SR of NCHs is extremely sensitive to changes in pH of medium. The SR of NCHs increased steadily as the pH rose from 2 to 9, owing to the ionization of polymer chain carboxyl groups into carboxylate anions. A further increase in pH above 9 resulted in a decrease in SR due to higher hydroxyl ion (OH$^-$) concentration and lower osmotic pressure. The presence of salt solution also influenced the SR of NCHs. The decrease in SR was clearly visible as the cross-linking density in the hydrogel framework increased due to coordination to several electron donor groups on the hydrogel matrix. Taken together, the nanocomposite hydrogels show a change in swelling rate with subsequent changes in solvent conditions, pH, magnetic field, electrical fields, etc. [62].

3.4.2 MECHANICAL PROPERTIES

Hydrogels usually have a low intrinsic mechanical strength and durability. To address the aforementioned issue, significant efforts have been raised to improve the stability features through incorporation of nanoparticles as filler materials. Ko *et al.* reported a highly compliant and resilient VPA hydrogel based on vinyl hybrid silica nanoparticles (VSNPs) with polyacrylamide and alginate chains. Specifically, the physically cross-linked polyacrylamide acts as a polyelectrolyte matrix for the VPA hydrogel, facilitating intrinsic toughness and network integrity, whereas the vinyl hybrid silica nanoparticle reinforces mechanical robustness by increasing cross-linking density points and decreasing average chain lengths. When compared to single-network hydrogels, the synthesized double-network hydrogel can synergistically balance energy dissipation to achieve high toughness and resiliency under repeated strain release processes. The developed hydrogel-based sensor responded well to strains ranging from 25% to 100% and regained its initial resistance upon strain release. Furthermore, the sensor demonstrated long-term durability for 2,500 strain cycles with a GF of 1.73 up to 100% strain, a response time of 0.16 s, an ultra-low electrical hysteresis of 2.43%, and a lower LOD of 0.4% (Figure 3.5). The stretching and releasing of more than 2,500 cycles without any permanent cracks suggested a reversible cross-linking interaction via energy dissipation and stress transfer [63].

Zhao *et al.* also reported the use of a triple network penetrating conductive hydrogels based on carbon nanotubes to improve inherent mechanical strength. The use of conductive fillers such as n-butylimidazolium bromide salt functionalized MWCNTs (CNT-Br) improved the uniform distribution of nanoparticles in the PAAS matrix as well as even dispersion of PEDOT: PSS, which helped to strengthen interfacial interactions with the polymer network, thereby improving the physical properties of nanomaterial-reinforced hydrogels. The stress-strain graphs clearly showed that as with increase in the content of CNT-Br, the tensile strength increased significantly (up to 8.02 MPa). On the other hand, the elongation at break decreased to 181%, resulting in lower toughness [64].

3.4.3 ELECTRICAL PROPERTIES

The electrical conductivity of a hydrogel can be significantly improved by incorporation of nanoparticles [65]. The strong interaction of graphene and its derivatives with hydrogel polymer chains linked via hydrogen bonds, ionic bonds, and hydrophobic interaction resulted in the formulation of nanocomposite hydrogels (NCHs) with exceptional electrical conductivity, mechanical ductility, and structural stability. The incorporation of reduced graphene oxide into a chitosan-based hydrogen matrix led to the enhancement of electrical conductivity from 0.57 mS/cm to 1.22 mS/cm [66]. In another study, Wang *et al.* produced a self-healable

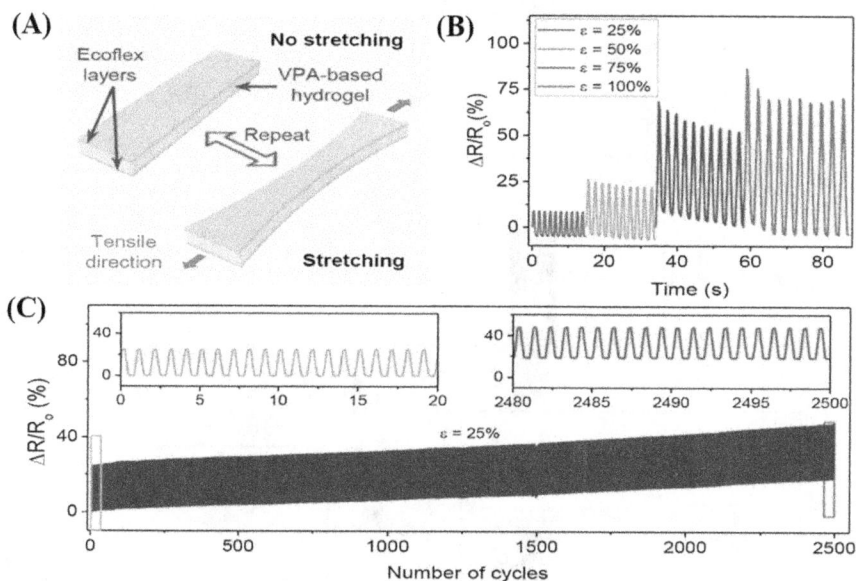

FIGURE 3.5 Electromechanical characterization of the VPA hydrogel-based strain sensor: (A) Schematic illustration of the straining mechanism of the VPA hydrogel-based strain sensor. (B) Cyclic response of VPA hydrogel measured at relatively higher strains. (C) Long-term stability of the sensor over 2,500 cycles. The insets represent the 20 cycles of the response at the beginning (left) and ending (right) of the durability test. Reproduced with permission from [63], Copyright (2022), American Chemical Society.

conductive hydrogel based on polydopamine-reduced Ag NPs and polypyrrole grafted gelatin-based hydrogel. The experimental results showed that the conductivity of the PPyGel-Fe hydrogel about 14 mS/cm, which increased from 24 to 36 mS/cm after the addition of different concentrations of Ag NPs into the above hydrogel matrix (Figure 3.6(A)). The enhanced electrical conductivity values further suggested that Ag NPs, polypyrrole, and Fe^{3+} made a significant contribution for the remarkable increase in conductivity [67].

3.4.4 MAGNETIC PROPERTIES

Incorporation of magnetic nanoparticles (MNPs) such as Fe_2O_3, Fe_3O_4 into the hydrogel matrix is one of the straightforward approaches to induce magnetism in hydrogels under the influence of applied external magnetic fields. For precise control over the hydrogel properties, the MNPs should be superparamagnetic in nature, since these can be positioned at designated locations and aggregated under the applied magnetic stimulus. To create magnetic nanocomposite hydrogels, pre-synthesized MNPs are blended with the hydrogel polymer precursors and subsequently cross-linked, allowing the formation of hybrid magnetic materials. For instance, a magnetically responsive hydrogel was prepared by modifying methacrylated gelatin (GelMA) anisotropic hydrogels with Fe_2O_3 MNPs. Two commercially available

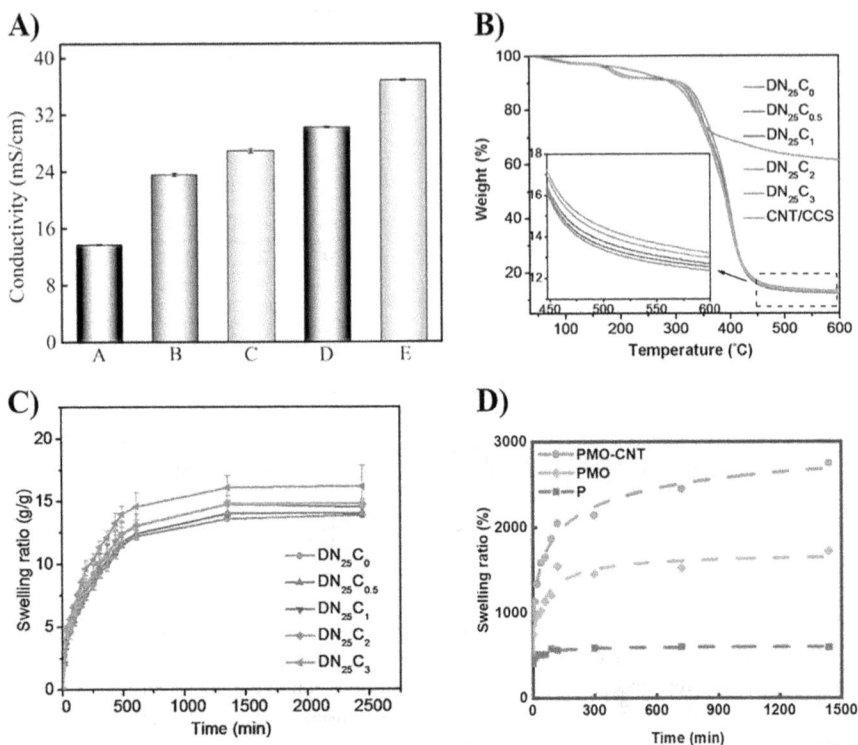

FIGURE 3.6 A) Electrical conductivity of the (A) PPyGel-Fe; (B) PDA@AgNPs0.5-PPyGel-Fe; (C) PDA@AgNPs1.0-PPyGel-Fe; (D) PDA@AgNPs1.5-PPyGel-Fe; and (E) PDA@AgNPs2.0-PPyGel-Fe. Reproduced with permission from ref [67]. Copyright from American Chemical Society. B) Thermal decomposition curve of hydrogels. C) Swelling rate of different hydrogels. Reproduced with permission from ref [72]. Copyright from American Chemical Society. D) Swelling kinetics of hydrogels in PBS (pH 7.4) at 20°C. Reproduced with permission from [74], Copyright (2022), American Chemical Society.

permanent magnets (~20 mT) were used to generate an external magnetic field and applied to Fe_2O_3, containing a hydrogel precursor solution, followed by UV-mediated cross-linking of networks develop the anisotropic nanocomposite [68]. Similarly, pre-synthesized Fe_3O_4 was incorporated inside polyvinyl alcohol/type-II collagen hydrogel matrices by freeze-thawing method [69]. In a different approach, *in situ* preparation method for magnetic hydrogels was also explored [70]. The precursor solution consisting of iron(II) chloride was added to kappa-carrageenan and polyvinyl alcohol polymer solution. The ammonia solution was added to form MNPs, while freeze-thawing and potassium chloride treatment was done to cross-link the polymer chains. Taleb *et al.* investigated the magnetic characteristics of Fe_3O_4 nanoparticles embedded N-Vinyl pyrrolidon/chitosan copolymer (NVP/CS) based hydrogels using a vibrating sample magnetometer. It was observed that the saturated magnetization (Ms) of a nanocomposite hydrogel (12 emu/g) was lower than that of bare Fe_3O_4 nanoparticles (40 emu/g). The quenching of surface moment caused by the entrapment of Fe_3O_4 nanoparticles into a hydrogel matrix led to lower Ms value. Despite the lower Ms value, the hydrogels nonetheless demonstrated superparamagnetic effect [71].

3.4.5 THERMAL PROPERTIES

The rapid response to changes in temperature is an important parameter of thermoresponsive nanocomposite hydrogels that can alter between swelling/deswelling kinetics and volume of hydrogels. Lin *et al.* developed a novel temperature-sensitive hydrogel by using carboxylated chitosan-modified carbon nanotubes (CNT/CCS) as a filler, diethylacrylamide (DEA) and *N*-hydroxymethylacrylamide (NHMAA) as a temperature-sensitive monomer, and a hydrophilic co-monomer, respectively. The initial weight loss in the CNT/CCS was between 43–127°C due to evaporation of bound and free water in the sample material, while the second weight loss over 160°C is ascribed to the chain disintegration of CCS, a low stability amino polysaccharide. However, three weight loss phases became obvious with the addition of CNT/CCS to the polymer matrix of hydrogels. The maximum weight loss of about 80% was seen between 266–490°C and was attributed to the fracture in the side groups of polymer or breakdown of the polymer skeleton. Nevertheless, it was clearly apparent from the graphs that an increase in CNT/CCS content causes a subsequent decline in the weight loss, thereby providing the thermal stability to the hydrogel (Figure 3.6(B)). According to the experimental findings, the inclusion of CNT/CCS increased the pore size and strengthen the thermal stability without altering the lower critical solution temperature (ca. 37.9°C) of the hydrogel [72]. Abudabbus *et al.* created silver-doped poly(vinyl alcohol)/graphene composite hydrogels by freezing and thawing the PVA hydrogel. The incorporation of graphene into the PVA hydrogel improved the thermal stability of the nanocomposite hydrogel, while *in situ* immobilization of silver nanoparticles denoted antibacterial properties. The PVA hydrogel demonstrated a three-step weight loss due to water molecule evaporation, dehydration, and PVA backbone disintegration. However, weight loss was observed at higher temperatures after the incorporation of Ag NPs into the PVA matrix. In comparison to PVA/Gr (96 wt.%) and Ag/PVA (97.6 wt.%), the incorporation of graphene into the Ag/PVA matrix resulted in a smaller weight loss of approximately 94.6 wt.%. The above experimental results indicated increased stability and stronger interactions between graphene, AgNPs, and PVA matrix molecules. Furthermore, graphene could increase the temperature required for polymer backbone thermal degradation. The addition of graphene at a low concentration (5 wt.%) could improve the thermal stability of nanocomposite hydrogels by interacting PVA chains with graphene oxygen-containing moieties, resulting in decreased mobility of the polymer chains and, as a result, a shift in the onset of degradation to higher temperatures [73].

3.4.6 SWELLING PROPERTIES

One of the key characteristics of hydrogels is their capacity to absorb and retain water, which is often assessed using an SR study. The swelling equilibrium can be obtained after immersion of hydrogels into deionized water for 24 h. Due to the hydrophilic effect of carboxyl, hydroxyl, and amide groups, as well as the difference in internal and external osmotic pressure, water molecules quickly diffuse inside the polymer matrix from the outside, causing the swelling rate to increase dramatically. The swelling rate increased slowly and eventually remained unchanged as the osmotic pressure difference decreased. Lin *et al.* observed an increase in swelling ratio with increasing CNT/CCS concentration over a 190-minute immersion time (Figure 3.6(C)). The swelling rate of CNT/CCS-incorporated hydrogels was determined to be 9.68 g/g, which is higher than the swelling rate of unmodified hydrogel (7.81 g/g). The hydrophilicity of CNT/CCS could explain the increased swelling rate. Furthermore, as the CNT/CCS content increased, the pore size of the hydrogel increased, allowing for more space for water molecules. Furthermore, the developed hydrogel could accelerate swelling without reducing cross-linking density [72]. In another work, Shi *et al.* studied the swelling performance of hydroxylated multiwalled carbon nanotubes embedded in a hydrogel. The swelling kinetics were assessed by immersing the hydrogels in PBS until the swelling equilibrium was attenuated. It was discovered that PEGDA hydrogel had a constant swelling rate of 600%. On the other hand, hydroxylated multiwalled carbon nanotubes incorporated poly(MEO2MA-co-OEGMA) hydrogel exhibited a swelling rate of 2748% at 20°C (Figure 3.6(D)). Within 180 minutes, the modified hydrogel swelled dramatically, and swelling equilibrium was reached after 360 minutes. The experimental results indicated that the modified hydrogel had high water absorption and swelling ability, which was attributed to the hydrogels' increased hydrophilic ethylene glycol side chains and lower cross-linking degree [74].

Ghanbari *et al.* investigated the swelling performance of a nanocomposite hydrogel reinforced with oxidized alginate (OA), gelatin (GEL), and ceramic silica nanoparticles (SiO_2). The aldehyde groups of OA were cross-linked with the amino groups of GEL through 1-ethyl-3-(3-dimethylaminopropyl) carbodiimide (EDC) and N-hydroxysuccinimide (NHS) as chemical cross-linkers resulted in the formation of a nanocomposite hydrogel. After 24 hours, the swelling properties of freeze-dried hydrogels were evaluated in a PBS solution at 37°C. Because of their hydrophilicity and high porosity, hydrogels have an excellent water uptake capability with a swelling rate of 838.2%. The hydrogel with a high SiO_2 content (3.0%) exposed more compressed pore size and cross-linked networks than the hydrogel with a low SiO_2 content (0.125%), resulting in less water absorption and lower swelling degrees. It can be concluded from the results that the swelling characterization of hydrogels is primarily determined by the density of the cross-linker. The formation of more covalent bonds in the hydrogel networks resulted in an increase in cross-linking density. Therefore, the addition of SiO_2 to the hydrogel matrix caused less movement of the hydrogel network chains, subsequently resulted into decreased swelling capability [75].

3.5 CONCLUSION AND FUTURE TRENDS

Hydrogels are soft polymeric materials with porous and hydrated network structures. Contrary to native hydrogels, nanocomposite hydrogels can overcome the limitations in physico-chemical, electrical, mechanical, and biological properties. By having a variety of methods to undertake the synthesis of nanocomposite hydrogels, the addition of different nanoparticles/cross-linkers has made feasible the design of multicomponent hydrogel networks with suitable functionality. Highly elastic network, increased swelling, light actuation, and electron transfer are some of the characteristics that have been associated with nanocomposite hydrogels. Designing the next generation of nanocomposite hydrogels will focus on developing large-scale production methods, which currently remains a challenge. Moreover, we believe new fabrication and design strategies for nanocomposite hydrogels will have to be devised for imparting novel stimuli-responsive behavior and mechanical stability for pursuing innovative applications.

REFERENCES

[1] S. Merino, C. Martín, K. Kostarelos, M. Prato, E. Vázquez, Nanocomposite hydrogels: 3D polymer-nanoparticle synergies for on-demand drug delivery, ACS Nano. 9 (2015) 4686–4697.

[2] S. Bashir, M. Hina, J. Iqbal, A.H. Rajpar, M.A. Mujtaba, N.A. Alghamdi, S. Wageh, K. Ramesh, S. Ramesh, Fundamental concepts of hydrogels: Synthesis, properties, and their applications, Polymers (Basel). 12 (2020) 1–60.

[3] L. Brannon-Peppast, Equilibrium swelling hydrogels of pH-sensitive, Chem. Eng. 46 (1991) 715–722.

[4] D. Roy, J.N. Cambre, B.S. Sumerlin, Future perspectives and recent advances in stimuli-responsive materials, Prog. Polym. Sci. 35 (2010) 278–301.

[5] A. Mostafavi, J. Quint, C. Russell, A. Tamayol, Nanocomposite hydrogels for tissue engineering applications, In Biomaterials for Organ and Tissue Regeneration (Elsevier) (2020) 499–528, Elsevier.

[6] G. Sharma, B. Thakur, M. Naushad, A. Kumar, F.J. Stadler, S.M. Alfadul, G.T. Mola, Applications of nanocomposite hydrogels for biomedical engineering and environmental protection, Environ Chem Lett. 16 (2018) 113–146.

[7] Y. Zhang, Q. Chen, Z. Dai, Y. Dai, F. Xia, X. Zhang, Nanocomposite adhesive hydrogels: From design to application, J. Mater. Chem. B. 9 (2021) 585–593.

[8] A.K. Gaharwar, N.A. Peppas, A. Khademhosseini, Nanocomposite hydrogels for biomedical applications, Biotechnol. Bioeng. 111 (2014) 441–453.

[9] V. Pardo-Yissar, R. Gabai, A.N. Shipway, T. Bourenko, I. Willner, Gold nanoparticle/hydrogel composites with solvent-switchable electronic properties, Adv. Mater. 13 (2001) 1320–1323.

[10] N. Moini, A. Jahandideh, G. Anderson, Inorganic nanocomposite hydrogels: Present knowledge and future challenge (Springer), In Sustainable Composites and Nanocomposites (2019) 805–853, Springer.

[11] P. Thoniyot, M.J. Tan, A.A. Karim, D.J. Young, X.J. Loh, Nanoparticle–hydrogel composites: Concept, design, and applications of these promising, multi-functional materials, Adv. Sci. 2 (2015) 1–13.

[12] L. Sheeney-Haj-Ichia, G. Sharabi, I. Willner, Control of the electronic properties of thermosensitive poly(N-isopropylacrylamide) and Au-nanoparticle/poly(N-isopropylacrylamide) composite hydrogels upon phase transition, Adv. Funct. Mater. 12 (2002) 27–32.

[13] F. Wahid, C. Zhong, H.S. Wang, X.H. Hu, L.Q. Chu, Recent advances in antimicrobial hydrogels containing metal ions and metals/metal oxide nanoparticles, Polymers (Basel). 9 (2017) 636.

[14] C. Wang, N.T. Flynn, R. Langer, Controlled structure and properties of thermo-responsive nanoparticle-hydrogel composites, Adv. Mater. 16 (2004) 1074–1079.

[15] A.J. Clasky, J.D. Watchorn, P.Z. Chen, F.X. Gu, From prevention to diagnosis and treatment: Biomedical applications of metal nanoparticle-hydrogel composites, Acta Biomater. 122 (2021) 1–25.

[16] K. Haraguchi, Stimuli-responsive nanocomposite gels, Colloid Polymer Sci. 289 (2011) 455–473.

[17] H.L. Tan, S.Y. Teow, J. Pushpamalar, Application of metal nanoparticle–hydrogel composites in tissue regeneration, Bioengineering. 6 (2019) 1–17.

[18] T. Chen, K. Hou, Q. Ren, G. Chen, P. Wei, M. Zhu, Nanoparticle–polymer synergies in nanocomposite hydrogels: From design to application, Macromol. Rapid Commun. 39 (2018) 1–26.

[19] S. Das, P. Chakraborty, S. Mondal, A. Shit, A.K. Nandi, Enhancement of energy storage and photoresponse properties of folic acid-polyaniline hybrid hydrogel by in situ growth of Ag, ACS Appl. Mater. Interfaces. 8 (2016) 28055–28067.

[20] H. Wu, G. Yu, L. Pan, N. Liu, M.T. McDowell, Z. Bao, Y. Cui, Stable Li-ion battery anodes by in-situ polymerization of conducting hydrogel to conformally coat silicon nanoparticles, Nat. Commun. 4 (2013) 1943–1946.

[21] K.H. Shen, C.H. Lu, C.Y. Kuo, B.Y. Li, Y.C. Yeh, Smart near infrared-responsive nanocomposite hydrogels for therapeutics and diagnostics, J. Mater. Chem. B. 9 (2021) 7100–7116.

[22] M. Biondi, A. Borzacchiello, L. Mayol, L. Ambrosio, Nanoparticle-integrated hydrogels as multifunctional composite materials for biomedical applications, Gels. 1 (2015) 162–178.

[23] L. Han, K. Liu, M. Wang, K. Wang, L. Fang, H. Chen, J. Zhou, X. Lu, Mussel-inspired adhesive and conductive hydrogel with long-lasting moisture and extreme temperature tolerance, Adv. Funct. Mater. 28 (2018) 1–12.

[24] I. Matai, G. Kaur, S. Soni, A. Sachde Vikas, S. Mishra, Near-infrared stimulated hydrogel patch for photothermal therapeutics and thermoresponsive drug delivery, J. Photochem. Photobiol. B Biol. 210 (2020) 111960.

[25] M. Zhong, Y.T. Liu, X.M. Xie, Self-healable, super tough graphene oxide-poly (acrylic acid) nanocomposite hydrogels facilitated by dual cross-linking effects through dynamic ionic interactions, J. Mater. Chem. B. 3 (2015) 4001–4008.

[26] V. Kumar, A. Sachdev, I. Matai, Self-assembled reduced graphene oxide-cerium oxide nanocomposite@cytochrome: C hydrogel as a solid electrochemical reactive oxygen species detection platform, New J. Chem. 44 (2020) 11248–11255.

[27] M.C. Arno, M. Inam, A.C. Weems, Z. Li, A.L.A. Binch, C.I. Platt, S.M. Richardson, J.A. Hoyland, A.P. Dove, R.K. O'Reilly, Exploiting the role of nanoparticle shape in enhancing hydrogel adhesive and mechanical properties, Nat. Commun. 11 (2020) 1420.

[28] I. Gholamali, M. Yadollahi, Bio-nanocomposite polymer hydrogels containing nanoparticles for drug delivery: A review, Regen. Eng. Transl. Med. 7 (2021) 129–146.

[29] J. Wang, H. He, R.C. Cooper, H. Yang, In situ-forming polyamidoamine dendrimer hydrogels with tunable properties prepared via Aza-Michael addition reaction, ACS Appl. Mater. Interfaces. 9 (2017) 10494–10503.

[30] I. Matai, A. Sachdev, P. Gopinath, Multicomponent 5-fluorouracil loaded PAMAM stabilized-silver nanocomposites synergistically induce apoptosis in human cancer cells, Biomater. Sci. 3 (2015) 457–468.

[31] I. Matai, P. Gopinath, Hydrophobic myristic acid modified PAMAM dendrimers augment the delivery of tamoxifen to breast cancer cells, RSC Adv. 6 (2016) 24808–24819.

[32] I. Matai, A. Sachdev, P. Gopinath, Self-assembled hybrids of fluorescent carbon dots and PAMAM dendrimers for epirubicin delivery and intracellular imaging, ACS Appl. Mater. Interfaces. 7 (2015) 11423–11435.

[33] I. Matai, P. Gopinath, Chemically cross-linked hybrid nanogels of alginate and PAMAM dendrimers as efficient anticancer drug delivery vehicles, ACS Biomater. Sci. Eng. 2 (2016) 213–223.

[34] R. Tutar, A. Motealleh, A. Khademhosseini, N.S. Kehr, Functional nanomaterials on 2D surfaces and in 3D nanocomposite hydrogels for biomedical applications, Adv. Funct. Mater. 29 (2019) 1–29.

[35] R. Esmaeely Neisiany, M.S. Enayati, P. Sajkiewicz, Z. Pahlevanneshan, S. Ramakrishna, Insight into the current directions in functionalized nanocomposite hydrogels, Front. Mater. 7 (2020) 1–8.

[36] P. Schexnailder, G. Schmidt, Nanocomposite polymer hydrogels, Colloid Polym. Sci. 287 (2009) 1–11.

[37] C. García-Astrain, C. Chen, M. Burón, T. Palomares, A. Eceiza, L. Fruk, M.Á. Corcuera, N. Gabilondo, Biocompatible hydrogel nanocomposite with covalently embedded silver nanoparticles, Biomacromolecules. 16 (2015) 1301–1310.

[38] T. Wang, J. Wang, R. Wang, P. Yuan, Z. Fan, S. Yang, Preparation and properties of ZnO/sodium alginate bi-layered hydrogel films as novel wound dressings, New J. Chem. 43 (2019) 8684–8693.

[39] P. Baei, S. Jalili-Firoozinezhad, S. Rajabi-Zeleti, M. Tafazzoli-Shadpour, H. Baharvand, N. Aghdami, Electrically conductive gold nanoparticle-chitosan thermosensitive hydrogels for cardiac tissue engineering, Mater. Sci. Eng. C. 63 (2016) 131–141.

[40] F. Lin, J. Zheng, W. Guo, Z. Zhu, Z. Wang, B. Dong, C. Lin, B. Huang, B. Lu, Smart cellulose-derived magnetic hydrogel with rapid swelling and deswelling properties for remotely controlled drug release, Cellulose. 26 (2019) 6861–6877.

[41] Y. Wang, B. Li, Y. Zhou, D. Jia, In situ mineralization of magnetite nanoparticles in chitosan hydrogel, Nanoscale Res. Lett. 4 (2009) 1041–1046.

[42] A. Fadakar Sarkandi, M. Montazer, T. Harifi, M. Mahmoudi Rad, Innovative preparation of bacterial cellulose/silver nanocomposite hydrogels: In situ green synthesis, characterization, and antibacterial properties, J. Appl. Polym. Sci. 138 (2021) 1–12.

[43] M.S. Marques, K.M. Zepon, J.M. Heckler, F.D.P. Morisso, M.M. da Silva Paula, L.A. Kanis, One-pot synthesis of gold nanoparticles embedded in polysaccharide-based hydrogel: Physical-chemical characterization and feasibility for large-scale production, Int. J. Biol. Macromol. 124 (2019) 838–845.

[44] A.K. Gaharwar, S.A. Dammu, J.M. Canter, C.J. Wu, G. Schmidt, Highly extensible, tough, and elastomeric nanocomposite hydrogels from poly(ethylene glycol) and hydroxyapatite nanoparticles, Biomacromolecules. 12 (2011) 1641–1650.

[45] M.I. Sujan, S.D. Sarkar, S. Sultana, L. Bushra, R. Tareq, C.K. Roy, M.S. Azam, Bi-functional silica nanoparticles for simultaneous enhancement of mechanical strength and swelling capacity of hydrogels, RSC Adv. 10 (2020) 6213–6222.

[46] Y. Si, L. Wang, X. Wang, N. Tang, J. Yu, B. Ding, Ultrahigh-water-content, superelastic, and shape-memory nanofiber-assembled hydrogels exhibiting pressure-responsive conductivity, Adv. Mater. 29 (2017) 1–7.

[47] X. Sun, C. Liu, A.M. Omer, W. Lu, S. Zhang, X. Jiang, H. Wu, D. Yu, X. kun Ouyang, pH-sensitive ZnO/carboxymethyl cellulose/chitosan bio-nanocomposite beads for colon-specific release of 5-fluorouracil, Int. J. Biol. Macromol. 128 (2019) 468–479.

[48] M. Rasoulzadehzali, H. Namazi, Facile preparation of antibacterial chitosan/graphene oxide-Ag bio-nanocomposite hydrogel beads for controlled release of doxorubicin, Int. J. Biol. Macromol. 116 (2018) 54–63.

[49] D. Garg, I. Matai, S. Agrawal, A. Sachdev, Hybrid gum tragacanth/sodium alginate hydrogel reinforced with silver nanotriangles for bacterial biofilm inhibition, Biofouling. 38 (2022) 965–983.

[50] K.M. Rao, A. Kumar, K.S.V. Krishna Rao, A. Haider, S.S. Han, Biodegradable tragacanth gum based silver nanocomposite hydrogels and their antibacterial evaluation, J. Polym. Environ. 26 (2018) 778–788.

[51] F. Li, J. Tang, J. Geng, D. Luo, D. Yang, Polymeric DNA hydrogel: Design, synthesis and applications, Prog. Polym. Sci. 98 (2019) 101163.

[52] Z. Shi, X. Gao, M.W. Ullah, S. Li, Q. Wang, G. Yang, Electroconductive natural polymer-based hydrogels, Biomaterials. 111 (2016) 40–54.

[53] V. Morya, S. Walia, B.B. Mandal, C. Ghoroi, D. Bhatia, Functional DNA based hydrogels: Development, properties and biological applications, ACS Biomater. Sci. Eng. 6 (2020) 6021–6035.

[54] P.K. Pandey, H. Ulla, M.N. Satyanarayan, K. Rawat, A. Gaur, S. Gawali, P.A. Hassan, H.B. Bohidar, Fluorescent MoS_2 quantum dot-DNA nanocomposite hydrogels for organic light-emitting diodes, ACS Appl. Nano Mater. 3 (2020) 1289–1297.

[55] E. Cheng, Y. Li, Z. Yang, Z. Deng, D. Liu, DNA-SWNT hybrid hydrogel, Chem. Commun. 47 (2011) 5545–5547.

[56] L. Zhou, C. Chen, J. Ren, X. Qu, Towards intelligent bioreactor systems: Triggering the release and mixing of compounds based on DNA-functionalized hybrid hydrogel, Chem. Commun. 50 (2014) 10255–10257.

[57] R.M. Jin, M.H. Yao, J. Yang, D.H. Zhao, Y. Di Zhao, B. Liu, One-step in situ synthesis of polypeptide-gold nanoparticles hybrid nanogels and their application in targeted photoacoustic imaging, ACS Sustain. Chem. Eng. 5 (2017) 9841–9847.

[58] Y. Piao, B. Chen, One-pot synthesis and characterization of reduced graphene oxide-gelatin nanocomposite hydrogels, RSC Adv. 6 (2016) 6171–6181.

[59] S. Mondal, S. Das, A.K. Nandi, A review on recent advances in polymer and peptide hydrogels, Soft Matter. 16 (2020) 1404–1454.

[60] T. Simon, C.S. Wu, J.C. Liang, C. Cheng, F.H. Ko, Facile synthesis of a biocompatible silver nanoparticle derived tripeptide supramolecular hydrogel for antibacterial wound dressings, New J. Chem. 40 (2016) 2036–2043.

[61] Z. Deng, T. Hu, Q. Lei, J. He, P.X. Ma, B. Guo, Stimuli-responsive conductive nanocomposite hydrogels with high stretchability, self-healing, adhesiveness, and 3D printability for human motion sensing, ACS Appl. Mater. Interfaces. 11 (2019) 6796–6808.

[62] G.R. Bardajee, N. Khamooshi, S. Nasri, C. Vancaeyzeele, Multi-stimuli responsive nanogel/hydrogel nanocomposites based on κ-carrageenan for prolonged release of levodopa as model drug, Int. J. Biol. Macromol. 153 (2020) 180–189.

[63] S. Ko, A. Chhetry, D. Kim, H. Yoon, J.Y. Park, Hysteresis-free double-network hydrogel-based strain sensor for wearable smart bioelectronics, ACS Appl. Mater. Interfaces. 14 (2022) 31363–31372.

[64] Z. Zhao, X. Yuan, Y. Huang, J. Wang, CNT-Br/PEDOT:PSS/PAAS three-network composite conductive hydrogel for human motion monitoring, New J. Chem. 45 (2021) 208–216.

[65] C.H. Lu, C.H. Yu, Y.C. Yeh, Engineering nanocomposite hydrogels using dynamic bonds, Acta Biomater. 130 (2021) 66–79.

[66] X. Jing, H.Y. Mi, B.N. Napiwocki, X.F. Peng, L.S. Turng, Mussel-inspired electroactive chitosan/graphene oxide composite hydrogel with rapid self-healing and recovery behavior for tissue engineering, Carbon. 125 (2017) 557–570.

[67] S. Wang, L. Yuan, Z. Xu, X. Lin, L. Ge, D. Li, C. Mu, Functionalization of an electroactive self-healing polypyrrole-grafted gelatin-based hydrogel by incorporating a polydopamine@AgNP nanocomposite, ACS Appl. Bio Mater. 4 (2021) 5797–5808.

[68] R. Tognato, A.R. Armiento, V. Bonfrate, R. Levato, J. Malda, M. Alini, D. Eglin, G. Giancane, T. Serra, A Stimuli-Responsive Nanocomposite for 3D anisotropic cell-guidance and magnetic soft robotics, Adv. Funct. Mater. 29 (2019) 1–10.

[69] W. Lan, M. Xu, X. Zhang, L. Zhao, D. Huang, X. Wei, W. Chen, Biomimetic polyvinyl alcohol/type II collagen hydrogels for cartilage tissue engineering, J. Biomater. Sci. Polym. Ed. 31 (2020) 1179–1198.

[70] G.R. Mahdavinia, H. Etemadi, In situ synthesis of magnetic CaraPVA IPN nanocomposite hydrogels and controlled drug release, Mater. Sci. Eng. C. 45 (2014) 250–260.

[71] M.F. Abou Taleb, F.I. Abou El Fadl, H. Albalwi, Adsorption of toxic dye in wastewater onto magnetic NVP/CS nanocomposite hydrogels synthesized using gamma radiation, Sep. Purif. Technol. 266 (2021) 118551.

[72] T. Lin, J. Zhang, H. Long, F. Yang, L. Huang, S.P. Deng, J. Zhang, X. Cai, Y. Yang, S. Tan, Temperature-sensitive hydrogels containing carboxylated chitosan-modified carbon nanotubes for controlled drug release, ACS Appl. Nano Mater. 5 (2022) 10409–10420.

[73] M.M. Abudabbus, I. Jevremović, K. Nešović, A. Perić-Grujić, K.Y. Rhee, V. Mišković-Stanković, In situ electrochemical synthesis of silver-doped poly(vinyl alcohol)/graphene composite hydrogels and their physico-chemical and thermal properties, Compos. Part B Eng. 140 (2018) 99–107.

[74] W. Shi, N. Song, Y. Huang, C. He, M. Zhang, W. Zhao, C. Zhao, Improved cooling performance of hydrogel wound dressings via integrating thermal conductivity and heat storage capacity for burn therapy, Biomacromolecules. 23 (2022) 889–902.

[75] M. Ghanbari, M. Salavati-Niasari, F. Mohandes, B. Dolatyar, B. Zeynali, In vitro study of alginate-gelatin scaffolds incorporated with silica NPs as injectable, biodegradable hydrogels, RSC Adv. 11 (2021) 16688–16697.

4 Synthesis of Hydrogels

Physical and Chemical Cross-Linking

Heidi Yánez-Vega and Karen Yánez-Vega
Universidad de las Fuerzas Armadas, Sangolquí, Ecuador

Moises Bustamante-Torres
Universidad de Buenos Aires, Buenos Aires, Argentina

4.1 INTRODUCTION

Hydrogels are 3D networks of cross-linked polymers with the ability to absorb a large amount of water without dissolving. Their capacity to retain water depends on the polymers and the fabrication method employed. According to their starting materials, they can be divided into natural polymer hydrogels and synthetic polymer hydrogels, as well as a mixture of both classes. Their ability to interact with solvents is due to the hydrophilic groups present in their structure, such as hydroxyl (-OH) and carboxyl (-COOH) [1].

Hydrogels are polymeric structures cross-linked forming a 3D networks. The hydrogel structure can be held together as water-swollen gels by primary covalent cross-links, ionic forces, hydrogen bonds, affinity or "bio-recognition" interactions, hydrophobic interactions, polymer crystallites, physical entanglements of individual polymer chains, or a combination of two or more of the above interactions [2].

Hydrogel properties will depend on the cross-linking way (physical or chemical) and the polymers employed (natural or synthetic). Cross-linking reactions involve the covalent or non-covalent bonding between macromolecules. During the cross-linking, the micro and macro properties of hydrogels can be modified. The physical synthesizing method includes ionic, hydrophobic interaction, and hydrogen bond formation, whereas the chemically cross-linked hydrogel preparation involves different polymerization techniques, such as chain-growth polymerization, irradiation polymerization, and step-growth polymerization [3].

Physical methods rely on ionic and hydrophobic forces and hydrogen bonds to maintain the hydrogel structure [4]. The main characteristic of this method does not require a cross-linking agent, forming a reversible and non-uniform hydrogel. Besides, these hydrogels are not stable under physiological conditions [5]. On the contrary, chemical hydrogels build cross-linked networks from covalent bonds;

DOI: 10.1201/9781003340485-4

here, the hydrophobic monomers become hydrophilic polymers [4]. This method requires either a cross-linking agent or high-energy irradiation to form permanent hydrogels. These kind of cross-links are required for in vivo applications [5].

Hydrogels have been studied for many years for multiple applications. Nonetheless, the cross-linking way determines the gist properties of the hydrogels. This chapter will provide information about the chemical and physical techniques to get cross-linking hydrogels.

4.2 CROSS-LINKING

Cross-linking corresponds to the formation of polymeric networks from polymers. A vast amount of polymers can be cross-linked through different mechanisms. During cross-linking, the polymer can react, forming intermolecular or intra-molecular linkages. Cross-linking agents tie together carbon atoms from different polymer chains, transforming what were once viscous linear segments into an insoluble gel network. Those insoluble gel networks are known as hydrogels.

Depending on the cross-linking degree, the hydrogel can have different properties. The density of the gel determines the degree of cross-linking. The cross-linked materials exhibit better mechanical (elastic modulus, ultimate tensile strength, and ultimate elongation percent), chemical and thermal properties, which are tied directly to the quality of bond types between molecules. Generally, uncross-linking polymers display viscous flow components, while cross-linked and highly cross-linked networks exhibit a rigid structure [1].

4.3 CHEMICAL CROSS-LINKING

Chemical cross-linking refers to the intermolecular or intramolecular joining of two or more molecules by irreversible bonds. These chemical mechanisms rely on the cross-linking of functional groups of the polymers (-COOH, -OH, or -NH$_2$, etc.) with cross-linkers and high-energy irradiation [6]. Hydrogels irreversible are obtained through a chemical reaction. Actually, chemical cross-linking hydrogels show better mechanical stability (tensile, shear, bending, etc.) than physical cross-linking hydrogels. These chemical hydrogels fabricated through chemical cross-linking cannot be dissolved in solvents unless the covalent bonds formed in the cross-linking process are cleaved [7]. Figure 4.1 illustrates hydrogels obtained through chemical cross-linking.

The cross-linkers are reagents that join two or more polymers, forming a permanent hydrogel. Table 4.1 represents some of the most common cross-linker agents. The cross-link formation is carried out by adding small cross-linkers molecules. The main factors that affect chemical cross-linking are the concentration of the cross-linking agents and the reaction duration [8]. Covalent cross-linking hydrogels can be fabricated through several approaches, which are discussed below.

4.3.1 CHAIN GROWTH POLYMERIZATION OR ADDITION POLYMERIZATION

Chain-growth polymerized hydrogels are formed by rapid propagation of active centers through monomers containing multiple carbon–carbon double bonds which

Chemical cross-linking

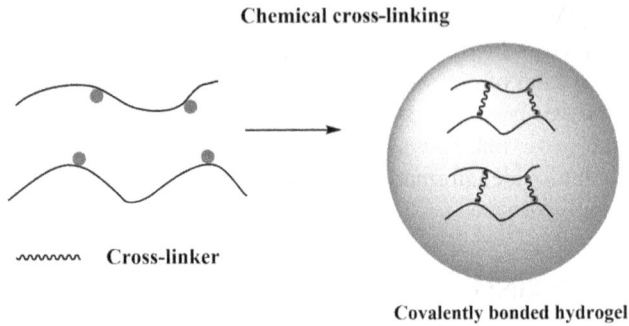

wwwww   Cross-linker

Covalently bonded hydrogel

FIGURE 4.1   Schematic representation of chemical cross-linking.

TABLE 4.1
Some Common Cross-Linker Agents

Cross-Linker	References
Glutaraldehyde	[9,10]
Epoxy compounds	[9]
Isocyanates	[9]
1,3-butadiene bicyclic oxidate	[10]
Pyridoxal 5′-phosphate	[10]
Amino resins	[11]

---

form a high-molecular-weight kinetic chain, serving as a cross-linking point [12]. This process involves initiation, addition, and termination. During initiation, the addition of an initiator, such as heat, light, or irradiation, can trigger the initiation process, where the monomer acquires an active site (free-radical) [7]. Subsequently, monomers are added with their structure unchanged onto a chain, creating active sites for the subsequent attachment [13]. Finally, the growing chain is neutralized, stopping the chain propagation. Some examples are polyethylene, polyvinyl chloride (PVC), acrylics, and poly(ethylene glycol) (PEG), among others.

Normally, polymers containing vinyl groups are produced by this way. Vinyl monomers contain an initiating and a propagating group in the same molecule. In contrast, anionic polymerizations are chain-growth processes in which the active center to which successive monomers are added is a negative ion that is associated with a positive counterion [14]. Anions that are active enough to cause polymerization through reactions of carbon–carbon double bonds are strong bases [14].

Lee and co-workers developed hydrogels from PEG through chain-growth polymerization for drug delivery [12]. Vats and collaborators synthesized hydrogels from PEG and poly(methyl methacrylate) or poly(acrylate) via chain-growth polymerization as a versatile platform for cell adhesion [15].

4.3.1.1 Condensation Polymerization or Step-Growth Polymerization

Condensation polymerization is also known as step-growth polymerization, which is a powerful technique that involves several condensation reactions between bi-functional or tri-functional monomers. The presence of at least two function groups is strictly required. As a result, condensation polymers are those polymers that are formed from polyfunctional monomers, which are typically prepared by the reaction between two kinds of bi-functional symmetric monomers [16]. This technique is usually employed to produce polyesters and nylons (Eq. 4.1). Below is an example of making nylons:

$$R - NH2 + R'COOH \rightarrow R'CONHR + H2O \tag{4.1}$$

The condensation reaction between two reactants yields one larger product and a second, smaller product, such as water [17], ammonia, and HCl as by-products.

4.3.2 RADIATION CROSS-LINKING

Radiation provides a cross-linking to the polymer, forming a 3D network structure. Besides, it confers to the cross-linked polymers a significant dimensional stability at extreme conditions. Figure 4.2 illustrates the radiation cross-linking technique. In particular, among the chemical cross-linking methods, the radiation cross-linking method has the advantage that a reaction is induced without using a chemical additive, such as a cross-linking agent or initiator [18].

When radiation is used, the excited molecules dissociate after absorption of energy, hydrogen is evolved, and radicals from the polymer form cross-links [19]. Nowadays, the formation of hydrogels employing controlled radiation is highly studied. However, the equipment is too expensive.

Several works have used radiation cross-linking to create polymeric networks. For example, Lee et al. developed blends of high-density polyethylene (HDPE)/ ethylene vinyl acetate (EVA)/polyurethane (PU) by radiation cross-linking to improve the thermal and mechanical properties of HDPE [18]. Bustamante-Torres and co-workers developed a hydrogel containing agar and acrylic acid (AAc) irradiated with gamma rays at 15, 20, and 25 kGy. The vinyl groups from AAc can efficiently react, forming inner cross-linking. Those hydrogels were characterized and evaluated as drug delivery systems of ciprofloxacin and silver nanoparticles. As increased the irradiation dose, the hydrogel became tough but lost elasticity. Finally,

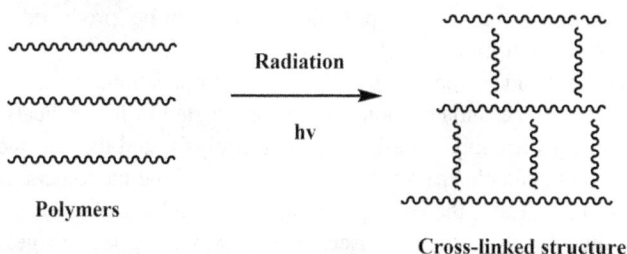

FIGURE 4.2 Representation of radiation cross-linking.

these hydrogels were tested against *Escherichia coli* and *Methicillin Resistant Staphylococcus Aureus* (MRSA), obtaining remarkable results [20].

4.3.2.1 Photo-Cross-Linking Radiation

Hydrogels are frequently cross-linked using photo-initiators and UV light to form permanent hydrogels. Photo-cross-linking requires the presence of reactive unsaturated groups in polymer chains and the presence of photo-initiators but may be accomplished under mild conditions (room temperature) [21]. The photo-cross-linking group activated by ultraviolet (UV) light generates a transient, high-energy intermediate that reacts through space to form a covalent bond [22]. The three most commonly used photo-cross-linking moieties are diazirines, benzophenones, and aryl azides [22].

4.3.3 GRAFT COPOLYMERIZATION

The graft copolymer is a type of copolymer in which one or more blocks of homopolymer are grafted as branches onto a main chain through covalent bonds. These covalent bonds are randomly distributed along the backbone chain. Three different methods can obtain graft copolymers: grafting from, grafting onto, and grafting through. Nonetheless, the most usual route for the preparation of a graft copolymer is to "activate" an initial "trunk polymer", which will then initiate the polymerization of a monomer corresponding to the second polymer [23]. Grafting reactions can be performed by chemical, radiation, or plasma discharge.

Graft copolymers are formed by growing a polymer as branches on another preformed macromolecule. If individual monomeric residues encode A and B, they can form a larger grafted structure from those monomers [14]. In a graft copolymer, the distinguishing feature of the side chains is constitutional, i.e., the side chains comprise units derived from at least one species of monomer different from those which supply the units of the main chain [24].

4.3.4 FREE-RADICAL POLYMERIZATION

Free-radical polymerization of low-molecular-weight monomers in the presence of a cross-linking agent or initiator can be used to produce hydrogels [25] from monomers containing carbon double bonds. The most common initiators are potassium persulfate (KPS), ammonium persulfate (APS), ceric ammonium nitrate, ferrous ammonium sulfate, 2-2′-azobisisobutyronitrile (AIBN), and benzoyl peroxide [26]. Besides, free-radical polymerization can be produced by thermal, photochemical, or with high-energy source.

This process includes four steps: initiation, propagation, chain transfer, and termination. Initially, the initiator molecules are converted to free radicals by different mechanisms, such as heating, photolysis, and electrolysis, and the free radicals, being highly active, can obtain electrons from the molecules of the monomers, making them highly reactive [27]. Then, the propagation and cross-linking of polymer chains will continuously increase the system's viscosity and eventually lead to gelation, which terminates the polymerization [28]. The polymer chain propagates to a high-molecular weight in a short time before its termination, obtaining the final hydrogel.

4.3.4.1   Photo-Cross-Linking Radiation

Hydrogels can be fabricated through light-sensitive functional groups. Photo-cross-linking requires the presence of reactive unsaturated groups in polymer chains and the presence of photo-initiators but may be accomplished under mild conditions (room temperature) [21]. The photo-cross-linking group activated by ultraviolet (UV) light generates a transient, high-energy intermediate that reacts through space to form a covalent bond [22]. This method offers numerous advantages over classical cross-linking approaches, including control of the interaction, which only forms upon UV exposure, and the precise mapping of the interaction interface, speedy preparation, and low cost of production [26]. The three most commonly used photo-cross-linking moieties are diazirines, benzophenones, and aryl azides [22].

Photo-cross-linkable hydrogels are frequently used in cartilage tissue engineering, with cross-linking systems relying on cytotoxic photo-initiators and UV light to form permanent hydrogels [29]. Reys and co-workers synthesized fucoidan hydrogels photo-cross-linked based on visible light. Fucoidan was functionalized by grafting methacrylic groups in the chain backbone, photo-cross-linkable under visible light, to obtain biodegradable structures for tissue engineering [30]. Bartnikowski et al. developed two photo-cross-linkable hydrogel systems: gelatin methacrylamide (GelMA) and gellan gum methacrylate (GGMA). The higher ratio of reactive groups to photo-initiator molecules protected chondrocytes but did not affect chondrocyte differentiation [29].

4.3.4.2   Graft Copolymerization

The graft copolymer is a type of copolymer in which one or more blocks of homopolymers are grafted as branches onto a main chain, meaning it is a branched copolymer with one or more side chains of a homopolymer attached to the backbone of the main chain [30]. In a graft copolymer, the distinguishing feature of the side chains is constitutional, i.e., the side chains comprise units derived from at least one species of monomer different from those which supply the units of the main chain [30].

Three different methods can obtain graft copolymers: grafting from, grafting onto, and grafting through [31]. Nonetheless, the most usual route for the preparation of a graft copolymer is to "activate" an initial "trunk polymer", which will then initiate the polymerization of a monomer corresponding to the second polymer [23].

Graft copolymers are formed by growing a polymer as branches on another preformed macromolecule. If individual monomeric residues encode A and B, they can form a larger grafted structure from those monomers [14]. Grafting can be carried out in such a way that the properties of the side chains can be added to those of the substrate polymer without changing the latter [32]. For example, cellulose fibers can be grafted with sodium polyacrylate while still maintaining their fibrous nature and most of their mechanical properties [32].

4.3.5   Free-Radical Polymerization

Free-radical polymerization of low-molecular-weight monomers in the presence of a cross-linking agent or initiator can be used to produce hydrogels [25]. The most common initiators are potassium persulfate (KPS), ammonium persulfate (APS), ceric ammonium nitrate, ferrous ammonium sulfate, 2-2'-azobisisobutyronitrile (AIBN), and benzoyl peroxide [26]. This process can be performed in a wide range of temperatures since it can be initiated thermally and photochemically [33] or with a high-energy source.

Several solvents have been studied for free-radical polymerization. In many cases, the influence of solvent is small. However, it is becoming increasingly evident that solvent effects can be used to control the polymerization reaction, both at the macroscopic and molecular levels [34]. There are four methods of radical polymerization: bulk polymerization, solution polymerization, suspension polymerization, and emulsion polymerization [35].

Nonetheless, in general, this method involves the chemistry of typical free-radical polymerization, the main steps being initiation, propagation, chain transfer, and termination [25]. Initially, the initiator molecules are converted to free radicals by different mechanisms such as heating, photolysis, and electrolysis, and the free radicals, being highly active, can obtain electrons from the molecules of the monomers, making them highly reactive [27]. After the initiation, the propagation and cross-linking of polymer chains will continuously increase the system's viscosity and eventually lead to gelation, which terminates the polymerization [27]. The polymer chain propagates to high-molecular weight in a relatively short time before its termination by adding monomer molecules to an active chain-end regenerates the active site at the chain-end [33]. The final step of the free-radical polymerization process is the termination of the growing chain and the realization of the final product [36].

Ma and co-workers reported tubular hydrogels with complex geometries by surface radical polymerization, in which an iron wire acts as both a catalyst and template for the formation of a gel layer with controllable thickness. The formed hydrogel layer can be easily removed from the template after secondary cross-linking to obtain hollow hydrogel tubes that exhibit extraordinary and tunable tensile strength, good elasticity, and pressure-bearing capability [37].

4.3.6 INTERPENETRATING NETWORKS

John Millar introduced the term *interpenetrating polymer networks (IPNs)* in 1960. IPNs combine two or more polymers in networks where a partial interlacing on the molecular scale is present in the matrix [38]. Many IPNs exhibit dual-phase continuity, meaning that two or more polymers in the system form continuous phases on a macroscopic scale, but are not covalently bonded to each other by any chemical bond [39]. For IPN synthesis, several parameters must be considered.

Since the IPNs are cross-linked, they swell but do not dissolve in the presence of solvents. Therefore, the mechanical strength and elasticity of the material are enhanced, while physical and chemical characteristics such as temperature sensitivity and interfacial compatibility are improved [39]. These properties depend strictly on the type of IPN. IPNs can be classified according to the chemistry preparation as simultaneous (polymers plus cross-linkers and activators of both networks are mixed simultaneously) or sequential (polymer network A is made, and

TABLE 4.2
Classification of IPNs

IPN Classification	Sub-Classification	References
Chemistry of preparation	Simultaneous IPN	[40]
	Sequential IPN	
Structure	Full IPNs	
	Homo-IPNs	
	Semi-IPNs	

then a second monomer B plus cross-linker and activator are added to A). Besides, IPNs can be classified depending on their structure as full, homo, or semi-IPNs. Table 4.2 represents the IPN classification.

Interpenetrating and semi-interpenetrating networks represent promising composite materials as they allow combining different polymers to enhance their components' characteristics or obtain materials with unique properties [40]. Feig and co-workers developed two interpenetrating hydrogel networks, one of which is formed by the gelation of the conducting polymer poly(3,4-ethylene dioxythiophene) polystyrene sulfonate (PEDOT: PSS) as promising soft electrode materials [41]. Hanyková et al. studied the temperature response of double network (DN) hydrogels composed of thermoresponsive poly(N, N'-diethylacrylamide) (PDEAAm) and hydrophilic polyacrylamide (PAAm). The second hydrophilic network in DN hydrogels influenced their thermal sensitivity significantly [40].

4.4 PHYSICAL CROSS-LINKING

Physical cross-linking of hydrogels is generated by reversible interactions, such as molecular interactions or secondary forces, which despite not being very strong or permanent, confer to hydrogels the ability to be insoluble in aqueous media [42]. Examples of this type of cross-linking are interactions through hydrogen bonds, ionic interactions, and hydrophobic interactions. These cross-links have a reversible response (sol-gel transition) and can be disrupted by changes in physical conditions or by exposure to external stimuli such as changes in pH, temperature, or ionic strength [43].

Physical methods of hydrogel cross-linking offer several biomedical advantages over chemical methods of cross-linking. For example, these methods do not require any covalent cross-linking agent in the synthesis of hydrogels; this provides safety in biomedical applications as it will not produce toxicity due to unreacted residues [26]. In addition, a higher integrity and stability of the bioactive substances (e.g., proteins, cells) that are anchored in the hydrogel are ensured, as they are not exposed to undesired reactions as in chemical cross-linking. The short lifetime of physical hydrogels in physiological environments is ideal for medical applications such as short-term drug delivery [43] (Figure 4.3).

Physical cross-linking

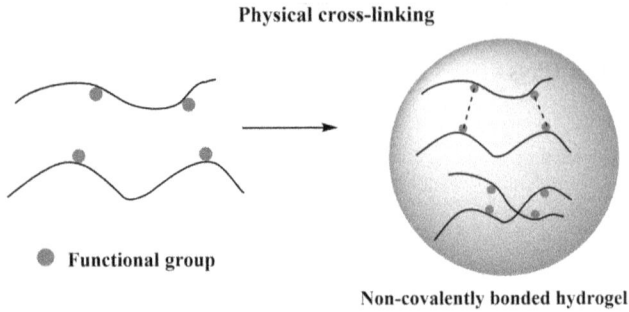

Functional group

Non-covalently bonded hydrogel

FIGURE 4.3   Schematic representation of physical cross-linking.

The formation of physically cross-linked hydrogels largely depends on hydro-colloid type, concentration, pH, salt type, temperature, and other thermodynamic parameters [7]. These features have made physically cross-linked hydrogels more interesting to study in recent years. Different physical cross-linking methods that have been reported for the synthesis of physical hydrogels are detailed.

4.4.1 CRYSTALLIZATION (FREEZE/THAW CYCLES)

When nonionic polymers are subjected to freeze/thaw cycles, crystallites are produced and used as cross-linking points to form the hydrogel network [43]. During freeze–thaw, cycling water freezes, causing phase separation and facilitating crystallization [44]. This cross-linking technique is known as crystallization. An example of this cross-linking is hydrogels made from polyvinyl alcohol (PVA) [43]. This compound can form a mechanically soft gel under ambient conditions, but when exposed to repeated freezing and thawing cycles, it acquires characteristics of a more resistant and elastic hydrogel [45]. Figure 4.4 illustrates the crystallization technique to obtain PVA-based hydrogels.

Hydrogen bonds in PVA are formed in the freezing process [42]. The low interaction between the frozen water and the hydroxyl groups of the polymer chains causes the PVA chains to bond together. Therefore, with a longer freezing time, a more significant number of hydrogen bonds will be formed, increasing the degree of cross-linking of the hydrogel [43].

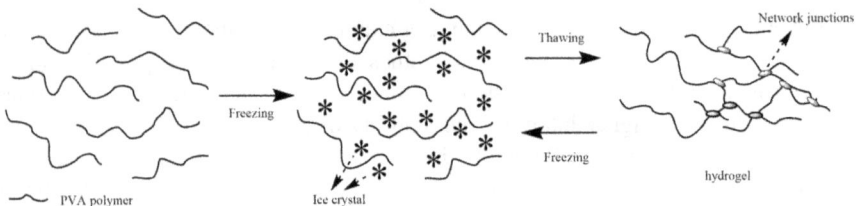

FIGURE 4.4   PVA crystallization processes by freeze/thaw cycles to form physical hydrogels.

Similarly, several researchers have reported that in this type of hydrogel, their properties such as swelling, degree of crystallinity, and crystal size depend on the number of cycles and freezing/thawing times, the PVA's molecular weight, the PVA solutions' concentration, and the minimum and maximum temperature used [45]. Hydrogels cross-linked by crystallization have been reported to have remained stable for up to 6 months under temperatures of 37°C, while at temperatures between 50°C and 60°C, the PVA crystallites melt, and the hydrogen bonds of the 3D network of the hydrogel are broken [43].

It has been reported that by synthesizing a PVA hydrogel by freeze/thawing, the hydrogel can self-repair autonomously at room temperature; this process does not require any stimuli or healing agents [46].

4.4.2 HYDROPHOBIC INTERACTIONS

The hydrophobic interaction term is used to describe the tendency of nonpolar groups or molecules to aggregate in a water solution [47]. For hydrogels, the hydrophobic cross-linking begins with water-soluble polymers, whose structure contains a hydrophobic or insoluble part (terminal groups, side chains, or monomers) responsible for forming physical cross-links [48,49]. The association between hydrophobic groups forms hydrophobic microdomains, and these serve as physical cross-linking points between two or more polymer chains to form the 3D network of the hydrogel [7]. Hydrogels synthesized with hydrophobic cross-linking have high mechanical strength, self-healing capacity, and environmental sensitivity [7].

The hydrophobic cross-linking depends on the amount of hydrophobic and hydrophilic structure in the polymers, and the components in the solution such as salts or surfactants. If the proportion of hydrophobic interactions is low, micelles with hydrophobic nuclei are formed. In contrast, if the proportion of hydrophilic parts is low, the polymers form loops with nuclei containing hydrophobic groups. The higher concentration of hydrophilic parts promotes the formation of links between the micelles, finally forming the hydrogel network [48]. Polysaccharides such as chitosan, dextran, pullulan, and carboxymethylcurdlan have also been reported to synthesize physically cross-linked hydrogels by the hydrophobic modification technique [50].

Several hydrogels have been reported using this technique. Sarmah and Karak synthesized a starch/poly(acrylamide) (PAM) hydrogel. Firstly, they incorporated starch into a hydrophobic domain (HA) of the PAM chain. The incorportation to starch enhanced the mechanical resistance of the hydrogel network when swollen, high flexibility, and high self-repair capacity [51]. Besides, Wang et al. developed a physically cross-linked double network hydrogel from hydrophobically associated PAM/physically cross-linked gellan gum. The authors reported that hydrogel could be used as a substitute for damaged cartilage because it exhibits characteristics similar to native cartilage, such as high mechanical resistance, self-regeneration, stability in physiological environments, and promotes chondrocyte proliferation [7].

4.4.3 AMPHIPHILIC COPOLYMERS

Amphiphilic copolymers are also called block polymers [50]. When found in an aqueous solution, these molecules can self-assemble in arrangements such as polymeric micelles. The hydrophilic structures of the polymer chains form loops while the hydrophobic groups remain in the center [52].

The synthesis of physical hydrogels is crucial to promote the attachment between the micelles through hydrophobic interactions formed by the thermal induction of the base polymers [50]. The induction temperature will depend on the properties of the polymers; some polymers are soluble below the lower critical solution temperature (LCST), but can become insoluble and hydrophobic when above this threshold [53], which is a characteristic of thermo-sensitivity. Smart polymers or stimuli-responsive polymers respond to external stimuli such as temperature [4].

Poly(lactic acid) (PLA) and polyethylene glycol (PEG) have been used to assemble block copolymers and synthesize hydrogels in various investigations. Due to their biodegradability and biocompatibility properties, these polymers are candidates as drug delivery hydrogels [54]. PEG and poly(N-isopropylacrylamide) (PNIPAM) hydrogels are also assembled by block copolymers and have allowed significantly high cell retention as stem cell carriers in in vivo studies. PEG is the water-soluble block in this synthesis, and PNIPAM is the thermosensitive terminal-bearing block [55]. Multiple triblock polymer systems with hydrophobic segments in between have been proposed; for example, PEG-(poly(lactic-co-glycolic acid))(PLGA)-PEG at low concentrations in water form micelles and high concentrations form thermoreversible hydrogels. Cross-linking is thought to occur by hydrophobic interactions [54].

4.4.4 CHARGE INTERACTIONS

Charge or ionic interaction also allows synthesizing of physically cross-linked hydrogels from two molecules with opposite electrical charges that interact to form polyelectrolyte complexes [42]. Figure 4.5 illustrates the charge interaction of two different charge monomers to form hydrogels; one positively charges with the amino group, and another negatively charges with the carboxyl group. Thus, ionic polymers can cross-link if divalent or trivalent counterions are added [56]. The properties of hydrogels synthesized from these polyelectrolyte complexes depend

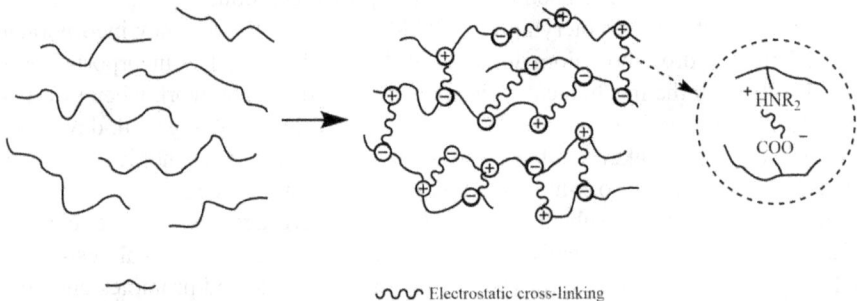

$+HNR_2$

COO^-

Electrostatic cross-linking

FIGURE 4.5 Formation of cross-linked physical hydrogels by charge interaction.

on factors such as the charge density in the polymers, the concentration of each polymer, and the polymer's solubility [57].

Over the years, strategies have been developed to enhance cross-linking of hydrogels through charge interaction. For example, alginate is a polymer that has mannuronic acid and glucuronic acid residues. Therefore, these residues can cross-link through ionic interactions with divalent cations, such as calcium (Ca^{2+}), barium (Ba^{2+}), and magnesium (Mg^{2+}) [43]. The cations only bind to the guluronate of the alginate chains. Subsequently, the guluronate blocks of the first polymer are attached to the guluronate blocks of the second polymer; thus, the hydrogel network is formed. These hydrogels can be destabilized by adding a chelating agent that extracts the ions [43].

4.4.5   INTERACTIONS BY HYDROGEN BONDS

Hydrogen bonds (H-bonds) are a specific type of electrostatic interaction between a proton attached to an electronegative atom (such as N or O) and a lone pair of electrons on an electronegative atom such as N, O, or F [58]. Hydrogen bonds often occur in networks frequently with water mediating. Water is especially facile at hydrogen bonding because it is both an acceptor and a donor [59]. Unlike covalent bonds, which vary in strength within a factor of ~4 (30–120 kcal mol^{-1}), hydrogen bonds are much less constrained in their geometry and physical properties, and they vary in strength by a factor of at least 20-fold (2–40 kcal mol^{-1}) [60]. Thus, hydrogels obtained from single hydrogen bonds are too weak. For this reason, employing multiple multivalent hydrogen bonds between the polymer chains is necessary to form a solid and insoluble 3D network [48]. These bonds have a high dependence on pH, which is why the properties of the hydrogel could be shaped based on pH [50].

Several investigations have been carried out to form hydrogels using hydrogen interaction. For example, Jing and co-workers synthesized hydrogen-bonded cross-linked sodium alginate/carboxymethylchitosan hydrogel beads to encapsulate proteins. These beads obtained excellent sensitivity to the pH of the different exposed environments; therefore, properties such as swelling and preventing premature protein release could be controlled [61].

4.4.6   STEREOCOMPLEX FORMATION

The stereocomplex is a complex of macromolecules that share an identical chemical composition but a different chiral. A typical example of cross-linking by stereo-complex formation occurs from stereoisomers of PLA. The formation of PLA stereocomplexes is generated by non-covalent interactions between the enantiomeric chains of poly(L-lactide) (PLLA) [i.e., poly(L-lactic acid)] and poly(D-lactide) (PDLA) [i.e., poly(D-lactic acid)] [62].

When the PLLA/PDLA stereocomplexes are found alone, they cannot swell like a hydrogel; therefore, it is necessary to establish mixtures of copolymers to obtain the characteristics of a suitable hydrogel [50,63]. For example, Liu et al. inserted PLLA and PDLA stereocomplexes into heat-sensitive PNIPAM to obtain enantio-meric copolymers of PNIPAM-g-PLLA and PNIPAM-g-PDLA that would be used later for the synthesis of the hydrogel. Physical cross-linking was generated between the hydrophobic domains of PLLA/PDLA for the formation of the 3D network of the hydrogel [63].

Besides, Lim and Park reported the manufacture of PLLA-PEG-PLLA and PDLA-PEG-PDLA triblock copolymers to form microspheres that would encapsulate bovine serum albumin (BSA). The authors compared the BSA release capacity between spheres made with the triblock copolymer and spheres made only with the PLLA stereocomplex. They determined that slightly more protein was released from the spheres formed by the copolymer of the stereocomplexes. This difference in releasing compared to the PLLA spheres was attributed to the fact that the copolymer spheres had a greater swelling capacity when absorbing water [64].

4.4.7 PROTEIN INTERACTIONS

Protein-conjugated hydrogels are used as a simplified mimic of the extracellular matrix (ECM) by incorporating various proteins in the scaffold structure and have more recently been appreciated as beneficial toward eliciting a desired cell response, degradability, or manipulating the mechanical and physical properties of the scaffold itself [65]. Protein aggregation into a gel network is the primary governing mechanism for the protein's gelation process, known as physical cross-linking [66]. The formed network maintains water within its structure. It can be stabilized through non-covalent cross-links, such as hydrophobic, van der Waals, electrostatic interactions, and hydrogen bonds [67]. Protein-based hydrogels are becoming increasingly attractive due to their infinite design options, substantial advantages such as biocompatibility, biodegradability, tunable mechanical properties, molecular binding capabilities; and intelligent responses to external stimuli such as pH, ionic strength, and temperature [68]. Physical cross-linking by protein interactions can be accomplished through genetically engineered proteins or antigen–antibody interactions [4].

4.5   CONCLUSION

Hydrogels are 3D polymeric structures with a great ability to interact with solvents such as water. Their capacity to uptake water will depend on the monomer concentration and the technique to form the cross-linking structure. There are two ways to form hydrogels: chemical and physical cross-linking. Nonetheless, some works have combined both processes. Chemical cross-linking consists of several techniques to form permanent hydrogels. These hydrogels require a cross-linking agent to start with the polymerization mechanism. Chemical cross-linking consists of several processes, such as condensation and addition polymerization, the two most common methods for polymer synthesis. These reactions occur between different functional groups. Moreover, controlled-radiation cross-linking is well-known as a convenient tool for modifying polymeric materials through cross-linking, grafting, and degradation. Free-radical polymerization employs cross-linking initiators to produce the polymerization and monomers that contain carbon double bonds.

Interpenetrating networks combine at least two polymers into networks, forming a rigid structure with enhanced mechanical, physical, and chemical properties.

Physical cross-linking consists of reversible interactions to form a hydrogel. The main characteristic is that cross-linking agents are required for these techniques. Crystallization employs freeze-thawing cycles to create crystallites that will be used as cross-linking points for hydrogels. Hydrophobic interactions consist of hydrophobic parts or non-soluble into the polymer structure that will develop microdomains that serve as physical cross-linking points. On the other hand, amphiphilic copolymers or block polymers form arrangements such as polymeric micelles containing either hydrophilic or hydrophobic segments. Charge interactions use an opposite electrical charge to generate hydrogels with physical cross-linking. Interactions by hydrogen bonds are too weak to form hydrogels; therefore, multivalent hydrogen bonds between polymers can create a solid polymeric structure. Stereocomplex formation uses macromolecules with identical chemical composition but a different chiral configuration, such as PLLA and PDLA. Finally, protein aggregation into a gel network is the primary governing mechanism for the protein's gelation process. This mechanism generally employs genetically engineered proteins.

REFERENCES

[1] Arcentales-Vera, B., Bastidas, L., Bustamante-Torres, M., Maldonado Pinos, P., Bucio, E. "Hyper-Crosslinked Polymers." *Porous Polymer Science and Applications*, 2022, 7–36. http://dx.doi.org/10.1201/9781003169604-2.

[2] Peppas, N. A., Hoffman, A. S. "Hydrogels." *Biomaterials Science*, 2013, 166–179. https://doi.org/10.1016/B978-0-08-087780-8.00020-6.

[3] Ranganathan, N., Bensingh, R. J., Kader, M. A., Nayak, S. N. In Mondal, Md., Ibrahim, H. (Eds.). *Polymers and Polymeric Composites: A Reference Series.* "Synthesis and Properties of Hydrogels Prepared by Various Polymerization Reaction Systems." 2019, Springer International Publishing.

[4] Bustamante-Torres, M., Romero-Fierro, D., Arcentales-Vera, B., Palomino, K., Magaña, H., Bucio, E. "Hydrogels Classification According to the Physical or Chemical Interactions and as Stimuli-Sensitive Materials." *Gels*, 7, 2021, 182/1–182/25.

[5] García, M. C., Cuggino, J. C. In Hamdy Mackhlouf, A. S., Abu-Thabit, N. Y. (Eds.). *Stimuli Responsive Polymeric Nanocarriers for Drug Delivery Applications.* "Stimulus-Responsive Nanogels for Drug Delivery." 2018, Woodhead Publishing.

[6] Ermis, M., Calamak, S., Calibasi Kocal, G., Guven, S., Durmus, N. G., Rizvi, I., Hasan, T. Hasirci, N., Hasirci, V., Demirci, U. In Conde, J. (Ed.). *Handbook of Nanomaterials for Cancer Theranostics.* "Hydrogels as a New Platform to Recapitulate the Tumor Microenvironment." 2018, Elsevier.

[7] Wang, M., Guo, L., Sun, H. In Narayan. (Ed.). *Encyclopedia of Biomedical Engineering.* "Manufacture of Biomaterials." 2019, Elsevier.

[8] Hasirci, V., Yilgor Huri, P., Endogan Tanir, T., Eke, G., Hasirci, N. In Ducheyne. (Ed.). *Comprehensive Biomaterials II.* "1.22 Polymer Fundamentals: Polymer Synthesis." 2017, Elsevier.

[9] Wagermaier, W., and Fratzl, P. In Matyjaszewski, K., Möller, M. (Eds.). *Polymer Science: A Comprehensive Reference.* "Collagen." 2012, Elsevier.

[10] Lu, X., Ma, G., Su, Z. In Moo-Young, M. (Ed.). *Comprehensive Biotechnology.* "Hemoglobin-Based Blood Substitutes – Preparation Technologies and Challenges." 2011, Academic Press.

[11] Pizzi, A., Ibeh, C. C. In Dodiuk, H., Goodman, S. H. (Eds.). *Handbook of Thermoset Plastics.* "Aminos." 2014, William Andrew Publishing.

[12] Lee, S., Tong, X., Yang, F. "Effects of the Poly(Ethylene Glycol) Hydrogel Crosslinking Mechanism on Protein Release." *Biomaterials Science*, 4, 2016, 405–411.

[13] Speight, J. G. In Speight, J. G. (Ed.). *Handbook of Industrial Hydrocarbon Processes.* "Monomers, Polymers, and Plastics." 2020, Gulf Professional Publishing.

[14] Rudin, A., Choi, P. In Rudin, A., Choi, P. (Eds.). *The Elements of Polymer Science & Engineering.* "Ionic and Coordinated Polymerizations." 2013, Elsevier.

[15] Vats, K., Marsh, G., Harding, K. Zampetakis, I. Waugh, R. E., Benoit, D. S. "Nanoscale Physicochemical Properties of Chain- and Step-Growth Polymerized Peg Hydrogels Affect Cell-Material Interactions." *Journal of Biomedical Materials Research Part A*, 105, 2017, 1112–1122.

[16] Fukukawa, K., Ueda, M. In Matyjaszewski, K., Möller, M. (Eds.). *Polymer Science: A Comprehensive Reference.* "Sequence Control in One-Step Polycondensation." 2012, Elsevier.

[17] Ouellette, R. J., Rawn, J. D. In Ouellette, R. J., Rawn, J. D. (Eds.). *Organic Chemistry.* "Synthetic Polymers." 2014, Elsevier.

[18] Lee, J., Jeong, J., Jeong, S., Park, S. "Radiation-Based Crosslinking Technique for Enhanced Thermal and Mechanical Properties of HDPE/EVA/Pu Blends." *Polymers*, 13, 2021, 2832.

[19] Rado, R., Lazár, M. "Role of the Source of Free Radicals in Crosslinking of Polyethylene." *Journal of Polymer Science*, 53, 1961, 67–74.

[20] Bustamante-Torres, M., Pino-Ramos, V. H., Romero-Fierro, D., Hidalgo-Bonilla, S. P., Magaña, H., Bucio, E. "Synthesis and Antimicrobial Properties of Highly Cross-Linked Ph-Sensitive Hydrogels through Gamma Radiation." *Polymers*, 13, 2021, 2223.

[21] Parhi, R. "Cross-Linked Hydrogel for Pharmaceutical Applications: A Review." *Advanced Pharmaceutical Bulletin*, 7, 2017, 515–530.

[22] Castaldi, M. P., Zuhl, A., Ricchiuto, P., Hendricks, J. A. In Goodnow, R. A. (Ed.). *Annual Reports in Medicinal Chemistry.* "Chemical Biology in Drug Discovery." 2017, Elsevier.

[23] Chapiro, A. In Buschow, K., Cahn, R., Flemings, M., Ilschner, B., Kramer, E., Mahajan, S., Veyssière, P. (Eds.). *Encyclopedia of Materials: Science and Technology,* "Radiation Effects in Polymers." 2004, 1–8, Elsevier.

[24] Jenkins, A., Kurt, L. In Allen, G., Bevington, J. (Eds.). *Comprehensive Polymer Science and Supplements,* "Nomenclature." 1989, 13–54, Oxford: Pergamon Press.

[25] Varghese, S. A., Rangappa, S. M., Siengchin, S., Parameswaranpillai, J. In Chen Y. (Ed.). *Hydrogels Based on Natural Polymers.* "Natural Polymers and the Hydrogels Prepared from Them." 2020, Elsevier.

[26] Bashir, S., Hina, M., Iqbal, J., Rajpar, A. H., Mujtaba, M. A., Alghamdi, N. A., Wageh, S., Ramesh, K., Ramesh, S. "Fundamental Concepts of Hydrogels: Synthesis, Properties, and Their Applications." *Polymers*, 12, 2020, 2702.

[27] Das, S. K., Chakraborty, S., Naskar, S., Rajabalaya, R. In Ahmed, S., Annu. (Eds.). *Bionanocomposites in Tissue Engineering and Regenerative Medicine.* "Techniques and Methods Used for the Fabrication of Bionanocomposites." 2021, Woodhead Publishing.

[28] Wang, W., Narain, R., Zeng, H. In Narain, R. (Ed.). *Polymer Science and Nanotechnology.* "Hydrogels." 2020, Elsevier.

[29] Bartnikowski, M., N. J. Bartnikowski, Woodruff, M. A., Schrobback, K., Klein, T. J. "Protective Effects of Reactive Functional Groups on Chondrocytes in Photocrosslinkable Hydrogel Systems." *Acta Biomaterialia*, 27, 2015, 66–76.

[30] Reys, L., Silva, S., Soares da Costa, D., Oliveira, N., Mano, J., Reis, R., Silva, T. "Fucoidan Hydrogels Photo-Cross-Linked with Visible Radiation as Matrices for Cell Culture." *ACS Biomaterials Science & Engineering*, 2, 2016, 1151–1161.

[31] Faust, R., Schlaad, H. In Craver, C., Carraher, C. (Eds.). *Applied Polymer Science: 21st Century*. "Ionic Polymerization." 2000, Applied Polymer Science: 21st Century.

[32] Stannett, V. T., Fanta, G. F., Doane, W. M., Chatterjee, P. K. In Chatterjee, P. K., Gupta, B. S. (Eds.). *Absorbent Technology*. "Polymer Grafted Cellulose and Starch." 2002, Elsevier.

[33] Lamaoui, A., García-Guzmán, J., Amine, A., Palacios-Santander, J., Cubillana-Aguilera, L. In Sooraj, M. P. Archana, S. N., Beena, M., Sabu, T. (Eds.). *Molecularly Imprinted Polymer Composites*. "Synthesis Techniques of Molecularly Imprinted Polymer Composites." 2021, Woodhead Publishing.

[34] Coote, M. L., Davis, T. P. In Wypych, G. (Ed.). *Handbook of Solvents*. "Solvent Effects on Free Radical Polymerization." 2014, ChemTec Publishing.

[35] Mantha, S., Sangeeth, P., Parisa, K., Akshaya, U., Yuli, Z., Tao, O., Pham, H., Tran, S. "Smart Hydrogels in Tissue Engineering and Regenerative Medicine." *Materials*, 12, 2019, 3323.

[36] Baruah, C., Sarmah, J. K. In Jana, S., Jana, S. (Eds.). *Micro- and Nanoengineered Gum-Based Biomaterials for Drug Delivery and Biomedical Applications*. "Guar Gum-Based Hydrogel and Hydrogel Nanocomposites for Biomedical Applications." 2022, Elsevier.

[37] Ma, S., Mingming, R., Peng, L., Bao, M., Xie, J., Xiaolong, W., Huck, W., Zhou, F., Weimin, L. "Fabrication of 3D Tubular Hydrogel Materials through on-Site Surface Free Radical Polymerization." *Chemistry of Materials*, 30, 2018, 6756–6768.

[38] Karak, N. In Karak, N. (Ed.). *Vegetable Oil-Based Polymers*. "Fundamentals of Polymers." 2012, Woodhead Publishing.

[39] Bongiovanni, R., Vitale, A. In Montemor, M. F. (Ed.). *Smart Composite Coatings and Membranes*. "Smart Multiphase Polymer Coatings for the Protection of Materials." 2016, Woodhead Publishing.

[40] Hanyková, L., Krakovský, I., Šestáková, E., Šťastná, J., Labuta, J. "Poly(N,N′-Diethylacrylamide)-Based Thermoresponsive Hydrogels with Double Network Structure." *Polymers*, 12, 2020, 2502.

[41] Feig, V. R., Tran, H., Lee, M., Bao, Z. "Mechanically Tunable Conductive Interpenetrating Network Hydrogels That Mimic the Elastic Moduli of Biological Tissue." *Nature Communications*, 9, 2018, 2740/1–2740/9.

[42] Hu, W., Wang, Z., Xiao, Y., Zhang, S., Wang, J. "Advances in Crosslinking Strategies of Biomedical Hydrogels." *Biomaterials Science*, 7, 2019, 843–855.

[43] Ullah, F., Othman, M., Javed, F., Ahmad, Z., Akil, H. "Classification, Processing and Application of Hydrogels: A Review." *Materials Science and Engineering: C*, 57, 2015, 414–433.

[44] Holloway, J., Lowman, A., Palmese, G. "The Role of Crystallization and Phase Separation in the Formation of Physically Cross-Linked PVA Hydrogels." *Soft Matter*, 9, 2013, 826–833.

[45] Boran, F. "The Influence of Freeze-Thawing Conditions on Swelling and Long-Term Stability Properties of Poly(Vinyl Alcohol) Hydrogels for Controlled Drug Release." *Polymer Bulletin*, 78, 2021, 7369–7387.

[46] Zhang, H., Xia, H., Zhao, Y. "Poly(Vinyl Alcohol) Hydrogel Can Autonomously Self-Heal." *ACS Macro Letters*, 1, 2012, 1233–1236.

[47] Sill, E., Arnau, A., Tuñón, I. In Wypych, G. (Ed.). *Handbook of Solvents*. "Fundamental Principles Governing Solvents Use." 2014, ChemTec Publishing.

[48] Voorhaar, L., Hoogenboom, R. "Supramolecular Polymer Networks: Hydrogels and Bulk Materials." *Chemical Society Reviews*, 45, 2016, 4013–4031.

[49] Wang, Y., Yu, W., Liu, S. "Physically Cross-Linked Gellan Gum/Hydrophobically Associated Polyacrylamide Double Network Hydrogel for Cartilage Repair." *European Polymer Journal*, 167, 2022, 111074.

[50] Augustine, R., Alhussain, H., Zahid, A. A., Raza Ur Rehman, S., Ahmed, R., Hasan, A. In Jose, J., Thomas, S., Thakur, V. K. (Eds). *Nano Hydrogels. Gels Horizons: From Science to Smart Materials*. "Crosslinking Strategies to Develop Hydrogels for Biomedical Applications." 2021, Springer.

[51] Sarmah, D., Karak N. "Physically Cross-Linked Starch/Hydrophobically-Associated Poly(Acrylamide) Self-Healing Mechanically Strong Hydrogel." *Carbohydrate Polymers*, 289, 2022, 119428.

[52] Förster, S., Markus, A. "Amphiphilic Block Copolymers in Structure-Controlled Nanomaterial Hybrids." *Advanced Materials*, 10(3), 1999, 195–217.

[53] Lue, S., Chen, C., Shih, C. "Tuning of Lower Critical Solution Temperature (LCST) of Poly(N-Isopropylacrylamide-Co-Acrylic Acid) Hydrogels." *Journal of Macromolecular Science, Part B*, 50, 2011, 563–579.

[54] Jeong, B., Bae, Y., Kim, S. "Thermoreversible Gelation of Peg–PLGA–Peg Triblock Copolymer Aqueous Solutions". *Macromolecules*, 32, 1999, 7064–7069.

[55] Cai, L., Dewi, R., Heilshorn, S. "Injectable Hydrogels with in Situ Double Network Formation Enhance Retention of Transplanted Stem Cells." *Advanced Functional Materials*, 25, 2015, 1344–1351.

[56] Gulrez, S. K., Al-Assaf, S. O. G. In A. Carpi (Ed.). *Progress in Molecular and Environmental Bioengineering – From Analysis and Modeling to Technology Applications*, "Hydrogels: Methods of Preparation, Characterisation and Applications." 2011, IntechOpen.

[57] Hamman, J. "Chitosan Based Polyelectrolyte Complexes as Potential Carrier Materials in Drug Delivery Systems." *Marine Drugs*, 4, 2010, 1305–1322.

[58] Abelian, A., Dybek, M., Wallach, J., Gaye, B., Adejare, A. In Adejare, A. (Ed.). *Remington: The Science and Practice of Pharmacy*, "Pharmaceutical Chemistry." 2021, Academic Press.

[59] Mcree, D. In Mcree, D. (Ed.). *Practical Protein Crystallography*, "Computational Techniques." 1999, Academic Press.

[60] Frey, P. In Lennarz, W., Lane, D. (Eds.). *Encyclopedia of Biological Chemistry*, "Low Barrier Hydrogen Bonds." 2004, Elsevier.

[61] Jing, H., Huang, X., Du, X., Mo, L., Ma, C., Wang, H. "Facile Synthesis of pH-Responsive Sodium Alginate/Carboxymethyl Chitosan Hydrogel Beads Promoted by Hydrogen Bond." *Carbohydrate Polymers*, 278, 2022, 118993.

[62] Wu, J., Shi, X., Wang, Z., Song, F., Gao, W., Liu, S. "Stereocomplex Poly(Lactic Acid) Amphiphilic Conetwork Gel with Temperature and Ph Dual Sensitivity." *Polymers*, 11, 2019, 1940.

[63] Liu, K., Cao, H., Yuan, W., Bao, Y., Shan, G., Wu, Z., Pan, P. "Stereocomplexed and Homocrystalline Thermo-Responsive Physical Hydrogels with a Tunable Network Structure and Thermo-Responsiveness." *Journal of Materials Chemistry B*, 35, 2020, 7947–7955.

[64] Lim, D., Park, T. "Stereocomplex Formation between Enantiomeric PLA–PEG–PLA Triblock Copolymers: Characterization and Use as Protein-Delivery Microparticulate Carriers." *Journal of Applied Polymer Science*, 75, 2000, 1615–1623.

[65] Zustiak, S., Wei, Y. Leach, Y. "Protein–Hydrogel Interactions in Tissue Engineering: Mechanisms and Applications." *Tissue Engineering Part B: Reviews*, 19, 2, 2013, 160–171.

[66] Le, X., Rioux, L., Turgeon, S. "Formation and Functional Properties of Protein–Polysaccharide Electrostatic Hydrogels in Comparison to Protein or Polysaccharide Hydrogels." *Advances in Colloid and Interface Science*, 239, 2017, 127–135.

[67] Sato, N., Aoyama, Y., Yamanaka, J., Toyotama, A., Okuzono, T. "Particle Adsorption on Hydrogel Surfaces in Aqueous Media Due to Van Der Waals Attraction." *Scientific Reports*, 7, 2017, 6099/1–6099/10.

[68] Panahi, R., Baghban-Salehi, M. Protein-Based Hydrogels *"Polymers and Polymeric Composites: A Reference Series."* 2019, 1561–1600, Springer.

5 Fabrication Techniques of Hydrogels

Estefani Chichande-Proaño
Universidad Central del Ecuador, Quito, Ecuador

Moises Bustamante-Torres
Universidad de Buenos Aires, Buenos Aires, Argentina

5.1 INTRODUCTION

Hydrogels are attractive 3D cross-linking polymeric biomaterials with a remarkable ability to absorb and retain water due to the hydrophilic groups in their chemical network. These cross-linking structures have attracted scientists due to their biocompatibility, biodegradability, and possible interaction with other materials. The synergic union would enhance the hydrogel properties.

The hydrogel properties depend mainly on the nature of their precursors and fabrication techniques. Hydrogels can be made up of natural or synthetic monomers. These structures can hold together as water-swollen gels by chemical, physical, or both interactions. Figure 5.1 illustrates a hydrogel representation. The fabrication techniques play an essential role in these interactions and have evolved to form hydrogel structures. Over the years, new techniques have been developed, including electrospinning, gas foaming, sol–gel method, 3D printing, melt molding, freeze-drying, and grafting. Therefore, the properties of the hydrogels will depend on the nature of their monomers and the strategy to produce them.

Currently, hydrogel synthesis is a remarkable study by scientists. These polymeric materials were the first biomaterials designed for the human body and are finding widespread biomedical applications. Furthermore, their versatility makes them ideal for multiple combinations with several molecules possible to obtain adequate characteristics for each scope. This chapter describes the main techniques employed for the fabrication of hydrogels currently.

5.2 FABRICATION TECHNIQUES OF HYDROGELS

Hydrogel properties and precursors should include high biocompatibility, biodegradability, and low cytotoxicity. Hydrogels can be classified in multiple ways; some of the most important are based on their nature and the fabrication process. Natural hydrogels commonly employ natural polymers such as cellulose, carrageenan, hyaluronic acid, polysaccharides (alginate, starch, agarose), chitosan (CS), fibrin, and proteins (gelatin, collagen) [1]. In contrast, synthetic polymer-based

DOI: 10.1201/9781003340485-5

Cross-linking

Hydrogel

Chemical/Physicial Cross-linkers

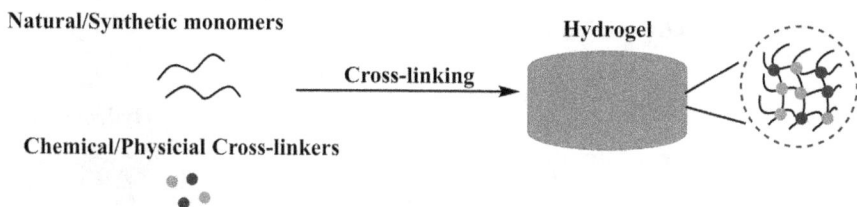

FIGURE 5.1 Schematic representation of hydrogel cross-linked from monomers from different natures.

TABLE 5.1
Fabrication Techniques to Produce Hydrogels

Techniques	Sub-Techniques	References
Electrospinning	Solution electrospinning	[5]
	Melt electrospinning	[6]
Gas foaming		[7]
Sol–gel method		[8]
3D printing	Laser printing	[9]
	Stereolithography	[10]
	Two-photon polymerization	[11]
	Laser-induced forward transfer	[12]
	Extrusion printing	[13]
	Inkjet printing	[14,15]
Melt molding		[16]
Freeze-drying		[17]
Grafting	Plasma treatment	[18,19]
	High-energy electron beam irradiation	[20,21]

hydrogels are obtained from polymers such as poly(acrylic acid) (PAAc), poly(N-isopropylacrylamide) (PNIPAm), poly(N-vinyl caprolactam) (PNVCL) [2–4], among others. The fabrication method contributes to the hydrogel properties as well. Table 5.1 summarizes the diverse fabrication techniques employed currently to synthesize hydrogels.

5.3 ELECTROSPINNING

Electrospinning is a straightforward and versatile technique to fabricate ultrafine continuous polymer fibers into different morphologies by applying an appropriate electric field on the viscose polymeric fluid [22]. This technique involves a syringe, feed pump, grounded collector, and power supply. The process of electrospinning involves applying high voltages to the emitter and collector electrodes to charge the polymer solution. When this polymer reaches the electric emitter, a droplet is held by its

Electrospining technique

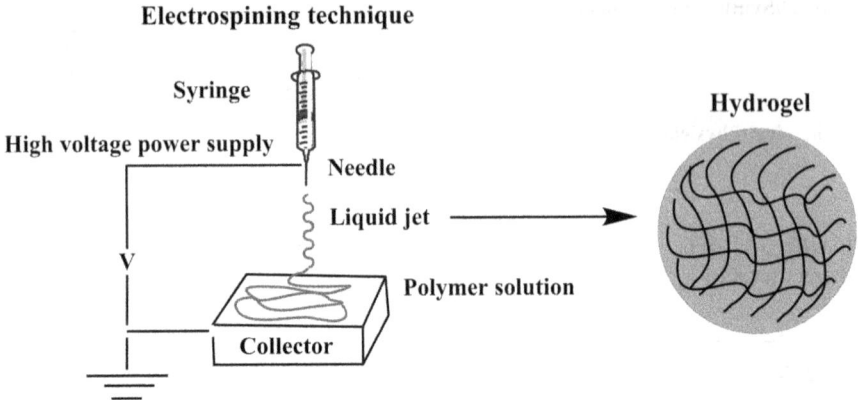

FIGURE 5.2 Schematic representation of electrospinning technique forming hydrogels.

surface tension, deforming into a cone-shaped structure called a "Taylor cone," from which the polymer solution emerges in the form of a jet from the emitter and accelerates towards the oppositely polarized collector. The difference in voltage between the nozzle or tip and the collector determines the strength of the electric field. Figure 5.2 illustrates the electrospinning technique. In case the temperature and the relative humidity must be controlled at a specific value, the 3D electrospinning system needs to include a heater, a humidifier, or a hot air blower [23].

Hydrogel fibers exhibit a 3D structure that provides a suitable microenvironment for cell proliferation, differentiation, and maintenance function, which allows cells to immerse and grow when treated by an electrospinning technique [24]. Kai and collaborators incorporated electrospun poly(ε-caprolactone) (PCL)/gelatin nanofiber into gelatin hydrogels to fabricate nanofiber-reinforced hydrogels, reaching the 20 Kpa, which can play a role similar to collagen fibers in the tissue [25]. Eom and co-workers developed a novel hydrogel based on the electrospinning method for drug delivery technology. They used a 3D hydrogel structure as a grounded collector instead of a metal collector in conventional electrospinning, thereby depositing electrospun nanofibers directly on its exterior surface [26]. However, the replication of complex shapes is a struggle due to the deposition of electrospun nanofibers being only partially effective. These disadvantages could reduce the functionalization and efficiency of the electrospinning technique. Moreover, this technique can be divided into two groups: solution electrospinning and melt electrospinning.

5.3.1 SOLUTION ELECTROSPINNING

The solution electrospinning technique does not need high temperatures and coagulation chemistry to yield nanofibers from a polymer solution [27]. The electrostatic force acting on the polymer solution overcomes the surface tension at a sufficient voltage differential. It generates a Taylor cone at the tip of the spinneret, from which a polymer jet is drawn [5]. Typically, 10–50 kV DC high voltages are applied between two electrodes.

5.3.2 MELT ELECTROSPINNING

Melt electrospinning is another technique to produce polymer fibers at high voltage and extreme temperatures. Melt electrospinning provides a lower-cost and safer approach to preparing ultrafine fibers without residual solvents [6]. The diameter of the spinneret used for melt electrospinning is higher than that for solution electrospinning [28]. Besides, the ambient humidity should be low enough to avoid air breakdown, and the spinning path temperature needs to be precisely controlled to obtain finer fibers [6]. The main reason for this is the need to use extreme voltage and temperature because this technique is easy to produce failure, such as the fiber diameter could be thick, and a reduction in fiber diameter has become a critical technical problem [29].

5.3.3 ELECTROSPINNING PARAMETERS

The fibers formed by electrospinning depend directly on the choice of solution, which is dictated by the viscosity, conductivity, and surface tension [30]. Viscosity is the resistance to flow in a liquid. The higher the viscosity, the higher the molecular weight, although the correlation is not direct. The relationship between intrinsic viscosity [h] and molecular weight (M) is given by the well-known Mark–Houwink–Sakurada equation (Eq. 5.1) as follows:

$$[h] = KM^a \tag{5.1}$$

where M is the viscosity-average molecular weight and K and a are constants for given solute-solvent system and temperature [31]. Conductivity can be defined as a measure of electrical conduction, which shows a material's ability to pass a current [32]. In electrospinning, the solution must be ionic so that the electric field between the needle and collector can flow through the solution. If the conductivity of the solution is insufficient, fibers cannot be formed. On another hand, surface tension is a contractive tendency of the surface of a liquid in contact with air, or another specific gas, that allows it to resist an external force [33]. Therefore, the electric field needs to overcome surface tension in energy to produce the solution jet for the electrospinning technique.

5.4 GAS FOAMING

The gas foaming process utilizes the nucleation and growth of gas bubbles (internal phase) dispersed throughout a polymer (continuous phase). The porous structure of polymer is formed when the dispersed gas phase (discontinuous phase) is removed from the continuous phase of polymer [34]. The first step of gas foaming is high-temperature compression, molding the polymer into a solid disc. After the disc is formed, the solid polymer rests in a high-pressure carbon dioxide (CO_2) chamber for several days [35]. Then, the pressure is lowered to the ambient pressure to induce thermodynamic instability of the gaseous phase producing a pore formation [36]. The release of pressure results in nucleation and growth of the air bubbles up to 100 µm; however, interconnectivity is still limited and is often combined with particulate leaching to obtain improved interconnectivity between pores [37]. The pores grow by the gas molecules diffusing to the pore nuclei formation [36].

Hydrogel foams are an essential sub-class of macroporous hydrogels. They are commonly obtained by integrating closely packed gas bubbles of 10–1,000 µm into a continuous hydrogel network, leading to more than 70% gas volume fractions in the wet state and close to 100% in the dried state [7] Hydrogel foams are mainly based on polysaccharides (starch, CS, alginates), vegetal (zein) and animal (gelatin) proteins, polyesters, PLA, poly(glycolic acid) (PGA), and their copolymers PLGA and PCL [38]. Nonetheless, this technique is poorly used for scaffold fabrication due to the difficulty in controlling the size and interconnectivity of pores, and large pores may be formed inside the polymer structure [39].

5.5   SOL–GEL METHOD

The sol–gel process is a wet chemical technique also known as chemical solution deposition, which involves several steps hydrolysis, polycondensation, gelation, aging, drying, densification, and crystallization [8]. The sol–gel process can fabricate hydrogels. The sol (or solution) evolves gradually towards forming a gel-like network containing both a liquid phase and a solid phase (continuous polymer networks). When the hydrogel is formed under physiological conditions and maintains its integrity for a desired time, it may provide various advantages over conventional hydrogels [40]. For example, the sol–gel technique was employed to form a structurally organized silica network in a hyaluronic acid (HA) hydrogel matrix rather than mixing discrete particles with the HA polymer matrix. The organized silica enhances the mechanical properties of the hydrogel [41]. Besides, the sol–gel method was used to synthesize polysiloxane hydrogel with encapsulated urease (immobilization degree in the range of 79–88%), which preserved the enzymatic activity at the level of 56–84% [42].

5.6   THREE-DIMENSIONAL PRINTING

3D printing is a versatile technology studied over the years. This technique relies on the successive deposition of the polymer layers to form the 3D structure with complex structures. This technique can construct hydrogels into complex artificial tissues with precise control. Hydrogel materials fabricated from 3D printing technology are determined by the rheological properties and cross-linking mechanisms, adopting complex shapes. Nonetheless, hydrogels obtained from this technique usually present poor mechanical properties (Table 5.2).

5.6.1   LASER PRINTING

Laser light is tightly focused and directly irradiated on the fabrication materials to fabricate structures along the designated laser light path [9]. During this process, designed 3D structures are directly materialized in the liquid vat, which means the hydrogel composites can be built within the photocurable organic-inorganic

TABLE 5.2

3D Techniques and Their Principles to Produce Hydrogels

3D Techniques	Principle	References
Laser printing	Laser beam	[9]
Stereolithography	UV light-based	[10]
Two-photon polymerization	Absorption of two photons	[11]
Laser-induced forward transfer	Laser direct-write technique based on ink	[12]
Extrusion printing	Heated print head to melt the polymers	[13]
Inkjet printing	Ink deposition	[14,15]

solution [43]. For example, there are three hydrogels in bioprinting (2% HA sodium salt, 1% methylcellulose, and 1% sodium alginate) through the laser printing process [44]. Similarly, an ultraviolet (UV) laser exposure on the surface of photocurable liquid causes the gel-formation of a single thin layer, and it is sequentially moved upward or downward with the sample stage to allow the next layer to form on top of a preformed structure [45].

5.6.2 STEREOLITHOGRAPHY

Stereolithography (SL), also known as vat photopolymerization, is a technique that utilizes a UV light-based approach for layer-by-layer polymerization of photo-sensitive resin [10]. It converts a liquid photoresin into a solid, which eventually will be removed when the process is completed. SL-based 3D printing has several advantages over traditional fabrication techniques, as it allows for the control of hydrogel synthesis at a very high resolution, making possible the creation of tissue-engineered devices with microarchitectures similar to the tissues they are replacing [46]. Several photosensitive poly(2-methyl-2-oxazoline) resins were synthesized as hydrogel precursors for SL photofabrication [47]. Annadakrishnan and collaborators presented a fast hydrogel SL printing within minutes. Through precisely controlling the photopolymerization condition, low suction force-driven, the high-velocity flow of the hydrogel prepolymer is established that supports the continuous replenishment of the prepolymer solution below the curing part and the constant part growth [48].

5.6.3 TWO-PHOTON POLYMERIZATION

Two-photon polymerization (TPP) is a versatile technique employed to generate photopolymerization. A typical photopolymerization process occurs through the following steps: (1) initiation, (2) propagation, and (3) termination. The polymerization is initiated by two-photon absorption in a high-intensity region that is limited to the focal volume of the laser beam [49]. With TPP, the light-matter interaction only takes place within the volume of a focused laser spot. The simultaneous absorption of two photons in the focused spot triggers the locally confined polymerization of an exposed photoresist.

The resolution of 3D hydrogels largely depends on the efficiency of TPP initiators [11]. Multiple polymeric, hydrogel, or composite materials were employed in combination with TPP to fabricate micro- and nano-patterns and 3D structures for in vitro cellular studies [50]. These compounds are mixed into a photoinitiator, an organic molecule responsible for absorbing the laser energy and starting the polymerization procedure [51].

5.6.4 LASER-INDUCE FORWARD TRANSFER

Laser-induce forward transfer (LIFT) is a versatile, non-contact, and nozzle-free printing technique with high resolution for multiple applications. LIFT consists of a donor material pre-coated on the bottom of a glass substrate (carrier). The carrier and the receiver are spaced at a distance, typically ranging from tenths of micrometers to a few millimeters, enabling contactless printing [12]. Laser pulses are focused on the metal layer via the glass slide, evaporating the laser-absorbing layer locally [52]. The substrate, or the receptor, is mounted at a very small standoff distance (25–75 μm) from the donor, while the thin film faces the substrate [53]. Hydrogels have been reported to be fabricated through this technique. Unger and co-workers reported the printing mechanism of an alginate-based hydrogel LIFT, which is investigated by spatial and temporal high-resolved stroboscopic imaging [54]. Yusupov et al. developed three hydrogels based on 2% hyaluronic acid sodium salt, 1% methylcellulose, and 1% sodium alginate based on LIFT. It was also shown that maintaining a stable temperature (±2°C) allowed for neglecting the temperature-induced viscosity change of hydrogels [44].

5.6.5 INKJET PRINTING

Inkjet printing, also called drop-by-drop bioprinting, is one of the most important bioprinting techniques to form 3D structures. For a thermally-induced inkjet printer, the droplets are generated through heating, while for a piezoelectrically-induced inkjet printer, ink is extruded from the chamber using piezoelectric actuators when a pulse is applied [14], into a specific location of the substrate. Furthermore, it requires only small amounts of functional materials, ranging from a simple polymer solution to advanced NP dispersions [55]. The materials can also be drawn to the tip of the nozzle by capillary forces without applying external pressures. Figure 5.3 illustrates an example of an inkjet printing technique.

Optimizing the reaction at the single-droplet level enables wet hydrogel droplets to be stacked, thus overcoming their natural tendency to spread and coalesce [15], as the gelation must be rapid. Nakawaga and co-workers fabricated 3D inkjet printing of ionically cross-linked star block copolymer hydrogels. They employed a dendritic polyester core, a poly(oligo(ethylene glycol) methyl ether acrylate) inner layer, and a PAAc outer layer. This solution formed a homogeneous hydrogel upon adding metallic ions, such as zinc, copper(II), aluminum, and ferric ions [56].

Inkjet Printing

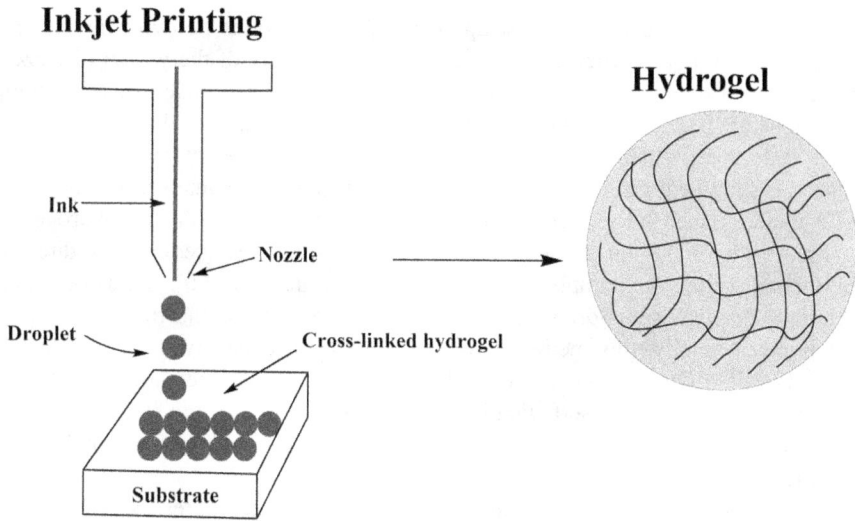

FIGURE 5.3 Schematic representation of inkjet printing technique forming hydrogels.

5.6.6 EXTRUSION PRINTING

Extrusion printing is the most common and versatile ink-3D printing technique for hydrogel materials. This technique uses a heated print head in which the thermoplastic filaments get melted and extruded through the nozzle and build the part in a layer-by-layer pattern [57]. The heated material is forced to flow from the nozzle in extruded form and deposited on the build platform layer-wise according to the generated path [13]. The travel speed of the deposition head typically varies depending on the material's desired material throughput, melt pressure, and rheological characteristics [58].

5.7 MELT MOLDING

Melt molding is a remarkable technique involving molds producing complex 3D external shapes like hydrogels without harsh chemical solvents. Firstly, the melted solution is placed into a mold. After the reorganization of the polymer, the material is removed from the mold, cooled, and soaked in an appropriate liquid to leach out the porogen. The resulting porous scaffold has the exact external shape as the mold [16]. Despite being one of the most employed techniques, it presents several limitations, such as high operating temperatures. Besides non-porous layers on the surface, it likely leaves behind porogen compounds in the scaffold due to the difficulty in leaching out the particles that can be created [39].

5.7.1 FREEZE-DRYING

Freeze-drying (FD) is a combination of freezing and drying processes. This technique is considered one of the simplest methods for forming pores in hydrogels.

The process is based on the freezing of the product, followed by removing the solvent by sublimation (below the triple-point temperature of the water) of frozen solution or dispersion of bioactive compound and carrier under high vacuum conditions [59] at reduced pressure. By doing this, the water will change from a liquid state to a crystalized state. Furthermore, because of the low temperature, this process is suitable for microencapsulating heat-sensitive compounds [59]. The dried solid materials can be termed "cryogels" when the water inside the hydrogel is sublimated by FD, under vacuum sublimation, where the ice crystals directly sublimated from the samples, leaving porous structures [60]. However, this technique displays some drawbacks. First, it is an expensive and time-consuming technique. Secondly, this method is limited to dry, heat-sensitive, and high-value products [61]. On the other hand, the conventional FD technique results in hydrogels that are not self-standing, presumably due to too high porosity. Besides, the pore formation worsens the mechanical properties of the hydrogels, limiting their application [17].

FD hydrogels are used to create 3D interconnected micropores. Several works have been published through this technique. For example, Guastaferro and co-workers prepared agarose gels based on the FD technique. The agarose cryogels were prepared by freeze-drying using two cooling rates: 2.5°C/min and 0.1°C/min. A more uniform macroporous structure and a decrease in average pore size were achieved when a fast cooling rate was adopted; however, when a slower cooling rate was performed instead, cryogels were characterized by a macroporous and heterogenous structure [62].

5.8   OTHER METHODOLOGIES

5.8.1   Grafting

High-energy radiation interacts with the matter, producing free radicals on polymers. These radicals can form bonds with other free radicals comprising graft copolymers. These graft copolymers can be obtained by a different technique that needs energy radiation, such as plasma treatment and high-energy electron beam radiation.

5.8.1.1   Plasma Treatment

Plasma treatment is a technique for inducing cross-linking and polymerization. This technique consists of the simultaneous grafting and polymerizing of functionalized monomers on the material surface. It is not only famous for not involving any toxic chemicals but also for the fact that all the modifications occur at the top few nanometers of the interfacial region, leaving the bulk properties unaffected [63].

Plasma treatment is a promising method to increase the surface hydrophilicity and wettability, and polar group can be introduced to the surface [18]. Plasma treatment involves using gases such as nitrogen (N_2), oxygen (O_2), and others. Using N_2 as a working gas for preparing a CS-AAc hydrogel, there was an increase in the protonated amines in the CS backbone, especially at high CS concentration, aside from the inductive effect in the C-O species [19].

Nowadays, several hydrogels have been studied through this technique. For example, Liu et al. employed plasma-activated water instead of water in the hydrogel polymerization process to prepare plasma-activated hydrogels based on poly-acrylamide (PAAm) for delivering antifungal agents [64]. Labay and collaborators investigated the generation of reactive oxygen and nitrogen species (RONS) in alginate hydrogels by comparing two atmospheric pressure plasma jets at various plasma treatments. The physico-chemical properties of the hydrogels remain unchanged by the plasma treatment. In contrast, the hydrogel shows a several-fold larger capacity for generating RONS than a typical isotonic saline solution [65].

5.8.2 High-Energy Electron Beam Irradiation

High-energy electron beam irradiation is a technique for penetrating a polymer layer, breaking down the bonds in order to create free radicals. Due to the higher doses, less exposure time is required, reducing the polymer degradation. Radiation cross-linking by electron beam irradiation has been studied for several polymers, using ^{60}Co and ^{137}Cs as a source of gamma irradiation. During gamma-ray irradiation, the polymers cross-link such that the backbone chains of polysaccharide polymers form chemical bonds and form a 3D network structure [20], known as hydrogels. Hydrogel fabrication and sterilization through electron beam technology in a single step is a rapid and convenient method because the cross-linking process completes in a short time at ambient temperature. There is no production of radioactive waste and process control is easy [21]. Bustamante-Torres and co-workers synthesized hydrogel from AAc and agar through cross-linked using gamma radiation (^{60}Co). These hydrogels were evaluated, characterized, and loaded with ciprofloxacin and silver NPs. Finally, the antimicrobial activity of biocidal-loaded hydrogel was tested against E. coli and methicillin-resistant Staphylococcus aureus (MRSA) in in vitro conditions [2]. Jeong et al. developed a metronidazole (MD) loaded PAA hydrogel with different MD content (0.1, 0.25, 0.5, and 1 wt%) using varying doses (25, 50, and 75 kGy) and the radiation doses (25, 50, or 75 kGy) in a one-step gamma-ray irradiation process. Then, these hydrogels were tested against some bacteria strains E. coli, Staphylococcus aureus (S. aureus), and Streptococcus mutans) showing good antibacterial activity [66].

5.8.3 Hydrogel Nanocomposites Formation

Hydrogels are hydrophilic materials that can be fabricated through the techniques mentioned previously. Nonetheless, despite the broad applicability of hydrogels, their mechanical properties limit their use. Therefore, the synergic union with other compounds can enhance it. Some of the most common are nanoparticles (NPs), which serve as a filler or reinforcement material, and the hydrogel as a matrix structure for the deposition of NPs. Incorporating NPs in 3D polymeric structures is an innovative means for obtaining multicomponent systems with various functions within a hybrid hydrogel network [67]. These systems, called nanocomposites, have been highly studied for multiple applications. The most common methods to get hydrogel nanocomposites are blending and in situ.

5.8.3.1   Blending Method

Blending is the most straightforward method for synthesizing polymer nanocomposites. This method can be achieved through two ways: (1) Preformed NPs to the polymer solution, causing polymer chains to cross-link and encapsulate the NPs, and (2) NPs and network hydrogel are made separately, and after, the NPs are trapped into the network by physical interactions [68,69].

5.8.3.2   In Situ Method

This method involves the interaction of highly dispersed NPs into preformed hydrogels. Subsequently, these hydrogels are immersed against but now in precipitating agents. Whereby permeated ions within the hydrogel structures are reacted by drastically increasing the pH, and nanocrystals are nucleated on the functional groups of the polymer chains within the hydrogels [69,70].

5.9   CONCLUSION

Hydrogels are attractive biomaterials with remarkable properties to interact with water and other solvents based mainly on the monomer precursors and the fabrication process. In the fabrication process, specific properties can confer on the material, such as biocompatibility, biodegradability, improved mechanical properties, and high thermal and chemical resistance, which are used depending on its application. Electrospinning techniques use high voltages on the polymeric fluid to create hydrogels. Gas foaming produces 3D polymeric structures based on bubble formation without organic solvents at low-temperatures process. 3D printing is a remarkable technique that relies on laying down successive polymer layers to form the 3D scaffold. It can be classified according to different ways to develop this deposition: laser printing, SL, TPP, LIFT, inkjet printing, and extrusion printing. On another melt, molding requires a high temperature to melt the solution before being placed into a mold. After the reorganization of the polymer, the mold is removed, resulting in a porous hydrogel that will depend on the monomeric solution. Meanwhile, FD is based on the freezing of the product, followed by removing the solvent by sublimation (below the triple-point temperature of the water) of a frozen solution under high vacuum conditions. Finally, grafting is a high-energy technique to synthesize hydrogels. This technique works on producing free radicals on the polymers (independently of the polymer). Plasma treatment (N_2, O_2, and other gases), high-energy electron beam irradiation (^{60}Co and ^{137}Cs), and UV irradiation (UV light) can produce those free radicals into the polymers.

Despite their countless favorable properties, the main disadvantage of hydrogels is their poor mechanical properties. This drawback can be improved by incorporating other compounds, such as NPs, as reinforcement material. Some main techniques to produce hydrogel composites include blending and in situ methods. These techniques are employed to incorporate the NPs after the hydrogel was performed previously.

REFERENCES

[1] Varghese, S. A., S. M. Rangappa, S. Siengchin, and J. Parameswaranpillai. "Natural Polymers and the Hydrogels Prepared from Them." *Hydrogels Based on Natural Polymers*, 2020; 17–47. 10.1016/b978-0-12-816421-1.00002-1.

[2] Bustamante-Torres, M., V. H. Pino-Ramos, D. Romero-Fierro, S. P. Hidalgo-Bonilla, H. Magaña, and E. Bucio. "Synthesis and Antimicrobial Properties of Highly Cross-Linked Ph-Sensitive Hydrogels through Gamma Radiation." *Polymers*, 2021;13(14): 2223. 10.3390/polym13142223.

[3] López-Barriguete, J. E., T. Isoshima, and E. Bucio. "Development and Characterization of Thermal Responsivehydrogel Films for Biomedical Sensor Application." *Materials Research Express*, 2018;5(4):45703.

[4] Rao, M. K., S. K. Rao, and C-S. Ha. "Stimuli Responsive Poly(Vinyl Caprolactam) Gels for Biomedical Applications." *Gels*, 2016;2(1):6–24.

[5] Goldstein, A. S., and P. S. Thayer. "Fabrication of Complex Biomaterial Scaffolds for Soft Tissue Engineering by Electrospinning." *Nanobiomaterials in Soft Tissue Engineering*, 2016; 299–330. 10.1016/b978-0-323-42865-1.00011-8.

[6] Yang, W., H. Li, and X. Chen. "Melt Electrospinning." *Electrospinning: Nanofabrication and Applications*, 2019; 339–361. 10.1016/b978-0-323-51270-1.00011-x.

[7] Djemaa, I. B., S. Auguste, W. Drenckhan-Andreatta, and S. Andrieux. "Hydrogel Foams from Liquid Foam Templates: Properties and Optimisation." *Advances in Colloid and Interface Science*, 2021;294: 102478. 10.1016/j.cis.2021.102478.

[8] Sakka, S. "Sol–Gel Process and Applications." *Handbook of Advanced Ceramics*, 2013; 883–910. 10.1016/b978-0-12-385469-8.00048-4.

[9] Serien, D., and Koji Sugioka. "Laser Printing of Biomaterials." *Handbook of Laser Micro- and Nano-Engineering*, 2021; 1767–1798. 10.1007/978-3-030-63647-0_52.

[10] Agrawal, A. A., K. A. Pawar, V. N. Ghegade, A. A. Kapse, and V. B. Patravale. "Nanobiomaterials for Medical Devices and Implants." *Nanotechnology in Medicine and Biology*, 2022; 235–272. 10.1016/b978-0-12-819469-0.00008-3.

[11] Xing, J.-F., M.-L. Zheng, and X.-M. Duan. "Two-Photon Polymerization Microfabrication of Hydrogels: An Advanced 3D Printing Technology for Tissue Engineering and Drug Delivery." *Chemical Society Reviews*, 2015;44(15): 5031–5039. 10.1039/c5cs00278h.

[12] Mikšys, J., G. Arutinov, M. Feinaeugle, and G.-W. Römer. "Experimental Investigation of the Jet-on-Jet Physical Phenomenon in Laser-Induced Forward Transfer (Lift)." *Optics Express*, 2020;28(25): 37436. 10.1364/oe.401825.

[13] Mahamood, R. M., T. C. Jen, S. A. Akinlabi, S. Hassan, K. O. Abdulrahman, and E. T. Akinlabi. "Role of Additive Manufacturing in the Era of Industry 4.0." *Additive Manufacturing*, 2021; 107–126. 10.1016/b978-0-12-822056-6.00003-5.

[14] Li, H., C. Tan, and L. Li. "Review of 3D Printable Hydrogels and Constructs." *Materials & Design*, 2018;159: 20–38. 10.1016/j.matdes.2018.08.023.

[15] Pataky, K., T. Braschler, A. Negro, P. Renaud, M. P. Lutolf, and J. Brugger. "Microdrop Printing of Hydrogel Bioinks into 3D Tissue-like Geometries." *Advanced Materials*, 2011;24(3): 391–396. 10.1002/adma.201102800.

[16] Murphy, M. B., and A. G. Mikos. "Polymer Scaffold Fabrication." *Principles of Tissue Engineering*, 2007; 309–321. 10.1016/b978-012370615-7/50026-3.

[17] Sornkamnerd, S., M. K. Okajima, and T. Kaneko. "Tough and Porous Hydrogels Prepared by Simple Lyophilization of LC Gels." *ACS Omega*, 2017;2(8): 5304–5314. 10.1021/acsomega.7b00602.

[18] MacLean-Blevins, M. T. "Process Selection—Which Plastics Process to Use?" *Designing Successful Products with Plastics*, 2018; 51–77. 10.1016/b978-0-323-44501-6.00003-9.

[19] Taaca, K. L., M. J. De Leon, K. Thumanu, H. Nakajima, N. Chanlek, E. I. Prieto, and M. R. Vasquez. "Probing the Structural Features of a Plasma-Treated Chitosan-Acrylic Acid Hydrogel." *Colloids and Surfaces A: Physicochemical and Engineering Aspects*, 2022;637: 128233. 10.1016/j.colsurfa.2021.128233.

[20] Raza, M. A., J.-O. Jeong, and S. H. Park. "State-of-the-Art Irradiation Technology for Polymeric Hydrogel Fabrication and Application in Drug Release System." *Frontiers in Materials*, 2021;8. 10.3389/fmats.2021.769436.

[21] Rosiak, J. "Radiation Formation of Hydrogels for Drug Delivery." *Journal of Controlled Release*, 1994;31(1): 9–19. 10.1016/0168-3659(94)90246-1.

[22] Moazeni, N., and M. Sadrjahani. "Stimuli-Responsive Nanofibrous Materials in Drug Delivery Systems." *Engineered Polymeric Fibrous Materials*, 2021; 171–189. 10.1016/b978-0-12-824381-7.00007-x.

[23] Radacsi, N., and W. Nuansing. "Fabrication of 3D and 4D Polymer Micro- and Nanostructures Based on Electrospinning." *3D and 4D Printing of Polymer Nanocomposite Materials*, 2020; 191–229. 10.1016/b978-0-12-816805-9.00007-7.

[24] Li, Y., J. Wang, Y. Wang, and W. Cui. "Advanced Electrospun Hydrogel Fibers for Wound Healing." *Composites Part B: Engineering*, 2021;223: 109101. 10.1016/j.compositesb.2021.109101.

[25] Kai, D., M. P. Prabhakaran, B. Stahl, M. Eblenkamp, E. Wintermantel, and S. Ramakrishna. "Mechanical Properties and *in Vitro* Behavior of Nanofiber–Hydrogel Composites for Tissue Engineering Applications." *Nanotechnology*, 2012;23(9): 095705. 10.1088/0957-4484/23/9/095705.

[26] Eom, S., S. M. Park, H. Hong, J. Kwon, S.-R. Oh, J. Kim, and D. S. Kim. "Hydrogel-Assisted Electrospinning for Fabrication of a 3D Complex Tailored Nanofiber Macrostructure." *ACS Applied Materials & Interfaces*, 2020;12(46): 51212–51224. 10.1021/acsami.0c14438.

[27] Bagbi, Y., A. Pandey, and P. R. Solanki. "Electrospun Nanofibrous Filtration Membranes for Heavy Metals and Dye Removal." *Nanoscale Materials in Water Purification*, 2019; 275–288. 10.1016/b978-0-12-813926-4.00015-x.

[28] Nayak, R., R. Padhye, and L. Arnold. "Melt-Electrospinning of Nanofibers." *Electrospun Nanofibers*, 2017; 11–40. 10.1016/b978-0-08-100907-9.00002-7.

[29] Liu, Y., K. Li, M. M. Mohideen, and S. Ramakrishna. "The Device of Melt Electrospinning." *Melt Electrospinning*, 2019; 7–19. 10.1016/b978-0-12-816220-0.00002-6.

[30] Robb, B., and B. Lennox. "The Electrospinning Process, Conditions and Control." *Electrospinning for Tissue Regeneration*, 2011; 51–66. 10.1533/9780857092915.1.51.

[31] Kasaai, M. R., G. Charlet, and J. Arul. "Intrinsic Viscosity–Molecular Weight Relationship for Chitosan." *Journal of Polymer Science Part B Polymer Physics*, 2000;38(19): 2591–2598. 10.1002/1099-0488(20001001)38:19<2591::AID-POLB110>3.0.CO;2-6.

[32] Sharaf, S. M. "Smart Conductive Textile." *Advances in Functional and Protective Textiles*, 2020; 141–167. 10.1016/b978-0-12-820257-9.00007-2.

[33] Camuffo, D. "Physics of Drop Formation and Micropore Condensation." *Microclimate for Cultural Heritage*, 2014; 165–201. 10.1016/b978-0-444-63296-8.00006-8.

[34] Lips, P. A. M., I. W., Velthoen, P. J., Dijkstra, M., Wessling, J., Feijen. "Gas Foaming of Segmented Poly(ester amide) Films." *Polymer*, 2005;46: 9396–9403.

[35] Gorth, D., and T. J., Webster. "Matrices for Tissue Engineering and Regenerative Medicine." *Biomaterials for Artificial Organs*, 2011; 270–286. 10.1533/9780857090843.2.270.

[36] Carter, P., and N. Bhattarai. "Bioscaffolds: Fabrication and Performance." *Engineered Biomimicry*, 2013; 161–188. 10.1016/b978-0-12-415995-2.00007-6.

[37] Hutmacher, D. W., T. B. F. Woodfield, and P. D. Dalton. "Scaffold Design and Fabrication." *Tissue Engineering*, 2014; 311–346. 10.1016/b978-0-12-420145-3. 00010-9.

[38] Iannace, S., L. Sorrentino, and E. Di Maio. "Biodegradable Biomedical Foam Scaffolds." *Biomedical Foams for Tissue Engineering Applications*, 2014; 163–187. 10.1533/9780857097033.1.163.

[39] Chung, S., and T. J. Webster. "Antimicrobial Nanostructured Polyurethane Scaffolds." *Advances in Polyurethane Biomaterials*, 2016; 503–521. 10.1016/b978-0-08-100614-6.00017-2.

[40] Jeong, B., S. W. Kim, and Y. H. Bae. "Thermosensitive Sol–Gel Reversible Hydrogels." *Advanced Drug Delivery Reviews*, 2002;54(1): 37–51. 10.1016/s0169-409x(01)00242-3.

[41] Lee, H.-Y., J. Kim, C.-H. Hwang, H.-E. Kim, and S.-H. Jeong. "Strategy for Preparing Mechanically Strong Hyaluronic Acid-Silica Nanohybrid Hydrogels via in Situ Sol-Gel Process." *Macromolecular Materials and Engineering*, 2018;303(9): 1800213. 10.1002/mame.201800213.

[42] Pogorilyi, R. P., V. P. Honcharyk, I. V. Melnyk, and Y. L. Zub. "Application of Sol-Gel Method for Synthesis of a Biosensitive Polysiloxane Matrix." *Sol-Gel Methods for Materials Processing*, 2008; 383–389. 10.1007/978-1-4020-8514-7_32.

[43] Bertsch, A., S. Jiguet, P. Bernhard, and P. Renaud. "Microstereolithography: A Review." *MRS Proceedings*, 2003;758. 10.1557/proc-758-ll1.1.

[44] Yusupov, V., S. Churbanov, E. Churbanova, K. Bardakova, A. Antoshin, S. Evlashin, P. Timashev, and N. Minaev. "Laser-Induced Forward Transfer Hydrogel Printing: A Defined Route for Highly Controlled Process." *International Journal of Bioprinting*, 2020;6(3). 10.18063/ijb.v6i3.271.

[45] Jang, T.-S., H.-D. Jung, H. M. Pan, W. T. Han, S. Chen, and J. Song. "3D Printing of Hydrogel Composite Systems: Recent Advances in Technology for Tissue Engineering." *International Journal of Bioprinting*, 2018;4(1). 10.18063/ijb.v4i1.126.

[46] Burke, G., D. M. Devine, and I. Major. "Effect of Stereolithography 3D Printing on the Properties of PEGDMA Hydrogels." *Polymers*, 2020;12(9): 2015. 10.3390/polym12092015.

[47] Brossier, T., B. T. Benkhaled, M. Colpaert, G. Volpi, O. Guillaume, S. Blanquer, and V. Lapinte. "Polyoxazoline Hydrogels Fabricated by Stereolithography." *Biomaterials Science*, 2022;10(10): 2681–2691. 10.1039/d2bm00138a.

[48] Anandakrishnan, N., H. Ye, Z. Guo, Z. Chen, K. I. Mentkowski, J. K. Lang, N. Rajabian, et al. "Fast Stereolithography Printing of Large-Scale Biocompatible Hydrogel Models." *Advanced Healthcare Materials*, 2021;10(10): 2002103. 10.1002/adhm.202002103.

[49] Ovsianikov, A., M. Malinauskas, S. Schlie, B. Chichkov, S. Gittard, R. Narayan, M. Löbler, K. Sternberg, K.-P. Schmitz, and A. Haverich. "Three-Dimensional Laser Micro- and Nano-Structuring of Acrylated Poly(Ethylene Glycol) Materials and Evaluation of Their Cytotoxicity for Tissue Engineering Applications." *Acta Biomaterialia*, 2011;7(3): 967–974. 10.1016/j.actbio.2010.10.023.

[50] Sharaf, A., B. Roos, R. Timmerman, G.-J. Kremers, J. J. Bajramovic, and A. Accardo. "Two-Photon Polymerization of 2.5D and 3D Microstructures Fostering a Ramified Resting Phenotype in Primary Microglia." *Frontiers in Bioengineering and Biotechnology*, 2022;10. 10.3389/fbioe.2022.926642.

[51] Otuka, A. J., N. B. Tomazio, K. T. Paula, and C. R. Mendonça. "Two-Photon Polymerization: Functionalized Microstructures, Micro-Resonators, and Bio-Scaffolds." *Polymers*, 2021;13(12): 1994. 10.3390/polym13121994.

[52] Zhu, W., J. G. Ock, X. Ma, W. Li, and S. Chen. "3D Printing and Nanomanufacturing." *3D Bioprinting and Nanotechnology in Tissue Engineering and Regenerative Medicine*, 2015; 25–55. 10.1016/b978-0-12-800547-7.00002-3.

[53] Alemohammad, H., and E. Toyserkani. "Laser-Assisted Additive Fabrication of Micro-Sized Coatings." *Advances in Laser Materials Processing*, 2010; 735–762. 10.1533/9781845699819.7.735.

[54] Unger, C., M. Gruene, L. Koch, J. Koch, and B. N. Chichkov. "Time-Resolved Imaging of Hydrogel Printing via Laser-Induced Forward Transfer." *Applied Physics A*, 2010;103(2): 271–277. 10.1007/s00339-010-6030-4.

[55] Perelaer, J., and U. S. Schubert. "Ink-Jet Printing of Functional Polymers for Advanced Applications." *Polymer Science: A Comprehensive Reference*, 2012; 147–175. 10.1016/b978-0-444-53349-4.00205-3.

[56] Nakagawa, Y., S. Ohta, M. Nakamura, and T. Ito. "3D Inkjet Printing of Star Block Copolymer Hydrogels Cross-Linked Using Various Metallic Ions." *RSC Advances*, 2017;7(88): 55571–55576. 10.1039/c7ra11509a.

[57] Velu, R., D. K. Jayashankar, and K. Subburaj. "Additive Processing of Biopolymers for Medical Applications." *Additive Manufacturing*, 2021; 635–659. 10.1016/b978-0-12-818411-0.00019-7.

[58] Kishore, V., and A. A. Hassen. "Polymer and Composites Additive Manufacturing." *Additive Manufacturing*, 2021; 183–216. 10.1016/b978-0-12-818411-0.00021-5.

[59] Rostamabadi, H., S. R. Falsafi, S. Boostani, I. Katouzian, A. Rezaei, E. Assadpour, and S. M. Jafari. "Design and Formulation of Nano/Micro-Encapsulated Natural Bioactive Compounds for Food Applications." *Application of Nano/ Microencapsulated Ingredients in Food Products*, 2021; 1–41. 10.1016/b978-0-12-815726-8.00001-5.

[60] Yuan, Y., L. Wang, R.-J. Mu, J. Gong, Y. Wang, Y. Li, J. Ma, J. Pang, and C. Wu. "Effects of Konjac Glucomannan on the Structure, Properties, and Drug Release Characteristics of Agarose Hydrogels." *Carbohydrate Polymers*, 2018;190: 196–203. 10.1016/j.carbpol.2018.02.049.

[61] Punathil, L., and T. Basak. "Microwave Processing of Frozen and Packaged Food Materials: Experimental." *Reference Module in Food Science*, 2016. 10.1016/b978-0-08-100596-5.21009-3.

[62] Guastaferro, M., L. Baldino, E. Reverchon, and S. Cardea. "Production of Porous Agarose-Based Structures: Freeze-Drying vs. Supercritical CO_2 Drying." *Gels*, 2021;7(4): 198. 10.3390/gels7040198.

[63] Kasoju, N., L. T. B. Nguyen, A. R. Padalhin, J. F. Dye, Z. Cui, and H. Ye. "Techniques for Modifying Biomaterials to Improve Hemocompatibility." *Hemocompatibility of Biomaterials for Clinical Applications*, 2018; 191–220. 10.1016/b978-0-08-100497-5.00015-x.

[64] Liu, Z., Y. Zheng, J. Dang, J. Zhang, F. Dong, K. Wang, and J. Zhang. "A Novel Antifungal Plasma-Activated Hydrogel." *ACS Applied Materials & Interfaces*, 2019;11(26): 22941–22949. 10.1021/acsami.9b04700.

[65] Labay, C., I. Hamouda, F. Tampieri, M.-P. Ginebra, and C. Canal. "Production of Reactive Species in Alginate Hydrogels for Cold Atmospheric Plasma-Based Therapies." *Scientific Reports*, 2019;9(1). 10.1038/s41598-019-52673-w.

[66] Jeong, J.-O., J.-S. Park, E. J. Kim, S.-I. Jeong, J. Y. Lee, and Y.-M. Lim. "Preparation of Radiation Cross-Linked Poly(Acrylic Acid) Hydrogel Containing Metronidazole with Enhanced Antibacterial Activity." *International Journal of Molecular Sciences*, 2019;21(1): 187. 10.3390/ijms21010187.

[67] Merino, S., C. Martín, K. Kostarelos, M. Prato, and E. Vázquez. "Nanocomposite Hydrogels: 3D Polymer–Nanoparticle Synergies for On-Demand Drug Delivery." *ACS Nano*, 2015;9(5): 4686–4697. 10.1021/acsnano.5b01433.

[68] Ganguly, S., and S. Margel. "Design of Magnetic Hydrogels for Hyperthermia and Drug Delivery." *Polymers* 2021;13(23): 4259. 10.3390/polym13234259.

[69] Bustamante-Torres, M., B. Arcentales-Vera, J. Estrella-Nuñez, H. Yánez-Vega, and E. Bucio. "Antimicrobial Activity of Composites-Based on Biopolymers." *Macromol*, 2022;2(3): 258–283. 10.3390/macromol2030018.

[70] Chen, S., T. S. Jang, H. M. Pan, H. D. Jung, M. W. Sia, S. Xie, Y. Hang, S. Chong, D. Wang, and J. Song 3D Freeform Printing of Nanocomposite Hydrogels through in situ Precipitation in Reactive Viscous Fluid. *International Journal of Bioprinting*, 2020;6: 258. 10.3390/polym14040752.

6 Hydrogel-Based Sensors with Advanced Properties

David Romero-Fierro, Lorena Duarte-Peña,
Y. Aylin Esquivel-Lozano, and Emilio Bucio
Instituto de Ciencias Nucleares, UNAM, Ciudad de México, México

6.1 INTRODUCTION

A sensor is a device capable to capture changes in environmental parameters, as an input, and transforming them into electrical signals to output, that can be interpreted by either a human or a machine. Currently, sensors are occupied in virtually everything, including robotics, electronics, artificial intelligence, security, health, and biomedicine, triplicating their use within the last two decades. For biomedical purposes, a wide variety of sensors have been extensively developed taking into account a key factor, which is the type of interface that they could form with the medium to be applied [1]. From this interaction between sensor and surface, good or bad, slow or fast responses can be obtained. In this sense, flexible sensors depict one of the most versatile types of sensors due to their potential to be applied on regularly and irregularly shaped surfaces, including skin and textile fabrics. Flexible sensors overcome some limitations of the surface on detection devices allowing an expansion in the application of sensors in human health and biomedical industry.

Hydrogels are an important type of biomaterial, which are composed of cross-linked polymer networks forming 3D structures that support a high proportion of water. They can swell without dissolution up to 99% (w/w) water of their dry weights. Hydrogels can be classified depending on the type of cross-linking to obtain them, that is, chemical and physical crosslinked hydrogels. Chemically cross-linked hydrogels are bonded by covalent networks and do not dissolve in water without breaking their covalent bonds. However, physically cross-linked hydrogels are formed by dynamic cross-linking of synthetic or natural building blocks by non-covalent interactions such as electrostatic interactions. Hydrogels can be formulated in a wide variety of physical forms with many typologies of methodologies and polymer functionalization [1,2]. Hydrogels are flexible and highly compatible with most biological molecules and substrates, due to their high water content, porosity, and consistency, making them attractive as a component of sensor systems where necessary to have versatile chemical/biochemical materials. Common flexible sensors based on flexible elastomer components like

DOI: 10.1201/9781003340485-6

flexible metal, polymer films, and polymer elastomers combined with graphene, carbon nanotubes, and metal particles, display limited stretchability, limited sensitivity, and poor fatigue, due to their inherent properties. Conversely, hydrogel-based sensors include some characteristics like high stretchability, moldability, and compatibility, but, above all, their excellent flexibility, which has attracted the attention of many researchers around the world. In fact, according to the Scopus database, there is an exponential growth of scientific production over the years, reflected in articles, passing from 1,458 documents in 2010 to 4,570 in 2022. This reflects the importance and impact of this research topic in our daily life.

The sensing mechanism of a hydrogel-based sensor is governed by the action of active sensing materials, which will suffer changes in their conformation, structure, or composition, derived from their flexibility which generates a signal, as Figure 6.1 shows. This mechanism can be explained in terms of two concepts, such as hydrophilicity and swelling. Hydrophilicity of the polymers constituting the polymer network produces an osmotic pressure in the hydrogel that generates a swelling process in the matrix under exposure to water. Generally, this process occurs in three steps that are a diffusion of water along the polymer matrix, relaxation of polymer chains, and expansion of the polymer network upon relaxation. The hydrogel reaches a maximum degree of water absorption, defined as equilibrium water content. This maximum degree is defined as the balance between the elastic retractive forces of the polymer chains in the 3D network and the osmotic force generated by water. When the polymer chains are stretched, elastic retractive forces appear as a counteraction for the network expansion. Then, when there is a balance between both forces, expansion is stopped and the system reaches equilibrium. If there are changes in either osmotic pressure or in cross-linking density, equilibrium is broken and the hydrogel shows a change in the degree of swelling. One of the most theoretical approaches used to explain this behavior is the equilibrium swelling theory proposed by Flory and Rehner. Swellings in hydrogel-based flexible sensors are induced by stimuli, like mechanical force, which induces phase transition in the hydrogel and simultaneously converts this sense into a macroscopic event. Finally, the electrical output or signal is possible by the use of some techniques, including light transmission measurement, conductometry, and pressure induced by the change in gel swelling [2].

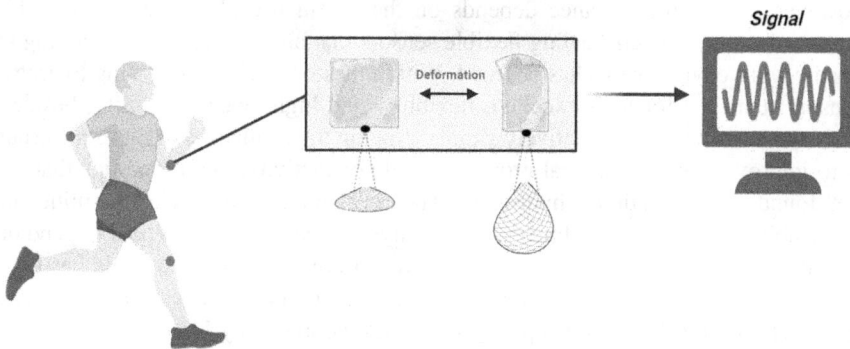

FIGURE 6.1 Operation mode of hydrogel-based flexible sensors.

The sensing behavior in hydrogels, primarily attributed to their flexibility and swelling, can be provided or enhanced by the action of other interesting properties. Polymer hydrogels can evidence unique and diverse performances and functionalities by their inherent properties or by the addition of different fillers, dopants, cross-linking, or hydration states to adapt and apply to different conditions and environments. Smart hydrogel-based flexible sensors with striking multi-functions could be easily obtained by appropriate design, inclusion, and optimization of the 3D network or composition of polymers [3].

This chapter aims to describe the most relevant advances in hydrogel-based flexible sensors, exhibiting some novel characteristics including self-healing, anti-freezing, shape memory, adhesive, hydrophobicity, superhydrophobicity, conductivity, and high stretchability. These advanced characteristics have been chosen due to the vast scientific production related to these topics and the potential that they have shown for their application as sensing systems. These properties are used for some diverse sensing applications starting from detection of chemical components to detecting human motion. In each section, it is mentioned a brief description of the contribution that each property provides to obtain functional hydrogel-based flexible sensors. Then, principles of performance, synthetic procedures, and/or classification of each category are summarized with the most remarkable examples given in that area. Finally, a section of discussion of deficiencies, challenges, and future trends will allow to the reader to generate new ideas and research opportunities to overcome them and produce advances in a very interesting field as it is hydrogel-based flexible sensors.

6.2 HYDROGELS WITH SELF-HEALING PROPERTIES

Hydrogels are currently the ideal matrix for manufacturing sensors in the biomedical field since they are elastic and easily adhere to different surfaces, constituting the so-called flexible sensors. Sensors are devices that capture changes in environmental parameters (temperature, pressure, humidity, and chemical changes) and convert them into electrical signals, being essential for the development of technologies that allow the treatment and timely diagnosis of many diseases.

Flexible sensors have a greater susceptibility to environmental changes; however, proper performance depends on the useful life of the sensor. For this reason, the study of self-healing flexible sensors that allow adequate functioning in hostile work environments has boomed. Within these devices, self-healing hydrogel sensors stand out for their excellent flexibility and high biocompatibility, besides, their synthesis process is relatively simple. Among the challenges of these materials is to improve their mechanical properties and conductivity, two properties that are not found in conventional hydrogels. The mechanical properties determine the applicability of the device while the sensitivity and stability of the sensor depend on the conductivity. For this reason, the manufacture of composite materials has been used in which conductive polymers or metallic nanoparticles are incorporated into a hydrogel with a self-healing capacity to acquire conductivity [4].

The self-healing capacity allows for extending the useful life of hydrogels that, due to their structure, have mechanical properties that make them prone to the appearance of microcracks, weakening the integrity of the system. Self-healing hydrogels can be divided into extrinsic or intrinsic, depending on how the repair occurs. In the case of extrinsic materials, a self-healing agent is added to the polymer, for example, a polymeric matrix with monomers or solvents encapsulated disperse, which are released in response to damage, or systems with integrated vascular networks. These methods are widely used for the self-healing of metallic systems, urea-formaldehyde microcapsules have been used to transport a liquid gallium-indium alloy, which allowed the automatic repair of a gold-based electrical circuit, almost completely recovering its conductivity [5].

On the other hand, intrinsic self-healing hydrogels are systems with dynamic structures, such as non-covalent or dynamic covalent bonds that break and form continuously, allowing the material to be restructured, the advantage of these systems is that it can be repaired multiple times since they are not limited by the self-healing agent. Non-covalent bonds are characterized because they do not share electrons. They are weak but have numerous interactions whose cumulative effect can stabilize a system; some of these bonds are van der Waals forces, hydrophobic interactions, hydrogen bonding, and host-guest interactions. Below are some examples of this type of system that use hydrophobic interactions and hydrogen bonds.

Hydrophobic interactions are formed because when water molecules come into contact with a nonpolar molecule, they try to organize themselves without damaging the network of hydrogen bonds, generating an unusually strong attraction between hydrophobic molecules, which is why they tend to aggregate and isolate. Hydrogels of this type can be achieved by copolymerizing a small amount of hydrophobic monomer with hydrophilic monomers using the micellar polymerization technique, this leads to the formation of hydrophobic domains in the structure that strongly interact and give stability to the material [6]. Yang et al. developed a conductive polyacrylamide hydrogel, incorporating multi-walled carbon nanotubes through cellulose nanofibers to improve the dispersion of the nanotubes, their composition provided greater mechanical properties and electrical conductivity to the hydrogel, which in turn has the self-healing capacity due to the interaction of hydrophobic domains, recovering its structure at room temperature and without external stimulus. These materials would have applications in different sectors, including the development of electronic skins and protection materials against electromagnetic interference [7].

Hydrogen bonds are a special type of high-energy dipole-dipole interaction, 10–40 kJ/mol, which occurs between two atoms with high electronegativity (N, O, or F) and hydrogen covalently bonded to one of them. These interactions stabilize a large number of biological structures, such as the DNA, and can be used for the development of self-healing materials; for example, Jie-Cao et al. synthesized a highly sensitive flexible sensor with a regeneration capacity of approximately 93% even after the third structural damage, in approximately 15 seconds, for use in intelligent systems that allow human-machine interaction. The sensor is formed by the interaction of chitosan, natural rubber latex, and carboxyl cellulose nanocrystals through multiple interactions of hydrogen bonds. In addition, the system contains carbon nanotubes that generate a conductive network sensitive to tension [8].

On the other hand, dynamic covalent bonds are reversible covalent bonds; that is, bonds that are in equilibrium but can be reversed in response to a stimulus. This type of bond allows the formation of hydrogels with higher resistance and rigidity. Some of the most commonly used dynamic covalent bonds for this type of structure are Schiff base bonds, disulfide bonds, and Diels-Alder reactions. Using Schiff base bonding, Hafeez et al. developed elastic and self-healing hydrogels for bioprinting applications that have potential use in the generation of smart bioinks. These hydrogels are stabilized by imine dynamic bonds between oxidized alginates and hydrazide or semicarbazone, which award the material self-healing capability [9].

A relevant problem in intrinsic self-healing hydrogels is that the mechanical properties are usually poor, so it is common to use systems that combine two or more of the aforementioned interactions, in order to have a more stable structure. Liu et al. prepared flexible sensors with a dual self-healing system, to monitor organ movements through tension. The hydrogel was formed as a dual network, which has host-guest interactions between cyclodextrin and ferrocene and borate dynamic bonds between poly(vinyl alcohol) and borax, the electrical conductivity of the material was given by dispersion of carbon nanotubes. The sensors showed a recovery capacity of approximately 95%, high mechanical resistance, and stretch-ability close to 450% [10]. On the other hand, Su et al. synthesized conductive self-healing hydrogels, cross-linking a conductive polyaniline network with a self-healing matrix of polyacrylic acid with a hydrophobic association that gives it its regenerative capacity. In addition, the self-healing capacity is enhanced by hydrogen bonds and electrostatic interactions with polyaniline. The resulting material presented high stretchability, tenacity, and mechanical resistance, due to which it could be used for the manufacture of flexible sensors that allow the monitoring of physiological signals and advances in soft robotics [11].

6.3   HYDROGELS WITH SHAPE MEMORY PROPERTIES

Hydrogel-type shape memory materials are 3D interpenetrating networks composed of a hydrophilic polymer, which is also sensitive to stimuli, and therefore, under the influence of an external change, such as temperature, pH, light, magnetic field, or electrical, the hydrogel will undergo a temporary structural change. This change gives a momentary shape that after removing the stimulus for a while returns to its original shape, which can be carried out in a different number of repetitive cycles. Such changes may include twisting or bending in response. Some materials have only one form of change; however, it is possible to find materials with two reversible forms of memory [12]. When there is a thermo-response in a temperature-sensitive hydrogel, the shapes depend on the transition temperatures (T_{trans}); for example, the glass transition temperature (T_g) or the melting temperature (T_m) in case the hydrogels count on it.

At the same time, shape memory hydrogels have had a great impact on the scientific community due to their properties similar to human muscles with great flexibility and softness, in addition to their high biocompatibility; also, the change of form that they can adopt while they are under the influence of the stimulus in question. For this reason, a large amount of research has focused on manufacturing flexible sensors with this type of material. Taking advantage of the properties that hydrogels can provide on their own and more specifically, shape memory–type hydrogels that have shown great precision in monitoring different analytes. This makes them useful in some applications in different areas such as biomedicine, biosensors, and aerospace, among others.

One of the disadvantages of this type of material is its deficiency in mechanical properties. One of the solutions for this problem is the incorporation of reinforcements, such as nanoparticles or metal ions; with this it is possible to modify and improve the mechanical properties. In this way, it is possible to have shape memory hydrogels that meet the necessary requirements for the desired application. In 2021, the design of artificial muscles was reported using hydrogels with shape memory of acrylic acid and acrylamide, reinforced with cellulose nanocrystals, which presented good stretching, twisting, and rolling properties in water; thus, demonstrating the improvement of mechanical properties with the introduction of the reinforcement of said nanocrystals [13].

The most recent research in the field of biomedicine and tissue engineering is based on the development of soft robots; that is, devices capable of imitating some mechanical movements of living organisms, being the result of intensive and rigorous research and development of biosensors. As reported by Jung Gi Choi et al., they used poly N-isopropylacrilamide and polycaprolactone (PNIPAM/PCL) composites to obtain a shape memory material that responds to a change in temperature, which also presents tension and compression responses, which can subsequently provide more complexes [14]. This research offers a chance to manufacture the mentioned soft robots, since, like human beings, it is necessary to have more than one movement to make it work, and with this material said movement will be modified.

In general, the synthesis of hydrogels is based on obtaining materials with cross-linking that can be carried out by the addition of an external cross-linking agent or by high-energy radiation. More specifically, the synthesis can be carried out by different forms of polymerization in which radical polymerization is the most used to obtain hydrogels with varied properties and structures. The monomers used directly influence the design of the biosensor, because by introducing new and diverse functional groups, these will allow the immobilization of some biomolecule having larger recognition areas due to the nature of the hydrogel structure. One of the hydrogel synthesis techniques with potential application in biosensors is coating them with a substance that is capable of providing the recognition sites for the biomolecules of interest. The other technique consists of the encapsulation of the enzyme or protein, generating the conservation of the native protein and with it a greater enzymatic activity and producing a greater and more effective response signal [15].

Regarding the synthesis for shape memory hydrogels, there are molecular imprinting polymers (MIPs), a technique widely used in the synthesis of flexible sensors that in turn involves various synthetic methods such as the synthesized polymer containing the template of the molecule to interact or remove the molecule from the template, leaving a cavity with the shape of interest, which can later be reunited [16]. The main objective of MIPs, which also makes them so attractive, is the fact that they allow systems with high specificity and selectivity to biological receptors analogous to natural ones to be obtained with greater applicability and duration.

Another important feature is that the synthesis of the polymers can be carried out by conventional radical polymerization methods, for example, either by the use of an external chemical or physical initiator. In 2019, the synthesis of a flexible biosensor based on a shape memory hydrogel with sensitivity to near-infrared radiation (NIR) and temperature provided by N-isopropylacrylamide (NIPAAm) was reported. The material was obtained with the incorporation of graphene oxide/polyaniline and NIPAAm in a glassy carbon electrode. As reported, the synthesized hydrogel has a response to bovine serum albumin (BSA), allowing it to be lodged in the biosensor holes and after irradiating at a wavelength of 808 nm, it releases the protein, leading to cleaning [17].

On the other hand, sensors with a biological response of DNA hydrogels have been designed, which exhibit certain limitations in terms of a slow response to the stimulus and high design costs. However, shape memory DNA hydrogel films were developed, and these manage to solve the problems of bulk hydrogels. For example, Chunyan Wang and his group synthesized a micron-thick hydrogel film for 2D photonic crystal regulation with a target recognition unit such as Ag^+/Cysteine. Such DNA hydrogels contain a composite between a pH-responsive polymer and DNA [18].

Thus, shape memory hydrogels represent a very advantageous alternative to other materials, mainly due to their similarity with muscle tissues and their high biocompatibility, because this provides favorable characteristics for their possible applications in biosensors or related to them. In recent years, valuable and successful advances have been made in this field with the use of reversible shape change properties that adequately allow the recognition of molecules of biological origin with good sensitivities and speed to be applied in this field that is in continuous development.

6.4 HYDROGELS WITH HYDROPHOBIC AND SUPERHYDROPHOBIC PROPERTIES

Hydrophobic and superhydrophobic materials have great biomedical applications, due to the null adherence of water and other substances such as bacteria as well as its self-cleaning capacity. This fact has positioned them in a highly valued place for the manufacture of new medical devices or their improvement. Lately, research in the field of biomedicine and electronics has focused on the development of biosensors, specifically for use in the human body, for this reason the idea of applying hydrophobic properties to them arises, to promote a good application regardless of environmental conditions or those generated by the human body itself, such as humidity and sweat.

It is important to clarify that hydrophobic surfaces are those that have a contact angle greater than 90° and superhydrophobic when the contact angle is greater than 150° [19] this is determined by the angle formed between the smooth surface of the sample and a drop of water deposited on it. It is also known that these surface wettability properties can be exploited to create materials that have characteristics very similar to those found in nature, for example, lotus leaves that have a natural superhydrophobicity. In addition, the possibility of controlling the hydrophilic and hydrophobic character in hydrogel-type samples has been reported, and with this, it is mandatory to direct research towards the development of biosensors where properties related to wettability [20].

Recently, the hydrophobic and superhydrophobic hydrogels have been represented on the sensors developing the biologic environment, given already described characteristics of water repellence and adding the same properties of hydrogels as their biocompatibility, softness, and flexibility. Combining each of the properties opens up a wide range of possibilities for the development of new devices that are more adaptable to the human body and current needs. In addition, it seeks to save production costs that at the same time are reflected in a response capacity by these, as will be described below.

It is worth mentioning that great improvements have been achieved in the operation and design of sensors for the human body; for example, in the electrodes used for electroencephalograms (EEGs). This was carried out with the design of a double-layer hydrogel, one of which is hydrophobic, using the layer-by-layer technique and a photoinitiated polymerization. The sensor consists of a hydrophobic layer with a contact angle of 133.87° on the outside, followed by a conductive layer and an adhesive layer. The function of the hydrophobic hydrogel is such that it does not allow malfunction due to sweat or dust contamination, while the adhesive layer ensures effective contact between the skin and the electrode, as can be seen in Figure 6.2. Compared with commercial electrodes, the manufactured hydrogel shows improved signal uptake. This opens a new branch for the design of biosensors that can be applied in the diagnosis of neurodegenerative diseases; for example, those that are neurological [21].

As in the previous sections, to obtain advanced and innovative properties with numerous applications in current requirements, it is common to resort to composites. The synergy that comes from the combination of polymeric materials with materials

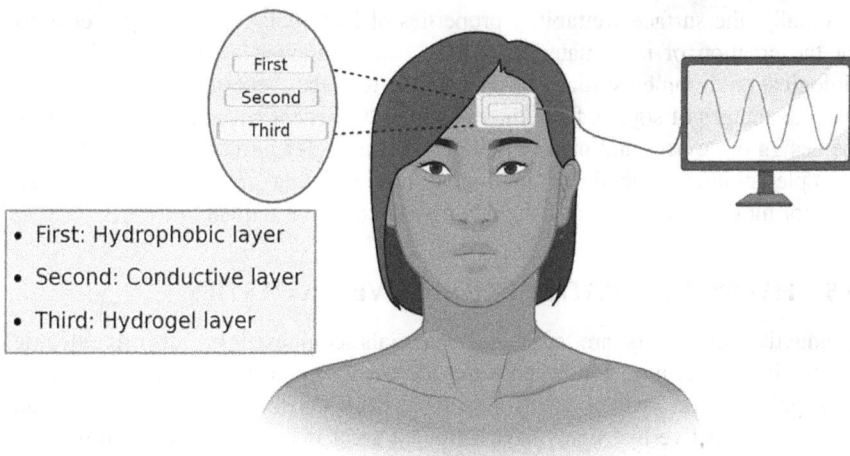

First
Second
Third

• First: Hydrophobic layer
• Second: Conductive layer
• Third: Hydrogel layer

FIGURE 6.2 Mode of action of hydrogel-based flexible sensor with hydrophobic properties.

of a metallic nature in bulk or in small sizes such as nanometers has resulted in a good opportunity in materials engineering and specifically for this case, in the design and development of flexible biosensors. Using materials other than polymers can significantly contribute to the final properties of the device since materials that favor or amplify the signals generated from the analyte of interest can be introduced into hydrogels. An improvement in the mechanical properties is even achieved, which in some cases is affected in hydrogels by absorbing a large amount of water.

Mengnan Qu and their research group recently reported obtaining a triboelectric nanogenerator made of polydimethylsiloxane (PDMS) film modified with poly-tetrafluoroethylene (PTFE) and a hydrogel of AgNWs/PVA that has the character-istic of a superhydrophobic material with a contact angle of 152°, which they call PP/AgH-TENG, using 3D printing technique and a spraying process. This material was applied as a flexible biosensor in a bracelet, in which the hydrophobic part provides high stability in the conductive interface and high performance in excessively humid environments. The results for this flexible device demonstrate a good ability to capture the energy generated by the movement carried out by the human body, as well as a high sensitivity to this. As reported, the collected energy is capable of turning on 360 commercial LED lights, in addition to charging capacitors with a power density of 3.07 W/m^2 [22]. This translates to a device capable of self-powering with good efficiency and effectiveness.

The synthesis of hydrophobic and superhydrophobic hydrogels can be carried out by various techniques, as has been discussed in the examples presented in previous paragraphs. Some of these include innovative synthesis with techniques such as 3D printing, spray, layer-by-layer technique, and coating, among others. However, these methods are applied for the generation of the hydrophobic and superhydrophobic parts found in the hydrogel. For the synthesis of hydrogels, traditional techniques are used, which can be photoinduced polymerizations or generated by some external agent; in any case, it is necessary to use a cross-linking agent that generates the networks.

Finally, the surface wettability properties of hydrogels result in a powerful tool for the creation of new materials. Those can be applied as detection devices in biological environments; that is, they are able to detect, record, and, in some cases, take advantages of signals from the human body. Recent research seeks to improve the design of devices and incorporate new technologies into everyday products, for example, clothing. Obtaining intelligent products that are capable of firsthand monitoring of movements and signals generated in the human body.

6.5 HYDROGELS WITH CONDUCTIVE PROPERTIES

Conductive hydrogels are synthetic materials composed of intrinsic electrical conductive materials, which provide electrical conductivity, and cross-linked hydrogel networks, which act as a scaffold [23]. On the development of flexible sensors, conductive hydrogels have attracted great interest due to the combination of advantages coming from traditional polymeric materials and organic conductors that include high conductivity, flexible modification, easy synthesis, and a high surface area. The sensing mechanism of conductive hydrogels is transforming mechanical stimulus into an electrical signal when the hydrogel sensor is deformed. When pressure is applied to the hydrogel matrix, the conductive backbone becomes denser and its resistance decreases. The ratio between initial and final resistance is used as an indicator of pressure. Also, when the conductive hydrogel is stretched or compressed, there is a change in the resistance or conductivity due to the deformation of the conductive network. These aspects make hydrogel-based sensors ideal for muscle movement sensing applications. In this way, conductive hydrogels show excellent biocompatibility, which is relevant to emulate many key aspects of biological tissues water-rich nature, elastic modulus, high modulation, and 3D network structure [24].

To achieve conductive properties in a hydrogel, it can be done by two approaches that are by providing electronic conduction or ionic conduction. In the first strategy, the inclusion of materials with electrical conductivities such as metals, MXene compounds, graphene, carbon nanotubes into conventional polymer network or the development of hydrogels based on conducting polymers, generate materials with attractive electron conductive properties. For the second case, conduction is mainly associated with directional transportation of free moving ions such as Li^+, Na^+, Fe^{3+}, and Al^{3+}. In this way, dispersing directly soluble inorganic salts into hydrogels is a facile and versatile way to prepare conductive hydrogels [25]. One of the most relevant achievements from this approach is that the obtained conductive hydrogel is usually transparent, which is necessary for some practical visualization usages [24]. Table 6.1 summarizes the main advantages of each approach.

Electronic conduction in hydrogels can be achieved by hybridizing conductive fillers with conductive networks. Even conductive networks can work alone, incorporation of some conductive particles greatly improves this property. The most common cases of reported electronic conductive hydrogels are composed by

TABLE 6.1

Summary of Advantages and Limitations of Electronic and Ionic Approaches to Providing Conductive Properties to Hydrogels

Conductive hydrogel	Advantages	Limitations
Electronic conductive hydrogel	• Excellent electrical properties and conductivities of filler • Hydrogels with improved mechanical properties	• Incompatibility and weak interfacial interaction between conductive filler and hydrogel • Improper contact between electronic device and skin or tissue, giving unreliable signal acquisition
Ion conductive hydrogel	• Ionic conductive hydrogels are transparent • Hydrogels show adhesion, self-healing, self-recovery, and biocompatibility	• Lower sensitivity and conductivity than electronic conductive hydrogels • High salt concentration could damage cells

conductive phase (which provides conductivity) joined to a hydrogel network (flexible and deformable matrix). In this way, some conductors such as carbon nanotubes, graphene/graphene oxide, metal particles, and conductive polymers can be directly cross-linked forming hydrated 3D networks by chemical bonding, ionic cross-linking, and self-assembling. Commonly, those networks show very high conductivity but low stretchability [26].

Electron conductivity in conductive hydrogels depends on the type of conductive material used for that purpose. In conductive polymer-based hydrogels, the conjugate structure of localized σ and π bonds is responsible to impart conductivity to the system. During polymerization process, the p-orbitals between π-bonds overlap each other activating electron distribution. Then, delocalization occurs and delocalized electrons move freely along the polymer backbone, inducing conductivity [27]. Some common conductive polymers are shown in Figure 6.3. An example of the application of those materials is exposed by Jin et al., who developed a stretchable and conductive hydrogel based on graphene oxide/chitosan, polyaniline, and polyacrylamide showing enhanced properties such as tensile strength, elongation, and conductivity (0.80 MPa, 360% and 2.88 S/m, respectively). Furthermore, this hydrogel displayed skin affinity, which is promising to its use as flexible wearable electronic devices [28]. In the case of hydrogels based on carbon nanotubes, the p electrons from carbon atoms create a wide range of delocalized π bonds, which create a bride along the structure and the electrons can move easily [27]. This prominent delocalization is well used for some special electrical properties. Lv et al., in an interesting research, reported the obtaining of a novel hydrogel based on multi-walled carbon nanotubes, which exhibits high conductivity (9×10^{-3} S/cm), antibacterial properties, and self-healing ability (GF = 12.71 at 100–1100%, quick response 500 ms), owing to the synergistic effect between matrix and carbon material [29].

On the other hand, ionic conductive hydrogels have a large number of free ions along their three-dimensional polymer network, conferring an extraordinary flexibility very similar to biosystems. Compared to most electronic conductive hydrogels, ionically conductive hydrogels display common characteristics like transparency, tissue adhesion, and self-healing. Also, the large amount of water supported by the hydrogel network facilitates ion transmission, forming current without losing the solid integrity during the stretching process. Those features are used to create soft, flexible devices [26]. As examples of a promising flexible sensor, Din et al. developed a semi-interpenetrated network of carboxymethyl

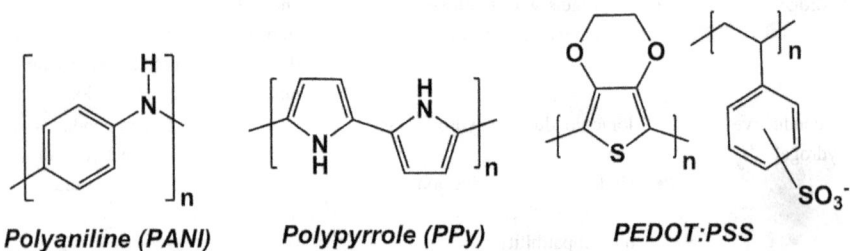

Polyaniline (PANI)   Polypyrrole (PPy)   PEDOT:PSS

FIGURE 6.3   Common conductive polymers used as matrices in flexible sensors.

chitosan, polyacrylamide, and sodium chloride. The resulting hydrogels exhibit transparency, high ionic conductivity (up to 6.44 S/m), elastic modulus (9–47 kPa), anti-freezing properties, and good sensitivity in monitoring human motion [30]. Also, Zhao et al. prepared a double-network hydrogel based on $AlCl_3$, acrylic acid, oxide sodium alginate, and aminated gelatin. The final hydrogel showed good self-adhesiveness, transparency, stretchability, ionic conductivity (1.85 S/m), and high sensitivity (GF = 7.091 at a strain of 419%). Further investigation on this hydrogel demonstrated that as a sensor it is capable of monitor various human movements and object deformations [31].

Both approaches, electronic and ionic conduction, can coexist in a simultaneous system, improving together their conductive properties. This purpose could be beneficious to obtain a high sensitivity and faster response in sensors for monitoring human motions. Wang et al. combined the flexibility of the polyacrylamide hydrogel with ionic conduction of sodium caseinate and electron conduction of reduced graphene oxide. This sensor exhibited excellent rapid response (190 ms), a high sensitivity (GF = 13.14), a large strain range (0.1–1100%), and high toughness of 104 kJ/m^3, demonstrating its enormous potential for wide applications, such as sports monitoring, e-skin, and artificial intelligence [32].

Conductive hydrogel-based flexible sensors can display unique mechanical flexibility, biocompatibility, anti-freezing properties, adequate durability, high sensitivity, and fast response in motion applications if the conductive network is properly designed. For that reason, efforts must be directed to optimize the design process of conductive hydrogels to obtain high degrees of similarity to natural soft tissues for sensing applications induced by normal touch, object manipulation, human motions, and physiological activity. The future of flexible conductive hydrogel sensors is very promising and it will be dependent on how we can balance the flexibility, ionic or electronic conductivity, and stability, as well as the cost of production.

6.6 HYDROGELS WITH MAGNETIC PROPERTIES

Magnetic hydrogels are an alternative to improve the active response of conventional hydrogels since they offer control over system activation through magnetic fields. These are composite materials formed by the dispersion of magnetic micro- or nanoparticles, generally iron oxides, in conventional hydrogels, which is why they are also known as ferrogels. The composite characteristics depend on each component and its percentage in the material. Three methods for the fabrication of magnetic hydrogels are highlighted: the blending method, the *in situ* precipitation method, and the grafting method. Figure 6.4 shows a schematization of the different methods of preparing hydrogels with magnetic properties.

Magnetic hydrogels have many applications in the biomedical field; due to their properties, they can be used in tissue engineering for the development of complex tissues, such as those of the heart or the head, acquiring functionality due to the presence of a magnetic field. Also, it has been demonstrated that 3D scaffolds of these materials allow the growth of biological agents due to the presence of nanoparticles. In addition, these hydrogels can be used for controlled drug delivery

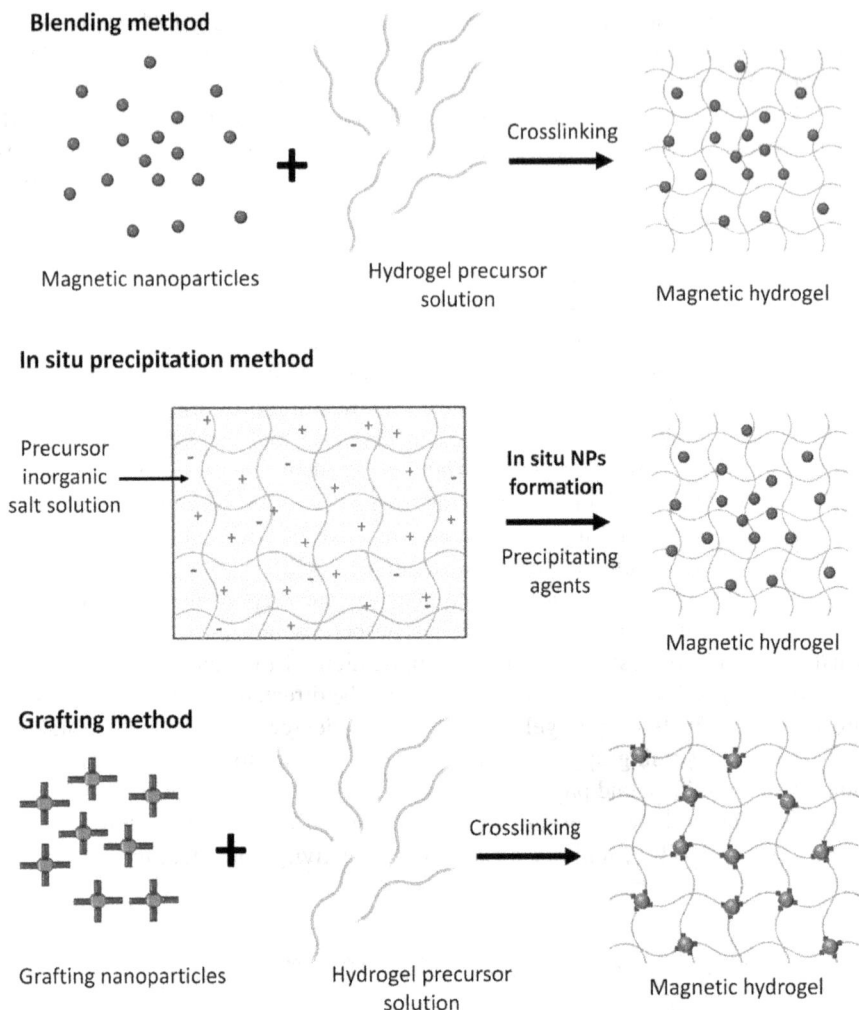

FIGURE 6.4   Preparation methods of hydrogels with magnetic properties.

via magnetic pulses or by employing temperature-sensitive hydrogels and the phenomenon of magnetic hyperthermia to control release. Finally, as flexible sensors, the hydrogels with magnetic properties allow control of mechanical properties changes and water adsorption processes, which are ideal for joint muscles and actuator construction [33].

In the blending method, the magnetic micro- or nanoparticles are dispersed in the hydrogel precursor solution, which is subsequently polymerized. Among the most widely used magnetic nanoparticles are those of iron oxides, which are usually obtained by coprecipitation methods. This method allows a homogeneous dispersion of the magnetic particles, regardless of whether they are micro or nano; it is also a fast and practical method. Brunsen et al. manufactured a magnetic hydrogel using a dextran solution as a polymer base and mixing it with iron oxide nanoparticles to subsequently obtain the hydrogel by photocross-linking with ultraviolet rays. The final material showed magnetic characteristics that make it promising for use in magneto-optic sensors in the biomedical field [34]. These systems have also been studied for artificial muscles and flexible sensor production because structural changes can be induced by impinging on a magnetic field. Szabo et al. synthesized a polyvinyl alcohol hydrogel loaded with magnetite nanoparticles using the mixing method; the synthesized material showed an abrupt reversible change when exposed to a magnetic field [35].

The *in situ* precipitation method is characterized because the formation of magnetic nanoparticles takes place in the hydrogel networks. It starts with a solution of precursor inorganic salts that are added to the hydrogel, allowing it to swell to subsequently react with precipitating agents that will allow the formation of micro- or nanoparticles *in situ*. Among the advantages of this method are the high dispersion and low cost. However, the base hydrogels must have high stability to avoid their degradation during the process. Wang et al. synthesized superparamagnetic nanoparticles *in situ* in a chitosan hydrogel to produce a composite with magnetic properties. It was observed that the pH in the saline solutions significantly affected the magnetic properties of the final material; due to the structural changes in chitosan when varying the pH, chitosan controlled the diffusion of iron ions by generating nucleation sites that prompted the formation of the nanoparticles and their uniformity in the polymeric matrix [36].

Finally, the grafting method consists of modifying the surfaces of the metallic particles by grafting functional groups that allow cross-linking with the hydrogel networks. This synthesis method provides a covalent union between the magnetic micro- or nanoparticles and the hydrogel, guaranteeing the stability of the composite. Barbucci et al. grafted aminopropyl silane onto nanoparticles of cobalt iron oxide ($CoFe_2O_4$) to allow cross-linking between the nanoparticles and the networks of a carboxymethylcellulose hydrogel. The grafted polysaccharide contained amino groups that react with the cellulose carboxyls, forming amide bonds that give high stability to the system, the composite showed susceptibility to being exposed to a magnetic field, having control over the movement of the material [37].

6.7 FUTURE TRENDS ON HYDROGELS WITH ADVANCED PROPERTIES

The impact of hydrogels in the development of flexible sensors is evidenced in the scientific interest and applications available in many different fields. The wide range of properties of these polymeric networks is clearly an enormous incentive for their application as sensing materials. One of them is the high water content that is able to support in their structure, which is a guarantee of biocompatibility and allows the diffusion of many compounds of interest along the polymeric matrix. Also, their capacity to be responsive, to tune their structure, and the possibility to integrate into their composition other sensing compounds, determine a unique versatility of this type of sensor. The impact is verified by the wide range of applications of hydrogel-based flexible sensors to detect some variable that includes temperature, pressure, motion, pH, electric current, magnetic signals, as well as enzymes, nucleotides, cell damage, etc. Although great efforts to achieve functional hydrogel-based flexible sensors have demonstrated their great potential, there are some challenges to be approached in future research.

This chapter has demonstrated the ability of hydrogel-based flexible sensors to detect two or more signals during the same process. In this way, the first approach should be addressed to obtain polymer networks by novel cross-linking techniques to achieve self-regulation of functions that include selectivity, mechanical flexibility, conductivity, biological or chemical responsiveness, and properties detailed in this chapter. In other words, a sensing mechanism related with each function should act independently, enhancing response times and generating improved signals. This should be accompanied by optimization of another critical problem, such as the improvement of reliability and stability of those flexible sensors. These parameters will determine and guarantee long-term operation in different environments, such as the biological one.

Since the development of hydrogel-based flexible sensors is still in its early stage, one of the most important lacks in the development of hydrogel-based flexible sensors is the gap between basic research and commercialization of final products. This lack arose from the strict requirements of commercialization since preparation, mechanical structure, and performances are far to meet them. To achieve this goal, the miniaturization and light weight of hydrogel-based flexible sensors are crucial aspects. In this nanotechnology era, it has become a necessity to develop functional and minimally invasive materials with the aim of generating comfort in the target group. Furthermore, design and preparation of hydrogel sensors at the nanoscale should not affect the properties achieved at the macroscale, which is still a challenge for further development of hydrogel flexible sensors. All this must be accompanied by not only production at a large scale but also with a specific design at relatively low manufacturing costs.

Another effort to be addressed is the correct use of computational tools to predict how hydrogel-based flexible sensors could behave in a real environment. Currently, computational modeling and virtual simulation techniques can help to obtain a better interpretation of the relationship between components, structure, dynamics, interactions of hydrogels, and performance of hydrogels with new network structures, and enhance the structural design of hydrogel-based sensors. Generally, molecular dynamics (MD) simulations are taken as the first option to study the molecular mechanism of hydrogels [38]. This technological approach can also help to solve issues related to manufacturing previously described in the last paragraph, providing conclusions about optimization of industrial processes [39].

Although the flexibility of hydrogels represents the main advantage of the use of those networks, water loss in hydrogels leads to a decrease in mechanical properties, affecting their sensing function. To improve water retention and sensitivity, some fillers and dopants are added to a hydrogel matrix, but possibly these components are incompatible with the surrounding flexible network, weakening mechanical strength of the network. For this reason, it is necessary to direct efforts to find the ideal filler that guarantees simultaneously high sensitivity and improved mechanical properties, using all available tools to that.

Although all efforts and attention to improve the use of hydrogels are not only to sensing applications, much more attention would be paid to environmental concerns. The development of functional and environmentally friendly hydrogel-based sensors would be the most important goal in the coming years. Even hydrogels display high biocompatibility attributed to their high water content, but very little has been studied about their impact on the environment. In this way, the synthesis and manufacturing of the hydrogels without the use of toxic cross-linking agents or organic solvents appear as a promissory direction. Another important direction should be the evaluation of how the novel properties acquired by the hydrogels for human welfare affect other life forms. Also, the manufacturing of reusable and biodegradable hydrogel-based flexible sensors could help to minimize the impact of these networks as disposal materials in soils.

Integrating all aspects described in this section including high sensitivity, novel properties, enhanced mechanical properties, environmental stability, design, and optimization of adequate manufacturing processes, low cost, commercialization, and low environmental impact will promote a promising future on the development of hydrogel-based flexible sensors for some practical applications.

6.8 CONCLUSIONS

For biomedical purposes, precisely in the design and development of flexible sensors, hydrogels have gained a lot of attention due their excellent characteristics, including biocompatibility, stretchability, and flexibility. Hydrogels have been applied to various flexible sensors to detect, in real time, variations in movement, facial expression, tiny physiological signals, strain, pressure, pulses, breathing, speaking, and even changes in touch. Hydrogel-based flexible sensors have displayed unique performances and functionalities by the addition of different fillers or dopants and by the development of novel synthetic methods to create new conditions of adaptability to diverse and extreme environments. Smart hydrogel-based flexible sensors with multi-functions can be achieved by the correct designing and optimizing of the three-dimensional network exhibiting properties described in this chapter; that is, self-healing, anti-freezing, shape memory, magnetic, hydro-phobic, adhesive, or highly stretchable capacity. Their progressive use in biomedical high-performance areas, such as tissue engineering, smart diagnosis, drug delivery, and implantable bioelectronics has demonstrated the potential of these materials and has encouraged to the scientific community to join efforts to understand the fundamental background, accelerating process, and applications of these smart hydrogel-based flexible sensors. Although there are still many things to do and many challenges that must be overcome, the road looks quite promising in this field. Maybe in the future, with the progress of science, technology, and decided support from economic groups, in the future, hydrogel-based flexible sensors can be available that are able to re-create extreme conditions of the human body and that fully comply with their sensing purposes.

ACKNOWLEDGMENTS

Thanks to CONAHCyT for the doctoral scholarship provided for Lorena Duarte Peña (887494), and master scholarship provided for David Romero Fierro (1175725) and Yeeimi Aylin Esquivel Lozano (952777). This work was supported by Dirección General de Asuntos del Personal Académico, Universidad Nacional Autónoma de México under Grant IN 204223.

DECLARATION OF CONFLICT OF INTEREST

The authors declare that there is no conflict of interest.

REFERENCES

[1] Pinelli F., Magagnin L., Rossi F. Progress in Hydrogels for Sensing Applications: A Review. Mater Today Chem 2020;17:100317.

[2] Buenger D., Topuz F., Groll J. Hydrogels in Sensing Applications. Prog Polym Sci 2012;37:1678–1719.

[3] Wang L., Xu T., Zhang X. Multifunctional Conductive Hydrogel-Based Flexible Wearable Sensors. TrAC Trends Anal Chem 2021;134:116130.

[4] Zhang J., Wang Y., Wei Q., Wang Y., Lei M., Li M., et al. Self-Healing Mechanism and Conductivity of the Hydrogel Flexible Sensors: A Review. Gels (Basel, Switzerland) 2021;7:1–33.

[5] Blaiszik B.J., Kramer S.L.B., Grady M.E., McIlroy D.A., Moore J.S., Sottos N.R., et al. Autonomic Restoration of Electrical Conductivity. Adv Mater 2012;24:398–401.

[6] Tuncaboylu D.C., Sari M., Oppermann W., Okay O. Tough and Self-Healing Hydrogels Formed via Hydrophobic Interactions. Macromolecules 2011;44:4997–5005.

[7] Yang W., Shao B., Liu T., Zhang Y., Huang R., Chen F., et al. Robust and Mechanically and Electrically Self-Healing Hydrogel for Efficient Electromagnetic Interference Shielding. ACS Appl Mater Interfaces 2018;10:8245–8257.

[8] Cao J., Lu C., Zhuang J., Liu M., Zhang X., Yu Y., et al. Multiple Hydrogen Bonding Enables the Self-Healing of Sensors for Human–Machine Interactions. Angew Chemie 2017;56:8795–8800.

[9] Hafeez S., Ooi H.W., Morgan F.L.C., Mota C., Dettin M., Van Blitterswijk C., et al. Viscoelastic Oxidized Alginates with Reversible Imine Type Crosslinks: Self-Healing, Injectable, and Bioprintable Hydrogels. Gels 2018;4:1–19.

[10] Liu X., Ren Z., Liu F., Zhao L., Ling Q., Gu H. Multifunctional Self-Healing Dual Network Hydrogels Constructed via Host–Guest Interaction and Dynamic Covalent Bond as Wearable Strain Sensors for Monitoring Human and Organ Motions. ACS Appl Mater Interfaces 2021;13:14612–14622.

[11] Su G., Yin S., Guo Y., Zhao F., Guo Q., Zhang X., et al. Balancing the Mechanical {,} Electronic{,} and Self-Healing Properties in Conductive Self-Healing Hydrogel for Wearable Sensor Applications. Mater Horiz 2021;8:1795–1804.

[12] Peponi L., Navarro-Baena I., Kenny J.M. Shape Memory Polymers: Properties, Synthesis and Applications. In: Aguilar M.R., San Román J., editors. Smart Polym. their Appl., Cambridge: Elsevier; 2014, pp. 204–236.

[13] Cui Y., Li D., Gong C., Chang C. Bioinspired Shape Memory Hydrogel Artificial Muscles Driven by Solvents. ACS Nano 2021;15:13712–13720.

[14] Choi J.G., Spinks G.M., Kim S.J. Mode Shifting Shape Memory Polymer and Hydrogel Composite Fiber Actuators for Soft Robots. Sensors Actuators A Phys 2022;342:113619.

[15] Herrmann A., Haag R., Schedler U. Hydrogels and Their Role in Biosensing Applications. Adv Healthc Mater 2021;10:1–25.

[16] Belbruno J.J. Molecularly Imprinted Polymers. Chem Rev 2019;119:94–119.

[17] Wei Y., Zeng Q., Wang M., Huang J., Guo X., Wang L. Near-Infrared Light-Responsive Electrochemical Protein Imprinting Biosensor Based on a Shape Memory Conducting Hydrogel. Biosens Bioelectron 2019;131:156–162.

[18] Wang C., Li F., Bi Y., Guo W. Reversible Modulation of 2D Photonic Crystals with a Responsive Shape-Memory DNA Hydrogel Film. Adv Mater Interfaces 2019;6:1–8.

[19] Zhao J., Gao X., Chen S., Lin H., Li Z., Lin X. Hydrophobic or Superhydrophobic Modification of Cement-Based Materials: A Systematic Review. Compos Part B Eng 2022;243:110104.

[20] Sidorenko A., Krupenkin T., Aizenberg J. Controlled Switching of the Wetting Behavior of Biomimetic Surfaces with Hydrogel-Supported Nanostructures. J Mater Chem 2008;18:3841–3846.

[21] Yang G., Zhu K., Guo W., Wu D., Quan X., Huang X., et al. Adhesive and Hydrophobic Bilayer Hydrogel Enabled On-Skin Biosensors for High-Fidelity Classification of Human Emotion. Adv Funct Mater 2022;32:2200457.

[22] Qu M., Shen L., Wang J., Zhang N., Pang Y., Wu Y., et al. Superhydrophobic, Humidity-Resistant, and Flexible Triboelectric Nanogenerators for Biomechanical Energy Harvesting and Wearable Self-Powered Sensing. ACS Appl Nano Mater 2022;5:9840–9851.

[23] Rong Q., Lei W., Liu M. Frontispiece: Conductive Hydrogels as Smart Materials for Flexible Electronic Devices. Chem – A Eur J 2018;24:chem.201886461.

[24] Chen Z., Chen Y., Hedenqvist M.S., Chen C., Cai C., Li H., et al. Multifunctional Conductive Hydrogels and Their Applications as Smart Wearable Devices. J Mater Chem B 2021;9:2561–2583.

[25] Tang L., Wu S., Qu J., Gong L., Tang J. A Review of Conductive Hydrogel Used in Flexible Strain Sensor. Materials (Basel) 2020;13:3947.

[26] Wang Z., Cong Y., Fu J. Stretchable and Tough Conductive Hydrogels for Flexible Pressure and Strain Sensors. J Mater Chem B 2020;8:3437–3459.

[27] Zhou C., Wu T., Xie X., Song G., Ma X., Mu Q., et al. Advances and Challenges in Conductive Hydrogels: From Properties to Applications. Eur Polym J 2022;177:111454.

[28] Jin X., Jiang H., Li G., Fu B., Bao X., Wang Z., et al. Stretchable, Conductive PAni-PAAm-GOCS Hydrogels with Excellent Mechanical Strength, Strain Sensitivity and Skin Affinity. Chem Eng J 2020;394:124901.

[29] Lv X., Tian S., Liu C., Luo L-L, Shao Z-B, Sun S-L. Tough, Antibacterial and Self-Healing Ionic Liquid/Multiwalled Carbon Nanotube Hydrogels as Elements to Produce Flexible Strain Sensors for Monitoring Human Motion. Eur Polym J 2021;160:110779.

[30] Ding H., Liang X., Wang Q., Wang M., Li Z., Sun G. A Semi-Interpenetrating Network Ionic Composite Hydrogel with Low Modulus, Fast Self-Recoverability and High Conductivity as Flexible Sensor. Carbohydr Polym 2020;248:116797.

[31] Zhao L., Ke T., Ling Q., Liu J., Li Z., Gu H. Multifunctional Ionic Conductive Double-Network Hydrogel as a Long-Term Flexible Strain Sensor. ACS Appl Polym Mater 2021;3:5494–5508.

[32] Wang Y., Gao G., Ren X. Graphene Assisted Ion-Conductive Hydrogel with Super Sensitivity for Strain Sensor. Polymer (Guildf) 2021;215:123340.

[33] Li Y., Huang G., Zhang X., Li B., Chen Y., Lu T., et al. Magnetic Hydrogels and Their Potential Biomedical Applications. Adv Funct Mater 2013;23:660–672.

[34] Brunsen A., Utech S., Maskos M., Knoll W., Jonas U. Magnetic Composite Thin Films of Fe_xO_y Nanoparticles and Photocross-Linked Dextran Hydrogels. J Magn Magn Mater 2012;324:1488–1497.

[35] Szabó D., Szeghy G., Zrínyi M. Shape Transition of Magnetic Field Sensitive Polymer Gels. Macromolecules 1998;31:6541–6548.

[36] Wang Y., Li B., Zhou Y., Jia D. Chitosan-Induced Synthesis of Magnetite Nanoparticles via Iron Ions Assembly. Polym Adv Technol 2008;19:1256–1261.

[37] Barbucci R., Pasqui D., Giani G., De Cagna M., Fini M., Giardino R., et al. A Novel Strategy for Engineering Hydrogels with Ferromagnetic Nanoparticles as Cross-Linkers of the Polymer Chains. Potential Applications as a Targeted Drug Delivery System. Soft Matter 2011;7:5558–5565.

[38] Fu J., in het Panhuis M. Hydrogel Properties and Applications. J Mater Chem B 2019;7:1523–1525.

[39] Wang B., Xu W., Yang Z., Wu Y., Pi F. An Overview on Recent Progress of the Hydrogels: From Material Resources, Properties, to Functional Applications. Macromol Rapid Commun 2022;43:2100785.

7 Chemical-Responsive Reversible Hydrogels

Vijay Kumar Pal
Institute of Nano Science and Technology, Punjab, India

Sangita Roy
Institute of Nano Science and Technology, Mohali, India

7.1 INTRODUCTION

Hydrogels are cross-linked networks of both synthetic and natural polymers, which upon chemical or physical stimulus, undergo polymerization and are extensively used for biomedical purposes because of their porous nature, loading capacity, and swelling behavior.[1] Stimuli-responsive polymeric hydrogels have gained immense attention in the field of material science and molecular engineering due to their application in bioelectronics, biosensors, drug delivery, and biomedical applications.[2] Three-dimensional polymer network hydrogels have the capability to absorb a large content of water to mimic the extracellular matrix.[3] Thus, hydrogels are generally regarded as soft and biocompatible materials, which have aroused increasing attention in various biomedical fields, such as drug delivery, wound dressings, three-dimensional cell culture, and tissue engineering, due to their biomimetic 3D nanofibrous structure and improved biocompatibility.[4] The swelling behavior of these polymeric hydrogels shows chemical responsiveness towards external stimuli, such as pH, ionic strength, chemical or biochemical change in the environment, molecular recognition, solvent composition, change in structural conformations, etc.[5] The hydrophilic functional groups connected to the backbone of the polymeric chain provide them the capacity to hold water in their interstitial spaces, while the cross-linking polymer prevents them from getting dissolved in water. Physically cross-linked hydrogels are created when polyelectrolytes of opposite charges and multivalent ions/surfactants interact and show a reversible degradation nature,[6] whereas the chemically cross-linked hydrogels are formed by covalently cross-linked polymeric network with permanent binding.[7] In order to improve the performances of hydrogels in a wide area, particularly for biomedical applications, their chemical structure, composition, biological functions, biodegradability, and a variety of physicochemical properties, including mechanical and rheological properties, along with spectral and pH stability, can be managed in an advanced formulation of the hydrogels.[8] It was evident that the physical cross-linking, such as entanglements and crystallites, as well as chemical cross-linking maintains the compact shape of the hydrogels in aqueous fluids. Construction of chemical-responsive hydrogel requires endogenous stimuli or contact stimuli to fabricate bio-functional materials. Advancement in the design of such

DOI: 10.1201/9781003340485-7

hydrogels with chemical-responsive cross-linker will be able to detect the abnormal signal in the pathological or wounded tissues, which leads to the activation of a targeted and on-demand drug release.[9] Stimuli-responsive hydrogels can have a variety of features, including controlled gelation, degradation, or diverse mechanical stiffness, depending upon the different gelator molecules and cross-linkers.[10] A sensible structuring of the endogenous and external stimuli-responsive components into a single hydrogel system may offer a wide variety of customized materials. It can be envisioned that the addition of multi-stimuli responsiveness makes it possible to combine several properties into a single hydrogel system.

Despite the extensively rich history of polymer hydrogel research, the field is still intriguing and hence still emerging. In particular, the current focus is dedicated towards exploring the ways to use polymeric hydrogels by utilizing the stimuli-responsive and biocompatible nature of these polymer-based hydrogels in a drug administrative domain.[11] Polymer hydrogels propose better mechanical characteristic properties due to their long covalent chains in comparison to the small molecule hydrogelator, whereas biomolecular hydrogels, especially peptide-based hydrogels formed using supramolecular self-assembly, have gained a lot of attention in recent years.[4] Peptide hydrogels feature a three-dimensional fibrillar network structure, just like polymer hydrogels.[12] The fibrils are related to coiled-coil, α-helical, β-sheet, micelle, vesicle, ribbon, tape, and tube-based morphologies. Furthermore, the most attractive features of these hydrogels lie in their biocompatible as well as biodegradable nature, which is further coupled with the multi-stimuli-responsive behavior when exposed to external stimuli. This intriguing property of peptide-based hydrogels has attracted the interest of many research groups over the last two decades. Due to their low toxicity, peptide hydrogels are currently widely employed in a variety of biomedical applications, including drug delivery, protein separation, biosensors, tissue engineering, and wound healing.[13] Thus, both polymer and peptide hydrogels show a different stimuli response towards a different environment. Here in this chapter, we will focus on the chemical-responsive behavior of polymer/peptide hydrogels and their applications.

Hydrogels responding to external triggers usually display a phase transition from gel to solid or gel to solution, which evidently are the signature of their stimuli-responsive behavior. To make reversible or one-cycle phase changes within a hydrogel system, several external physical or chemical stimuli can be used. While the physical trigger includes thermal, magnetic, acoustic, electrochemical, or light stimuli, the chemical triggers consider the changes in pH, ionic composition of overall ionic strength, redox reaction, solvent composition, molecular recognition, etc. Such chemical stimuli trigger the hydrogelation among these peptide/polymeric chains and further display an on-demand gel-to-sol transition for different energy and healthcare applications. It was claimed that stimuli-responsive hydrogels might be used for a variety of purposes, such as functional matrices for sensing, actuators, and several other biological purposes including controlled drug release, tissue engineering, and bioimaging.[14] Hydrogels that are stimulated have also been utilized to build catalytic switches, logic gates, surfaces for regulated cell growth, and more.[15] Surfaces with signal-triggered stiffness and switchable interfacial electron transport capabilities were created using techniques to immobilize and shape stimuli-responsive hydrogels on surfaces.[16] With the aid of stimuli-responsive hydrogels and synthetic advancements in developing nano- and micrometer-sized inorganic particles, hybrid materials for electrical and optical applications may be

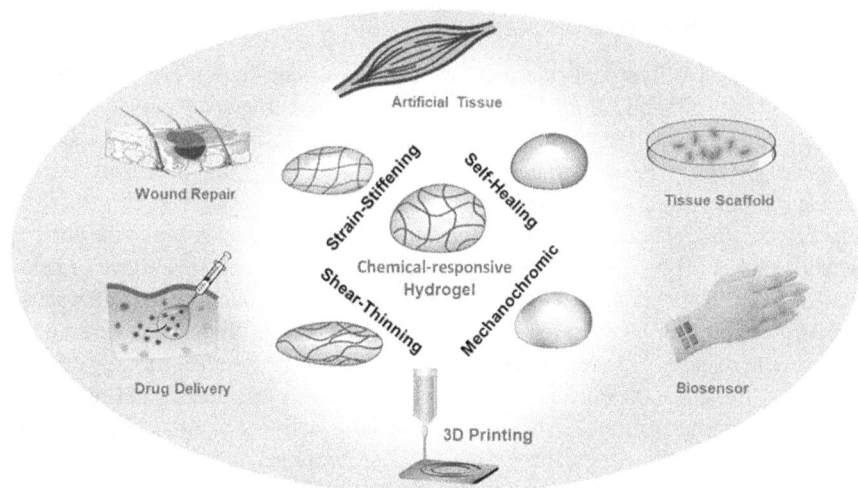

FIGURE 7.1 Chemically responsive hydrogel showing different applications under external chemical stimuli. Adapted with permission from reference [17]. Copyright (2020) American Chemical Society.

fabricated (Figure 7.1).[17] New ideas for regulated and targeted drug release were also provided by techniques for creating stimuli-responsive nano- or micro-hydrogels.[18] Significant research accomplishments have been made in the sector due to the quick development of stimuli-responsive hydrogels and their wide range of applications.[14]

In this chapter, we specifically concentrate on the chemically responsive hydrogels with different synthesis strategies, such as changes in pH, ionic strength of the gelator microenvironment, addition of chemical stimuli, like biological relevant ions, changes in solvent composition followed by supramolecular assembly, radical polymerization, use of different chemical reactions, using molecular recognition-based methodologies, etc. We will first discuss different chemical-responsive pathways of hydrogelation among hydrogelator molecules, followed by various applications in the field of tissue engineering and regenerative medicines. It is anticipated that the development of hydrogels with chemical responsiveness would advance this field of research and open new, interesting possibilities.

7.2 CHEMICALLY RESPONSIVE HYDROGELS

Recently, smart polymeric hydrogels have gained considerable interest due to their chemical responsiveness property. The stimuli-responsive hydrogels have the tendency to show reversible phase changes, along with the changes in the physical or chemical properties of the gelator molecules so as to respond to the chemical signals, such as pH, addition of anions or metal ions, solvent environmental changes, molecular recognition, changes in structural conformation, etc.[7] Chemically responsive hydrogels, which show differential structural or mechanical behavior toward environmental triggers, are among the most important class of hydrogels, especially for the development of drug carriers,[1] environment responsive gene delivery vehicles,[19] robotics,[12] regenerative medicines, and more.

7.2.1 pH-Responsive Hydrogels

The pH-responsive hydrogels are the most studied stimuli-responsive hydrogels for their wide range of applications in controlled and targeted drug/gene delivery.[1] Several chemical cross-links, like Schiff base cross-links,[20] acylhydrazone cross-links,[21] ionizable groups,[22] boronate-catechol ligands,[23] and DNA triple helixes tend to associate or dissociate at a specific pH value, which can be used to construct pH-responsive hydrogel networks.[19] The ionization of the functional group in the polymer network causes pH-responsive hydrogels to swell or contract at a pH above or below its pKa point. This phenomenon further leads to the absorption or loss of water; a similar principle can be used for developing therapies with pH-responsive drug and nanoparticle release.[24] pH-responsive hydrogels are mostly designed for drug delivery in cancer treatment.[25] For example, Sarwar et al. presented a pH-responsive nanoparticle-based hydrogel hybrid system for drug delivery to solid tumors.[26] The chitosan-based hydrogel contains acid-susceptible polymer nanoparticles with high drug encapsulation and loading efficiency.[26] The three-layered nanoparticles contain amphiphilic copolymers (polycaprolactone-polyethyleneimine-polyethylene glycol) with stealthy and acid labile linkage. The nanoparticles released from chitosan hydrogel in an acidic tumor environment were able to penetrate the tumor tissues to release the encapsulated drug inside the tumor cells.[26] Similarly, an injectable pH-responsive hydrogel was shown to effectively perform the combinatorial chemoimmunotherapy at the tumor site.[27] A silk-chitosan composite scaffold was fabricated, which showed a controlled local release of the chemotherapeutic drug (doxorubicin and JQ1 molecular inhibitor) at a weak acidic environment (Figure 7.2a). The released

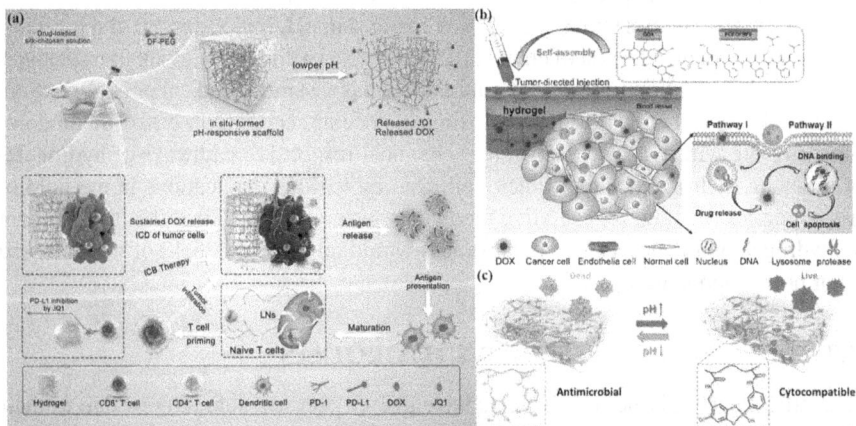

FIGURE 7.2   (a) Schematic representation of combinatorial therapy of the pH-responsive for the delivery of antitumor agents into tumor microenvironment. Adapted with permission from reference [27] Copyright (2022) Springer Nature. (b) Schematic representation of the injectable octapeptide hydrogel capable of releasing the anti-cancerous drug doxorubicin on target at acidic pH and further shows structural transition from nanofibers to nanosphere. Adapted with permission from reference [29] Copyright (2022) Elsevier. (c) Schematic illustration of pH reversible nature of the hydrogel to present antibacterial and cytocompatible application. Adapted with permission from reference [30] Copyright (2022) American Chemical Society.

drug at the target site could directly eliminate the cancerous cells or trigger the antitumor immunity, which leads to the cell death of tumor cells.[27] Another interesting building block for constructing pH-responsive self-assembled hydrogels is peptide- and protein-based biomolecules.[28] In this context, Liu et al. designed an injectable pH-responsive OE peptide hydrogel for the delivery of anticancer drugs, gemcitabine, and paclitaxel, which can be released simultaneously to achieve an antitumor effect.[25] The peptide hydrogel has demonstrated effective biological activities both *in-vitro* and *in-vivo* against cancerous cells. In particular, the drug gemcitabine was released in the first three days, while the hydrophobic drug paclitaxel showed sustained release up to seven days, leading to the prolonged action of the drug and further enhancing its therapeutic efficacy towards cancer.[25] The blank peptide hydrogels showed bio-compatible behavior towards cells and its proposed use is safe and non-toxic.

The use of a bioactive peptide as a biomaterial has a good developmental prospect as it is generally derived from natural proteins.[4] One such study by Zhang and co-workers demonstrated an injectable ionic-complementary octapeptide hydrogel loaded with doxorubicin that showed significantly high antitumor activity.[29] The octapeptide hydrogel was stable at physiological conditions, while showing phase transition through disassembly at an acidic environment (tumor microenvironment), leading to the structural transition of nanofibers to nanospheres, leading to the enhanced cellular uptake of doxorubicin (Figure 7.2b).[29] In this direction, Pat et al. have also demonstrated a peptide-based vehicle system for the delivery of antimicrobial agents at the wound site with microbial infection.[31] The bio-nano construct offers dual functionality and environmental tunability as the peptide nanoparticles demonstrate shape transition at a slightly basic environment, which leads to the release of an encapsulated hydrophobic drug molecule, ferulic acid. The spherical nanoparticles show structural transition into nanofibers, which provides an advanced synthetic scaffold for wound healing and closure.[31] Another report of a pH-responsive antibacterial hydrogel was developed by reversible linkage between bactericidal chlorinated catechol and phenylboronic acid.[30] At an acidic environment, the hydrogel tends to release the catechol-Cl, which shows excellent antibacterial properties against multiple bacterial strains. On the other hand, at a basic environment (pH 8.5), the antimicrobial property of the hydrogel is lost and becomes non-toxic due to reduced exposure of the catechol-Cl (Figure 7.2c). Upon adjusting the pH to acidic pH, the hydrogel showed reversible behavior by re-exposing the catechol-Cl and regaining its antibacterial property.[30] It was observed that these peptide-based hydrogels are capable of nurturing and protecting cells from mechanical membrane breakdown while being injected from a needle, which proposes their bright future in tissue engineering and biomedicines.

7.2.2 Ion-Responsive Hydrogels

Several complex biological phenomena and supramolecular associations are supported by cooperative non-covalent interactions, which require metal ions for their structural integrity and functional property.[32] Polymeric hydrogels cross-linked via different cations or anions have been extensively studied, which includes several synthetic polymers, natural peptides/protein, and DNA serving as the framework or template for this kind of responsive hydrogel.[33,34] Several reports with ion-responsive double network hydrogel systems involving alginate/gelatin, acrylamide, etc. were involved to provide the mechanical support and memory code.[35] Different metal cations were found to show ion-responsive shape memory property.[35] In this context, a robust strategy was used to prepare a single-component sodium alginate hydrogel sheet and further cross-linked with Ca^{2+} ions. This ionic cross-linking further displayed reversible deformation and was able to retain variable shapes as triggered by biologically relevant ions (Na^+/Ca^{2+}).[36] Various morphologies like tubular, 3D twisting, and plant-inspired architectures can be programmed by varying the orientation of the patterned microchannels (Na^+/Ca^{2+}).[36] Metal ions can also be used to induce various structural modifications in proteins, peptide derivates, and nucleic acids.[5] Recently, several studies have shown that the ionic cross-linking may lead to hydrogelation with enhanced mechanical strength of the hydrogel matrix, which can be used for several biomedical applications. Lin and co-workers designed a pentapeptide (FFRGD) with fluori-nated benzyl group at the N-terminal end of the peptide, which showed a propensity to form a stable supramolecular hydrogel at a physiological pH.[37] The addition of metal ions into these twisted nanobelt morphologies results in a variable gel with diversified storage modulus, where Ca^{2+}/Mg^{2+} showed enhanced mechanical stiffness, due to strong metal-ligand chelation and high cross-linked fibers. Another classical example in this direction displayed that the participation of Ba^{2+} ions leads to weak metal-ligand cross-linking and results in gels with weaker mechanical strength.[37] Recently, Roy and co-workers have demonstrated the effect of metal coordination into a collagen-inspired peptide (NapFFGDO), surpassing the limitation of the monomeric peptide moieties, which failed to induce higher-order self-assembly at a physiological pH, owing to charge repulsion (Figure 7.3a). The biologically relevant metal ions (Ca^{2+}/Mg^{2+}) were used to induce multiscale self-assembly among the peptide nanofibers to form a hydrogel, which otherwise remained as sol at a physiological pH.[38] A carboxyl group of aspartic acid (Asp) present at the peptide surface interacts with the divalent metal ions and thus masks the surface charge and leads to a mesh-like structure, which helps in stabilizing the hydrogel network. Such a metal coordinated peptide hydrogel network provides a suitable scaffold for enhanced cellular adhesion and proliferation.[38]

On the other hand, nucleic acids also show ion responsiveness, such as K^+ stabilized G-quadruplexes and other cooperative duplexes formed with Ag^+/Hg^{2+}, which results in the formation of DNA hydrogels.[5] Several synthetic and natural polymers (CMC-cellulose, polyacrylamide, PNIPAM, etc.) were introduced into these DNA hydrogels to provide the structural support to the scaffolds.[12] Hydrogels with flexible wearable property have shown remarkable promise in the field of health monitoring. The excellent mechanical/stretchable property, conductivity, and tissue-mimicking properties contribute to the development of biosensors.[13] For example, an ionic-conductive hydrogel was fabricated using polyvinylpyrrolidone (PVP)/tannic acid/ Fe^{3+} cross-linked into a N, N-methylene diacrylamide and poly (N-isopropylacrylamide-co-acrylamide) network (Figure 7.3b). By varying the concentration of metal ion and tannic acid, maximum stretchability (720%) and sensitive strain response (GF = 3.61) were achieved.[39] Another example of ionic gel

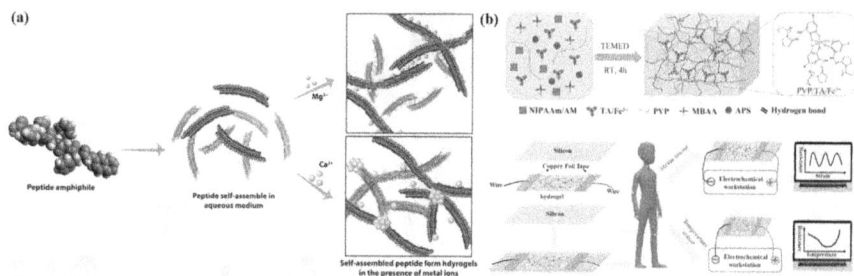

FIGURE 7.3 (a) Schematic representation of metal ion-induced hydrogelation in the collagen-inspired peptides at a physiological pH. Adapted with permission from reference [38] Copyright (2022) Wiley-VCH GmbH. (b) Schematic illustration of ion-responsive conductive hydrogel and its potential to be used as wearable strain sensor and temperature sensor. Adapted with permission from reference [39] Copyright (2022) American Chemical Society.

includes the design of a self-healable ionic gelatin/glycerol-based hydrogel for the application of soft-sensing. The hydrogel showed strain capability of 454% and stability over a long period of time, with a biocompatible and biodegradable nature.[40] The use of metal ions in polymeric hydrogels has shown great potential to develop diverse nanostructured, structurally stable, and biocompatible hydrogels with several advantages for futuristic applications in biomedicine.

7.2.3 CHEMICAL- AND BIOCHEMICAL-RESPONSIVE HYDROGELS

Apart from pH- and ion-responsive hydrogels, other stimuli like redox and exchange reactions are among the processes used to induce cross-linking to form hydrogels and to dissociate the monomers to release the encapsulated load inside the nanostructures. A dynamic redox reaction leads to hydrogelation of an ionic-conductive poly (hydroxyethyl methacrylate), which overcomes the limitations of the weak mechanical strength, lack of adhesiveness, and poor performance stability.[41] As discussed, stimuli-responsive hydrogels have been used to trigger the release of the therapeutic agents or act as a scaffold for wound recovery. However, an essential parameter for an ideal material lies in the ease of removal of these hydrogels from the wound site, for which an on-demand degradation or a facile method of dissolution is required. Such an important feature of degradability can be achieved in redox-sensitive hydrogels. Redox-responsive hydrogels have emerged as an important design criterion for the release of encapsulated drug molecule and degrading the biomaterial used for the hydrogelation. One such example is the use of cyclic thiosulfinate-based cross-linker or macromonomers for the formation of a hydrogel.[42] The most common approach was the use of free-radical polymerization (disulfide-containing bis-methacrylamides), which undergoes post-polymerization via interchain cross-linking.[43] Sanyal and co-workers have proposed a facile fabrication technique for a hydrogel using the thiol-disulfide exchange reaction, which showed sustained and on-demand release of the encapsulated biomacromolecules (Figure 7.4a).[43] In particular, a functionalization of the pyridyl disulfide group at the end chain of polyethylene glycol yielded hydrogels

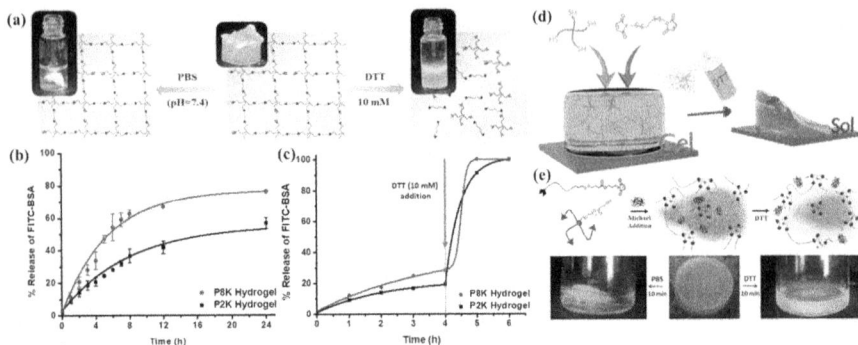

FIGURE 7.4 (a) Degradation study of the dye-conjugated hydrogel (P8K) in PBS and DTT (10 mM). (b) Release profile of the dye-conjugated (FITC-BSA) hydrogels P8K and P2K. (c) Forced release profile of FITC-BSA with DTT (10 mM) after 4 h of passive release at 37°C. Adapted with permission from reference [43] Copyright (2022) American Chemical Society. (d) Schematic representation of the gel to sol transition with the addition of DTT. (e) Fabrication and encapsulation of the protein in side hydrogel and its fast release in the presence of DTT and degradation behavior of the disulfide-containing hydrogel after incubating for 10 minutes in PBS and DTT. Adapted with permission from reference [44] Copyright (2022) American Chemical Society.

upon mixing with tetra-arm thiol containing PEG of two different molecular weights (M_n: 2 and 8KDa). The FITC-BSA trapped inside the hydrogel showed a slower release in a P2K hydrogel compared to P8K hydrogels, and the on-demand release was checked by adding DTT into the solution medium (PBS) (Figure 7.4b,c). One more example of the exchange reaction deals with the combination of a thiol-maleimide conjugate and thiol-disulfide, which leads to a fast-forming hydrogel and shows gel to sol responsiveness towards dithiothreitol (DTT) (Figure 7.4d).[44] A conjugation between two PEG derivatives; one with thiol-terminated tetra-arm poly(ethylene glycol) (PEG) and the other with disulfide-containing maleimide-terminated polymers was used to synthesize the hydrogels. The conjugate hydrogels were studied for their sustained release of the fluorescently labeled dextran polymers and bovine serum albumin (BSA) via a stimuli-responsive release mechanism (Figure 7.4e).[44]

Recently, a firefly luciferin-inspired hydrogel matrix was developed, exhibiting rapid gelation rate, tunable mechanical and biological property, and enhanced cell encapsulation efficiency along with injectable property. The luciferin-inspired hydrogel has redox-triggering capability, which allows the precise control over the gelation onset time and kinetics.[45] In a similar line, a transient hydrogel has been reported by Thordarson et al., which showed a controlled lifetime in the presence of a reducing agent, tris(2-carboxyethyl)-phosphine TCEP.[46] In this approach, a redox-responsive gelator with disulfides was used to self-assemble into hydrogel. Due to the dynamic reduction of disulfide cross-links, the system exhibited a dynamic sol-gel-sol transition with a range of lifetimes at various starting TCEP concentrations.[46] Another report of on-demand degradation of a dendritic thioester hydrogel has shown applications as a sealant system for wound closure.[47] The rapid and on-demand deformation of the hydrogel was achieved by thio-thioester exchange reaction in the presence of cysteine methyl ester. The hydrogel sealant closes an *ex-vivo* venous puncture, attaches firmly to tissues, and withstands intense pressure applied to a lesion.[47] When exposed to thiolates, the hydrogel sealant can be entirely removed, allowing for progressive wound re-exposure during definitive surgical care. Such a system with a redox and exchange reaction dilates the current toolkit of chemically responsive hydrogels.

7.2.4 MOLECULAR RECOGNITION

The non-covalent interactions play a major role in stabilizing a supramolecular hydrogel and induce stimuli responsiveness in the molecular domain. The hydrogels formed by a "smart" polymer are frequently described in the literature as being pH or temperature sensitive, which somehow limits the tuning of the polymeric backbone,[5] as the environmental stimuli play a major role in the sensitivity of the arrangement of the monomers that may lead into a nanostructure. On the other hand, molecular recognition can be a powerful strategy, as it allows us to adjust and control more factors like density and location of the binding sites during the synthesis.[48] More importantly, molecular interactions leading to environmental changes often alter the kinetics and thermodynamics of the molecular recognition.[48] Apart from the sol-gel transition, molecular recognition can be seen in various applications like self-healing materials, sensors, drug delivery vehicles, etc. The self-healing behavior of the supramolecular hydrogel can be shown by joining two different pieces of the hydrogel and by sufficiently healing the crack.[48] A suitable example to this end represents a polyacrylate-based supramolecular hydrogel that has been reported as containing cyclodextrins as a host and ferrocene as a guest polymer. Ferrocene (Fc) is oxidized to Fc^+ by sodium hypochlorite, which hinders the self-healing behavior. Upon reduction of the oxidized Fc^+ by glutathione, self-healing can be achieved (Figure 7.5a). This is a clear example of the molecular recognition playing crucial role in the self-process.[48] Several reports have indicated that the supramolecular hydrogels fabricated by a molecular recognition strategy showed self-healing and self-repairing ability in the presence of glutathione (Figure 7.5b). A self-assembled hydrogel provides a semi-wet microenvironment for developing a molecular recognition chip, which allows the immobilization of the artificial receptors and miniaturization of the functionalized hydrogel for high-throughput assays (Figure 7.5c).[49] Hamachi and co-workers reported a fluorescently labeled lectin array, which has the tendency to bind saccharides independently. A hydrogel containing glycosylated amino acid showed entangled nanofibers with micrometer-sized cavities, which can be used for the molecular recognition (Figure 7.5d). At first, an array of hydrogels was formed on a plate and incubated with lectin, which renders fluorescence to the hydrogel. A quencher was used to decrease the fluorescent intensity of the lectin. A strong competitive interaction between saccharide and lectin removes the quencher and recovers the fluorescence of the hydrogel (Figure 7.5e).[50]

Supramolecular hydrogels containing cyclodextrin could lead to the delivery of drug molecules, where the molecular recognition can be used to transport the drug and sustained release can be achieved with high precision. A nanogel reported by

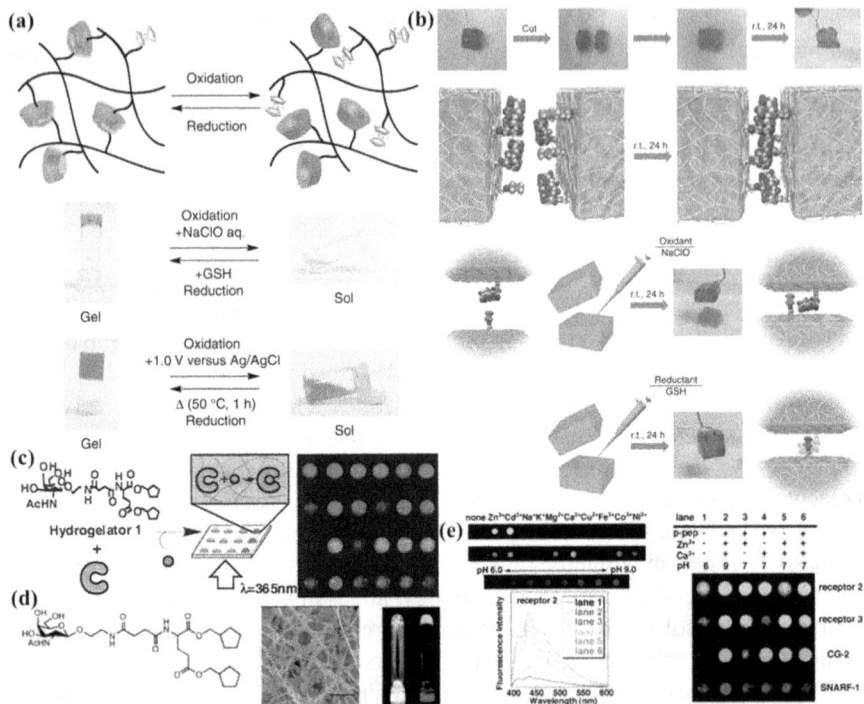

FIGURE 7.5 (a) Schematic representation of sol-gel transition via change in the oxidized and reduced environment. (b) Self-healing ability of the pAA-Fc hydrogel using oxidizing/reducing agent. Adapted with permission from reference [48] Copyright (2011) Nature Publishing Group. (c) Schematic representation of a semi-wet sensor chip probed with self-assembled hydrogel. (d) Nanofibrous network of the hydrogel and its UV-responsive nature in presence of phenyl phosphate. (e) Sensing pattern of semi-wet chemo-sensor chip at different pH in presence of different metal ions. Adapted with permission from reference [49,50] Copyright (2004, 2006) American Chemical Society.

the Akiyoshi group was shown to be capable of bearing cholesterol-pullulan and showed sustained release.[51] This nanogel was able to prevent the denaturation of horse radish peroxidase (HRP) by complexation and can only release when cholesterol groups on the pullulan make an inclusion complex with the cyclodextrin.[51] All these advanced functional materials further corroborated the fact that the structural conformation, molecular recognition, or lock and key-based synthesis are among the very important chemically responsive methods for hydrogelation, which serve multiple applications in the fabrication of biodevices and in biomedical applications.

7.3 FUNCTIONALITY AND APPLICATIONS

Hydrogels developed from chemical responsiveness provide multiple functionalities to the monomers of bioinspired moieties and advance their applications in the field of biomedical engineering. As discussed in the above sections, the polymeric hydrogels showed advancement in the mechanical stiffness and biocompatible nature in response to the changes in the bulk environmental stimuli, like pH, ionic strength, temperature, solvent, etc. The ability of the hydrogels to deliver through many implantation or injection methods (including subcutaneous, transdermal, nasal, pulmonary, ophthalmic, vaginal, and rectal routes) has increased the importance of their utilization. Intelligent hydrogel polymers offer distinctive properties, such as swelling-deswelling behavior, biocompatibility, hydrophilicity, and biodegradability; and support their ability to work effectively in a variety of biomedical applications, such as tissue engineering, targeted distribution, biosensors, and actuators.[13] The polymer/peptide-based hydrogels show chemical responsiveness and extend their applications in the field of biomedicines and bioelectronics.[12] By combining chemical functionality and external stimulus, emergent properties can be achieved, which play crucial roles in drug delivery, self-healing, antibacterial, antitumor property, robotics, etc. Due to their superior quality, the polymeric biomaterials have made significant advancements over the past few decades.[14] Several synthetic polymers and bioinspired biomolecules have shown an outstanding potential for different biomedical applications like tissue engineering, regenerative medicines, drug delivery, wound healing, *ex-vivo* and *in-vivo* cell culture, etc. Smart/stimuli-responsive hydrogels were able to exhibit distinctive qualities that led to their widespread use in tissue engineering applications. They provide a platform for cell migration to the site of wound, have a remarkable ability to re-create the ECM's surrounding environment, and can successfully modify their mechano-physical characteristics to fulfill the need.[5] One of the major applications of these stimuli-responsive biomaterials is in the field of drug delivery. For some pharmacological molecules, notably hydrophobic ones, the loading/release capacity of traditional hydrogels is still subpar.[15] Following that, several studies using polyampholyte, a two-fold system, and nanocomposites were carried out to enhance their mechanical strength and drug discharge capabilities as well as their electrical conductivity behaviors. Due to its potential 3D carrier property, biocompatibility, minimally invasiveness, together with conforming shape for administration, injectable hydrogels are emerging as futuristic material for numerous biomedical applications, including drug delivery and tissue engineering.[7] In the clinical setup, hydrogel-based drug release devices can deliver an effective therapeutic result. To bring success in a medical domain, traditional drug delivery techniques sometimes need high doses or repeated administrations. As a result, negative qualities like toxic effects and negative impacts occur more frequently. Hydrogels can provide a suitable platform for a number of physicochemical interactions with the loaded pharmaceuticals because of their adaptable physical, mechanical, and biodegradation assets. They can therefore provide rigorous control of the release rate of the medicine to the intended delivery location. When exposed to solvents, the aqueous structure of hydrogels plays a critical role in reducing the likelihood of drug buildup and degradation. They can therefore get a large loading of water-soluble medications. They can also encapsulate medicines that are water-insoluble. On the other hand, biosensors are extremely sensitive analytical tools that have several benefits, such as high sensitivity, specificity, affordable production, simplicity, portability, and quick reaction times. They are created by combining materials that respond to nearby stimuli like light, ionic strength, temperature, and pH. There are three major ways that smart/stimuli-responsive hydrogels are used in biosensors: signal recognition, signal transmission to the gauging electrode, and signal response customization. In recent years, biomimetic hydrogels have been widely used to construct biochemical sensors because of their exceptional receptivity to environmental stimuli. Because they can use phase-alteration mechanics and transmit biological information to these sensors, there has been a substantial breakthrough in the control of many illnesses.

7.4 CONCLUSION AND OUTLOOK

Hydrogels present an extremely diverse group of materials with applications in a wide field of energy and healthcare. Smart hydrogels exhibit chemical responsiveness and have led to the development of novel materials with an array of several chemical stimuli, including pH, ionic, species, solvent, conformational triggers, etc., which enables the controlled and precise assemblies of the biomolecules. Researchers have explored a range of materials from synthetic molecules to natural polymers cross-linked via covalent or non-covalent interactions, to study the sol-gel transition or stiffness of the hydrogels via chemical responsiveness. Yet, the potential toxicity of the degradation product, the automatic activation of the cargo release, or differential stiffness in these hydrogels might be overlooked during the material selection, which may need sincere attention. Overall, chemical-responsive hydrogels provide a powerful new paradigm in stimuli-responsive nanomaterials.

REFERENCES

[1] Suhail, M.; Shao, Y. F.; Vu, Q. L.; Wu, P. C., Designing of pH-sensitive hydrogels for colon targeted drug delivery; Characterization and in vitro evaluation. *Gels (Basel, Switzerland)*, 8 (3), 2022, 155.

[2] Pourjavadi, A.; Heydarpour, R.; Tehrani, Z. M., Multi-stimuli-responsive hydrogels and their medical applications. *New Journal of Chemistry*, 45 (35), 2021, 15705–15717.

[3] Fennell, E.; Huyghe, J. M., Chemically responsive hydrogel deformation mechanics: A review. *Molecules (Basel, Switzerland)*, 24 (19), 2019, 3521.

[4] Sharma, P.; Pal, V. K.; Roy, S., An overview of latest advances in exploring bioactive peptide hydrogels for neural tissue engineering. *Biomaterials Science*, 9 (11), 2021, 3911–3938.

[5] Li, Z.; Zhou, Y.; Li, T.; Zhang, J.; Tian, H., Stimuli-responsive hydrogels: Fabrication and biomedical applications. *VIEW*, 3 (2), 2022, 20200112.

[6] Zhu, T.; Sha, Y.; Yan, J.; Pageni, P.; Rahman, M. A.; Yan, Y.; Tang, C., Metallo-polyelectrolytes as a class of ionic macromolecules for functional materials. *Nature Communications*, 9 (1), 2018, 4329.

[7] Zhang, S.; Ge, G.; Qin, Y.; Li, W.; Dong, J.; Mei, J.; Ma, R.; Zhang, X.; Bai, J.; Zhu, C.; Zhang, W.; Geng, D., Recent advances in responsive hydrogels for diabetic wound healing. *Materials Today Bio*, 18, 2023, 100508.

[8] Yang, Y.; Ren, Y.; Song, W.; Yu, B.; Liu, H., Rational design in functional hydrogels towards biotherapeutics. *Materials & Design*, 223, 2022, 111086.

[9] Correia, C. R.; Nadine, S.; Mano, J. F., Cell encapsulation systems toward modular tissue regeneration: From immunoisolation to multifunctional devices. *Advanced Functional Materials*, 30 (26), 2020, 1908061.

[10] El-Husseiny, H. M.; Mady, E. A.; Hamabe, L.; Abugomaa, A.; Shimada, K.; Yoshida, T.; Tanaka, T.; Yokoi, A.; Elbadawy, M.; Tanaka, R., Smart/stimuli-responsive hydrogels: Cutting-edge platforms for tissue engineering and other biomedical applications. *Materials Today Bio*, 13, 2022, 100186.

[11] Xu, G.; Li, J.; Wu, J.; Zhang, H.; Hu, J.; Li, M.-H., Tough polymeric hydrogels formed by natural glycyrrhetinic acid-tailored host–guest macro-cross-linking toward biocompatible materials. *ACS Applied Polymer Materials*, 1 (10), 2019, 2577–2581.

[12] Erol, O.; Pantula, A.; Liu, W.; Gracias, D. H., Transformer hydrogels: A review. *Advanced Materials Technologies*, 4 (4), 2019, 1900043.

[13] Cao, H.; Duan, L.; Zhang, Y.; Cao, J.; Zhang, K., Current hydrogel advances in physicochemical and biological response-driven biomedical application diversity. *Signal Transduction and Targeted Therapy*, 6 (1), 2021, 426.

[14] Roy, A.; Manna, K.; Pal, S., Recent advances in various stimuli-responsive hydrogels: From synthetic designs to emerging healthcare applications. *Materials Chemistry Frontiers*, 6 (17), 2022, 2338–2385.

[15] Willner, I., Stimuli-controlled hydrogels and their applications. *Accounts of Chemical Research*, 50 (4), 2017, 657–658.

[16] Esmaeilzadeh, P.; Groth, T., Switchable and obedient interfacial properties that grant new biomedical applications. *ACS Applied Materials & Interfaces*, 11 (29), 2019, 25637–25653.

[17] Chen, J.; Peng, Q.; Peng, X.; Han, L.; Wang, X.; Wang, J.; Zeng, H., Recent advances in mechano-responsive hydrogels for biomedical applications. *ACS Applied Polymer Materials*, 2 (3), 2020, 1092–1107.

[18] Devi, L.; Chopra, H.; Gaba, P., ed Barhoum, A.; Jeevanandam, J.; Danquah, M. K., Chapter 10 – Nanohydrogels for targeted drug delivery systems, in Bionanotechnology: Emerging Applications of Bionanomaterials, 2022, Elsevier.

[19] Hu, Y.; Gao, S.; Lu, H.; Ying, J. Y., Acid-resistant and physiological pH-responsive DNA hydrogel composed of A-Motif and i-Motif toward oral insulin delivery. *Journal of the American Chemical Society*, 144 (12), 2022, 5461–5470.

[20] Dai, T.; Wang, C.; Wang, Y.; Xu, W.; Hu, J.; Cheng, Y., A nanocomposite hydrogel with potent and broad-spectrum antibacterial activity. *ACS Applied Materials Interfaces*, 10 (17), 2018, 15163–15173.

[21] Li, L.; Gu, J.; Zhang, J.; Xie, Z.; Lu, Y.; Shen, L.; Dong, Q.; Wang, Y., Injectable and biodegradable pH-responsive hydrogels for localized and sustained treatment of human fibrosarcoma. *ACS Applied Materials Interfaces*, 7 (15), 2015, 8033–8040.

[22] Chen, Y.; Pang, X. H.; Dong, C. M., Dual Stimuli-responsive supramolecular polypeptide-based hydrogel and reverse micellar hydrogel mediated by host–guest chemistry. *Advanced Functional Materials*, 20 (4), 2010, 579–586.

[23] Yesilyurt, V.; Webber, M. J.; Appel, E. A.; Godwin, C.; Langer, R.; Anderson, D. G., Injectable self-healing glucose-responsive hydrogels with pH-regulated mechanical properties. *Advanced Materials*, 28 (1), 2016, 86–91.

[24] Rizwan, M.; Yahya, R.; Hassan, A.; Yar, M.; Azzahari, A. D.; Selvanathan, V.; Sonsudin, F.; Abouloula, C. N., pH sensitive hydrogels in drug delivery: Brief history, properties, swelling, and release mechanism, material selection and applications. *Polymers*, 9 (4), 2017, 137.

[25] Liu, Y.; Ran, Y.; Ge, Y.; Raza, F.; Li, S.; Zafar, H.; Wu, Y.; Paiva-Santos, A. C.; Yu, C.; Sun, M.; Zhu, Y.; Li, F., pH-Sensitive peptide hydrogels as a combination drug delivery system for cancer treatment. *Pharmaceutics*, 14 (3), 2022, 652.

[26] Sarwar, S.; Bashir, S.; Asim, M. H.; Ikram, F.; Ahmed, A.; Omema, U.; Asif, A.; Chaudhry, A. A.; Hu, Y.; Ustundag, C. B., In-depth drug delivery to tumoral soft tissues via pH responsive hydrogel. *RSC Advances*, 12 (48), 2022, 31402–31411.

[27] Gu, J.; Zhao, G.; Yu, J.; Xu, P.; Yan, J.; Jin, Z.; Chen, S.; Wang, Y.; Zhang, L. W.; Wang, Y., Injectable pH-responsive hydrogel for combinatorial chemoimmunotherapy tailored to the tumor microenvironment. *Journal of Nanobiotechnology*, 20 (1), 2022, 372.

[28] Pal, V. K.; Jain, R.; Roy, S., Tuning the supramolecular structure and function of collagen mimetic ionic complementary peptides via electrostatic interactions. *Langmuir*, 36 (4), 2020, 1003–1013.

[29] Li, J.; Wang, Z.; Han, H.; Xu, Z.; Li, S.; Zhu, Y.; Chen, Y.; Ge, L.; Zhang, Y., Short and simple peptide-based pH-sensitive hydrogel for antitumor drug delivery. *Chinese Chemical Letters*, 33 (4), 2022, 1936–1940.

[30] Liu, B.; Li, J.; Zhang, Z.; Roland, J. D.; Lee, B. P., pH responsive antibacterial hydrogel utilizing catechol–boronate complexation chemistry. *Chemical Engineering Journal*, 441, 2022, 135808.

[31] Pal, V. K.; Roy, S., Bioactive peptide nano-assemblies with pH-triggered shape transformation for antibacterial therapy. *ACS Applied Nano Materials*, 5 (8), 2022, 12019–12034.

[32] Zou, R.; Wang, Q.; Wu, J.; Wu, J.; Schmuck, C.; Tian, H., Peptide self-assembly triggered by metal ions. *Chemical Society Reviews*, 44 (15), 2015, 5200–5219.

[33] Nakamura, T.; Takashima, Y.; Hashidzume, A.; Yamaguchi, H.; Harada, A., A metal–ion-responsive adhesive material via switching of molecular recognition properties. *Nature Communications*, 5 (1), 2014, 4622.

[34] Xie, T.; Ding, J.; Han, X.; Jia, H.; Yang, Y.; Liang, S.; Wang, W.; Liu, W.; Wang, W., Wound dressing change facilitated by spraying zinc ions. *Materials Horizons*, 7 (2), 2020, 605–614.

[35] Bercea, M., Bioinspired hydrogels as platforms for life-science applications: Challenges and opportunities. *Polymers*, 14 (12), 2022, 2365.

[36] Du, X.; Cui, H.; Zhao, Q.; Wang, J.; Chen, H.; Wang, Y., Inside-out 3D reversible ion-triggered shape-morphing hydrogels. *Research*, 2019, 2019, 6398296.

[37] Mohammed, M.; Chakravarthy, R. D.; Lin, H.-C., Influence of metal ion cross-linking on the nanostructures, stiffness, and biofunctions of bioactive peptide hydrogels. *Molecular Systems Design & Engineering*, 7 (10), 2022, 1336–1343.

[38] Pal, V. K.; Roy, S., Cooperative metal ion coordination to the short self-assembling peptide promotes hydrogelation and cellular proliferation. *Macromolecular Biosciences*, 22 (5), 2022, 2100462.

[39] Pang, Q.; Hu, H.; Zhang, H.; Qiao, B.; Ma, L., Temperature-responsive ionic conductive hydrogel for strain and temperature sensors. *ACS Applied Materials & Interfaces*, 14 (23), 2022, 26536–26547.

[40] Hardman, D.; George Thuruthel, T.; Iida, F., Self-healing ionic gelatin/glycerol hydrogels for strain sensing applications. *NPG Asia Materials*, 14 (1), 2022, 11.

[41] Wang, F.; Chen, C.; Wang, J.; Xu, Z.; Shi, F.; Chen, N., Facile preparation of PHEMA hydrogel induced via Tannic Acid-Ferric ions for wearable strain sensing. *Colloids and Surfaces A: Physicochemical and Engineering Aspects*, 658, 2023, 130591.

[42] Aluri, K. C.; Hossain, M. A.; Kanetkar, N.; Miller, B. C.; Dowgiallo, M. G.; Sivasankar, D.; Sullivan, M. R.; Manetsch, R.; Konry, T.; Ekenseair, A., Cyclic thiosulfinates as a novel class of disulfide cleavable cross-linkers for rapid hydrogel synthesis. *Bioconjugate Chemistry*, 32 (3), 2021, 584–594.

[43] Kilic Boz, R.; Aydin, D.; Kocak, S.; Golba, B.; Sanyal, R.; Sanyal, A., Redox-responsive hydrogels for tunable and "on-demand" release of biomacromolecules. *Bioconjugate Chemistry*, 33 (5), 2022, 839–847.

[44] Altinbasak, I.; Kocak, S.; Sanyal, R.; Sanyal, A., Fast-forming dissolvable redox-responsive hydrogels: Exploiting the orthogonality of thiol–maleimide and thiol–disulfide exchange chemistry. *Biomacromolecules*, 23 (9), 2022, 3525–3534.

[45] Jin, M.; Gläser, A.; Paez, J. I., Redox-triggerable firefly luciferin-bioinspired hydrogels as injectable and cell-encapsulating matrices. *Polymer Chemistry*, 13 (35), 2022, 5116–5126.

[46] Wojciechowski, J. P.; Martin, A. D.; Thordarson, P. J., Kinetically controlled lifetimes in redox-responsive transient supramolecular hydrogels. *Journal of the American Chemical Society*, 140 (8), 2018, 2869–2874.

[47] Konieczynska, M. D.; Villa-Camacho, J. C.; Ghobril, C.; Perez-Viloria, M.; Tevis, K. M.; Blessing, W. A.; Nazarian, A.; Rodriguez, E. K.; Grinstaff, M. W., On-demand dissolution of a dendritic hydrogel-based dressing for second-degree burn wounds through thiol-thioester exchange reaction. *Angewandte Chemie (International ed. in English)*, 55 (34), 2016, 9984–9987.

[48] Nakahata, M.; Takashima, Y.; Yamaguchi, H.; Harada, A., Redox-responsive self-healing materials formed from host–guest polymers. *Nature Communications*, 2 (1), 2011, 511.

[49] Yoshimura, I.; Miyahara, Y.; Kasagi, N.; Yamane, H.; Ojida, A.; Hamachi, I., Molecular recognition in a supramolecular hydrogel to afford a semi-wet sensor chip. *Journal of the American Chemical Society*, 126 (39), 2004, 12204–12205.

[50] Koshi, Y.; Nakata, E.; Yamane, H.; Hamachi, I., A fluorescent lectin array using supramolecular hydrogel for simple detection and pattern profiling for various glycoconjugates. *Journal of the American Chemical Society*, 128 (32), 2006, 10413–10422.

[51] Sawada, S.-i.; Sasaki, Y.; Nomura, Y.; Akiyoshi, K., Cyclodextrin-responsive nanogel as an artificial chaperone for horseradish peroxidase. *Colloid and Polymer Science*, 289 (5), 2011, 685–691.

8 Hydrogels with Electrical Properties

José García-Torres and Carlos Alemán
Universitat Politècnica de Catalunya-Barcelona Tech,
Barcelona, Spain

8.1 INTRODUCTION

Among hydrogels, which are three-dimensional (3D) network structures able to imbibe large amounts of water, conducting hydrogels (CHs) deserve consideration due to their charge (ionic and/or electronic) transport properties. CHs are excellent candidates for *the fabrication of flexible (bio)electronics* owing to their good conductivity, adjustable mechanical properties (*i.e.,* stretchability, compressibility, and elasticity), multiple stimuli-responsive properties, and biocompatibility [1,2]. Among others, these materials have become essential for manufacturing energy storage devices (batteries and supercapacitors for energy conversion) [3]; interfaces for biomedical platforms for biosensing (health diagnosis), drug delivery, and tissue regeneration (health therapies) [4]; actuators able to transform electrochemical energy into mechanical energy (soft robotics) [5]; touch panels (pressure sensors) [6]; and electrochromic displays [7].

The conductivity of CHs, which synergizes the advantageous features of hydrogels and (semi)conductors, originates from the ion or/and electron transport across the 3D cross-linked porous polymer networks [8]. This can be achieved through the chemical structure of the polymer chains, which contain ionized groups and the corresponding counter ions are transported through the water that fills the pores (ion CHs) or contain π-bonded electrons delocalized over the conjugated backbone structures (electronically CHs made of intrinsically conducting polymers; ICPs), or by incorporating conducting compounds (*e.g.,* metal nanoparticles (NPs), conducting polymers (CP) forming a semi-interpenetrating network, and graphene sheets or carbon nanotubes (CNT) as electron conductors) into insulating hydrogels to form conducting (nano)composites. This variability in the chemical composition is consistent with an enormous preparation versatility, a large number of synthetic routes being used to prepare CHs that exhibit facile processability, excellent conductivity, and high electrochemical activity.

This chapter will discuss recent achievements on the synthesis, properties, and potential applications of different types of CHs. For this purpose, CHs have been categorized in five types: 1) ion CHs prepared with polymers bearing charged functional groups, in which the measured conductivity comes from the electrolyte (counter ions) that swells them; 2) CHs obtained through the direct gelation of ICPs; 3) CHs obtained by entrapping conducting polymer chains in the hydrogel matrix (*i.e.,* semi-interpenetrated conducting hydrogels); 4) CHs made by combining the

  DOI: 10.1201/9781003340485-8

properties of inherently insulating hydrogels and metal NPs, which are well known to possess good electrical conductivity and are also easier to process; and 5) carbon-based CHs, the electrical, thermal, and mechanical properties of hydrogels are reinforced by adding carbon-based materials such as CNT, carbon black (CB), and graphene.

8.2 ION-CONDUCTING HYDROGELS

Water-rich ion-containing hydrogels have ionic conductivity, which makes them ideal systems as biointerfaces because in biological systems electric signals are mainly carried on by ions [9]. Ion CHs consist of cross-linked polyelectrolytes (fixed ions) with repeated negative charges (polyanions), positive charges (polycations), or both (zwitterionic gels), and mobile ions (Figure 8.1).

Although many neutral hydrogels are able to conduct ions when a salt is properly loaded into their aqueous environment (doped hydrogels), better conducting properties are usually obtained using the polyelectrolyte hydrogels bearing fixed charges. The ionic conductivity (σ) corresponds to the sum of the contributions of all of charge carriers (mobile ions), charge carrier density (n_i), and the mobility of such charged species (μ_i):

$$\sigma = \sum_i Z_i \cdot e \cdot n_i \cdot \mu_i \qquad (8.1)$$

where Z_i is the absolute value of the ion charge and e is the fundamental charge. Besides, the mobility (μ_i) of a hydrated ion depends on the temperature (T), the diffusion coefficient (D_i), and the charge of the ion (q_i):

$$\mu_i = \frac{D_i \cdot q_i}{k_b T} \qquad (8.2)$$

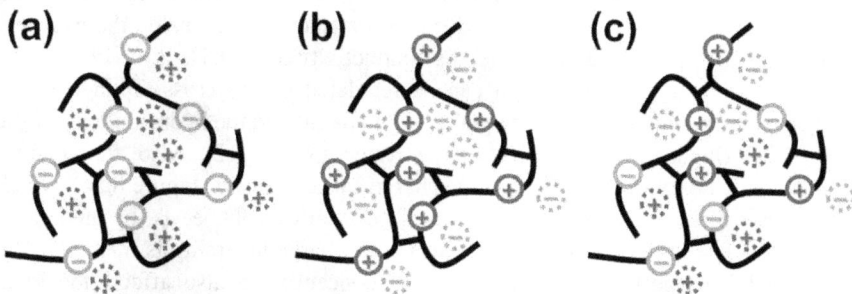

FIGURE 8.1 Schematic of the structure of ion CHs made of (a) polyanion, (b) polycation, and (c) zwitterionic polyelectrolyte, with a large number of fixed ions. Polyanion, polycation, and zwitterionic hydrogels can conduct cations, anions, and both, respectively. Charges in solid and dashed circles refer to fixed and mobile ions, respectively.

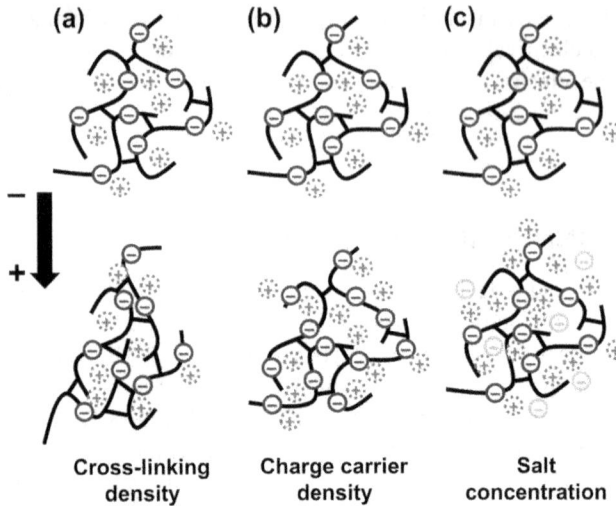

FIGURE 8.2 Schematic of the structure of ion CHs with different (a) cross-linking density, (b) charge carrier density, and (c) salt concentration.

where k_b is the Boltzmann constant. In porous 3D network structures, D_i is smaller than for the same ion in water (D_o). Therefore, D_i is adapted to the topography of the hydrogel [10]:

$$D_i = D_o \frac{\varepsilon}{\tau} \tag{8.3}$$

where ε is the porosity and τ is the tortuosity of the path followed by the ion in the hydrogel.

The conductivity of ion CHs increases with water uptake, as water molecules not directly involved in the hydration of fixed and mobile ions contribute to enhance ion transport by dynamics (mobility and diffusion) of mobile ions inside the hydrogel [11]. The water uptake and, therefore, the conductivity of ion CHs is affected by the hydrogel's porosity, which in turn can be regulated by the cross-linking density (higher cross-linking, lower water uptake; Figure 8.2a) and the hydrophilicity of the cross-linker (higher hydrophilicity, higher water uptake) [12]. Also, the conductivity of ion CHs largely depends on the charge carrier density (Figure 8.2b), which corresponds to the amount of fixed charges per unit of volume, as the amount of mobile ions increases with such parameters (*i.e.*, charge neutrality is maintained in the overall hydrogel) [13]. In addition, salt concentration also affects the ionic conductivity of these systems. As the concentration of salt increases, the conductivity of ion CHs increases since both mobile cations and anions are provided in the system (Figure 8.2c) [12].

Ion CHs have been extensively used in flexible (bio)electronics due to their high stretchability, transparency, and biocompatibility. For example, ion CHs have been successfully employed as effective organic electronic ion pumps [14] and ion sensors [15] instead of planar polymer-based membranes. Thus, 3D structured ion pumps made of CHs enable the transport and deliver of larger compounds upon an electric bias than ion-exchange membranes [14]. Also, ion CHs are being used as selective ion sensors in the fields of environmental monitoring and healthcare [15]. Other relevant applications of ion CHs are related with developments in prosthetics and soft robotics as, for example, stretchable artificial skin for hand-motion monitoring [16] and artificial muscles to transform electrical energy into mechanical energy [17]. Application of ion CHs in sensors, actuators, biomedicine, and soft electronics has been recently reviewed by Wang and coworkers [18].

8.3   ELECTRONICALLY CONDUCTING HYDROGELS

ICPs are a class of organic polymers (π-conjugated polymers) with electrical properties similar to metals and semiconductors, while retaining the most relevant properties of common polymers (*e.g.,* ease of synthesis and flexibility). Charge propagation in ICPs is mainly explained by the free movement of delocalized π-electrons along more or less extended parts of the conjugated systems, constructing an electrical pathway for mobile charge carriers under the applied electric potential [19,20]. The electrical conductivity of ICPs depends strongly on the incorporation of ionic dopants to further increase the degree of oxidation or reduction, which is usually named the degree of doping, by removing or donating electrons, respectively. Such ionic doping (*i.e.,* formation of polarons and bipolarons) implies a re-organization of the bonds and gives place to the creation of positively and negatively charged delocalized states (p-doping and n-doping, respectively) [21,22]. For charge compensation, anions or cations from the medium solution move into the ICP matrix during the p-doping or n-doping processes, respectively.

The conductivity of ICPs is frequently explained using the Schaefer–Siebert–Roth model, which takes into account the localization length with conjugation length. When the ICP is doped with p dopants or n dopants, breakage of a π-bond and charge carriers like polarons and bipolarons occur, and the formation of these charge mobilizers causes the formation of different conjugation lengths within the backbone [23]:

$$\sigma = \sigma_0 \exp[-(T/T_0)^{-\gamma}] \qquad (8.4)$$

where σ_0 and T_0 are parameters that depend on the localization length and γ is a function of the density of states at the Fermi level.

The electrical conductivity of ICPs is due to two important factors: the number of carriers (electrons or holes) and charge carrier mobility. Higher mobility will occur with more crystalline, better oriented, defect-free materials. Increasing the doping level will increase the density of charge carriers. The conductivity decreases with falling temperature, just like that of semiconductors. In addition, conductivity is supported by electron hopping between chains (inter-chain conduction) and defects. The intra- and inter-chain electron transport mechanisms have been the subject of studies in the last decade [22,24].

The co-presence of ionic and electronic conductivity at the polymer chains makes ICPs particularly attractive for bioelectronic applications. It should be noted that the penetration of ionic dopants into the bulk ICP matrix is significantly enhanced in water-rich hydrogels with respect to other formats (*e.g.,* films and fibers). The addition of this feature to the intrinsic electronic conductivity along the ICP backbone results in a significant volumetric capacitance since the accumulation of charged ions gives place to the formation of an electric double layer (EDL) at practically the molecular level (*i.e.,* in the whole volume of the sample), as is schematically illustrated in Figure 8.3a [25,26]. This represents an important difference with respect to metallic substrates, which exhibit high superficial capacitance (rather than volumetric) because the EDL is created at the surface of the sample (Figure 8.3b). On the other hand, the electronic conductivity of CP-based CHs is lower than that of dry CPs, which is attributed to the loss of percolation with swelling. Indeed, the electronic conductivity of CHs was found to decrease with increasing degree of swelling [27]. This feature is schematized in Figure 8.4, which sketches the progressive loss of percolation when dry ICPs undergo hydration, swelling, and gelation.

Among ICPs, polyaniline (PAni), polypyrrole (PPy), and polythiophene (PTh) are most frequently investigated. The most famous PTh derivative is poly(3,4-ethylenedioxythiophene) (PEDOT), which is immiscible in water when it is pure state but water soluble when doped with poly(styrene sulfonate) (PSS), facilitating the preparation of ICP-based CHs. In a recent work, Lu *et al.* [27] used PEDOT:PSS nanofibrils, which were connected by adding dimethyl sulfoxide (DMSO) and, subsequently, applying dry-annealing and re-hydration. Although PEDOT:PSS CHs showed outstanding electrochemical stability during the charge storage and injection processes in wet physiological environments, its mechanical stability of was a matter of some controversy and debate [28] their great electrical properties and flexibility CHs have encouraged widespread usages in a number of

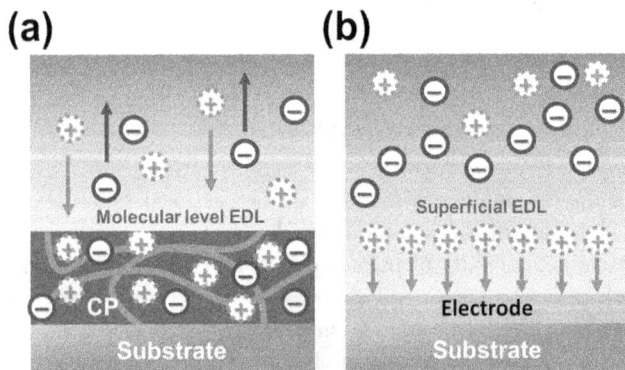

FIGURE 8.3 (a) ICP hydrogel whose capacitive properties rely on the formation of EDL at the molecular level and, therefore, capacitance is proportional to the hydrogel volume. (b) Metallic electrode whose capacitive properties rely on the superficial EDL and, therefore, capacitance is proportional to the area of the metal sheet.

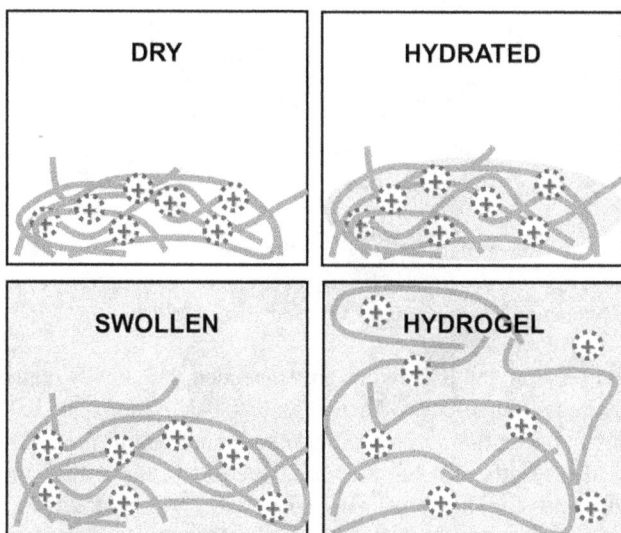

FIGURE 8.4 Sketch showing how dry ICPs loss percolation upon hydration, swelling, and gelation.

applications, such as neural implants, implantable sensors, prosthetic interfaces, and controlled drug delivery [29].

A different approach is the utilization of redox polymers for preparing CHs able to transport electrons through their hopping between neighboring redox sites [30]. These materials typically consist on polymers bearing organic or organometallic redox sites covalently attached, as for example poly(vinylferrocene). However, the electronic conductivity of redox polymers is orders of magnitude lower than that of doped ICPs. In general, electron-conducting redox hydrogels are used to electrically connect the redox centers of enzymes to electrodes, enabling their use whenever leaching of electron-shuttling diffusional redox mediators must be avoided, which is the case in subcutaneously implanted biosensors for diabetes management and in miniature, potentially implantable, glucose-O_2 biofuel cells.

8.4 SEMI-INTERPENETRATED CONDUCTING HYDROGELS

While interpenetrated polymer networks (IPNs) consist of a class of polymer mixture in which polymer chains are partially interlaced at the molecular scale but not covalently bonded to each other (*i.e.,* they cannot be separated unless chemical bonds are broken; Figure 8.5a) [31], semi-interpenetrated polymer networks (semi-IPNs) are found when the chains of the second polymer are only dispersed into the network defined by the first one (*i.e.,* without defining another interpenetrated network; Figure 8.5b) [31].

Semi-IPN CHs have been extensively prepared by incorporating ICPs into an already formed 3D hydrogel network. Two basic approaches have been mainly used for such a purpose. In the first one, which unfortunately leads to hydrogels with very

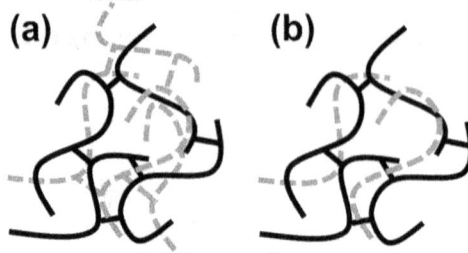

FIGURE 8.5 Sketch representing an (a) IPN and a (b) semi-IPN.

low conductivities [32], the ICP chains are embedded into the 3D matrix during the hydrogel polymerization process. In the second basic approach, ICP chains are grown by polymerizing the corresponding monomers within the hydrogel matrix. This strategy also results in some drawbacks, as for example the inhomogeneous electrical properties and heterogeneous structures, which affect the mechanical properties [32]. However, some improvements have been recently successfully developed to overcome the limitations of such basic approaches. Among them, three refinements deserved special consideration [33,34]. One such refinement was specifically designed to produce conducting semi-IPC hydrogels with good and homogeneous mechanical properties. This was achieved using templates to synthesize the hydrogel surrounding the ICP chains and the resulting semi-IPN CHs exhibited high chemical and mechanical stability, and outstanding electrical and electrochemical properties [33]. The second refinement was based on the utilization of CPs doped with polyelectrolytes (*i.e.*, charged polymer chains), as, for example, PEDOT:PSS, and the substitution of such a polyelectrolyte by a new polymeric dopant able to form hydrogels. This strategy has been successfully used by replacing the PSS of PEDOT:PSS by alginate chains, which is a thermodynamically favored process [28]. After cross-linking, the resulting semi-IPN CHs exhibited excellent conducting, piezoelectric, and electrochemical properties, as well as self-healing behavior [28,34].

The third improved approach, which also allowed the formation of semi-IPNs CHs with excellent mechanical properties and usually involved the utilization of two CPs (CP1 and CP2), is schematically depicted in Figure 8.6 [35]. Firstly, CP1 films were prepared by anodic polymerization on supporting electrodes (*e.g.*, steel electrodes). The resulting films were removed from the electrodes and, subsequently, processed into microparticles (MPs) by mechanical stirring (Figure 8.6a). Then, CP1 MPs were introduced in a viscous aqueous solution (paste) containing the polymer chains used to prepare the hydrogel, as for example poly-γ-glutamic acid, cellulose, and alginic acid (Figure 8.6b) [35,36]. The cross-linker was applied to transform the CP1 MPs-loaded paste into a hydrogel through chemical or physical processes (Figure 8.6c). After this, the conducting performance of the hydrogel was enhanced by incorporating the second CP (CP2) inside the hydrogel through *in situ* anodic polymerization. To that end, the CP1 MPs-loaded hydrogel was attached to an electrode and immersed in a CP2 monomer aqueous solution (Figure 8.6d), which was kept under stirring to ensure the penetration of the

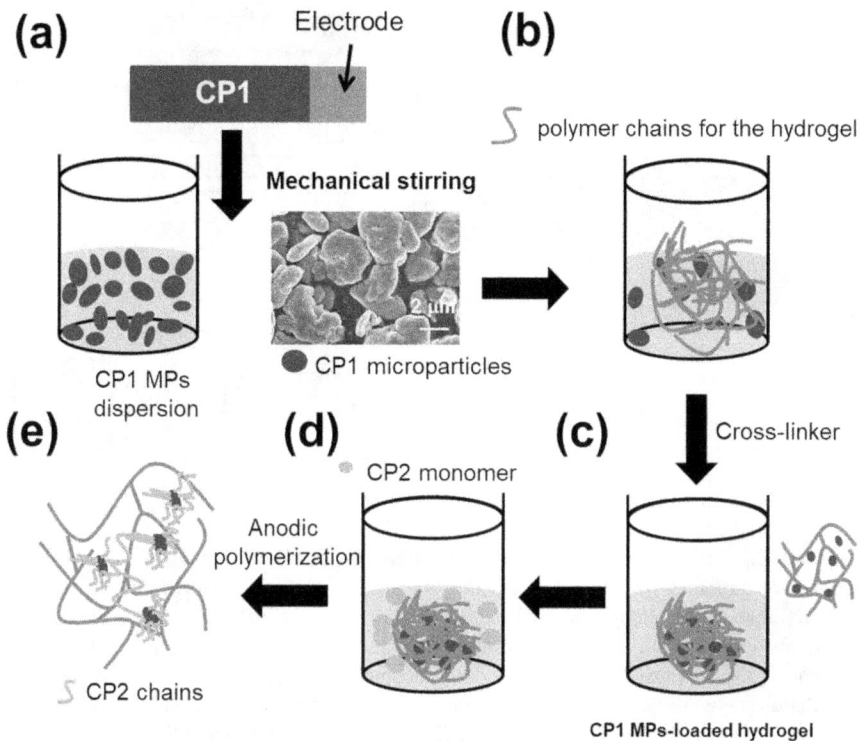

FIGURE 8.6 Process used to prepare semi-IPN CHs using two CPs (CP1 and CP2): (a) CP1 MPs are prepared and dispersed; (b) the polymer used to prepare the hydrogel is added to create an MPs-containing paste; (c) the addition of the cross-linker results in the formation of the CP1 MPs-loaded hydrogel; (d) the latter hydrogel is introduced in an aqueous solution containing the CP2 monomer; and (e) after an anodic polymerization process in which MPs act as polymerization nuclei, the CP1 MPs percolate because of the formation of CP2 connecting them.

monomer into the matrix. This step was crucial to ensure the access of the CP2 monomer to the CP1 MPs, which act as polymerization nuclei. The anodic polymerization process allowed the connection among CP1 MPs through the polymerized CP2 (*i.e.*, the percolation of the CPs), as shown in Figure 8.6e. Although CP1 and CP2 can be same, they are usually different since CP1 films can be prepared using an organic environment, which is easier, while the solvent used in the reaction medium of CP2 is water.

8.5   METALLIC NANOMATERIALS

Another widely explored strategy to confer electrical conductivity to hydrogels is the incorporation of conductive nanomaterials, including metallic- (*e.g.,* Au, Ag, Cu) and carbon-based nanomaterials (*e.g.,* graphene, CNT, CB) into the polymeric network [37–42]. The conductivity of such nanocomposite hydrogels arises due to

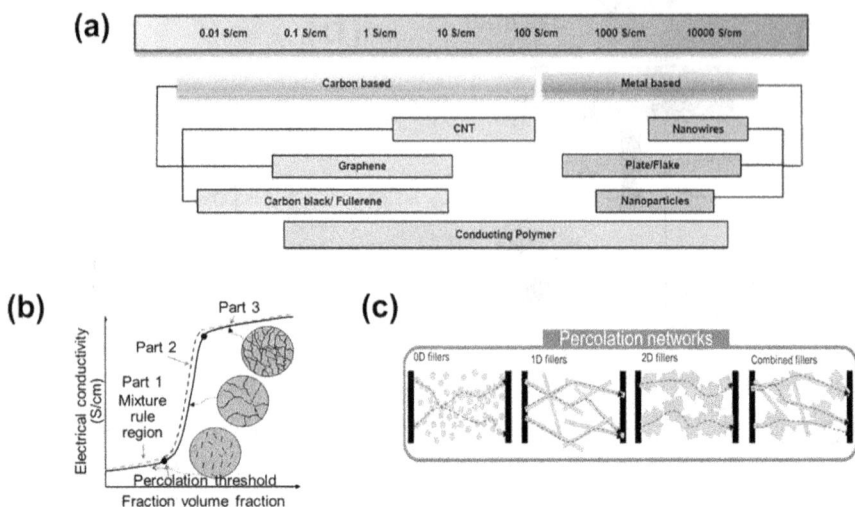

FIGURE 8.7 (a) Chart showing the electrical conductivity range for the different conductive nanomaterials based on metals, carbons, and CPs. Adapted with permission from reference [42], Copyright (2019), The Royal Society of Chemistry. (b) Variation of electrical conductivity of composites vs filler volume fraction. Adapted with permission from reference [43], Copyright (2014), The Electrochemical Society. (c) Scheme showing the percolation network mechanism of different aspect ratio nanomaterials. Adapted with permission from reference [41], Copyright (2021), American Chemical Society.

the formation of electronic conductive paths of the nanomaterials within the matrix. Thus, many factors play a role on the formation of such paths and therefore on the composite electrical properties like the intrinsic conductivity of the materials used, the content of the nanomaterials, their size and shape, and their dispersion within the matrix. Figure 8.7a shows a flow chart showing how the electrical conductivity of nanomaterials varies depending on the material employed to synthesize them (*e.g.,* metals, carbons, CPs) or their shape (*e.g.,* NPs, nanowires (NWs), nanoplatelets). Another critical parameter influencing conductivity is the nanomaterial's content inside the hydrogel, as it must surpass a critical value, the so-called percolation threshold, to create conducting pathways. Beyond this threshold, conductivity starts increasing rapidly by several orders of magnitude until reaching a plateau with a conductivity value approaching that of the bulk material as described by the percolation theory (Figure 8.7b) [43].

Among the conductive nanomaterials, noble metal NPs (0D) have received more attention than any other shapes to fabricate conductive hydrogels for three main reasons. First, 0D NPs show very interesting properties, like high surface-to-volume ratio, making them ideal for highly functional and stable nanocomposite hydrogels without structurally disrupting a polymer network. Second, NPs of different metals and alloys can be synthesized following simple procedures, leading to stable NP's dispersions. And third, the high electrical conductivity and oxidation resistance of noble metals (*e.g.,* Au and Ag). However, one of the main problems is that the percolating threshold occurs at high loadings (> 15%), impacting other properties' performance (*e.g.,* mechanical properties) [37–39]. At such high contents, phase separation between the matrix and NPs causes weak mechanical toughness, poor fatigue resistance, or low stretchability, leading to nanocomposite hydrogels with limited performance— hence, their applicability in fields like wearable and flexible bioelectronics is limited. One way to keep the percolation threshold low is to use high-aspect ratio nanomaterials, like NWs, nanorods (NRs), or nanofibers (NFs) [44–46]. 1D nanomaterials, with one dimension higher than the other two, allow for creating larger and more effective percolated networks for carrier transport and therefore, nanocomposite hydrogels with higher electrical conductivity compared to 0D or 2D nanomaterials (Figure 8.7c). The reason can be explained as follows. On the one hand, NWs with a given diameter and different lengths can lead to more effective percolation networks as the NW length increases. The reason is that a lower number of junctions are formed among NWs; therefore, lower resistance. On the other hand, when composites with the same volume fraction present a larger number of thinner NWs than thicker ones (with the same length), the number of percolation pathways is also higher and resistance decreases, as it is inversely proportional to thickness.

2D nanomaterials have also been shown to have great potential to create conducting pathways inside hydrogels due to their high-aspect ratio compared to 0D nanostructures [47,48]. Thus, Majidi's group reported the fabrication of a polyacrylamide-alginate hydrogel containing Ag flakes, which showed high conductivity ($\sim$350 S cm^{-1}) and, simultaneously, maintained softness (Young's modulus < 10 kPa) and deformability. Thus, the high-aspect ratio of the flakes allowed a low percolation threshold ($\sim$6 vol%), which, in turn, did not affect the mechanical properties. They also observed that the nanocomposite hydrogel electrical properties were not altered under mechanical deformations since the contact points between the nanomaterials were successfully sustained [48]. Another way to keep percolation loadings low is to mix with nanomaterials of different aspect ratios (*e.g.,* 0D-1D, 0D-2D, 1D-2D) (Figure 8.7c). Such combinations favor the formation of bridges between the different nanomaterials and thus, enhance the conductivity of the nanocomposite at low loadings; all, at no expense of sacrificing mechanical properties. For example, it was observed that 0D NPs can strengthen the contact in a 2D nanoflake network, improving electrical performance of the nanocomposites [49].

Two important points to consider during nanocomposite hydrogel synthesis are as follows: (i) aggregation due to the physicochemical interactions (*e.g.,* van der Waals forces and electrostatic forces) between the nanomaterials; and (ii) matrix-nanomaterial interactions. Both have a significant impact on the electrical performance of the composite [41]. Thus, a homogenous distribution of the conductive nanomaterials inside the hydrogel matrix is strongly necessary since nanomaterials aggregation affects the quality of the electrical junction contact between them, hindering the electrical conductivity. One effective way to control aggregation is through the chemical modification of the nanomaterials' surface to regulate their distribution within the hydrogel and maximize the conducting properties [41]. Surface chemistry is again essential to establish a good interaction between hydrogels and fillers to also avoid aggregation and favor dispersion.

The nanocomposite synthesis procedure also has an effect on the dispersion and interaction of the nanomaterials within the matrix. Thus, different methodologies have been followed to study blending, *in situ* precipitation, and covalent bonding among the most relevant [50]. Blending is the most widely employed approach due to its simplicity and the high number of different nanomaterials that can be incorporated into the hydrogel. This approach consists of mixing the hydrogel precursors and the colloidal nanomaterials suspension followed by hydrogel cross-linking to entrap the nanostructures. For example, a chitosan hydrogel containing Au NPs was successfully developed for cardiac tissue regeneration (Figure 8.8a) [51]. The Au NPs precursors (HAuCl$_4$, sodium citrate) were added to a chitosan solution, where the reduction of the gold salt took place. After that, chitosan was cross-linked with β-glycerophosphate. The electrical conductivity of the Au NPs-containing hydrogel was higher than that of the bare hydrogel. The conductive composite supported differentiation of the mesenchymal stem cells (MSC) into the cardiac lineage, as the increase of the cardiac markers revealed [51]. Nevertheless, such a strategy does not allow ensuring a proper dispersion of the nanomaterials inside the hydrogel network. Thus, *in situ* precipitation has been used as an alternative. In such an approach, nanomaterials (*e.g.,* metallic salts) and hydrogel precursors are homogeneously mixed. Then, the metallic ions are reduced into the metallic nanomaterials and, finally, the hydrogel formation favors a major nanostructure-hydrogel interaction and therefore minimizes aggregation. For example, Dolya *et al.* [52] did not observe aggregation when, after exhaustive mixing of hydrogel precursors (acrylamide, N, N′-methylene bisacrylamide and ammonium persulfate) and Au NPs precursors (HAuCl$_4$ and poly(ethyleneimine)), the hybrid hydrogel was formed by heating the solution at 80°C for 30 min (Figure 8.8b).

The third approach to improve even more the nanomaterial-hydrogel interaction is the covalent bonding that, in fact, is less used due to the more complex procedures. Here, nanomaterials are functionalized with different (bio)molecules to form a covalent bond with polymeric chains aiming to stabilize the nanomaterials inside the hydrogel network but also to enhance hydrogel chemical and biological properties. Skardal and collaborators [53] incorporated thiol-functionalized and non-functionalized Au NPs inside hyaluronic acid/gelatin hydrogels (Figure 8.8c). Due to the covalent interaction between the biopolymers and the thiol-functionalized NPs, they observed higher stiffness compared to the non-functionalized NPs. All those hydrogels have been successfully employed for applications like (bio)sensors, drug delivery systems, energy storage devices, or tissue engineering.

8.6 CARBON NANOMATERIALS

Carbon nanomaterials have been at the center of research for preparing electrically conductive hydrogels, despite their electrical conductivities are lower than those of metallic nanomaterials. However, carbon nanomaterials show some interesting advantages like relatively low production cost, high mechanical properties, or chemical and thermal stability. Among them, CB, CNT, and graphene have been the most widely investigated [41,50]. CB, the oldest carbon allotrope, has been frequently employed as a conductive filler, despite their lower conductivity compared to other carbon forms and the high percolation threshold. Its main

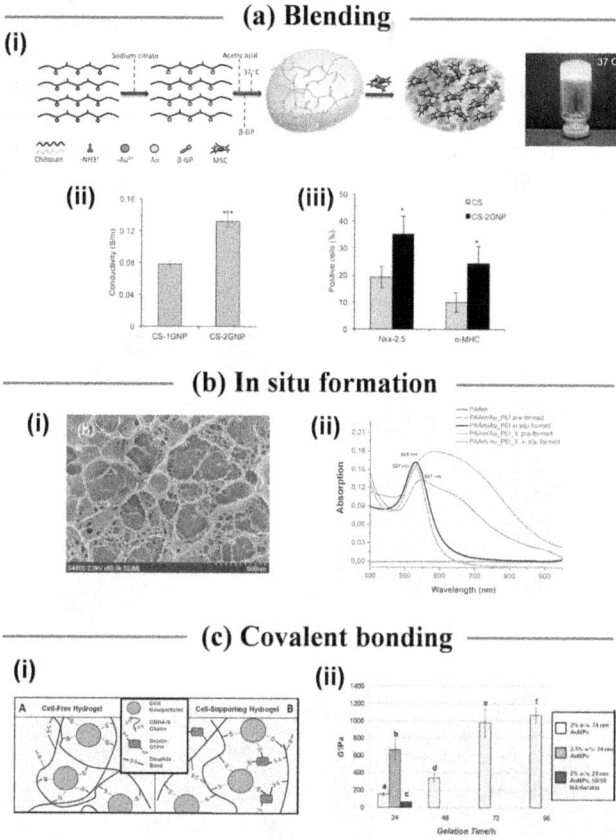

FIGURE 8.8 Different methodologies to incorporate metallic nanomaterials into CHs: (a) Blending: (i) Scheme showing the preparation of the chitosan-Au NPs hydrogel and further seeding. Photograph of the hydrogel. (ii) Variation of the electrical conductivity of the hydrogel with the content of Au NPs. (iii) Variation of cardiac markers (Nkx-2.5, α-MHC) content for chitosan and chitosan-Au NPs hydrogels. Adapted with permission from reference [51], Copyright (2016), Elsevier. (b) *In situ* formation: (i) SEM image and (ii) absorption spectra of the different PAM-Au NPs hydrogels. Adapted with permission from reference [52], Copyright (2013), Wiley-VCH Verlag GmbH. (c) Covalent bonding: (i) Thiol-functionalized NPs inside HA and HA/gelatin hydrogels and (ii) variation of the hydrogel stiffness with the concentration of HA and cross-linking time. Adapted with permission from reference [53], Copyright (2010), Wiley-VCH Verlag GmbH.

advantages are the low cost and the low density of the composite hydrogels. For example, CB has been employed for energy storage (*e.g.,* supercapacitors) and sensing applications [54,55].

Recently, CNT and graphene have gained interest owing to their better electrical conductivity and mechanical properties, and the reasons why they have been widely employed to fabricate composites materials for soft and flexible electronics (*e.g.,* sensors, actuators, and supercapacitors) and other biomedical applications like regenerative medicine [56–59]. CNT and graphene are more effective in creating conductive pathways at lower contents than 0D CB due to, again, the high-aspect ratio. For example, Xiao *et al.* fabricated a poly(vinyl alcohol) (PVA)/poly(ethylene glycol) (PEG)/graphene hydrogel by blending method (Figure 8.9a) [57]. First, PVA and PEG were dissolved in water at 90°C, followed by cooling the mixture at room temperature. Second, an aqueous graphene dispersion was added to the previous PVA/PEG solution and mixed until getting a homogeneous dispersion.

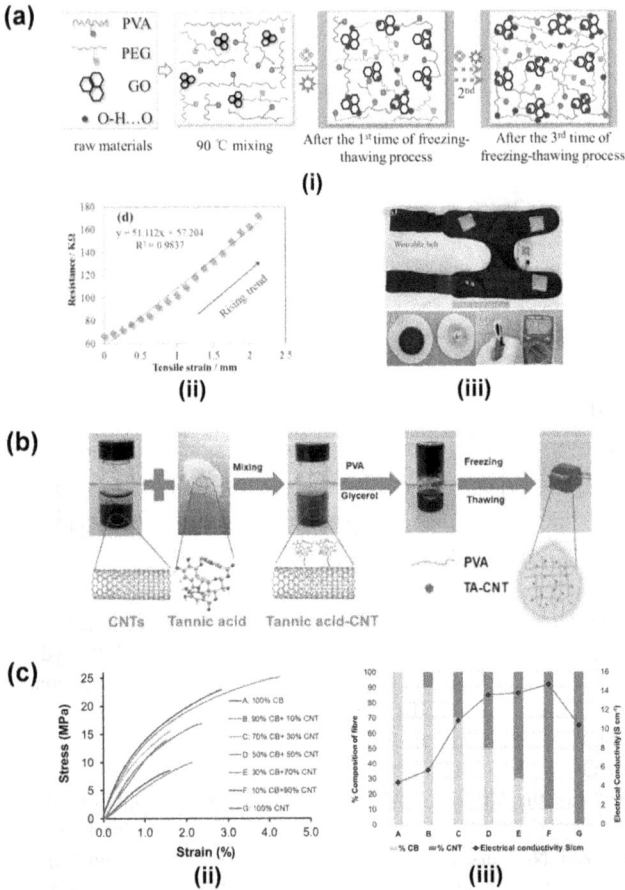

FIGURE 8.9 (a) (i) Scheme showing the preparation of the PVA/PEG/graphene hydrogel. (ii) Variation of the electrical resistance with the applied tensile strain. (iii) Photographs of the electrodes fabricated with the hydrogel, tested resistance of a pair of electrodes and self-fabricated belt with the electrodes and integrated circuits for electrocardiogram recording. Adapted with permission from reference [57], Copyright (2017), mdpi. (b) Scheme showing the fabrication of the PVA hydrogel containing TNA-functionalized CNT. Adapted with permission from reference [59], Copyright (2020), The Royal Society of Chemistry. (c) Variation of the (i) mechanical and (ii) electrical properties of chitosan hydrogels containing variable amounts of CB and CNT. Adapted with permission from reference [54], Copyright (2018), Elsevier.

Finally, hydrogel cross-linking was obtained by the freezing-thawing method with excellent conductivity and successfully employed as an electrode for electrocardio-gram recording. J. Garcia-Torres *et al.* [54] successfully prepared CNT-chitosan hydrogels as electrodes for supercapacitor applications, showing better response than the same hydrogels containing CB. The Ali Khademhosseini group incorpo-rated graphene into a methacryloyl-substituted recombinant human tropoelastin hydrogel to confer electrical conductivity for cardiac tissue regeneration [58]. Those authors observed that the good electrical conductivity of the resulting hydrogel improved the function of cardiomyocytes seeded on the hydrogel, as a higher expression of cardiac markers (*e.g.,* connexion-43, troponin I) revealed.

However, one of the main challenges of carbon-based hydrogels is the preparation of stable dispersion in solution and/or homogeneous mixtures within the hydrogel matrix. Due to strong non-covalent interactions (*e.g.,* hydrogen bonding, π-π stacking, and electrostatic interactions), carbon nanomaterials form aggregates or bundles. Thus, one viable strategy is the use of surfactants (*e.g.* sodium dodecylbenzensulfo-nate) to avoid such interactions and disperse them [41,54,55]. However, the content of the surfactants must be minimized as they hinder the electrical conductivity of the composite hydrogel due to their insulating nature. Alternatively, the use of oxidized CNT or reduced graphene oxide, which are similar to CNT or graphene but with oxygen containing polar functional groups (*e.g.,* hydroxyl groups), has also been common to improve dispersability [59,60]. Thus, He and coworkers [59] were able to disperse CNTs functionalized with tannic acid (TA) for a long time compared to bare CNT (Figure 8.9b). These TA-functionalized CNTs were successfully dispersed into a PVA hydrogel and it turned into a higher electrical conductivity (from 0.189 to 5.13 S m^{-1}) and improved mechanical properties.

On the other hand, the use of different aspect ratio nanocarbons, as in the case of metallic nanomaterials, has also been proved to improve dispersability. For example, Chen and coworkers [61] successfully dispersed CNT into polyacrylamide (PAM) hydrogels by mixing carbon nanofibers (CNFs) and CNTs in an acrylamide (AM) solution, followed by the free-radical polymerization using N, N′-methylene bisacrylamide and potassium peroxydisulfate as the cross-linker and initiator, respectively. The authors observed that the incorporation of 1 wt.% CNFs doubled the electrical conductivity of PAM/CNT hydrogels (from 0.041 to 0.085 S cm^{-1}), due to the dispersion effect of CNFs [61]. Also, Garcia-Torres *et al.* [54] mixed CB and CNT into a chitosan hydrogel, where CB was an effective dispersant for CNTs by acting as spacers and creating gaps between them (Figure 8.9c). Moreover, CB also acted as electrical contact between the CNTs. Overall, electrical conductivity and mechanical properties were improved compared to either the CB- or CNT-based hydrogels. Moreover, the authors also observed that the CB/CNT hydrogel electrodes had a higher capacitance due to the presence of gaps, meaning a higher surface area.

8.7 OTHER CONDUCTING NANOMATERIALS: MXENES AND CONDUCTING POLYMERS

Other nanomaterials based on MXenes and CP have been also successfully employed to fabricate flexible and/or stretchable electronic devices. MXenes are two-dimensional transition metal carbides, nitrides, or carbonitrides that have emerged in the field of bioelectronics owing to their high electrical conductivity, excellent electrochemical and optoelectronic properties, large specific surface area, high-aspect ratio, and easy surface tuneability. Those properties make them useful for the healthcare system, flexible electronics, and energy devices [62–64].

The improved coupling of MXenes with other materials at the nanoscale is expected to witness tremendous progress in flexible electronics [65]. MXenes show several interesting advantages to avoid aggregation within the hydrogel matrix. On one hand, they are electrically charged, originating electrostatic repulsion among individual nanosheets. Furthermore, their hydrophilicity is suitable for establishing a large number of interactions with the hydrogel. Also, they are easily functionalized with, for example, anticancer molecules. Finally, the nanosheets act as entanglement points for the polymeric chains to be densely packed. Overall, MXene-based hydrogels show enhanced electrical and mechanical properties compared to the bare electrodes, making them excellent candidates for supercapacitors, sensors, drug delivery systems, or catalytic systems [66,67]. For example, Zhang et al. [68] developed a hydrogel composed of PVA and $Ti_3C_2T_x$, showing outstanding sensitivity towards tensile strains (i.e., it was ten times higher than that of the bare hydrogel), along with self-healing, stretchability (~3400%), conformability, and adhesiveness to different surfaces. However, before being a reality, some issues related to a precise control of the MXene synthesis and the scalability of those processes, or the metastable character of the MXene-based hydrogels, as they are kinetically trapped materials that can undergo a nanoscale structuring during time, should be solved [67].

CPs have also been incorporated into hydrogels in the form of nanomaterials (e.g., NP, NW) as they are miscible with hydrogels able to confer low electrical impedance and stability, reducing the mechanical mismatch compared to inorganic nanomaterials. For example, Gan et al. [69] developed an electrically conductive PAM/chitosan by the in situ incorporation of PPy NWs. The bare hydrogel was immersed in a Py solution followed by the addition of the oxidizing agent (e.g., ferric chloride). The presence of the PPy NWs improved the mechanical and electrical properties and they were successfully used for skin regeneration. The electroactivity of the hydrogels allowed the closing of the wound with new skin tissue in fewer days than with the bare hydrogel [69].

8.8 CONCLUSIONS

In this chapter, the authors have shown the origin of the electrical properties of CHs and have also reviewed recent developments in CHs from the perspective of materials, properties, and applications. CHs with ionic conductivity transmit electrical signals via ion transport within the porous network, while the design of CHs with electronic conductivity is based on incorporating electron-conductive components such as metal NPs, carbon-based nanomaterials, and ICPs into a flexible polymer network. CHs can be modulated with a wide range of conductivity thanks to the vast range of (nano)materials that can be employed as well as the versatility in the synthesis procedures. Thus, CHs are extensively explored as flexible supercapacitors, electrodes for solar cells, biosensors for human health monitoring, and implantable platforms for cell regeneration and drug delivery. We can envision that CHs will facilitate the design of next-generation electronic systems requiring 3D hierarchical nanostructured morphological control.

REFERENCES

[1] H. Dechiraju, M. P. Jia, L. Luo, M. Rolandi. On-conducting hydrogels and their applications in bioelectronics. *Adv. Sustain. Syst.* 2021, *6*, 2100173.

[2] X. P. Hao, C. Y. Li, C. W. Zhang, M. Du, Z. M. Ying, Q. Zheng, Z. L. Wu. Self-shaping soft electronics based on patterned hydrogel with stencil-printed liquid metal. *Adv. Funct. Mater.* 2021, *31*, 2105481.

[3] J. Y. Nan, G. T. Zhang, T. Y. Zhu, Z. K. Wang, L. J. Wang, H. S. Wang, F. X. Chu, C. P. Wang, C. B. Tang. A highly elastic and fatigue-resistant natural protein-reinforced hydrogel electrolyte for reversible-compressible quasi-solid-state super-capacitors. *Adv. Sci.* 2020, *7*, 2000587.

[4] S. G. Alamdari, A. Alibakhshi, M. de la Guardia; R. Baradaran, R. Mohammadzadeh, M. Amini, P. Kesharwani, A. Mokhtardadeh, F. Oroojalian, A. Sahebkar. Conductive and semiconductive nanocomposite-based hydrogels for cardiac tissue engineering. *Adv. Healthc. Mater.* 2022,*11(18)*, 2200526.

[5] M. L. Guo, X. Yang, J. Yan, Z. J. An, L. Wang, Y. P. Wu, C. X. Zhao, D. Xiang, H. Li, Z. Y. Li, H. W. Zhou. Anti-freezing, conductive and shape memory ionic glycerol-hydrogels with synchronous sensing and actuating properties for soft robotics. *J. Mater. Chem. A* 2022,*10*, 16095–16105.

[6] J. Yu, M. Wang, C. Dang, C. Z. Zhang, X. Feng, G. X. Chen, Z. Y. Huang, H. S. Qi, H. C. Liu, J. Kang. Highly stretchable, transparent and conductive double-network ionic hydrogels for strain and pressure sensors with ultrahigh sensitivity. *J. Mater. Chem. C* 2021, *9*, 3635–3641.

[7] F. Y. Ding, Y. Zou, S. P. Wu, X. B. Zou. Self-healing and tough hydrogels with conductive properties prepared through an interpenetrating polymer network strategy. *Polymer* 2020, *206*, 122907.

[8] X. C. Wang, Z. X. Bai, M. H. Zheng, O. Y. Yue, M. D. Hou, B. Q. Cui, R. R. Su, C. Wei, X. H. Liu. Engineered gelatin-based conductive hydrogels for flexible wearable electronic devices: Fundamentals and recent advances. *J. Sci. Adv. Mater. Dev.* 2022, *7*, 100451.

[9] T. Someya, Z. Bao, G. G. Malliaras. The rise of plastic bioelectronics. *Nature* 2016, *540(7633)*, 379.

[10] W. Barros. Solvent self-diffusion dependence on the swelling degree of a hydrogel. *Phys. Rev. E* 2019, *99(5–1)*, 052501.

[11] S. De, C. Cramer, M. Schönhoff. Humidity dependence of the ionic conductivity of polyelectrolyte complexes. *Macromolecules* 2011, *44*, 8936–8943.

[12] C.-J. Lee, H. Wu, Y. Hu, M. Young, H. Wang, D. Lynch, F. Xu, H. Cong, G. Cheng. Ionic conductivity of polyelectrolyte hydrogels. *ACS Appl. Mater. Interfaces* 2018, *10*, 5845.

[13] G. Ruano, J. I. Iribarren, M. M. Pérez-Madrigal, J. Torras, C. Alemán. Electrical and capacitive response of hydrogel solid-like electrolytes for supercapacitors. *Polymers* 2021, *13*, 1337.

[14] M. Seitanidou, R. Blomgran, G. Pushpamithran, M. Berggren, D. T. Simon. Modulating inflammation in monocytes using capillary fiber organic electronic ion pumps. *Adv. Healthcare Mater.* 2019, *8*, 1900813.

[15] M. Parrilla, M. Cuartero, G. A. Crespo. Wearable potentiometric ion sensors. *TrAC Trends Anal. Chem.* 2019, *110*, 303.

[16] G. Gu, H. Xu, S. Peng, L. Li, S. Chen, T. Lu, X. Guo. Integrated soft ionotronic skin with stretchable and transparent hydrogel–elastomer ionic sensors for hand-motion monitoring. *Soft Rob.* 2019, *6*, 368.

[17] C. Keplinger, J.-Y. Sun, C. C. Foo, P. Rothemund, G. M. Whitesides, Z. Suo. Stretchable, transparent, ionic conductors. *Science* 2013, *341*, 984.

[18] K. Liu, S. Wei, L. Song, H. Liu, T. Wang. Conductive hydrogels—A novel material: Recent advances and future perspectives. *J. Agricult. Food Chem.* 2020, *68* (28), 7269.

[19] S. Brazovskii, N. Kirova. Physical theory of excitons in conducting polymers. *Chem. Soc. Rev.* 2010, *39*, 2453.

[20] U. Riaz, N. Signh, S. Banoo. Theoretical studies of conducting polymers: A mini review. *New J. Chem.* 2022, *46*, 4954.

[21] S. T. Keene, V. Guiskine, M. Berggren, G. G. Malliaras, K. Tybrandt, I. Zozoulenko. Exploiting mixed conducting polymers in organic and bioelectronic devices. *Phys. Chem. Chem. Phys.* 2022, *24*, 19144.

[22] J. Torras, J. Casanovas, C. Alemán. Reviewing extrapolation procedures of the electronic properties on the π-conjugated polymer limit. *J. Phys. Chem. A* 2012, *116*, 7571.

[23] P. Chandrasekhar. Conducting polymers, Fundamentals and applications, 1999, 143–172. ISBN: 978-1-4615-5245-1.

[24] S.-J. Sun, M. Menšík, P. Toman, C.-H. Chung, C. Ganzorig, J. Pfleger. Gate voltage impact on charge mobility in end-on stacked conjugated oligomers. *Phys. Chem. Chem. Phys.* 2020, *22*, 8096.

[25] A. V. Volkov, K. Wijeratne, E. Mitraka, U. Ail, D. Zhao, K. Tybrandt, J. W. Andreasen, M. Berggren, X. Crispin, I. V. Zozoulenko. Understanding the capacitance of PEDOT:PSS. *Adv. Funct. Mater.* 2017, *27*, 1700329.

[26] J. Rivnay, P. Leleux, M. Ferro, M. Sessolo, A. Williamson, D. A. Koutsouras, D. Khodagholy, M. Ramuz, X. Strakosas, R. M. Owens. High-performance transistors for bioelectronics through tuning of channel thickness. *Sci. Adv.* 2015, *1*, e1400251.

[27] B. Lu, H. Yuk, S. Lin, N. Jian, K. Qu, J. Xu, X. Zhao. Pure PEDOT:PSS hydrogels. *Nat. Commun.* 2019, *10*, 1043.

[28] I. Babeli, G. Ruano, J. Casanovas, M. P. Ginebra, J. García-Torres, C. Aleman. Conductive, self-healable and reusable poly(3,4-ethylenedioxythiophene)-based hydrogels for highly sensitive pressure arrays. *J. Mater. Chem. C* 2020, *8*, 8654.

[29] H. Shi, Z. X. Dai, X. Sheng, D. Xia, P. Shao, L. Yang, X. Luo. Conducting polymer hydrogels as a sustainable platform for advanced energy, biomedical and environmental applications. *Sci. Total Environ.* 2021, *786*, 147430.

[30] N. Casado, G. Hernandez, H. Sardon, D. Mecerreyes. Current trends in redox polymers for energy and medicine. *Prog. Polym. Sci.* 2016, *52*, 107.

[31] A. D. Jenkins, P. Kratochvíl, R. F. T. Stepto, U. W. Suter. Glossary of basic terms in polymer science (IUPAC recommendations). *Pure Appl. Chem.* 1996, *68*, 2287.

[32] S. P. Ansari, A. Anis. Conducting polymer hydrogels. In *Polymeric Gels*; K. Pal, I. B. T. Banerjee, Eds.; Woodhead Publishing, 2018; pp. 467–486. DOI: 10.1016/B978-0-08-102179-8.00018-1.

[33] D. Myung, D. Waters, M. Wiseman, P.-E. Duhamel, J. Noolandi, C. N. Ta, C. W. Frank. Progress in the development of interpenetrating polymer network hydrogels. *Polym. Adv. Technol.* 2008, *19*, 647.

[34] A. Puiggalí-Jou, E. Cazorla, G. Ruano, I. Babeli, M.-P. Ginebra, J. García-Torres, C. Alemán. Electroresponsive alginate-based hydrogels for controlled release of hydrophobic drugs. *ACS Biomater. Sci. Eng.* 2020 *6* (11), 6228.

[35] M. C. G. Saborío, S. Lanzalaco, G. Fabregat, J. Puiggalí, F. Estrany, C. Alemán. Flexible electrodes for supercapacitors based on the supramolecular assembly of biohydrogel and conducting polymer. *J. Phys. Chem. C* 2018, *122*(2), 1078.

[36] M. G. Saborío, P. Svelic, J. Casanovas, G. Ruano, M. M. Pérez-Madrigal, L. Franco, J. Torras, F. Estrany, C. Alemán. Hydrogels for flexible and compressible free standing cellulose supercapacitors. *Europ. Polym. J.* 2019, *118*, 347.

[37] R. Ma, B. Kang, S. Cho, M. Choi, S. Baik. Extraordinarily high conductivity of stretchable fibers of polyurethane and silver nanoflowers. *ACS Nano* 2015, *9*, 10876.

[38] I. You, B. Kim, J. Park, K. Koh, S. Shin, S. Jung, U. Jeong. Stretchable E-skin apexcardiogram sensor. *Adv. Mater.* 2016, *28*, 6359.

[39] F. Curry, A. M. Chrysler, T. Tasnim, J. E. Shea, J. Agarwal, C. M. Furse, H. Zhang. Biostable conductive nanocomposite for implantable subdermal antenna. *APL Mater.* 2020, *8*, 101112.

[40] I. Babeli, A. Puiggalí, J. J. Roa, M. P. Ginebra, J. Garcia-Torres, C. Aleman. Hybrid conducting alginate-based hydrogel for hydrogen peroxide detection from enzymatic oxidation of lactate. *Int. J. Biol. Macromolec.* 2021, *193*, 1237.

[41] K. W. Cho, S.-H. Sunwoo, Y. J. Hong, J. H. Koo, J. H. Kim, S. Baik, T. Hyeon, D.-H. Kim. Soft bioelectronics based on nanomaterials. *Chem. Rev.* 2022,*122*(5), 5068.

[42] S. Choi, S. I. Han, D. Kim, T. Hyeon, D.-H. Kim. High-performance stretchable conductive nanocomposites: Materials, processes, and device applications. *Chem. Soc. Rev.* 2019, *48*, 1566.

[43] R. Taherian. Development of an equation to model electrical conductivity of polymer-based carbon nanocomposites. *ECS J. Solid State Sci. Technol.* 2014, *3*, M26.

[44] T. Gurunathan, C. R. K. Rao, R. Narayan, K. V. S. N. Raju. Polyurethane conductive blends and composites: Synthesis and applications perspective. *J. Mater. Sci.* 2013, *48*, 67.

[45] D. Untereker, S. Lyu, J. Schley, G. Martinez, L. Lohstreter. Maximum conductivity of packed nanoparticles and their polymer composites. *ACS Appl. Mater. Interf.* 2009, *1*, 97.

[46] F. Curry, T. Lim, N. S. Fontaine, M. D. Adkins, H. Zhang. Highly conductive thermoresponsive silver nanowire PNIPAM nanocomposite for reversible electrical switch. *Soft Matter* 2022, *18*, 7171.

[47] Y. R. Jeong, S. Y. Oh, J. W. Kim, S. W. Jin, J. S. A. Ha. Highly conductive and electromechanically self-healable gold nanosheet electrode for stretchable electronics. *Chem. Eng. J.* 2020, *384*, 123336.

[48] Y. Ohm, C. Pan, M. J. Ford, X. Huang, J. Liao, C. Majidi. An electrically conductive silver–polyacrylamide–alginate hydrogel composite for soft electronics. *Nat. Electr.* 2021, *4* 185.

[49] N. Matsuhisa, D. Inoue, P. Zalar, H. Jin, Y. Matsuba, A. Itoh, T. Yokota, D. Hashizume, T. Someya. Printable elastic conductors by in situ formation of silver nanoparticles from silver flakes. *Nat. Mater.* 2017, *16*, 834.

[50] J. Garcia-Torres. Hybrid hydrogels with stimuli-responsive properties to electric and magnetic fields. In *Hydrogels – From Tradition to Innovative Platforms With Multiple Applications* (ISBN: 978-1-80355-583-6); L. Popa, Ed.; 2022, IntechOpen.

[51] P. Baei, S. Jalili-Firoozinezhad, S. Rajabi-Zeleti, M. Tafazzoli-Shadpour, H. Baharvand, N. Aghdami. Electrically conductive gold nanoparticle-chitosan thermosensitive hydrogels for cardiac tissue engineering. *Mater. Sci. Eng. C* 2016, *63*, 131.

[52] N. Dolya, O. Rojas, S. Kosmella, B. Tiersch, J. Koetz, S. Kudaibergenov. "One-pot" in situ formation of gold nanoparticles within poly(acrylamide) hydrogels. *Macromol. Chem. Phys.* 2013, *214*, 1114.

[53] A. Skardal, J. Zhang, L. McCard, S. Oottamasathien, G. D. Prestwich. Dynamically crosslinked gold nanoparticle-hyaluronan hydrogels. *Adv. Mater.* 2010, *22(42)*, 4736.

[54] J. Garcia-Torres, C. Crean. Ternary composite solid-state flexible supercapacitor based on nanocarbons/manganese dioxide/PEDOT:PSS fibres. *Mater. Des.* 2018, *155*, 194.

[55] J. Garcia-Torres, A. J. Roberts, R. C. T. Slade, C. Crean. One-step wet-spinning process of CB/CNT/MnO₂ nanotubes hybrid flexible fibres as electrodes for wearable supercapacitors. *Electrochim. Acta* 2019, *296*, 481e490.

[56] D. Gan, Z. Huang, X. Wang, L. Jiang, C. Wang, M. Zhu, F. Ren, L. Fang, K. Wang, C. Xie, X. Lu. Graphene oxide-templated conductive and redox-active nanosheets incorporated hydrogels for adhesive bioelectronics. *Adv. Funct. Mater.* 2020, *30*, 1907678.

[57] X. Xiao, G. Wu, H. Zhou, K. Qian, J. Hu. Preparation and property evaluation of conductive hydrogel using poly(vinylalcohol)/polyethylene glycol/graphene oxide for human electrocardiogram acquisition. *Polymers* 2017, *9*, 259.

[58] N. Annabi, S. R. Shin, A. Tamayol, M. Miscuglio, M. A. Bakooshli, A. Assmann, P. Mostafalu, J.-Y. Sun, S. Mithieux, L. Cheung, X. Tang, A. S. Weiss, A. Khademhosseini. Highly elastic and conductive human-based protein hybrid hydrogels. *Adv. Mater.* 2016, *28(1)*, 40.

[59] P. He, J. Wu, X. Pan, L. Chen, K. Liu, H. Gao, H. Wu, S. Cao, L. Huang, Y. Ni. Anti-freezing and moisturizing conductive hydrogels for strain sensing and moist-electric generation applications. *J. Mater. Chem. A* 2020, *8*, 3109.

[60] C. M. Palicpic, R. Khadka, J.-H. Yim. An effectively enhanced vapor phase hybridized conductive polymer based on graphene oxide and glycerol influence for strain sensor applications. *New J. Chem.* 2022, *46*, 22162.

[61] C. Chen, Y. Wang, T. Meng, Q. Wu, L. Fang, D. Zhao, Y. Zhang, D. Li. Electrically conductive polyacrylamide/carbon nanotube hydrogel: Reinforcing effect from cellulose nanofibers. *Cellulose* 2019, *26*, 8843.

[62] S. P. Sreenilayam, I. U. Ahad, V. Nicolosi, D. Brabazon. MXene materials based printed flexible devices for healthcare, bio-medical and energy storage applications. *Mater. Today* 2021, *43*, 99.

[63] R. Qin, G. Shan, M. Hu, W. Huang. Two-dimensional transition metal carbides and/or nitrides (MXenes) and their applications in sensors. *Mater. Today Phys.* 2021, *21*, 100527.

[64] K. Li, M. Liang, H. Wang, X. Wang, Y. Huang, J. Coelho, S. Pinilla, Y. Zhang, F. Qi, V. Nicolosi, Y. Xu. 3D MXene architectures for efficient energy storage and conversion. *Adv. Funct. Mater.* 2020, *30(47)*, 1.

[65] L. Gao, C. Li, W. Huang, S. Mei, H. Lin, Q. Ou, Y. Zhang, J. Guo, F. Zhang, S. Xu, H. Zhang. MXene/polymer membranes: Synthesis, properties, and emerging applications. *Chem. Mater.* 2020, *32(5)*, 1703.

[66] T. Someya, M. Amagai. Toward a new generation of smart skins. *Nat. Biotechnol.* 2019, *37*, 382.

[67] Y.-Z. Zhang, J. K. El-Demellawi, Q. Jiang, G. Ge, H. Liang, K. Lee, X. Dong, H. N. Alshareef. MXene hydrogels: fundamentals and applications. *Chem. Soc. Rev.* 2020, *49*, 7229.

[68] Y.-Z. Zhang, K. H. Lee, D. H. Anjum, R. Sougrat, Q. Jiang, H. Kim, H. N. Alshareef. MXenes stretch hydrogel sensor performance to new limits. *Sci. Adv.* 2018, *4*, eaat0098.

[69] D. Gan, L. Han, M. Wang, W. Xing, T. Xu, H. Zhang, K. Wang, L. Fang, X. Lu. Conductive and tough hydrogels based on biopolymer molecular templates for controlling in situ formation of polypyrrole nanorods. *ACS Appl. Mater. Interf.* 2018, *10*, 36218.

9 Hydrogels with Magnetic Properties

Nuraina Anisa Dahlan
Universiti Malaya, Kuala Lumpur, Malaysia

Brianna
Sunway University, Selangor Darul Ehsan, Malaysia

Fatimah Ibrahim
Universiti Malaya, Kuala Lumpur, Malaysia

Sin-Yeang Teow
Wenzhou-Kean University, Wenzhou, China

9.1 INTRODUCTION

Sensors, biomarkers, wound dressings, contact lenses, dental care products, biomedical implants, and capsule-based medications have one material in common: *hydrogels*. Three-dimensional polymeric networks of hydrogels are capable of imbibing water up to 95% of the polymer mass. Hydrogels are considered promising biomaterials, owing to their water-enriched and biocompatible properties, which can serve as an ideal interface between human bodies and biomedical products [1,2]. The strong polymer networks allow reversible swelling and deswelling behaviors, leading to their widespread use in various translation applications. The rising popularity of hydrogel-based products, especially in health-focused technologies, contributes to the establishment of multibillion-dollar industries. Magnetic hydrogels have been making inroads in various research and development fields with a phenomenal growth of scientific publications indexed in the Web of Science database since 1983. Up to now, more than 3,000 publications on magnetic hydrogels have been reported, as shown in Figure 9.1a. On top of that, the enthusiasm towards magnetic hydrogels as flexible sensors is evident by the exponential increase of publications since 1997, despite it still being at its infancy stage, as depicted in Figure 9.1b.

A sensor constitutes three main components: (a) receptor, (b) transducer, and (c) electronic system with the ability to transmit quantifiable signals. The versatility and resourceful origin of hydrogels gained widespread attention from the scientific community and industrial experts to step up their game in advancing magnetic hydrogel-based sensors for point-of-care (POC) devices, wearable electronics, food safety determination, environmental monitoring, and soft robotics. The strong magnetic field could influence hydrogel's structural integrity leading to performance

DOI: 10.1201/9781003340485-9

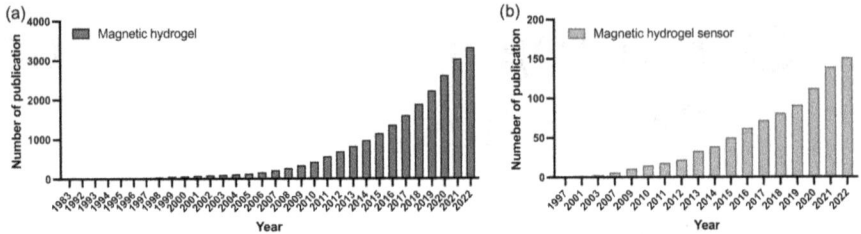

FIGURE 9.1 Scientific publications on (a) magnetic hydrogel and (b) magnetic hydrogel sensors since 1983, indexed in Web of Science.

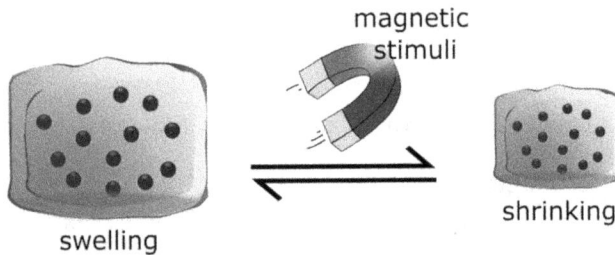

FIGURE 9.2 Mechanism of swelling/shrinking of magnetic hydrogels influenced by magnetic stimulation.

enhancement and making it an exciting material for various contactless applications [3–5]. The fundamental sensing behavior of magnetic hydrogel-based sensors lies in the physical changes of hydrogels (*e.g.,* swelling or shrinking behavior) influenced by the magnetic fields (Figure 9.2). To date, the lack of sufficient information on magnetic sensing principles indicates that this is a rather new research niche awaiting major breakthroughs. This chapter discusses the innovation and developmental trends of magnetic hydrogels in cancer theranostics, sensing applications in wearable devices, and gas detection systems as well as sensors and actuators for soft robotics.

9.2 PREPARATION OF MAGNETIC HYDROGELS

Designing hydrogels with specific functionalities (*e.g.,* magnetic, electroconductive, thermoresponsive) requires tailored preparative methods. This can be done by exploring a series of different *in situ* and *ex situ* synthesis routes with consideration on the properties of hydrogels and their magnetic components. An embodiment of hydrogels with magnetic properties can be achieved by incorporating or immobilizing magnetic materials (*e.g.,* magnetic nanoparticles or magnetic ferrofluids). Stability and hardness are important criteria to make a successful sensing component. In this instance, appropriate modification strategies are commonly considered to achieve desirable hydrogel functionalities.

9.2.1 *IN SITU* SYNTHESIS

In lieu of the conventional co-precipitation method to prepare magnetic nanoparticles, *in situ* co-precipitation involves the introduction of magnetic nanoparticle precursors to

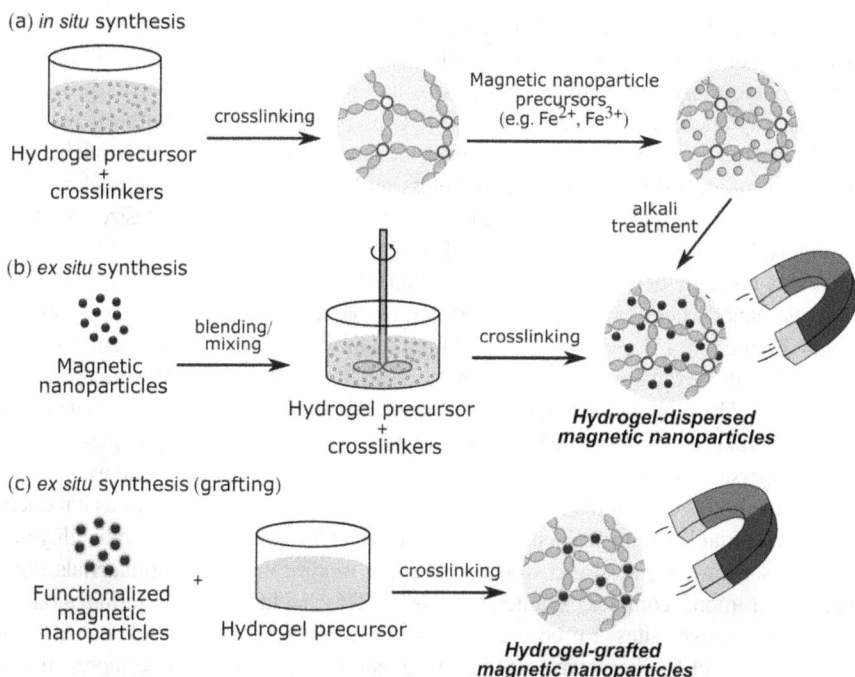

FIGURE 9.3 The synthesis routes of magnetic hydrogels for various sensing applications: (a) *in situ* synthesis, (b) *ex situ* synthesis, and (c) *ex situ* grafting methods.

hydrogel or hydrogel precursor solution (Figure 9.3a). An example is an *in situ* co-precipitation method in which Fe^{2+} and Fe^{3+} starting iron precursors were co-precipitated in as prepared polypyrrole/polyvinyl alcohol (PPy/PVA) hydrogel matrix [6]. Despite offering a straightforward and simple method, an *in situ* co-precipitation method is limited to hydrogels with stable polymeric networks against high alkaline conditions during synthesis. Different types of metal have been explored for facile *in situ* synthesis of magnetic hydrogels apart from Fe-based magnetic nanostructures. For instance, Dumitrescu and colleagues synthesized magnetic nickel ferrite ($NiFe_2O_4$) crystallites with polyacrylamide-based hydrogels as the template matrices using a sol-gel autocombustion method. Interestingly, as the size of $NiFe_2O_4$ crystallite increased, the research team observed increasing magnetic properties [7].

9.2.2 *Ex situ* Synthesis

In this approach, the magnetic components (typically magnetic nanomaterials) are prepared separately. The as-prepared magnetic nanoparticles are subsequently mixed with polymer solutions followed by *in situ* cross-linking to polymerize the polymeric networks (Figure 9.3b). Physical blending/mixing using sonication or mechanical stirring provides a straightforward and cost-effective synthesis under mild reaction conditions [8]. The *ex situ* approach poses great advantages for improving biocompatibility and surface functionalization of nanoparticles with targeted ligands, drugs, or markers, as well as providing additional cross-linking sites for the polymeric chains. In a recent work, Zhou *et al.* introduced the prepared magnetic iron oxide (Fe_3O_4) nanoparticles in a "one-pot" free radical polymerization of *N*-isopropylacryalamide (NIPAAm) and acryalamide (Am) cross-linked by *N, N'*-methylenebisacrylamide (MBA) [9]. The "one-pot" *ex situ* synthesis resulted in a flexible magnetic hydrogel-based sensor with controllable mechanical properties as wearable sensors with wireless and non-contact sensing mechanisms.

The *ex situ* grafting method is another renowned strategy to synthesize magnetic hydrogels. This involves the formation of covalent bonds between magnetic nanoparticles and polymer networks to form cross-linked magnetic hydrogels. Typically, the magnetic nanoparticles are functionalized to promote interaction with targeted polymer chains (Figure 9.3c). An example of a grafting method is magnetic Fe_3O_4 nanoparticles modified with hexamethylene diisocyanate (HMDI) to carry diisocyanate functional groups [10]. The modified magnetic Fe_3O_4 nanoparticles were subsequently reacted with hydroxyl groups of modified PVA. The resultant grafted magnetic hydrogels showed better dispersion of doxorubicin drugs within the polymeric magnetic networks. The grafting method is considered to be superior to other *ex situ* synthesis routes as it reduces the nanomaterial leaching problem, which could affect the overall sensor's reliability and robustness. However, it is also worth noting that modification of nanomaterials often leads to a more complex synthesis method. This includes the determination of appropriate active sites to ensure successful grafting and potential increase of cytotoxicity level by the newly modified magnetic nanoparticles. Furthermore, studies showed that functionalization or modification of polymers with specific functional groups (*e.g.,* carboxyl, hydroxyl, vinyl, thiol, amine functional groups) imparted new physicochemical properties with minimal changes to their functionality and final outcome [11,12]. Polymer modification with appropriate functional groups holds a great advantage to control grafting between two mutually incompatible segments.

9.3 THE FUTURE OF MAGNETIC NANOMATERIALS

Magnetic nanomaterials garnered significant interest due to their superparamagnetic behavior in nanoscale ranges. Metal-based magnetic nanostructures such as copper ferrite ($CuFe_2O_4$), neodymium-iron-boron (NdFeB), cobalt ferrite ($CoFe_2O_4$), and iron oxide nanoparticles (*e.g.,* magnetite Fe_3O_4 and hematite α-Fe_2O_3) with superior magnetic and biocompatible properties are the primary focus in the field of magnetic nanomaterials [13,14]. Sensing devices with magnetic fields can easily penetrate thick biological tissues and fluids compared to conventional optical signal-based sensing devices, allowing rapid non-contact or wireless sensing. Additionally, the ability of magnetic-based sensing devices to sense changes in temperature, stress, or chemical reaction without power excitation opens a wider spectrum for passive sensing mechanisms [15,16]. Apart from the favorable magnetic properties, magnetic-based sensing devices are economical, lightweight due to their small nanoscale size, robust, and easy to synthesize [17]. The progressive development of nanoscience allows controlled syntheses of numerous magnetic-based nanomaterials with specific morphology, shape, size distribution, and properties for desired applications.

Magnetic iron oxide (*e.g.,* Fe_3O_4, α-Fe_2O_3, γ-Fe_2O_3) nanoparticles are the most widely studied magnetic nanoparticles with immense potential in multifarious applications from environmental remediation, agriculture, catalysis, food industry, and biomedicine [10,14,18,19]. Bare magnetic Fe_3O_4 nanoparticles are also ideal for various surface functionalization strategies, which sparked the interest of many researchers to further expand the dynamic functionalities of the nanoparticles. Surface functionalization can be achieved via surface coating (*e.g.,* natural/synthetic polymers, polyethylene glycol (PEG) or peptide), conjugation of targeted functional groups via chemical means, or metal doping (*e.g.,* Co, Fe, Mn, Zn) [14,20]. Surface-functionalized nanoparticles offer improved biocompatibility and pharmacokinetic properties, which are useful for various biomedical applications, particularly cancer theranostics and drug delivery systems [12]. To date, a series of synthesis routes have been developed to synthesize magnetic Fe_3O_4 nanoparticles, such as co-precipitation, thermal decomposition, microemulsion, hydrothermal, sol-gel, polyol, and microwave-assisted [20–22]. Below are descriptions of the typical synthesis methods.

1. **Co-precipitation.** The co-precipitation method is considered a break-through in the synthesis of magnetic Fe_3O_4 nanoparticles. This conventional synthesis method involves the vigorous mixing of Fe^{2+} and Fe^{3+} precursors in the presence of a base at ambient or high temperatures [14]. The outcome of resultant magnetic nanoparticles such as size distribution, shape, and composition is highly influenced by pH of the mixture, reaction temperature, molar ratio of Fe^{2+}: Fe^{3+}, and type of iron salts [21,22]. Total control over reaction parameters often leads to the reproducible formation of magnetic nanoparticles. Unlike other preparative methods, co-precipitation offers an environmental friendly approach with water-based routes for a greener magnetic nanoparticle synthesis [1].

2. **Thermal decomposition.** In this approach, the synthesis of magnetic nanoparticles requires high temperature and long reaction time to decompose organometallic precursors in organic solvents and in the presence of stabilizing surfactants (*e.g.,* fatty acids, hexadecylamine, and oleic acid). Although thermal decomposition produces highly mono-dispersed magnetic nanoparticles at a larger scale, the formation of toxic by-products remains a serious limitation [20]. Furthermore, the use of organic solvents at higher temperatures and pressures under anaerobic conditions is health and work safety concerns [21].

3. **Microemulsion.** The use of microemulsion strategy often results in nanometer-sized and uniform magnetic nanoparticles. For instance, Fe core/iron oxide shell magnetic nanoparticles were successfully synthesized in water-octane microemulsions with cetyltrimethylammonium bromide and butanol as surfactants, aqueous solutions of iron (III) chloride ($FeCl_3$) and sodium borohydride ($NaBH_4$) as the starting precursors. The microemulsion strategy used for the magnetic nanoparticle synthesis yielded spherical nanoparticles with a core diameter ranging between 8 to 16 nm and a shell diameter of 2 to 3 nm [23].

4. **Hydrothermal.** Scaling up magnetic Fe_3O_4 nanoparticles for commercialization may benefit from the hydrothermal method. In this method, the synthesis of magnetic nanoparticles and ultrafine powders with specific size distribution can be achieved at higher temperatures (125–250°C) and

pressures (0.3–4 MPa) [20]. Despite the high yield production, corrosive hydrothermal slurries in high-pressure conditions pose a great threat to accidental explosions and hence require proper regulation and preventive measures.

5. **Sol-gel technique.** The sol-gel process involves (i) preparation of colloidal solution (sol) containing starting precursors dissolved in water or solvents, (ii) hydrolysis and gelation of sol, and (iii) removal of solvents. The sol-gel technique proved to be advantageous in producing a myriad of metal oxide nanostructures (*e.g.*, nanoparticles, nanotubes, nanorods) at lower reaction temperatures [17]. However, sol-gel is notoriously known for its slow processing time, making it less profitable for large-scale production.

6. **Polyol.** The polyol method offers control over scalability, production cost, and a wider range of morphologies from smooth spheres to complex "nanoflower" structures. In this approach, polyol derivatives such as ethylene glycol act as both surfactant and solvent in high-temperature conditions to provide control over nanoparticle growth, size, and distribution. Polyol derivates are stable in high temperatures due to their highly polar characteristics [21]. This allows the combination of polyol method with other synthesis methods such as microwave-assisted to up-scale the production of magnetic nanoparticles within a shorter period of time. Hemery *et al.* reported that their attempt to control the addition of water during polyol synthesis of magnetic Fe_3O_4 nanoparticles resulted in diameters ranging from 3–37 nm [24]. It was also found that exploiting the reaction conditions such as natural mixing (smooth spheres) *versus* mechanical stirring (nanoflowers) resulted in different nanoparticle morphologies.

7. **Microwave-assisted.** The need for rapid synthesis of nanoparticles with uniform products has led to the discovery of microwave-assisted technology. This technology relies on the use of electromagnetic waves for non-contact heating to heat iron precursor mixtures under a shorter reaction time. Uniform heating supplied by the electromagnetic wave improves physical and chemical properties of the resultant magnetic nanoparticles. On top of that, microwave-assisted technology is easily coupled with other nanoparticle fabrication strategies such as co-precipitation, polyol, and hydrothermal due to its simplicity and flexibility [21].

9.4 RECENT ADVANCES OF MAGNETIC HYDROGELS

Magnetic hydrogels are considered valuable tools in a myriad of applications including sensing and soft technologies owing to their foreseeable capabilities to perform structural or mechanical changes triggered by external magnetic stimuli. These changes can be converted into specific signaling cues (*e.g.*, optical, electrical, or chemical) in the fabrication of sensors, biological probes, or other biomedical devices [3,8,25].

9.4.1 CANCER THERANOSTICS

Cancer theranostics is a growing field of medical interest that combines diagnostic technique and treatment into a single integrated system, enabling the diagnosis, treatment, and monitoring of cancer treatment to be done simultaneously. This allows clinicians to develop individualized therapies for patients with the aim to increase the safety and effectiveness of treatment administered. The potential cancer theranostic use of magnetic hydrogels is currently being researched due to their unique characteristics such as high biocompatibility, enhanced magnetic properties for imaging purposes, mechanical properties, and ability to absorb and release drugs in response to predetermined stimuli (*e.g.,* heat, pH change, reactive oxygen species (ROS) level, enzymatic changes, and other external stimuli) [26,27]. Therefore, magnetic hydrogels are able to selectively deliver anticancer drugs to the tumor microenvironment and provide real-time monitoring of cancer status along with therapeutic response when localized at a targeted area, as depicted in Figure 9.4.

With the aim to investigate the theranostic potential of magnetic hydrogels as drug carriers, Jaiswal and colleagues developed a thermoresponsive magnetic hydrogel using Fe_3O_4 magnetic nanostructures encapsulated with poly(N-isopropyl acrylamide), along with PEG and polyhedral oligomeric silsesquioxane (POSS) [28]. Their study showed that the designed magnetic hydrogels are capable of performing diagnostic imaging of cancer due to the incorporation of Fe_3O_4, and undergo conformational change in response to the increased temperature and magneto-mechanical vibration when exposed to radiofrequency, which leads to the enhanced release of doxorubicin (anticancer drug). Moreover, their magnetic hydrogel-based delivery system functionalized with PEG and loaded with doxorubicin achieved high cell death efficacy (more than 80%) in cervical carcinoma (HeLa) cells with radiofrequency exposure for an hour, supporting its potential as a cancer theragnostic

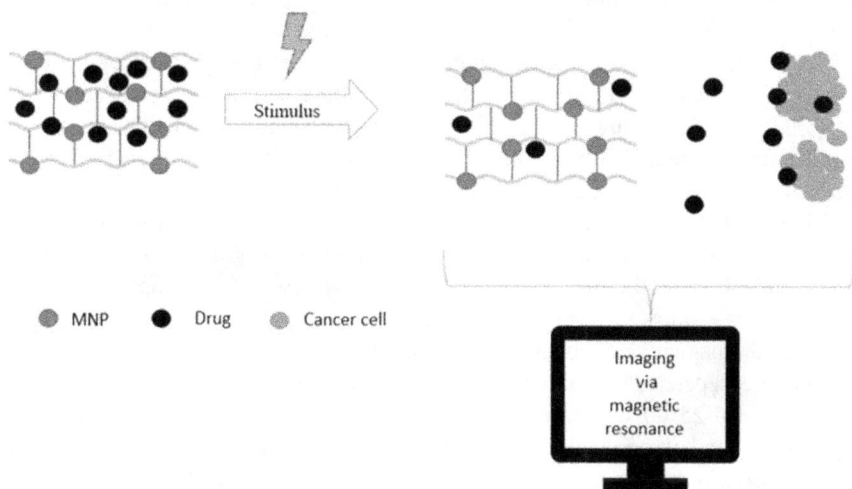

FIGURE 9.4 The dual function use of magnetic hydrogel in cancer theranostics, furnished with magnetic nanoparticles for imaging and drugs with anticancer effects.

agent. In another study, glycol chitosan and difunctional telechelic PEG (DT-PEG) functionalized hydrogels were instead incorporated with a low concentration of ferromagnetic vortex-domain iron oxide (FVIOs) with magnetic properties to optimize the therapeutic results [29]. Results demonstrated that the magnetic hydrogels loaded with doxorubicin were capable of significantly suppressing the recurrence of breast cancer in mice as compared to when treated with chemotherapy or hyperthermia. With the aim to increase the efficacy of current cancer-targeted drug delivery systems, Kim *et al.* developed a gelatin/PVA hydrogel-based microrobot constructed from magnetic nanoparticles and poly lactic-co-glycolic acid particles loaded with doxorubicin (PLGA-DOX) [30]. Using electromagnetic actuation and near-infrared stimulation, the microrobot will be guided to the tumor site and, upon reaching the site, the microrobot will decompose and release the magnetic nanoparticles and PLGA-DOX complexes. As magnetic nanoparticles are toxic if they remain in the body for a long duration, the magnetic nanoparticles will be retrieved using electromagnetic actuation. Gradually, the PLGA-DOX complexes will also biodegrade and release doxorubicin at controlled levels to exert anticancer effects. *In vitro* results showed that liver cancer (Hep3B) cells treated with the microrobot had significantly higher cell death compared to the control group, proving the efficacy of the novel microrobot. With the continuous development of magnetic hydrogels that act as a multifunctional imaging system and a controlled drug delivery carrier at the same time, magnetic hydrogels-based theragnostic will be the future of cancer diagnosis and treatment.

9.4.2 DIABETES

Diabetes is a lifelong condition that requires proper management to prevent further complications. Patients with uncontrolled diabetes could experience life-threatening complications such as kidney failure, cardiovascular diseases, and blindness [5]. The exponential growth in popularity of magnetic hydrogel-based glucose biosensors has contributed to relentless collaborations between researchers and industrial players in an effort to produce low-cost and reliable biosensors for glucose monitoring. Song *et al.* made a positive contribution to glucose biosensing applications by employing green synthesized magnetic Fe_3O_4 nanoparticles as mimetic enzymes in an amphiphile hydrogel system [1]. Impressively, their calorimetric biosensor exhibited high sensitivity towards glucose with a detection limit of 0.37 µmol/L. Huang and colleagues immobilized glucose oxidase (GOD) via chemical cross-linking on $Fe_3O_4@SiO_2(F)@meso-SiO_2$ complex magnetic nanoparticles followed by *in situ* immobilization of the prepared GOD-conjugated magnetic nanoparticles in thermoresponsive hydrogels [5]. The temperature-dependent detection system showed satisfactory detection at ambient temperature (~25°C) compared to an elevated temperature of 38°C. Favorable repeatability and glucose selectivity characteristics further emphasized the potential of proposed glucose biosensors for practical POC monitoring [5].

Currently, there are limited patents on magnetic hydrogel-based glucose biosensors filed in hope to improve current biosensors for future end users. For example, invention US9737244B2 refers to a sensor comprising magnetic particles dispersed within a ferrogel network [31]. According to the inventors, the sensor detected changes in the magnetic fields to transmit electrical signals corresponding to the blood glucose level. The invention is anticipated to track blood glucose levels in a timely manner. Seamless glucose monitoring could be a useful tool for physicians to obtain quality data on glucose levels from their patients and plan effective diabetes management.

9.4.3 WEARABLE DEVICES

In addition to the development of magnetic hydrogel-based biosensors as highlighted in Sections 9.4.1 and 9.4.2, the research on magnetic hydrogel-based sensors as wearable sensors is also eminently executed within the wearable sensing community. Wearable sensors offer affordability and ergonomic capability due to miniaturized electronics, continuous health monitoring data to meet unmet medical needs, increased awareness of health and physical fitness, as well as the advancement of mobile technology (*e.g.,* smartphones and Bluetooth). Wearable sensors mandate reversible and rapid response with interference-free characteristics to selectively detect target analytes [32]. Current technologies are moving from rigid sensors made from metal or semiconductor materials to flexible substrates such as hydrogels and elastomers. Magnetically responsive hydrogels are considered as ideal materials to magnetically control the shape and stiffness of hydrogels within a confined space of wearable devices. This concept is compelling in such a way that the actuation of rigid polymer chains is easily controlled in favor of magnetism which enhances sensing sensitivity [33]. Advances in healthcare wearable devices hitherto contribute to profound changes within society and medical healthcare. Increasing individual awareness on health monitoring has led to the blooming of wearable and wireless sensors to track physical activities and monitor chronic diseases (*e.g.,* monitoring of blood glucose levels, and neurological disorders such as seizure and epilepsy) [32,34]. Wearable sensors can be sewn into garments, affixed to clothing, or directly attached to the skin [34]. Figure 9.5 summarizes the integration of wearable sensors for specific biochemical (*e.g.,* biofluids) or physical (*e.g.,* pressure, motion, heart rate or pulse rate, temperature, and tactile) targets. Integration of IoT that includes web-enabled sensor platforms, software, and processors further paved the way for instant and constant data transmission. Information gathered from wearable and wireless sensors followed by instant transmission via an information gateway to patients and clinicians can be used to determine appropriate intervention and medical care.

To date, commercialized products such as FitBit, Apple Watch, Happy Ring (real-time mood tracking), and HeartGuide (blood pressure monitoring) have gained customer trust in the booming wearable technology industry. In this context, researchers have shifted to the exploration of magnetically enhanced wearable sensors with wireless signal transmission and data processing. For example, Song *et al.* explored the attachment of magnetic hydrogels "ferrogels" containing magnetic Fe_3O_4 nanoparticles to a planar inductor, which later modulates inductance created by the swelling and shrinking of hydrogel networks [3]. The ferrogel system coupled with a pH-sensitive hydrogel was a promising component to fabricate pH sensors for wireless monitoring. It was found that the use of carriers

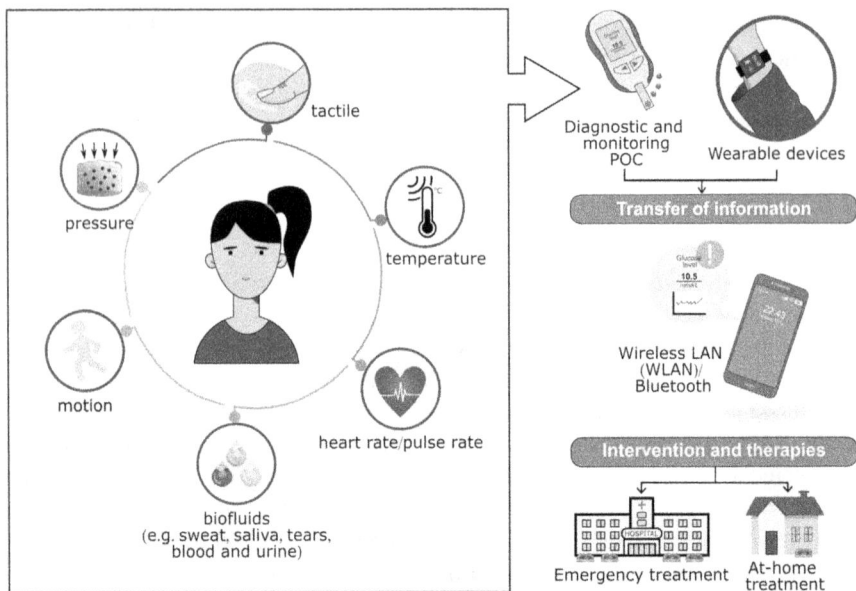

FIGURE 9.5 Schematic illustration of biochemical (*e.g.*, biofluids) and physical (*e.g.*, tactile, temperature, heart rate/pulse rate, motion, and pressure) sensing strategies for rapid diagnosis and monitoring of health. Digital information gathered from wearable sensors can provide better intervention and clinical outcomes.

such as PPy nanotubes to encapsulate magnetic nanoparticles acted not only as a protective barrier but also prevented nanoparticle agglomeration, a common problem associated with highly reactive magnetic nanoparticles. In this instance, a study showed that PPy nanotubes immobilized magnetic $MnFe_2O_4$ nanoparticles immobilized in hydrogel networks successfully produced ultra-stretchable and reusable hydrogel matrix useful for wireless and real-time monitoring [35]. van Bruggen and van Zon demonstrated a proof-of-concept study to demonstrate the potential of transducer miniaturization using magnetic hydrogels [36]. In their study, the transduction principle was explained based on a giant magneto resistance (GMR) sensing element used to detect magnetic stray field changes emitted by the superparamagnetic nanoparticles embedded in the hydrogel matrix. Their theoretical study suggested that the magnitude of hydrogel changes (swelling or shrinking behavior) was detected magnetically, which resulted in quantifiable electrical signals by the GMR element. Interestingly, they also found that thin magnetic hydrogels (~5 μm) can transmit maximum signals for fast sensor response, suggesting a bright future to miniaturize transducers.

9.4.4 SOFT ROBOTICS

Traditional rigid and bulky robotic machines limit the miniaturization of their mechanical components such as actuators (*e.g.*, motors), computational elements, power sources, and electronic circuits. In a conventional robotic system, actuators play an important role to convert energy (*e.g.,* hydraulic, thermal, magnetic, electrical, and pneumatic) into linear or rotational motions. These actuators, however, necessitate high-precision control to mobilize the rigid mechanical structures. As such, hydrogels are anticipated to generate smooth motions (*e.g.,* bending, elongation, and torsion) that mimic biological movements. For example, soft actuators containing magnetic nanoparticles (*e.g.,* Fe_3O_4 nanoparticles) could generate considerable forces under magnetic stimulation that change the fidelity and shape of the hydrogel matrix, resulting in smooth robotic movements [4,37]. These soft actuators could be very useful in the development of artificial limbs or fingers with native-like muscle flexibility to grasp and hold objects. Flexible materials like hydrogels are also anticipated to bridge the gap between bulky conventional robots and biological systems that are known to be soft, flexible, and tough. Miniaturized soft robotics with magnetic properties offer promising potentials to be transported within confined spaces of the human body. Considering the applications of soft robotics in enclosed and uncontrolled biological environments, the alignment of magnetic isotropic or anisotropic hydrogels can be tuned under magnetic stimulation to cause changes to the microstructure of the composite matrix for specific applications. For example, recent innovations in magnetic hydrogels offer technological foundations to control soft robotics such as surgical catheters or untethered magnetic drug delivery robots (from millirobots to microrobots) using a magnet or programmable magnetic stimulation. Unlike other types of smart hydrogels, magnetic hydrogels offer fast response with remote control ability using magnetic fields. This technology is useful to execute non-invasive or minimally invasive delivery of medications to deep tissues with narrow passages. Future development of soft robots includes the ability to measure *in situ* clinical conditions such as temperature, biomarkers, and pressure [13,37].

9.4.5 GAS DETECTION SYSTEM

Integrated sensing devices with a gas detection system have the potential to detect specific hazardous and flammable gases (*e.g.,* methane (CH_4), nitrogen dioxide (NO_2), carbon monoxide (CO), carbon dioxide (CO_2), etc.) as well as volatile organic compounds (*e.g.,* ethanol). This is useful in many aspects of technology, such as in-door air monitoring (*e.g.,* factories and manufacturing plants), environmental monitoring (*e.g.,* mining sites or industrial pollution), public safety (*e.g.,* households and vehicles), and medical diagnosis for gas detection and monitoring [38,39]. Gas sensors are often developed as either portable or fixed systems, depending on their target application in integrated safety instrument systems. The gas sensing mechanism is well explained in terms of the interaction between materials (adsorbent) and target gases (adsorbate). Since several magnetic parameters such as M_s (saturation magnetization), M_r (remanence magnetization), and H_c (coercivity) are extremely sensitive to reducing or oxidizing gases, metal oxide-based sensors are widely studied for various gas sensing applications. Exposure of metal oxides to air leads to the adsorption of oxygen ions to the metal surfaces. In this case, the oxygen ions serve as electron acceptors to remove electrons from the conduction bands of oxides and dissociate into different types of oxygen anions (*e.g.*, O_2^-, O^-, and O^{2-}). The removal of electrons from the conduction bands creates an electron depletion layer rich in hole carriers on the material surfaces, leading to an elevated resistance level, as illustrated in Figure 9.6 [40,41]. Furthermore, physical interactions, such as hydrogen bond formation or weak van der Waals forces between hydrogels and target gases, play a vital role in the sensing process [41]. In view of this, Aldalbahi *et al.* developed electrically conductive magnetic methane gas sensors composed of magnetic Fe_3O_4 nanoparticles and conductive-based materials (multi-walled carbon nanotubes (MWCNTs) and polyaniline (PANI) [38]. It was found that the magnetic hydrogel nanocomposites containing Fe_3O_4/hydrogel/MWCNTs showed higher sensitivity towards CH_4 at elevated temperatures. Their study outlined the importance of magnetic Fe_3O_4 nanoparticles to enhance the gas sensing characteristics as an oxidizing agent. Another study demonstrated a change in the rheological behavior of magnetic ferrogels controlled via modulation of alternating current magnetization to detect ammonia gas [42]. In this study, they reported that the stimulation of alternating current magnetization induced rotational relaxation of embedded magnetic nanoparticles, causing the swelling of the gel microstructures within specific pH ranges (7–12). The combination of pH response and swelling magnitude under the influence of alternating current magnetization corresponded significantly to the sensitivity of their sensing label towards ammonia gas. On the contrary, conventional electrical-based gas sensors are exposed to higher explosion risks, especially in the presence of reactive chemicals and temperature fluctuations. In these instances, magnetic gas sensors are deemed attractive due to ease of operation at ambient temperature with a significantly lower explosion risk. Additionally, the magnetic gas sensors can be designed to work at substantially low or high temperatures by selecting magnetic materials with appropriate Curie temperatures [4].

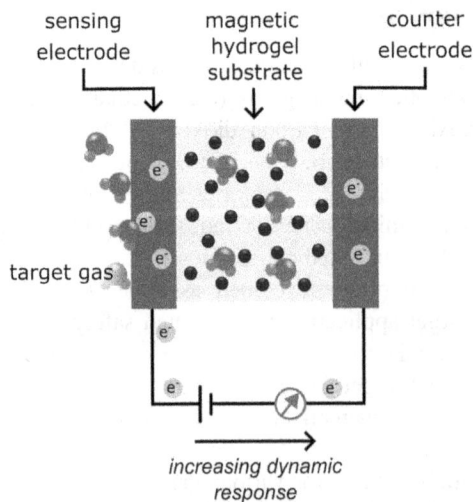

FIGURE 9.6 Illustration of the change of dynamic response upon exposure to target gases (*e.g.*, CH_4, CO, CO_2, etc.) for gas sensors incorporated with magnetic hydrogels.

9.5   CHALLENGES AND FUTURE PROSPECTS

Despite magnetic hydrogels being deemed as a prospective candidate in the field of sensors and biomedical devices, several challenges still need to be confronted for their realization in practical applications. Hitherto, integration of appropriate magnetic materials into the hydrogel networks is one of the bottlenecks faced by magnetic hydrogel-based sensors. Materials compatibility, gelation strategies, and hydrogel structures are the common problems to produce highly sensitive and robust magnetic hydrogel-based sensors. *In vivo*, biocompatibility without compromising mechanical strength is a grave problem in soft robotics. In this technology, mechanical considerations are of paramount importance to penetrate blood clots or deep tissues *in vivo* yet maintain appropriate softness to pass through narrow capillary channels [25].

Furthermore, the toxicity of magnetic nanoparticles used to magnetically control the direction of hydrogels and for imaging purposes is of concern. This is because, unlike hydrogels, magnetic nanoparticles are not biodegradable and have the potential to remain in the body post-treatment [43]. While magnetic nanoparticles may be excreted by the body incrementally, the remaining magnetic nanoparticles will attach to cell membranes or penetrate cells, leading to metabolic stress, generation of ROS, inflammation, disrupted cell proliferation, and eventually cell death [43]. This untoward reaction to magnetic nanoparticles greatly limits the clinical translation of magnetic hydrogels. Nevertheless, as previously discussed, there is a research group who circumvented this problem by retrieving the magnetic nanoparticles using a magnetic field once they have reached the target site, eliminating this grave cause for concern [30].

The rise of biomedical devices integrated with IoT could unleash the biggest future opportunities for healthcare technologies. IoT is a thriving technological frontier through its ability to connect physical sensors, computing devices, processing ability, software, and other technologies, thus eliminating the need for human intervention. For instance, Yu *et al.* introduced an automated cell manipulation biosensing platform integrated with magnetically responsive hydrogel actuators and optical imaging system to analyze cellular deformation under magnetic stimuli [16]. The intelligent biosensing platform could transmit accurate information and images to smartphone devices, proving their potential for rapid bedside testing. To date, IoT is changing the way biomedical and robotic products are designed for the comfort of end users by connecting a huge network ecosystem (*i.e.,* machinery, objects, humans, environment). Advances in IoT are anticipated to navigate the communication of reliable health information for monitoring healthcare at every stage. Biomedical devices integrated with IoT could be beneficial for patients with long-term conditions, thereby eliminating unnecessary hospital visits to constant daily monitoring. Despite the promising future of IoT, further challenges lie ahead in integrating IoT into the sensing platforms. Reliability of communication networks via wireless LAN could be one of the major challenges to ensure consistent connectivity of data transmission. In addition, data transmission is connected to the performance of various nanodevices and nanobatteries with an appropriate lifetime to guarantee uninterrupted access to essential information. Ideally, the IoT-integrated sensors should be self-powered with the ability to host large amounts of data and process accurate readings. Coexistence of different types of IoT-based devices is also a challenge to deliver seamless encrypted information between the devices. Therefore, multidisciplinary research combining material, biology, and intelligent technology is needed to inspire creativity and explore the relationship between sensing platforms and IoT [39,44].

9.6 CONCLUDING REMARKS

Magnetic hydrogels have tremendous potential in the application of health-focused and environmental technologies, including glucose biosensors, imaging, and treatment cancer theranostic systems; wearable sensors that provide continuous health monitoring; gas detecting systems; and soft robotics in intelligent systems. Magnetic hydrogels are small, robust, easy to manufacture, economical, biocompatible, and have a superior superparamagnetic ability, making them ideal sensing appliances. As reviewed in the above sections, the incorporation of magnetic hydrogels will improve the quality of healthcare, safety, and life. Nevertheless, for research translation of novel inventions to occur in the next decade, the aforementioned challenges must be taken into consideration and overcome in future research and development. It is also hopeful that scientists continue leveraging the magnetic properties of magnetic hydrogels in other progressive research fields such as hyperthermia-based treatments, disease biomarkers, and tissue engineering that are confronted with numerous difficulties to realize their potential from bench to bedside.

REFERENCES

[1] Shasha Song, Yang Liu, Aixin Song, Zengdian Zhao, Hongsheng Lu, and Jingcheng Hao. "Peroxidase Mimetic Activity of Fe3O4 Nanoparticle Prepared Based on Magnetic Hydrogels for Hydrogen Peroxide and Glucose Detection." *Journal of Colloid and Interface Science* 506, 2017, 46–57.

[2] Janarthanan Pushpamalar, Puviarasi Meganathan, Hui Li Tan, Nuraina Anisa Dahlan, Li-Ting Ooi, Bibi Noorheen Haleema Mooneerah Neerooa, Raahilah Zahir Essa, Kamyar Shameli, and Sin-Yeang Teow. "Development of a Polysaccharide-Based Hydrogel Drug Delivery System (DDS): An Update." *Gels* 7, 2021, 153.

[3] SH Song, JH Park, G Chitnis, RA Siegel, and B Ziaie. "A Wireless Chemical Sensor Featuring Iron Oxide Nanoparticle-embedded Hydrogels." *Sensors and Actuators B: Chemical* 193, 2014, 925–930.

[4] Pratik V Shinde, and Chandra Sekhar Rout. "Magnetic Gas Sensing: Working Principles and Recent Developments." *Nanoscale Advances* 3, 2021, 1551–1568.

[5] Jun Huang, Peipei Zhang, Mengshi Li, Pengfei Zhang, and Liyun Ding. "Complex of Hydrogel with Magnetic Immobilized GOD for Temperature Controlling Fiber Optic Glucose Sensor." *Biochemical Engineering Journal* 114, 2016, 262–267.

[6] Yanqin Wang, Yaping Zhu, Yanan Xue, Jinghui Wang, Xiaona Li, Xiaogang Wu, YiXian Qin, and Weiyi Chen. "Sequential In-situ Route to Synthesize Novel Composite Hydrogels with Excellent Mechanical, Conductive, and Magnetic Responsive Properties." *Materials & Design* 193, 2020, 108759.

[7] AM Dumitrescu, G Lisa, AR Iordan, F Tudorache, I Petrila, AI Borhan, MN Palamaru, C Mihailescu, L Leontie, and C Munteanu. "Ni Ferrite Highly Organized as Humidity Sensors." *Materials Chemistry and Physics* 156, 2015, 170–179.

[8] Hongying Su, Xiaodong Han, Lihua He, Lihua Deng, Kun Yu, Hai Jiang, Changqiang Wu, Qingming Jia, and Shaoyun Shan. "Synthesis and Characterization of Magnetic Dextran Nanogel Doped with Iron Oxide Nanoparticles as Magnetic Resonance Imaging Probe." *International Journal of Biological Macromolecules* 128, 2019, 768–774.

[9] Hongwei Zhou, Zhaoyang Jin, Yang Gao, Ping Wu, Jialiang Lai, Shuangli Li, Xilang Jin, Hanbin Liu, Weixing Chen, Yuanpeng Wu, and Aijie Ma. "Thermoresponsive, Magnetic, Adhesive and Conductive Nanocomposite Hydrogels for Wireless and Non-contact Flexible Sensors." *Colloids and Surfaces A: Physicochemical and Engineering Aspects* 636, 2022, 128113.

[10] Mojtaba Abasian, Vahid Hooshangi, and Peyman Najafi Moghadam. "Synthesis of Polyvinyl Alcohol Hydrogel Grafted by Modified Fe3O4 Nanoparticles: Characterization and Doxorubicin Delivery Studies." *Iranian Polymer Journal* 26, 2017, 313–322.

[11] Nuraina Anisa Dahlan, Sin Yeang Teow, Yau Yan Lim, and Janarthanan Pushpamalar. "Modulating Carboxymethylcellulose-based Hydrogels with Superior Mechanical and Rheological Properties for Future Biomedical Applications." *Express Polymer Letters* 15, 2021, 612–625.

[12] Marziyeh Fathi, Jaleh Barar, Hamid Erfan-Niya, and Yadollah Omidi. "Methotrexate-conjugated Chitosan-grafted pH- and Thermo-responsive Magnetic Nanoparticles for Targeted Therapy of Ovarian Cancer." *International Journal of Biological Macromolecules* 154, 2020, 1175–1184.

[13] Anna Puiggalí-Jou, Ismael Babeli, Joan Josep Roa, Justin O. Zoppe, Jaume Garcia-Amorós, Maria-Pau Ginebra, Carlos Alemán, and Jose García-Torres. "Remote Spatiotemporal Control of a Magnetic and Electroconductive Hydrogel Network via Magnetic Fields for Soft Electronic Applications." *ACS Applied Materials & Interfaces* 13, 2021, 42486–42501.

[14] Nuraina Anisa Dahlan, Anand Kumar Veeramachineni, Steven James Langford, and Janarthanan Pushpamalar. "Developing of a Magnetite Film of Carboxymethyl Cellulose Grafted Carboxymethyl Polyvinyl Alcohol (CMC-g-CMPVA) for Copper Removal." *Carbohydrate Polymers* 173, 2017, 619–630.

[15] Qi Zhang, Guannan Yang, Li Xue, Guohua Dong, Wei Su, Mengjie Cui, Zhiguang Wang, Ming Liu, Ziyao Zhou, and Xiaohui Zhang. "Ultrasoft and Biocompatible Magnetic-Hydrogel-Based Strain Sensors for Wireless Passive Biomechanical Monitoring." *ACS Nano* 16, 2022, 21555–21564.

[16] Le Yu, Longfei Chen, Yantong Liu, Jiaomeng Zhu, Fang Wang, Linlu Ma, Kezhen Yi, Hui Xiao, Fuling Zhou, Fubing Wang, Long Bai, Yimin Zhu, Xuan Xiao, and Yi Yang. "Magnetically Actuated Hydrogel Stamping-Assisted Cellular Mechanical Analyzer for Stored Blood Quality Detection." *ACS Sensors* 8, 2023, 1183–1191.

[17] M Hjiri. "Highly Sensitive NO2 Gas Sensor Based on Hematite Nanoparticles Synthesized by Sol–gel Technique." *Journal of Materials Science: Materials in Electronics* 31, 2020, 5025–5031.

[18] Li Zhou, Benzhao He, and Jiachang Huang. "One-Step Synthesis of Robust Amine- and Vinyl-Capped Magnetic Iron Oxide Nanoparticles for Polymer Grafting, Dye Adsorption, and Catalysis." *ACS Applied Materials & Interfaces* 5, 2013, 8678–8685.

[19] Yingyot Poo-arporn, Saithip Pakapongpan, Narong Chanlek, and Rungtiva P. Poo-arporn. "The Development of Disposable Electrochemical Sensor Based on Fe3O4-doped Reduced Graphene Oxide Modified Magnetic Screen-printed Electrode for Ractopamine Determination in Pork Sample." *Sensors and Actuators B: Chemical* 284, 2019, 164–171.

[20] Rutuja Prashant Gambhir, Sonali S. Rohiwal, and Arpita Pandey Tiwari. "Multifunctional Surface Functionalized Magnetic Iron Oxide Nanoparticles for Biomedical Applications: A Review." *Applied Surface Science Advances* 11, 2022, 100303.

[21] Saima Gul, Sher Bahadar Khan, Inayat Ur Rehman, Murad Ali Khan, and MI Khan. "A Comprehensive Review of Magnetic Nanomaterials Modern Day Theranostics." *Frontiers in Materials* 6, 2019, 179.

[22] Nene Ajinkya, Xuefeng Yu, Poonam Kaithal, Hongrong Luo, Prakash Somani, and Seeram Ramakrishna. "Magnetic Iron Oxide Nanoparticle (IONP) Synthesis to Applications: Present and Future." *Materials* 13, 2020, 4644.

[23] Katsiaryna Kekalo, Katherine Koo, Evan Zeitchick, and Ian Baker. "Microemulsion Synthesis of Iron Core/Iron Oxide Shell Magnetic Nanoparticles and Their Physicochemical Properties." *Materials Research Society Symposium Proceedings*, 2012, 1416.

[24] Gauvin Hemery, Anthony C. Keyes, Jr., Eneko Garaio, Irati Rodrigo, Jose Angel Garcia, Fernando Plazaola, Elisabeth Garanger, and Olivier Sandre. "Tuning Sizes, Morphologies, and Magnetic Properties of Monocore Versus Multicore Iron Oxide Nanoparticles through the Controlled Addition of Water in the Polyol Synthesis." *Inorganic Chemistry* 56, 2017, 8232–8243.

[25] Yi Ouyang, Gaoshan Huang, Jizhai Cui, Hong Zhu, Guanghui Yan, and Yongfeng Mei. "Advances and Challenges of Hydrogel Materials for Robotic and Sensing Applications." *Chemistry of Materials* 34, 2022, 9307–9328.

[26] Fangli Gang, Le Jiang, Yi Xiao, Jiwen Zhang, and Xiaodan Sun. "Multi-functional Magnetic Hydrogel: Design Strategies and Applications." *Nano Select* 2, 2021, 2291–2307.

[27] Bibi Noorheen Haleema Mooneerah Neerooa, Li-Ting Ooi, Kamyar Shameli, Nuraina Anisa Dahlan, Jahid M. M. Islam, Janarthanan Pushpamalar, and Sin-Yeang Teow. "Development of Polymer-Assisted Nanoparticles and Nanogels for Cancer Therapy: An Update." *Gels* 7, 2021, 60.

[28] Manish K Jaiswal, Mrinmoy De, Stanley S. Chou, Shaleen Vasavada, Reiner Bleher, Pottumarthi V. Prasad, Dhirendra Bahadur, and Vinayak P. Dravid. "Thermoresponsive Magnetic Hydrogels as Theranostic Nanoconstructs." *ACS Applied Materials & Interfaces* 6, 2014, 6237–6247.

[29] Fei Gao, Wensheng Xie, Yuqing Miao, Dan Wang, Zhenhu Guo, Anujit Ghosal, Yongsan Li, Yen Wei, Si-Shen Feng, Lingyun Zhao, and Hai Ming Fan. "Magnetic Hydrogel with Optimally Adaptive Functions for Breast Cancer Recurrence Prevention." *Advanced Healthcare Materials* 8, 2019, 1900203.

[30] Dong-in Kim, Hyoryong Lee, Su-hyun Kwon, Hyunchul Choi, and Sukho Park. "Magnetic Nano-particles Retrievable Biodegradable Hydrogel Microrobot." *Sensors and Actuators B: Chemical* 289, 2019, 65–77.

[31] Babak Ziaie, and Ronald A Siegel. Sensor Having Ferrogel with Magnetic Particles. US Patents US20150087945A1, 2017.

[32] J Heikenfeld, A Jajack, J Rogers, P Gutruf, L Tian, T Pan, R Li, M Khine, J Kim, J Wang, and J Kim. "Wearable Sensors: Modalities, Challenges, and Prospects." *Lab on a Chip* 18, 2018, 217–248.

[33] Wanxin Tang, Zhen Gu, Yao Chu, Jian Lv, Li Fan, Xinling Liu, Feng Wang, Ye Ying, Jian Zhang, Yuning Jiang, Jiaying Cao, Anni Zhu. "Magnetically-oriented Porous Hydrogel Advances Wearable Electrochemical Solidoid Sensing Heavy Metallic Ions." *Chemical Engineering Journal* 453, 2023, 139902.

[34] Yuji Gao, Longteng Yu, Joo Chuan Yeo, and Chwee Teck Lim. "Flexible Hybrid Sensors for Health Monitoring: Materials and Mechanisms to Render Wearability." *Advanced Materials* 32, 2020, 1902133.

[35] Duanli Wei, Jiaqing Zhu, Licheng Luo, Huabo Huang, Liang Li, and Xianghua Yu. "Ultra-stretchable, Fast Self-healing, Conductive Hydrogels for Writing Circuits and Magnetic Sensors." *Polymer International* 71, 2022, 837–846.

[36] MPB van Bruggen, and JBA van Zon. "Theoretical Description of a Responsive Magneto-hydrogel Transduction Principle." *Sensors and Actuators A: Physical* 158, 2010, 240–248.

[37] Nipping Deng, Jinghang Li, Hao Lyu, Ruochuan Huang, Haoran Liu, Chengchen Guo. "Degradable Silk-based Soft Actuators with Magnetic Responsiveness." *Journal of Materials Chemistry B* 37, 2022, 7650–7660.

[38] Ali Aldalbahi, Peter Feng, Norah Alhokbany, Tansir Ahamad, and Saad M Alshehri. "Synthesis, Characterization, and CH4-Sensing Properties of Conducting and Magnetic Biopolymer Nano-composites." *Journal of Environmental Chemical Engineering* 4, 2016, 2841–2847.

[39] Muhammad Khatib, and Hossam Haick. "Sensors for Volatile Organic Compounds." *ACS Nano* 16, 2022, 7080–7115.

[40] Tao Li, Wen Yin, Shouwu Gao, Yaning Sun, Peilong Xu, Shaohua Wu, Hao Kong, Guozheng Yang, and Gang Wei. "The Combination of Two-Dimensional Nanomaterials with Metal Oxide Nanoparticles for Gas Sensors: A Review." *Nanomaterials* 12, 2022, 982.

[41] Jin Wu, Zixuan Wu, Songjia Han, Bo-Ru Yang, Xuchun Gui, Kai Tao, Chuan Liu, Jianmin Miao, and Leslie K. Norford. "Extremely Deformable, Transparent, and High-Performance Gas Sensor Based on Ionic Conductive Hydrogel." *ACS Applied Materials & Interfaces* 11, 2019, 2364–2373.

[42] Ye Chen, Masataka Abe, Ikuyoshi Tomita, Yuta Kurashina, Yoshitaka Kitamoto. "Synthesis and Characterization of pH-responsive Ferrogels Comprising Sulfamethazine-based Polymer and Magnetic Nanoparticles for Sensing Ammonia Gas." *Journal of Magnetism and Magnetic Materials* 565, 2023, 170201.

[43] Morteza Mahmoudi, Heinrich Hofmann, Barbara Rothen-Rutishauser, and Alke Petri-Fink. "Assessing the In Vitro and In Vivo Toxicity of Superparamagnetic Iron Oxide Nanoparticles." *Chemical Reviews* 112, 2012, 2323–2338.

[44] Nuraina Anisa Dahlan, Aung Thiha, Fatimah Ibrahim, Lazar Milić, Shalini Muniandy, Nurul Fauzani Jamaluddin, Bojan Petrović, Sanja Kojić, and Goran M. Stojanović. "Role of Nanomaterials in the Fabrication of bioNEMS/MEMS for Biomedical Applications and Towards Pioneering Food Waste Utilisation." *Nanomaterials* 12, 2022, 4025.

10 Hydrogels with Thermal Responsiveness

Mirian A. González-Ayón,
Yadira D. Cerda-Sumbarda,
Arturo Zizumbo-López, and Angel Licea-Claverie
Tecnológico Nacional de México/Instituto Tecnológico de
Tijuana, Tijuana, México

10.1 INTRODUCTION

Hydrogels composed by tridimensional covalently cross-linked polymer chains have been extensively investigated. These materials are characterized by their ability to imbibe water and swell in aqueous environments, without deforming. Hydrophilic pendant groups, such as alcohols, carboxylic acids, and amides, are responsible for this behavior through the formation of hydrogen bonds polymer-water. Their physical-chemical properties depend on their constitution, grafting, cross-linking, coil conformation, morphology, and the nature of their constituents. Environment responsive/sensitive hydrogels, also named "smart" hydrogels, show generally a reversible sol-gel transition within their chains, accompanied by a phase separation of domains that can respond to external environment stimuli as temperature, pH, ionic strength, magnetic field, concentration of chemicals and solvents, and so on.[1] Temperature-responsive "smart" hydrogels have aroused great interest for potential applications in numerous fields, as controlled drug delivery systems, chemical separation, sensors, catalysts, and enzyme immobilization, due to its ability to swell and shrink reversibly as a function of a temperature stimulus.[2]

The main synthetic method for hydrogel preparation is a free-radical polymerization initiated by a redox method using a cross-linking agent and forming a stable tridimensional network structure. Other synthetic strategies explored make use of chemical or photochemical polymerization like click chemistry, grafting reactions, cross-linking of synthetic or natural polymers by esterification, amidation, or other catalyzed reactions.[3] Different structures were synthetized, like linear homopolymers forming the network, interpenetrated, and semi-interpenetrated copolymers, with ionizable or amphiphilic moieties. Also, physical cross-linking can be used to obtain non-covalent bonds like hydrophobic, ionic, and intermolecular hydrogen bonds in an amphiphilic block copolymer network.

Thermoresponsive hydrogels are particularly interesting because their response temperature can be easily designed and artificially tailored by varying some synthesis parameters. The response temperature in hydrogels is attributed to the balance between

DOI: 10.1201/9781003340485-10

the hydrophilic and hydrophobic polymeric components within the macromolecule network structure and by the interaction of free functional groups with the specific solvent. Based on these interactions, hydrogels have been divided into two categories: hydrogels that show a volume phase transition temperature (VPTT) upon heating based on the principle of lower critical solution temperature (LCST) and hydrogels that show phase separation upon cooling (upper critical solution temperature, UCST). Hydrogels exhibiting VPTT upon heating/cooling are thermoresponsive polymers, and are dominating in biomedical field and similar applications.[4]

In addition to swelling rate, mechanical properties, and extent of thermal response the addition of another response to a second stimulus, modulates the behavior and expand the applications of thermoresponsive "smart" hydrogels. A common approach with this goal in mind has been the preparation of interpenetrating polymer network (IPN) hydrogels, which consist of at least two networks, where one of them is being synthesized and/or cross-linked within the immediate presence of the other, with no covalent bonds between them. One network is thermoresponsive while the second one may be responsive to a second stimulus or could improve mechanical properties. On the other hand, the incorporation of nanoparticles to yield nanocomposite hydrogels may improve certain properties without affecting the thermal responsivity. In composite hydrogels, the mechanical properties are significantly influenced by the shape, orientation, size, continuity, and composition ratios of their reinforcing phases.[5] Both in the IPNs and composite hydrogels, the modification of polymer chains with functional groups that respond to stimuli or the addition of responsive fillers expand their properties and responses to stimuli such as pH, light, magnetic fields, chemical agents, and so on.

This chapter focuses on thermoresponsive hydrogels, IPNs, and nanocomposite hydrogels obtained from natural and/or synthetic polymers; their different classes; and the structural, chemical, and physical properties responsible for their responsive behavior, along with strategies to control their temperature responsiveness and to design switchable materials for applications in the biomedical field, biotechnology, sensors, and food packaging, among others.

10.2   THERMORESPONSIVE BEHAVIOR IN POLYMERS

Thermoresponsive polymers are a class of smart materials that show a unique sol-gel transition above a certain temperature in a specific solvent. According to their response to a change in temperature, they are classified as polymers that become insoluble above a critical temperature (LCST), representing a minimum in a concentration-temperature phase-curve and as polymers that precipitate and undergo a phase change below a critical temperature (UCST), representing a maximum in a concentration-temperature phase-curve. These transitions in covalently cross-linked polymer networks are known as volume phase transition temperature (VPTT) or transition temperature. In the case of linear polymers, this transition is shown as a change in the solubility of the polymer, changing reversibly from a completely soluble to an insoluble state (LCST by heating or UCST by cooling); while in polymer networks a change in volume is observed, changing from a swollen to a shrunk state by heating (LCST based), or by cooling (UCST based). In both cases, the phenomenon is reversible.

If the hydrogel contains structures with hydrophobic and hydrophilic interacting moieties, they undergo a collapse or expansion when they reach a specific temperature where their chains possess a LCST or UCST. Therefore, the polymer chains may be sensitive to temperature due to the overall contribution between the hydrophilic and hydrophobic moieties within the chains. For polymer chains with a LCST, when they are at temperatures below the LCST value, chains are unfolded, the osmotic pressure is high, and the corresponding hydrogel is in the swollen state mainly by hydrogen bonding interactions between the polar moieties and water. When the temperature is increased, the efficiency of hydrogen bonds of the polymeric chains with water decrease, chains gain a hydrophobic character causing that the hydrogel shrinks. In more detail, if the temperature is increased above the LCST of the chains, the coil conformation change to yield a phase separation by favoring intermolecular interactions between polymer chains instead of water-polymer chains, producing a collapse of the chains forming hydrophobic micro-domains, and releasing water; resulting in the overall shrinkage of the hydrogel (Figure 10.1).

While VPTT (from LCST or from UCST) is defined as a single temperature from a thermodynamic point of view, real hydrogels undergo a phase transition over a range of temperatures owing to inherent polydispersity coming from different chain lengths, branching, crystallinity, and so on within the polymer network.

In general, LCST and VPTT (from LCST) represent a point of minimum solubility and a minimum interaction polymer-water, respectively, and the polymers that show this behavior are also called "negative temperature-sensitive polymers", while UCST and VPTT (from UCST) refers to the point of maximum solubility of the polymer or the highest polymer-water interaction and are also referred to as "positive temperature-sensitive polymers". At LCST, an order-less arrangement of

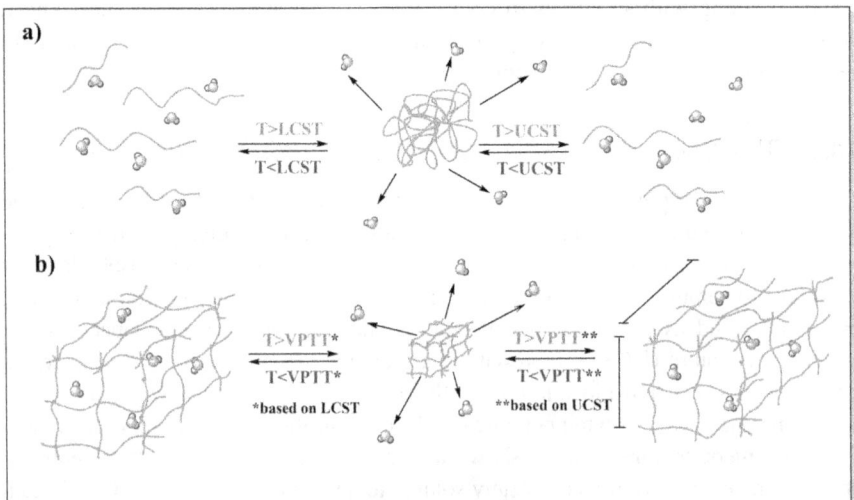

FIGURE 10.1 Schematic representation of thermoresponsive polymers (a) and thermoresponsive hydrogels (b) at lower and higher temperatures than the transition temperature.

water molecules causes an increase in entropy of the water in the system, which turns out to be more than the enthalpy observed by water hydrogen attached to the polymer. Therefore, LCST is governed by the entropy, whereas UCST is regulated by the enthalpy of the system.

When the temperature of an aqueous solution of a thermoresponsive polymer exhibiting LCST behavior is increased, the polymer chains show coil to aggregate transition (Figure 10.1). This transition is caused by a break in the hydrogen bonding interactions between the polymer and water that produce a phase separation. It has been demonstrated that both hydrogen bonding and hydrophobic interactions in a polymer-solvent system are responsible for phase transition and hydrated random coil to hydrophobic globule transition above a critical solution temperature. In systems that have a UCST transition, this behavior is inverse, that is when temperature increases above the UCST value, the polymeric systems have better interactions with the water through hydrophilic interactions.

In the case of polymers with UCST behavior, the transition temperature can be increased by copolymerization with hydrophobic comonomers. This can be attributed to the fact that polymers dissolve in a solvent when the Gibbs energy of dissolution is negative ($\Delta G < 0$). Polymer solutions show a UCST when both the enthalpy of dissolution (ΔH) and entropy of dissolution (ΔS) are positive. At the phase transition temperature, dissolution and phase separation are in equilibrium ($\Delta G = 0$); therefore, UCST = $\Delta H/\Delta S$.

Table 10.1 shows a selection of polymers with either LCST or UCST behavior in aqueous media.[6,7] These transitions are not restricted to an aqueous solvent environment, but only the aqueous systems are of interest for specific applications as biomedical materials and the vast majority of the reports are focused on the study of these transitions in water. Besides, polymers with LCST are more dominant in biomedical and similar applications, compared to polymers with UCST, mainly because of the abundant availability of LCST polymers, the obvious advantage of having LCST close to the body temperature, and the lack of stable UCST behavior. It has to be noted that the thermoresponsive behavior depends on the solvent interaction with the polymer and the hydrophilic/hydrophobic balance within the polymer molecules. The transition temperature can be strongly dependent on factors such as molecular weight, concentration, solvent quality, salt concentration, and so on.[8] Additionally, the UCST is affected too by pH, ionic strength, and the presence of electrolytes, so ionic radical initiators or carboxylic side or end groups can significantly suppress this transition. Obviously, the transition temperature has to be determined for the setting of the potential applications. Very few polymers are known that exhibit a UCST in water. The UCST behavior itself arises from markedly strong intra- and intermolecular polymer interactions, while entropic contributions are comparatively small.

Thermosensitive switchable hydrogels, physically or chemically cross-linked, have been synthesized using different polymerization techniques. It is known that chemically cross-linked polymer networks are more stable materials and have better mechanical properties, compared to hydrogels formed by physical bonds. Nevertheless, all hydrogels are characterized by poor mechanical strength due to their high water content and low efficiency of the dissipation process. In this sense, authors have proposed hybrid hydrogels, building molecular architectures as

TABLE 10.1
Phase Transition Temperatures for Polymers with LCST and UCST Behavior in Water[6,7]

LCST-type Polymers

Poly(diethylene glycol methyl ether methacrylate) (PDEGMA) (~27 °C)

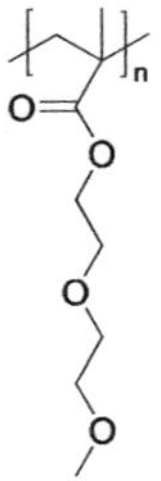

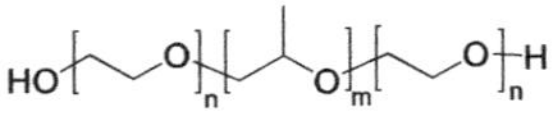

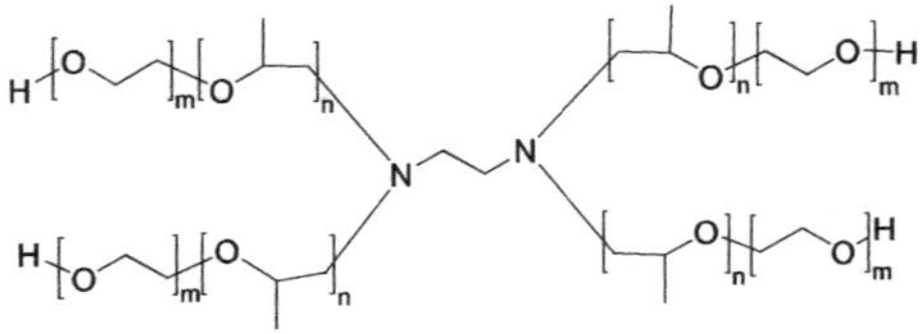

Poly(2-ethoxyethyl vinyl ether) (PEOVE) (~40 °C)

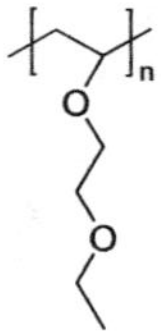

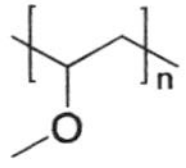

Polyethylene oxide (PEO) (100–180 °C)

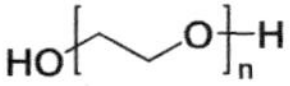

Ethers

*Pluronic® and Tetronic® (30–40 °C)

Poly(vinyl methyl ether) (PVME) (~40 °C)

Poly(2-methoxyethyl vinyl ether) (63 °C)

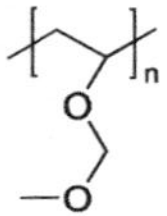

(Continued)

TABLE 10.1 (Continued)
Phase Transition Temperatures for Polymers with LCST and UCST Behavior in Water[6,7]

LCST-type Polymers

Poly(propylene glycol) (PPG)
(~ 50 °C)

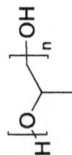

Hydroxypropyl cellulose (~ 55 °C)

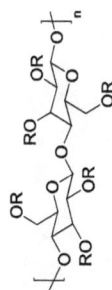

R= H, [CH₂CH(CH₃)O]H

Chitosan (CS)(~ 32 °C)

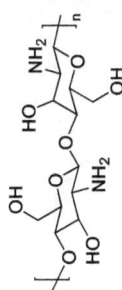

Poly(vinyl alcohol) (PVA)
(~125 °C)

Alcohols

Hydroxypropyl methylcellulose
(70 °C)

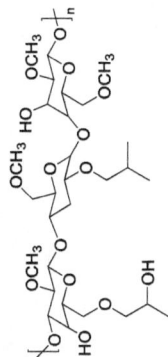

Methylcellulose (MC) (~ 80 °C)

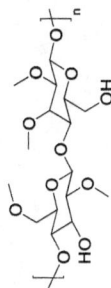

Xyloglucan (27–37 °C)

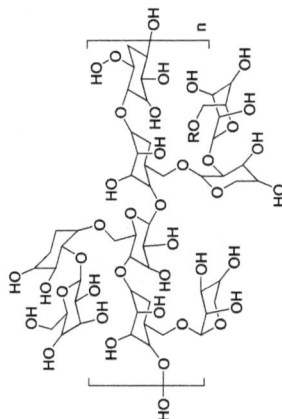

(Continued)

TABLE 10.1 (Continued)

Phase Transition Temperatures for Polymers with LCST and UCST Behavior in Water[6,7]

LCST-type Polymers

Poly(ethylene glycol) (PEG)
(~ 120 °C)

Poly(siloxethylene glycol)
(10–60 °C)

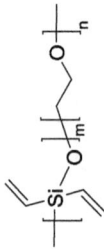

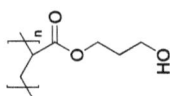

Hydroxypropyl acrylate (16 °C)

Poly(N-isopropylacrylamide)
(PNIPAM) (~32 °C)

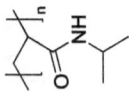

Poly(N,N-diethylacrylamide)
(PDEAM) (32–34 °C)

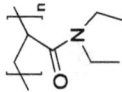

Amides

Poly(N-ethylacrylamide) (PEA)
(82 °C)

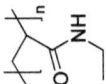

Poly(N-vinylcaprolactam)
(PNVCL) (30–50 °C)

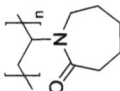

(Continued)

TABLE 10.1 (Continued)
Phase Transition Temperatures for Polymers with LCST and UCST Behavior in Water[6,7]

LCST-type Polymers

Poly(2-dimethylamino ethyl methacylate) (PDMAEM) 47 °C (pH 7), 35 °C (pH 10)

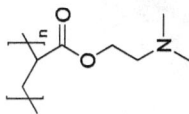

Poly(N-vinyl pyrrolidone) (PVP) (~ 160 °C)

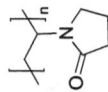

Poly(N-ethyl oxazoline) (~ 65 °C)

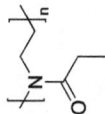

Poly(N-isopropyl oxazoline) (~ 36 °C)

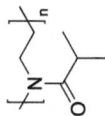

Poly(N-propyl oxazoline) (~ 24 °C)

Poly(silamine) (~ 37 °C)

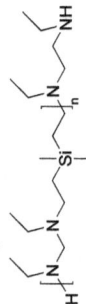

Poly(vinyl methyl oxazolidone) (PVMO) (~ 65 °C)

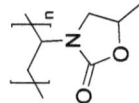

(Continued)

TABLE 10.1 (*Continued*)

Phase Transition Temperatures for Polymers with LCST and UCST Behavior in Water[6,7]

LCST-type Polymers

Poly(*N*-methacryloylasparagineamide) (33 °C)

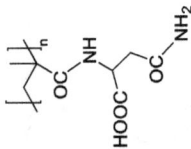

Poly(*N*-acryloylglutamineamide) (30 °C (in PBS))

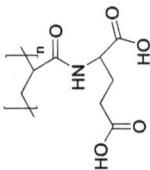

Poly(methacrylic acid) (PMAAc) (~75 °C)

Other synthetic polymers

Polyphosphazene derivatives (33–100 °C)

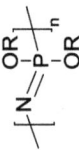

Gellan gum (50–60 °C)

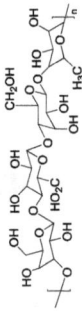

UCST-Type Polymers

Poly(*N*-acryloylasparagineamide) (22 °C)

Poly(ethylene oxide) (PEO) (296 °C)

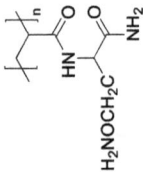

Poly(*N*-acryloylglycinamide) (PNAGA) (60 °C)

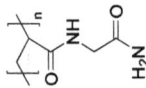

(Continued)

TABLE 10.1 *(Continued)*

Phase Transition Temperatures for Polymers with LCST and UCST Behavior in Water[6,7]

LCST-type Polymers

Poly(vinyl methyl ether) (PVME)**(−15 °C and −25 °C)

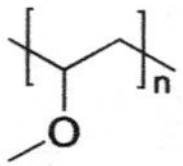

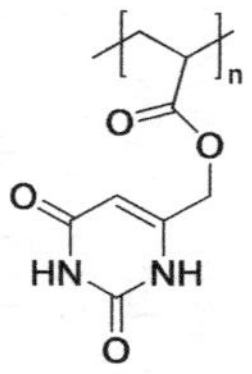

Poly(vinylalcohol) hydrophobically modified (>100 °C)

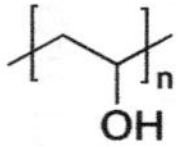

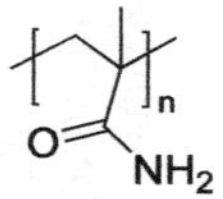

Poly(hydroxyethyl methacrylate) (PHEMA) (>100 °C)

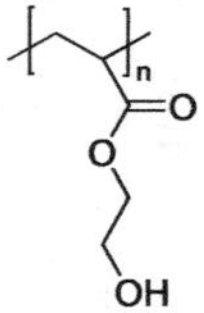

Poly(uracilacrylate) (5–60 °C)

Poly(methacrylamide) (10–80 °C)

Poly(3-dimethyl (methacryloyloxyethyl) ammonium propane sulfonate (PDMAPS)***(32 °C to 15 °C)

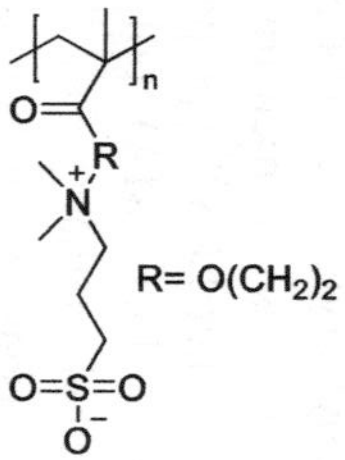

R= O(CH₂)₂

Notes

* diblocks and triblocks

** at low and high concentrations

*** at 0.1 wt% to 8 wt%

interpenetrating or semi-interpenetrating polymer networks (IPNs, or semi-IPNs) with different functionalities to generate polymer domains to induce opposite charges or phase separation of polymer chains which are driven to improve the mechanical properties. On the one hand, semi-IPNs are hydrogels with a second polymer within an already polymerized hydrogel either in the absence of a cross-linker and on the other hand, full-IPNs are carried out in the presence of a cross-linker. Typical semi-IPN- and full-IPN-hydrogels have been synthesized in a variety of compositions and formats for diverse applications in the biomedical field, biotechnological applications, sensors, and food packaging, among many others.

Poly(N-isopropylacrylamide) (PNIPAM)-based hydrogels are the most studied as single hydrogel and as interpenetrating network, with a LCST located around 32 °C.[2] Likewise, other LCST-type polymers such as poly(diethylene glycol methyl ether methacrylate) (PDEGMA), poly(N-vinylcaprolactam) (PNVCL), and poly(2-dimethylamino ethyl methacrylate) (PDMAEM) are the most widely used in the synthesis of thermosensitive hydrogels (Tables 10.2, 10.3, and 10.4). Some polymers chemically derived from natural polymers like methylcellulose, xyloglucan, and chitosan also exhibit LCST behavior (Table 10.1).

Typical UCST systems are based on a combination of synthetic polymers like acrylamide (AAm) and acrylic acid (AAc); polysaccharides like carrageenans, starch, agarose, and gellan gum; or proteins like gelatin, collagen, and elastin-like polypeptides have been explored in obtaining thermoresponsive hydrogels, too.[9]

On the other hand, polymers with both LCST and UCST transitions have been reported; only a handful of hydrogels with both LCST and UCST transitions are designed. LCST and UCST hydrogels have a weak mechanical strength, due to the existence of large amounts of embedded water molecules.

In general, both LCST (VPTT in hydrogels) and UCST (VPTT in hydrogels) can be modified using specific architectures, and combining comonomers with hydrophilic and hydrophobic characteristics as well as their concentration.

10.2.1 HYDROGELS WITH VPTT BASED ON LCST

In aqueous systems, polymerization with a hydrophobic comonomer will lower the LCST, and the addition of a hydrophilic comonomer will increase the LCST and therefore the LCST-based VPTT.[10]

Lehmann *et al.* report the synthesis of PNIPAM-based hydrogel films, using N,N'-methylenebisacrylamide (MBA), and poly(ethylene glycol) diacrylate (PEGDA) as cross-linkers. Results show changes in water content between swollen and collapsed state of the hydrogel increasing with decreasing cross-linking density and increasing size of the cross-linker, resulting in bigger meshes in the network. Since the elasticity of the hydrogels decreases upon collapse, the storage modulus (G'), and the loss modulus (G") increase when increasing the temperature. The hydrogels of PNIPAM-PEGDA displayed the greatest change in mechanical properties. The hydrogels with the highest deswelling ratio, PNIPAM-PEGDA (0.5 mol%) and PNIPAM-MBA (0.25 mol%), turned out to be materials with the potential for soft touch application due to its great variance in elasticity and water content when changing from the swollen to the collapsed state at around 35 °C.[11]

Thermo-/pH-responsive hydrogels based on TEMPO-oxidized nanofibrillated cellulose from wheat straw with food-grade cationic-modified poly(NIPAM-*co*-acrylamide) were reported by Shaghaleh *et al.* These materials revealed desirable enhancement in antimicrobial properties and pH/thermal-responsive behavior. The VPTT was between 30 and 40 °C and the *in vitro* delivery and release studies with natamycin revealed the faster release in preferred low-pH media and at temperatures close to the VPTT.[12]

Other interesting polymers have been investigated, like poly(ethylene oxide)-*b*-poly(propylene oxide)-*b*-poly(ethyelene oxide) (PEO-PPO-PEO) copolymers (also termed Pluronics®), PDMAEM-copolymers, natural polymers as cellulose, and chitosan grafted or copolymerized, which also show a LCST-based thermal response.

Dragan *et al.* reported the preparation of multi-stimuli-responsive semi-interpenetrating hydrogels by cross-linking polymerization of PDMAEM in the presence of potato starch or anionically modified potato starch. The last one allowed the decrease of the VPTT of the composite cryogels by its electrostatic interactions with the matrix. The VPTT obtained was situated in the range 36–39 °C, and is attributed to the PDMAEM concentration and the nature and content of the polysaccharide. To evaluate the capacity of these materials as drug delivery systems, diclofenac sodium was taken as a model acidic drug. The controlled delivery of diclofenac from the semi-interpenetrating hydrogel was optimized by the investigation of the effects of pH, temperature, and cycling changes in the release temperature. The results show that these semi-IPN hydrogels are promising for the sustained delivery of diclofenac in simulated intestinal fluid.[13]

Thermoresponsive semi-IPN hydrogels from poly(*N,N*,dimethylacrylamide) (PDMAM) and PNIPAM were obtained by Guo *et al.* In the PDMAM network interpenetrated by PNIPAM chains, the phase transition of PNIPAM occurs at a lower temperature compared to other hydrogels and gives rise to large micro-domains with a very weak percolation through the PDMAM network. The enhancement of the mechanical properties above the transition temperature is indeed very low with a threefold increase of the elastic modulus and little improvement of the fracture energy, with the isolated PNIPAM phase acting mainly like a filler in polymer materials. On the other hand, the PNIPAM network interpenetrated by PDMAM chains brings a large improvement of the mechanical properties at high temperatures with a ten-fold increase of the modulus and a very high fracture energy. Nevertheless, the absence of connectivity between the PNIPAM and PDMAM phase is responsible for the large scale of the phase separation, which induces a small collapse of the network at 60 °C when starting from the preparation stage.[14]

10.2.2 HYDROGELS WITH VPTT BASED ON UCST

A positive thermoresponsive IPN hydrogel of poly(acrylic acid)-*graft*-β-cyclodextrin (PAAc-*g*-CD) and polyacrylamide (PAM) was obtained by Wang *et al.*, the UCST-based VPTT was approximately 35 °C. Additionally, increases in the VPTT and non-sensitivity to changes in salt concentration were observed for the IPN hydrogel versus the normal IPN hydrogel PAAc/PAM without β-cyclodextrin (CD).[15]

The polymer-polymer interactions at UCST can derive from Coulombic interactions (polybetaines) or in non-ionic polymers from hydrogen bonding. Poly(N-acryloyl glycinamide) (PNAGA) is the most famous non-ionic UCST polymer; however, due to a growing interest in this field, an increasing variety of UCST polymers have been studied in the last years. Derivatives of PNAGA, poly(uracil acrylate), and urea-modified polymers show UCST behavior dependent on hydrogen bonding. In the case of poly(uracil acrylate), the UCST transition cannot be reversible due to the significant hydrolysis of the ester bond.[7] On the other hand, copolymers like poly(acrylamide-co-styrene), poly(NAGA-co-diacetone acrylamide), poly(N-[2-hydroxy-propyl]methacrylamide), and poly(acrylamide-co-acrylonitrile) have shown a UCST behavior in aqueous solution.[16]

Alazri *et al.* synthesized two different types of multi-stimuli-shrinking interpenetrating and terpolymeric hydrogels by copolymerization of PNIPAM, 2-hydroxyethyl methacrylate (HEMA), and 2-acrylamido-2-methylpropanesulfonic acid (AMPSA), using classical and frontal polymerization. The results show that some IPNs exhibit marked stimuli-shrinking properties, while some terpolymers present the opposite behavior. IPNs swell more than terpolymers and show a sharper stimuli response, with a larger swelling ratio variation. The first synthesized network derived from NIPAM polymerization showed a VPTT based on LCST at around 35 °C, while when the IPN of poly(HEMA-co-AMPSA) was first obtained, the materials were not thermoresponsive anymore. The thermoresponsive swelling was different for both materials. At a higher AMPSA concentration, terpolymers swell strongly above a UCST-based VPTT. This behavior underlines the crucial influence of the order in which the two networks are synthesized.[17]

10.2.3   HYDROGELS WITH VPTT BASED ON BOTH LCST AND UCST BEHAVIOR

Biobased hydrogels are very attractive, but an efficient approach to combine LCST and UCST is rare. As example, cellulose derivatives are well known to show LCST behavior in water, but UCST-type hydrogels produced by the interaction of cellulose with other substances have not been reported. However, the copolymerization of methyl cellulose (MC) with PAM provides well-defined polymer networks with high mechanical strength and a LCST- and UCST-based VPTT behavior.[18] Ding *et al.* obtained stimuli-responsive methylcellulose-$graft$-polyacrylamide (MC-g-PAM) hydrogels with tunable LCST- and UCST-based VPTT in one system (LCST based > UCST based) for a transparent "temperature window", which can be shifted via using varied cellulose derivatives or adding inorganic salts.[18]

Thermoresponsive polymer systems showing widely tunable UCST/LCST behaviors based on amino acid–derived vinyl polymers, were reported by Higashi *et al.* Four amino acids of Gly, Ala, Phe, and Val and their methyl esters were employed. The COOH-carrying Ala-based polymer hydrogels displayed an UCST behavior-based VPTT in water below pH 2.0 due to thermoreversible hydrogen bonding of the pendent COOH groups, while the Gly-based polymer did not show any phase separation. Methylation of COOH groups induced a LCST-based VPTT behavior. Widely tunable UCST/LCST VPTT behaviors were achieved from 18 °C to 73 °C by using copolymers from different monomer combinations in the hydrogel preparation.[19]

10.3 DUAL THERMO- AND PH-RESPONSIVE HYDROGELS

Many of the hydrogel applications depend on the interval of response, response time, porosity, and toxicity; different strategies have been investigated to decrease the response time, to increase the capacity of swelling, and the chemicals load, and to improve the mechanical behavior. Physical characteristics such as porosity or cavities in the materials are often responsible for a better response. Homogeneous or heterogeneous morphology can produce a regular or irregular internal structure. To get a better response to external stimuli, shape differences have been proposed: films, microparticles, microfibers, rods, spheres, membranes, microgels, nanogels, and fibers are among the physical structures tested.

Materials with cationic, anionic, or hydrophobic functionalities in their internal structure can have a dual pH and temperature response associated with the structure and morphology of the system. Then, swelling and deswelling properties depend on both temperature and pH. These functionalities can help to achieve better water retention, so gels can swell due to the diffusion of water into the polymer network by the formation of hydrogen bonds between the polar groups of polymer moieties and water.

In this section, we discuss the behavior of smart hydrogels capable of responding to two environment stimuli like temperature and pH. These kinds of polymeric materials should have some specific properties. Generally, they are macromolecular structures with both hydrophilic moieties, such as carboxylic or amine groups, and hydrophobic moieties as alkyl or aryl groups incorporated into the main chains. These smart polymeric materials are able to experience a small or large change in coil conformation, resulting in phase separation as a result of molecular interactions by the effect of both pH and temperature changes.

Figure 10.2 shows a reversible swelling and deswelling for acid, basic, or amphiphilic hydrogels. For acid hydrogels, variations of pH into the polymer

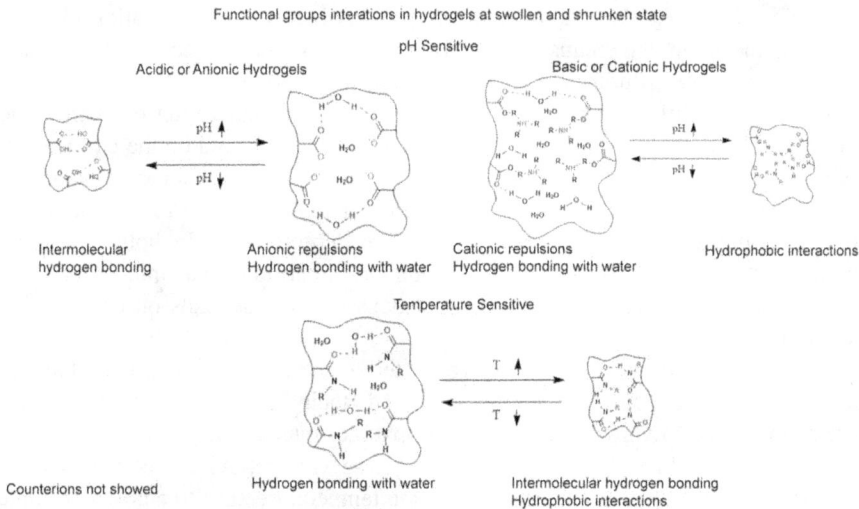

FIGURE 10.2 Swelling behavior of temperature- and pH-sensitive hydrogels.

network can induce a network ionization with a change in swelling capacity. The molecular structure of a cross-linked hydrogel is formed by chain lobules in swollen or shrunken state depending on molecular interactions between water and the polymeric chains. At a low pH, below the pK_a of the acid groups, chains are protonated and the polymer chains are folded and collapsed by hydrogen bonding intermolecular interactions; when the pH is increased, above the pK_a, the intermolecular interactions are broken by deprotonating of acid moieties and formation of carboxylate anions, chains begin to unfold forming a water hydrogen bonding with the polymeric chains, and anionic repulsion charges lead the hydrogel to swell. Water penetrates by diffusion into the polymer chains, increasing the osmotic pressure, gaining volume, and staying in a swollen state. This process is reversible when the pH is decreased below the pK_a, the intermolecular interactions are stronger, the chains begin to fold again, and the gel is contracted, releasing water molecules.

Otherwise, when a basic hydrogel is below the pK_a of its functional groups, the chains are protonated in the cationic state, the chains form eliminate hydrogen bonds with water producing an expansion by strong electrostatic repulsions with fixed cations, chains are unfolded and the osmotic pressure is high, and the pores are in the swollen state. If the pH is increased above the pK_a, the quaternary amines begin to deprotonate, forming amine groups until the concentration of cations are the same as those of the amine groups and the chains begin to collapse by intermolecular interactions between the polymeric chains. Above the pK_a, the chain moieties are deprotonated and the intermolecular hydrophobic interactions are higher and stronger; the pores are contracted and the hydrogel collapses. The process is reversible and the equilibrium depends on composition, ionic strength, hydrophobicity of the polymeric chains, and diffusion of water in a transient process.

When a polymer chain contains two ionizable groups with one of them with a pK_{a1} less than the other pK_{a2}, the swelling behavior depends on the concentration of ionizable groups. At about the isoelectric point, where the concentration of fixed groups is the same, the chains are mainly collapsed by intermolecular electrostatic interactions between the opposite ions fixed in the polymer chains. If the pH is decreased below pKa_1 or increased above pKa_2, the concentration of some ionic groups will become dominant and the swollen state will depend on the equilibrium forces and overall effects between ionic species and hydrophobic interactions. Below the pKa_1, the carboxylic acid and amine moieties are protonated, and above the pKa_2, the amine and carboxylic acids are deprotonated. In both cases, the hydrogen bonds with water and hydrophobic intermolecular interactions are responsible for the pH and temperature response, which depends on composition and ionic strength.

A dual pH and thermal-responsive material should have a critical balance between the intermolecular interactions and their medium. Van der Waals interactions, like hydrogen bonds, hydrophobic interactions, and electrostatic forces, can modulate the response of the phase change. As more units with a hydrophobic character are incorporated to the temperature-sensitive polymer, ionic units are more hindered between the chains, the interaction with the aqueous medium decreases, the diffusion of water is slower, and the phase transitions become wider, displacing the transition temperature by a local intermolecular aggregation.[20] Then, the phase separation by temperature and pH effect is determined by the overall hydrophilicity and balance of the ionizable charge of the copolymers.

Some polymeric materials with dual pH and temperature responses are presented in Table 10.2. The PDMAEM homopolymeric hydrogel was reported by Taktac *et al.*, showing that this hydrogel can present a dual response effect to pH due to their ionizable groups of tertiary amines and to temperature by an association between the hydrophobic backbone and ethylene tails with their hydrophilic substituent amide. This showed a phase transition at about pHs of 3 to 4 and temperatures at about 45 °C, with rapid swelling and deswelling. This behavior was associated with the high free volume by the substituted group, as well as the homogeneous and uniformly open porous for water diffusion, which were well distributed in the morphological structure. Quaternary amine substituents and electrostatic repulsion between polymeric chains increased volume and swollen rate of the gel.[21]

In order to avoid the polymer collapse, Guo *et al.* reported a polymer combination of PNIPAM and hydrophilic poly(*N,N*-dimethylacrilamide) (PDMAM) in the same cross-linked architecture to maintain the swollen state both below and above the VPTT of PNIPAM, exhibiting strong thermo-toughness and fracture properties.[31]

Other authors like Paris *et al.* have synthetized copolymers of poly(DMAEM-*co*-DEGMA) and they reported phase transitions between 26 to 53 °C and a pH transition at about 6 to 8 for compositions between 50 to 10 mol% of DMAEM.[22] Emileh *et al.* synthetized a copolymer of DMAEM and butyl methacrylate (BuMA) and poly(DMAEM-*co*-BuMA), and they reported that the phase transition depends on concentration of BuMA. While BuMA was increased, the phase transition at 25 °C was decreased to pH values of about 7 to 5. For pH = 7, the temperature for phase transition of PDMAEM was decreased when BuMA was added and disappeared for concentrations of BuMA greater than 20 mol%. In general, when DMAEM was used as temperature-responsive entity, the swollen state was present at low temperatures and the collapsed state was present at high temperatures. The same occurred for pH sensibility; the swollen state was present at a low pH where the amine of DMAEM is protonated and the shrunken state is observed at high pH values where the amines are deprotonated.[23]

Salgado *et al.* reported a dual pH and temperature-sensitive PNIPAM hydrogel copolymerized with potassium carboxybutyl methacrylate (KBuMA) and potassium carboxydecyl methacrylate (KDeMA). For a pH lower than the pK_a of the ionizable comonomer and when the amounts of hydrophobic moieties increased in the copolymer hydrogel network, the temperature of the phase transitions was displaced to a lower temperature, and when the pH was higher than the pK_a, the temperature of the phase transitions were increased in any composition of the hydrogel until a maximum pH and after that, the temperature began to decrease as a result of ionic strength shielding.[20]

Other authors have proposed different strategies using a natural polymer like carboxymethyl cellulose (CMC), carboxymethyl chitosan (CMCS), and chitosan copolymerized with synthetic copolymers to get a more controlled swelling behavior with dual pH and T response, among others. Gungor *et al.* described

TABLE 10.2

Some Polymer Hydrogels with a Dual T and pH Responsiveness

M1	M2	Physical State	Cross-linker	VPTT	Temperature Response (pH = 7)		pH Response (T=20–25 °C)		Ref.
					Swollen	Shrunken	Swollen	Shrunken	
PDMAEM	----	Cross-linked	MBA	~45	T<40	T>50	pH<2	pH>8	[21]
PDMAEM	DEGMA	Cross-linked	TEGDMA	53–26	T<45	T>65	pH<6	pH>9	[22]
PDMAEM	BuMA	Cross-linked	EGDMA	~30	T<25	T>50	pH<4	pH>7	[23]
PNIPAM	KBuMA	Cross-linked	MBA	40–73	T<20	T>35	pH>7	pH<5	[20]
	KDeMA			~30	T<15	T>35	pH>10	pH<5	
PNIPAM	TBMAC	Cross-linked	MBA	32–33	T<28	T>38	pH>4	pH<2	[24]
PNIPAM	PMAAc	Cross-linked	EGDMA	36.6	T<25	T>40	pH>7.4	pH<1	[25]
PNIPAM	PVP	Cross-linked	MBA	33–40	T<35	T>40	pH>5	pH<1	[26]
			MAAm	30–37	T<30	T>35	pH>5	pH<1	
CMC	PNIPAM/AM	IPN	ECH MBA	33–40	T<32	T>45	pH>7	pH<5	[27]
CS	PNIPAM, AAc	Cross-linked	MBA	~33	T<32	T>35	4.3<pH<7.5	4.3>pH>9	[28]
CS/PVA		Semi-IPN							
CMCS	PAM	Semi-IPN	MBA	----	T<25	T>40	pH>8	pH<4	[29]
MEA	PEGA	Gelled	Physical	63	T<45	T>60	pH>14.6	pH<12.4	[30]
MEA	PEGA-AAc			58,37					

IPN: interpenetrating polymer network; **semi-IPN:** semi-interpenetrating polymer network; **KBuMA:** potassium carboxybutyl methacrylate; **KDeMA:** potassium carboxydecyl methacrylate; **TBMAC:** N-ter-buthylmaleimic acid; **CMC:** Carboxymethylcellulose; **CMCS:** Carboxymethyl chitosan; **CS:** Carboxymethyl chitosan; **MEA:** 2-Methoxyethyl acrylate; **PEGA:** Poly(ethylenglycol)methyl ether acrylate; **TEGDMA:** Tetraethylene glycol dimethacrylate; **EGDMA:** Ethylene glycol dimethacrylate; **MAAm:** N,N-tris Acryloyl melamine; **ECH:** Epoxychloropropane.

the behavior of chitosan with PNIPAM and AAc as cross-linked hydrogels and as a semi-IPN mixing them first with polyvinyl alcohol (PVA), both presented phase transitions of around 33 °C at pH 7; however, they presented a swollen state at a pH between 4.3 and 8.0 and two contractions with one at a pH below 4.3 due to the presence of hydrogen bonding of chitosan intermolecular chains and a second contraction at a pH above 9 due mainly by deprotonation of ammonium because of gaining a hydrophophic character and separation of phases. The PVA effect was attributed to a decrease in the pore size, which was denser and decreased the availability of free volume in the network.[28]

Wei *et al.* proposed a semi-IPN polymer network of cross-linked PAM with a concentration of CMCS, which contains some free amine and carboxyl moieties for the dual effect of temperature and pH sensitivity. Therefore, concentrations of CMCS between 5–35 wt% were reported and the swelling behavior was studied at room temperature. The hydrogel collapsed under highly acidic conditions at pH values below 3.5; the amino and carboxylic groups were protonated and the hydrogels contracted because of the intra- and inter-molecular hydrogen bond interactions between PAM and CMCS. When the pH increased between 3.5 to 7.4, the concentration of amine and carboxyl groups changed due to partial deprotonation, increasing the electrostatic repulsion between the ionic species and increasing the osmotic pressure with a resulting increase in the swelling degree. With a pH higher than 7.4, the swelling decreased again and the behavior was attributed to the high charge screening effect as a result of the excess of the counterions. Hydrogels of PAM with concentrations higher than 20 wt% of CMCS showed a temperature response, whereas the hydrogels with low concentration did not show any evident response due to the absence of hydrophobic groups in the side chains of PAM.[29] Huang *et al.* synthetized a copolymer as a physical hydrogel of 2-methoxyethyl acrylate (MEA) and poly(ethyleneglycol) methyl ether acrylate (MEA-*co*-PEGA) and with AAc (MEA-*co*-PEGA-*co*-AAc), which presented a swollen state at temperatures lower than 45 °C, pH higher than 14.6, and a shrunken state for temperatures higher than 60 °C and pH lower than 12.6. The effect of swelling was attributed to the water bonding with the polymer in the swollen state and polymer chain segregation by intermolecular interaction polymer-polymer in the shrunken state. The swelling and shrinking behavior by pH effect was attributed to the protonation of the fixed carboxylic acid groups at low pHs, maintaining the hydrogel in the shrunken state due to the interaction of hydrogen bonds both intra- and inter-molecular and in the swollen state for high pHs by the electrostatic repulsion by deprotonation of fixed carboxylic acid groups of the polymer chains.[30]

10.4   THERMOSENSITIVE NANOCOMPOSITE HYDROGELS

Thermosensitive composite hydrogels are considered advanced materials due to their properties. A composite material comprises at least two materials of different nature, to obtain a material with which the set of performance characteristics is greater than those shown by the components separately, in which one of these is called the continuous phase or matrix and the other is the dispersed phase. This, taken in terms of nanotechnology, implies that the materials are made up of two or more phases, where at least one of their three dimensions is at the nanometer scale, i.e., between 1 and 100 nanometers (nm); these are known as nanocomposites.

The ease of processing polymers combined with the improvements offered by nanoparticles has led to the fabrication of several new materials. It is worth noting that one of the most attractive advantages of nanocomposite hydrogels is that a small proportion ($\sim$<10 wt%) of nanoparticles or nanostructures is needed to obtain a significant improvement compared to traditional composite materials (Figure 10.3). This type of materials ranges from thermogelling hydrogels to thermoresponsive hydrogels with enhanced properties, dual-responsive thermosensitive hydrogels, and thermoresponsive hydrogels formed by interpenetrating networks as a matrix. Some of these materials are described below. Table 10.3 lists thermogelling injectable nanocomposite hydrogels, meaning nanocomposite hydrogels that are formed *in situ* at body temperature (37 °C).

Thermoresponsive hydrogels have been previously modified to improve mechanical properties, or to have a multifunctional material for drug release or contrast agents among others. New strategies based on thermoresponsive hydrogels have been tested now with nanometric components to obtain thermoresponsive

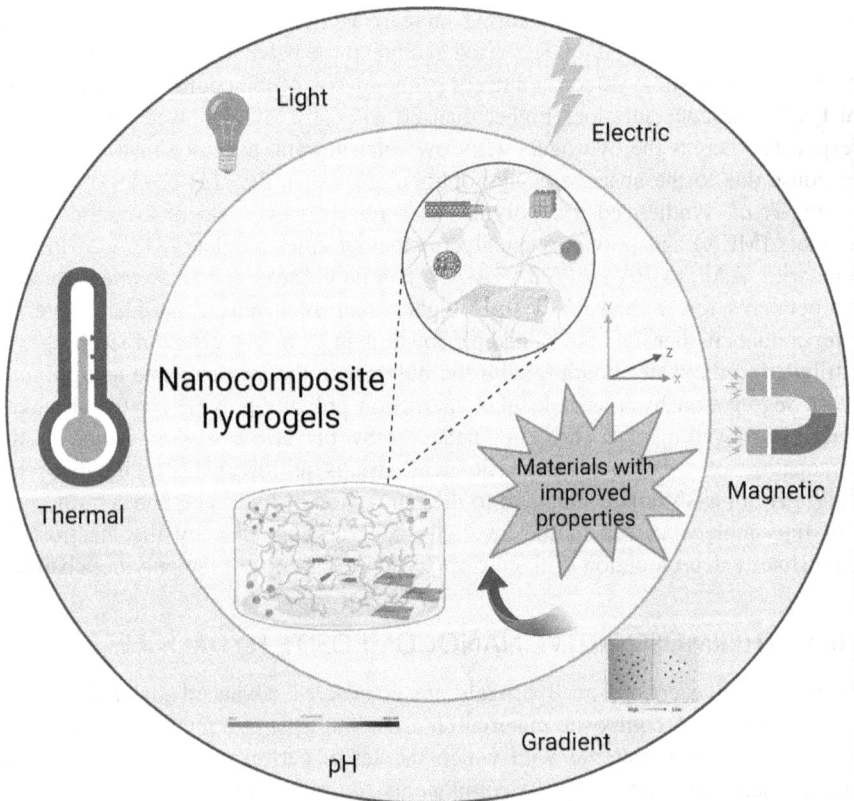

FIGURE 10.3 Scheme of responsive nanocomposite hydrogels.

TABLE 10.3

Examples of Thermogelling Nanocomposite Hydrogels Formulated in Recent Years[32]

Matrix Components		Filler Components
Chitosan	Nano-hydroxyapatite (nHAp)	CaCO$_3$ microspheres	Attapulgite
Combined	Collagen	Nano-bioactive glasses	GO nanosheets
β-glycerophosphate disodium salt/Collagen/ PMMA/ PVA/	CMCS	Borate-containing nano-	MWCNTs
Gelatin/ Oxidized alginate/ Hyaluronic acid/NaHCO$_3$/	Oxidized alginate	bioactive glasses	Copper
Silk fibroin/Telechelic difunctional PEG/Platelet-rich	TEMPO–oxidized cellulose	Heterotypic collagen	Nanodiamonds
plasma	nanofiber	type I and III	Fe$_3$O$_4$ NPs
		CS NPs	Chitin nano-whiskers
		Nano-fibrin and CaSO$_4$	
Collagen	nHAp		CNTs
Combined	β-tricalcium phosphate nano-bioactive glasses		AuNPs
Tyramine-functionalized PEG-*grafted* gelatin	silica nanoparticles/whitlockite nanoparticles		magnetic nanoparticles
	poly(L-lactide-*co*-ε-caprolactone) nanoyarns		Alendronate
Gelatin	Polyethyleneimine functionalized GO nanosheets complexed with DNA-VEGF Methacrylated gelatin/		
Combined	nanodiamonds/bFGF-loaded self-assembled heparin-pluronic nanogels		
Methacrylated/ Tyramine-functionalized PEG			
Alginate	Porous nHAp microspheres	Bioactive glass	Calcium silicate
Combined	Collagen–nHAp	Borosilicate	Sr-containing calcium
PVA/Gelatin/Keratin/ CMCS/Hyaluronate	microparticles	Ga-based glass	silicate
	nHAp mineralized on collagen	Bioactive glass 52S4 and	Tricalcium phosphate
	fibers	poly(caprolactone)	granules and PLGA
	CaF$_2$ nanoparticles	(PCL) microspheres	microspheres

(*Continued*)

TABLE 10.3 (Continued)

Examples of Thermogelling Nanocomposite Hydrogels Formulated in Recent Years[32]

Matrix Components	Filler Components	
• Biomineralized polymeric microspheres	• GO nanosheets • Chondroitin sulfate nanoparticles • Nano-curcumin using mPEG-PCL copolymer as a carrier	• Fluorenylmethoxycarbonyl-diphenylalanine self-assembling peptide • Nanocrystalline cellulose • CNTs • Nano-fibrin
Hyaluronic acid	Nano-hydroxyapatite/Nano-bioactive glasses/FITC-conjugated mesoporous silica nanoparticles/ GO nanosheets/PCL fibers grafted with PAAc and surface-modified with maleimide groups/HDI-Pluronic F127/Aldehyde modified cellulose nanocrystals	
Silated hydroxylpropylmethyl cellulose	Laponite nanoparticle	
Nanocellulose	Ultrafine fluorcanasite glass-ceramic particulates	
Cyclodextrins	Carboxymethyl cellulose-betulinic acid/Hydroxycamptothecine nanoparticles	
PNIPAM and methacrylated β-cyclodextrin-based macromer copolymer	Gold nanorods	
Hydrazide-functionalized poly(NIPAM-*co*-AAc) and aldehyde-functionalized dextrin	Starch nanoparticles	
Hydrazide-functionalized PNIPAM and aldehyde-functionalized dextran aldehyde-functionalized dextran	SPIONs/NIPAM-*co*-N-isopropylmethacrylamide microgels	

(Continued)

TABLE 10.3 (Continued)

Examples of Thermogelling Nanocomposite Hydrogels Formulated in Recent Years[32]

Matrix Components	Filler Components
Gellan gum	Various nano-bioactive glasses/nano-bioactive glasses
Fibrin	Rod-shaped PEG/SPION microgels
L-lysine-based amphiphilic hydrogelator/ dimethyl sulfoxide/phosphate buffer	Silver nanoparticles
PEG-PCL-PEG copolymer/collagen	nHAp
PEGDMA	nHAp/nano-bioactive glasses
4-armed star-shaped PEG with gold (III) chloride hydrate	Nano-bioactive glasses
Hydrophobically modified PEG	Hydrophilized silica nanoparticles
Dopamine-modified 4-armed PEG	Laponite
Thiol-modified hyaluronic acid/multi-armed PEGDA-melamine cross-linker	GO nanosheets
Poly(ethylene glycol) monoacrylate	CNTs
PEGDA	Poly(ethylene glycol) methacrylate-functionalized SWCNTs
4-armed PEG-acrylate/ PEG-dithiol cross-linker	CNTs
Vinyl sulfone functionalized 4-arm PEG	CNTs

(Continued)

TABLE 10.3 (Continued)

Examples of Thermogelling Nanocomposite Hydrogels Formulated in Recent Years[32]

Matrix Components	Filler Components
PEG-poly(β-aminoester urethane) triblock copolymer	Chondroitin sulfate nanogels
Oligo(poly(ethylene glycol) fumarate)	Gelatin microparticles
Poly(oligoethylene glycol methacrylate)	Cellulose nanocrystals
Laponite cross-linked, copolymer of PNIPAM and dimethyl acrylamide	nHAp
PNIPAM	Clay
Poly(NIPAM-co-DMAEM)	Sodium alginate modified magnetic GO nanosheets
Poly(NAGA-co-AAm)	Polydopamine (PDA) coated-gold nanoparticles
Poly(NIPAM-co-GMA-co-DBA-co-AAc)	Gelatin microparticles
Pluronic F127	Borate-based bioactive glass (13–93B3) microparticles/α-tricalcium phosphate/Nanoclay/Borate-based bioactive glass (13–93B3) microparticles/GO nanosheets or MWCNTs/GO nanosheets/ Alginate microparticles/ N,N-trimethyl chitosan-pluronic nanomicelles
PEG-PLGA-PEG	nHAp
PLGA dissolved in N-methyl pyrrolidone	nHAp/PGA nano- or micro-fibers
PLGA-PEG-PLGA	Nanoclay disk/laponite
PEG-PLGA-PNIPAM	Silica nanoparticles

(Continued)

TABLE 10.3 (Continued)

Examples of Thermogelling Nanocomposite Hydrogels Formulated in Recent Years[32]

Matrix Components	Filler Components
Chitin-PLGA	Whitlockite nanoparticles
PLA-pluronic L64-PLA	PLA-pluronic L35-PLA nanoparticles
PLA-pluronic 10R5-PLA/PVA	PLA-pluronic 10R5-PLA nanoparticles
PEG-PCL-PEG	nHAp/PLA-g-βTCP
mPEG-PCL-mPEG	GO nanosheets
PEG-PCL-PEG	Gold nanorods/acellular bone matrix granules
PVA-GMA	nHAp
PVA	Nano-bioactive glasses/nanocellulose
PHEMA	Rosette nanotubes and carbon nanofibers
Copolymer of 2-methacryloyloxyethyl phosphorylcholine	3-Aminopropyl triethoxysilane-modified
Aldehyde-functionalized monomer, 4-formylbenzoate ethyl methacrylate	Silica nanoparticles

CNTs: Carbon nanotubes; **MWCNTs:** Multiwalled carbon nanotubes; **SWCNTs:** Single wall carbon nanotubes; **AuNPs:** Gold nanoparticles; **GO:** Graphene oxide; **nHAp:** Nano-hydroxyapatite; **VEGF:** Vascular endothelial growth factor; **HDI:** Hexamethylene diisocyanate; **PLGA:** poly(lactic-co-glycolic acid); **PLA:** poly(lactic acid); **PLLA:** poly(l-lactide); **GMA:** glycidyl methacrylate; **SPIONs:** superparamagnetic iron oxide nanoparticles; **mPEG:** methoxy-polyethylene glycol; **β-TCP:** beta-tricalcium phosphate.

nanocomposite hydrogels. Here are some cases of such materials based on natural or synthetic polymeric matrices combined with the most popular nanomaterials as reinforcements, including nanoceramics, carbon-based nanostructures, metallic nanomaterials, and various nanosized polymeric materials.

10.4.1 NATURAL THERMOSENSITIVE POLYMERIC MATRIXES FOR NANOCOMPOSITE HYDROGELS

The most used polymeric matrices for nanocomposite hydrogels are based on chitosan, collagen, gelatin, alginate, hyaluronic acid, cellulose, and gellan gum, among others. These matrixes show thermogelling properties and/or thermoresponsive behavior when they are functionalized or grafted with other polymers. Some examples of thermogelling materials are described below.

The combination of natural polymers such as chitosan/β-glycerophosphate with disodium salt form thermogelling irreversible hydrogels at body temperature to which nanohydroxyapatite was incorporated and which proved to have good properties for bone applications. On the other hand, the incorporation of nano-bioactive glass into chitosan hydrogels results in the ability to bind to soft tissues and stimulate angiogenesis favorable for soft tissue repair. In the case of carbon-based materials, such as graphene oxide nano-sheets, reduce the gelation time of hydrogels, and cell adhesion and growth were demonstrated in this type of nanocomposite hydrogels. Nanodiamonds have also been tested as fillers in a mixture of chitosan and gelatin. As a result, the mechanical properties of the hydrogel are improved. Finally, for chitosan matrices, other polymeric fillers such as chitin nano-bigots can also be incorporated to improve mechanical properties.

Collagen has also been used as a matrix and combined with nano-bioactive glass has superior performance with respect to mineralization, angiogenesis, and cell penetration *in vivo*. The incorporation of whitlockite nanoparticles in a mixture of oxidized alginate and gelatin reduces the gelation time and increases the storage modulus. Moreover, the incorporation of carbon nanotubes into this type of hydrogel results in improved mechanical and electrical properties.

Alginate hydrogels that were filled with bioglass nanoparticles tend to form mechanically stronger materials due to ions that serve as additional physical cross-linking. In another study, an alginate nanocomposite hydrogel containing CaF_2 nanoparticles was evaluated and showed enhanced proliferation, fibroblast cell migration along with superior regeneration, and significantly higher antimicrobial activity than the matrix. Alginate hydrogels have also been tested with multi-walled carbon nanotubes (MWCNTs), which render the hydrogel more robust by increasing its stiffness. Among the carbon nanomaterials, we also find graphene oxide (GO), which was tested in alginate hydrogels and was found to increase the compressive elastic modulus up to a maximum of 300 KPa. On the other hand, the use of cellulose nanocrystals in alginate and gelatin hydrogels has been tested, which increased mechanical properties and resistance to biodegradation sought for some applications.

Part of the limitations of hyaluronic acid applications is its rapid resorption due to low stability in the body, with the intention of overcoming that limitation hydroxyapatite nanoparticles have been used and as a result had good volumetric persistence. The filling of GO nanosheets in hyaluronic acid hydrogels can be used to increase mechanical strength. There is also an increase in storage modulus by incorporating cellulose nanocrystals and a slower degradation rate.[32]

10.4.2 SYNTHETIC THERMOSENSITIVE POLYMERIC MATRIXES FOR NANOCOMPOSITE HYDROGELS

Synthetic thermoresponsive polymeric matrixes have been improved through the incorporation of nanomaterials and the most widely used ones are described. As mentioned before, PNIPAM is the most studied synthetic thermosensitive polymer; however, its limitations in terms of mechanical properties and non-biocompatibility remain a challenge, so the addition of nanomaterials to form nanocomposite hydrogels with improved properties is still widely explored. For example, nanomaterials, such as GO, carbon nanotubes (CNTs), Laponite XLG, nano starch, tetraethoxysilane, gold nanoparticles (AuNPs), and Fe_2O_3 NPs, have been used for the improvement of mechanical properties.[32]

Some other cases, such as chitosan nanohydrogel incorporated into PNIPAM hydrogel, improved the water-holding capacity of cotton fabric and achieved an antibacterial function against *Staphylococcus aureus* and *Escherichia coli* through modification with chitosan.[33] Finally, another example is the study and evaluation of a new thermogelling hydrogel based on poly(*N*-isopropylacrylamide-*graft*-chondroitin sulfate) (PNIPAM-*g*-CDS) for NP replacement. The hypothesis was tested with the addition of freeze-dried calcium cross-linked alginate particles to aqueous solutions of PNIPAM-*g*-CDS, which would allow the adjustment of the rheological properties as well as the bio-adhesive and mechanical properties of the injectable solution.[34]

Examples of hydrogels based on poly(ethylene glycol) (PEG) follow. The poly (ethylene glycol) dimethacrylate (PEGDMA) polymer is reported as an *in situ* photocross-linkable hydrogel in the presence of PVP; hydroxyapatite nanoparticles were embedded into the hydrogel to improve cell attachment, and it was found that the incorporation of nanohydroxyapatite (nHAp) could also reduce the temperature for the gelation process from 42.48 °C to 38.1 °C, which is advantageous. Moreover, by adding nHAp, the gelation time also increased from ~2 min to ~3–5 min. By adding silica nanoparticles in PEG hydrogels, both the sol-gel transition temperature and storage modulus of the nanocomposite hydrogel decreased and it was indicated that the refractive index of the nanocomposite could be increased up to 0.0667 compared to pure matrix without an increase in turbidity.[35]

Another of the most studied thermoresponsive polymers is Pluronic®, a commercially available ttiblockcopolymer PEO-*b*-PPO-*b*-PEO, however, thermoresponsive hydrogels based on the Pluronic PF127 have been found to have poor mechanical strength and stability. Research has been carried out with the intention of improving the mechanical resistance of thermoresposive hydrogels formed by PF127, such as mixed polymeric micelles with Pluronic P123, tween 80; conjugation with gold nanoparticles; and hydrophobically modified with carboxymethyl-chitosan (CMCS) and alginate.[33]

Poly(lactic-*co*-glycolic acid) (PLGA) can be converted into a thermosensitive hydrogel through its copolymerization with PEG blocks (PEG-PLGA-PEG). The hydrogel was embedded with nanohydroxyapatite and the resulting hydrogel was examined; the dynamic mechanical analysis of the nanocomposite showed two crossings for storage and loss moduli, revealing a gel-sol transition at 26 °C and a sol-gel transition at 33 °C. The results showed that the incorporation of alkaline nanoparticles could prolong the degradation rate of the matrix. In another study, fibers, ranging from 70 nm to 191 μm, were incorporated as a reinforcing component to improve the mechanical strength. A nanocomposite based on a highly sensitive amphiphilic copolymer was also fabricated by incorporating GO nanosheets into a thermosensitive poly(ethylene glycol) methyl ether (mPEG)-polycaprolactone-(PCL)-mPEG copolymer. The GO-loaded copolymer showed a light-activated sol-gel transition at UV intensities as low as 0.8 mW/cm^2, which is of particular interest since gelation has not occurred for the pure copolymer.[32]

10.5   THERMO- AND DUAL-RESPONSIVE HYDROGELS WITH NANOMATERIALS AS FILLERS

Hydrogels can respond to more than one stimulus, usually because they have polymers of different nature in their structure or because the nanostructures used as fillers are sensitive to some other type of electrical, magnetic, pH, or light change (Figure 10.3). Table 10.3 shows some of these composite materials.

10.5.1   THERMO- AND pH-RESPONSIVE HYDROGELS

Temperature- and pH-sensitive nanocomposite hydrogels are widely studied for their ease of loading and release of drugs or other important agents. An example of this type are monolithic nanocomposite hydrogels with dual and ultrafast response to temperature and pH stimuli based on PNIPAM with sodium methacrylate (SMA) comonomer with a strong pH response and physically cross-linked by clay nanoplatelets. In addition to their dual response, this hydrogel possesses extraordinarily high self-healing, mechanical, and tensile properties.[36]

10.5.2   THERMO- AND LIGHT-RESPONSIVE HYDROGELS

Nanocomposite hydrogels that, in addition to temperature respond to light, have applications ranging from drug delivery systems and therapeutic agents to contrast agents in imaging, whereby the light sensitivity can be in the range of NIR to UV light. As a sample of these systems, graphene oxide was functionalized as a thermosensitive vehicle for DOX release and embedded within a polymeric matrix of decellularized porcine small intestine submucosa methacrylate (SISMA) and chitosan methacrylate (CSMA), yielding a photoreactive material with tunable mechanical properties that is suitable for *in situ* deposition and room temperature polymerization.[37]

10.5.3   THERMO- AND ELECTRICAL RESPONSIVE HYDROGELS

For imparting nanocomposite hydrogels with electrical and temperature sensitivity, electrical/thermal conductors have been used as fillers, especially carbon-based materials such as graphite, graphene, and CNTs, since they are excellent electrical and thermal conductors. Such is the case of 3D hydrogels sensitive to electrical and thermal stimuli based on CNTs as core unit (electrical/thermal conductor), CS as capping unit (shell), and hydrophilic dispersant, poly(NIPAM-*co*-BBVIm) (BBVIm is 3,3′-(Butane-1,4-diyl) bis(1-vinyl-imidazol-3-ium) bromide cross-linker) as drug carrier and temperature-sensitive copolymer. By formulating CNT-core and CS-shell units and constructing a CNT/CS-sponge framework, the uniform distribution and 3D connectivity of CNTs were improved. The CNT/CS-sponge filled with cross-linked poly(NIPAM-*co*-BBVIm) formed a doubly responsive 3D hydrogel that delivered approximately 37% of a drug by 30% shrinkage after electrical and thermal on/off switches.[38]

10.5.4 THERMO- AND MAGNETIC-RESPONSIVE HYDROGELS

The development and property control of a thermo-magnetically sensitive and mechanically strong smart nanocomposite hydrogel remains a challenge; however, progress has been made in obtaining hydrogels with a PDMAM matrix, and as fillers of laponite nanoparticles, and Fe_3O_4 magnetite nanoparticles, which formed a material with physical and chemical cross-linking; as a result of its high hardness, ultra-stretch, and biocompatibility, it has great potential as a soft actuator remotely controllable for biomedical and pharmaceutical technologies.[39]

Some materials are more advanced and go from dual-response to multi-response, since they respond to more than two stimuli. For example, it has been possible to obtain triple-sensitivity hydrogels for drug delivery composed of PNIPAM for thermal response, PDMAEM for pH response, and magnetic graphene oxide (MGO) for magnetic response. For these materials, there is efficient drug loading and release and excellent biocompatibility.[32]

The systems described above need a lot of attention in the future to be applied, given their responses to multiple stimuli, could be used in different biomedical applications, specifically immunotherapy, drug delivery, and tissue engineering.

10.6 DUAL-RESPONSIVE TEMPERATURE-SENSITIVE HYDROGELS FOR SPECIFIC APPLICATIONS

Temperature-sensitive hydrogels have been engineered to respond to two or more stimuli when the thermal response is not good enough for a specific application. Sometimes the second responsiveness is only used to sharpen the thermal response to more specific conditions, but in some cases, the second or even third sensitive ability opens the door for new, more sophisticated applications. Table 10.4 summarizes a series of reports using temperature-sensitive hydrogels for a range of applications. In the majority of cases, the second responsivity is a result of the use of a second polymer in the hydrogel preparation. A variety of ways have been developed to attach the second polymer to the temperature-sensitive hydrogel: copolymerization using two monomers, grafting of the second polymer to a temperature-sensitive polymer backbone, semi- and fully interpenetrating polymer networks, growing of bilayers of hydrogels, and filling of hydrogels with microgels. Composite hydrogels have been also explored for more demanding applications where magnetic or photosensitive nanoparticles have been included. A detailed description follows.

TABLE 10.4

Dual-Responsive Temperature-Sensitive Hydrogels and Their Applications

40Temperature-Sensitive Polymer	Second (third) Responsivity/Polymer	Architecture	Application	Ref.
PNIPAM	pH/ PAAc grafted onto κ-carrageenan +magnetic/Iron oxide	Hydrogel filled with Nanogel + magnetic nanoparticles (NPs)	LDOPA controlled release system (CRS) for Parkinson	[41]
Poly(N-acryloyl-L-valine)	pH/poly(N-acryloyl-L-histidine)	Copolymeric hydrogel	Pilocarpine CRS for ocular delivery	[42]
PDMAEM	pH/PDMAEM + Salecan polysaccharide from *Agrobacterium sp. ZX09.*	Semi-interpenetrating polymer network (semi-IPN)	Insulin citocompatible CRS	[43]
PNIPAM	pH/poly(aminoethyl acrylate) + PAM + oxidized cellulose nanofiber	Cellulose nanofiber filled hydrogel	Fruit packaging with delivery of preservative	[12]
PNIPAM	Salt/zwitterionic sulfobetaine methacrylate	Copolymeric hydrogel	Tunable cell adhesion and detachment material	[44]
PNIPAM	pH/ PMAAc + PAM	Semi-IPN	Biomaterial with enhanced mechanical properties	[45]
PNIPAM	pH/CS + surface functionalized graphene oxide (GO)	GO filled semi-IPN	Biomaterial with superior mechanical properties	[46]
PDMAEM	pH/PDMAEM + PDEAM	Laponite filled hydrogel	Biomaterial with superior mechanical properties	[40]
PNIPAM	pH/PAAc + GO	GO filled double network hydrogels (full-IPN)	Biomaterial with superior mechanical properties	[47]
PNIPAM	Salt/PVBIPS	Bilayer hydrogels	Shape adaptable actuators	[48]
PNIPAM	Light/PNIPAM grafted AuNPs + PMAAc	Layer by layer hydrogels including AuNPs filled layers	Actuators	[49]
PNIPAM	Light/AuNPs + poly(NIPAM-*co*-AAc)	AuNPs filled hydrogels	Programable patterning	[50]

(Continued)

TABLE 10.4 (Continued)

Dual-Responsive Temperature-Sensitive Hydrogels and Their Applications

Temperature-Sensitive Polymer	Second (third) Responsivity/Polymer	Architecture	Application	Ref.
PNIPAM	pH (metal ions)/allylAlginate + Me^{2+}	Double network hydrogels	Shape memory actuators	[51]
Poly[oligo (ethylene glycol) methyl ether methacrylate] (POEGMA)	2nd Temperature/different POEGMA	Multilayer hydrogels	Shape adaptable actuators	[52]
PNIPAM	Ion/maleyl-Gelatin + Me^{2+} or Al^{3+}	Double network hydrogels	Smart Window	[53]
PNIPAM	Ion/maleyl-Chitosan + Al^{3+}	Double network hydrogels	Smart Window	[54]
PNIPAM	Light/polydopamine NPs + Clay	Filled nanocomposite double layer hydrogel	Electronic skin	[55]
PNIPAM	Magnetic/Iron oxide NPs + PAM + PVA + KCl	Filled semi-IPN hydrogel	Wearable sensor	[56]

Smart micro/nanogels have been extensively studied as controlled release systems (CRSs) for drugs in parenteral administration; this is out of the scope of the current book chapter. Smart macroscopic hydrogels can be used also as CRSs, however drug delivery, in this case, is oral or topical. Bardajee *et al.* reported the use of PNIPAM nanogels as fillers of PAAc-grafted κ-carrageenan hydrogels for the temperature and pH-controlled delivery of levo-3,4-dihydroxyphenylalanine (LDOPA) for Parkinson disease in oral administration.[41] The nanocomposite hydrogel was praised as a sustained release system for LDOPA ranging up to 12 days. Similarly, Wei *et al.* also reported a smart temperature- and pH-responsive hydrogel system for oral delivery of insulin.[42] In this case, the thermoresponsive polymer is PDMAEM and the hydrogel is a semi-IPN based on a biocompatible polysaccharide Salecan. A hydrogel system designed for topical delivery of drugs is the one reported by Casolaro *et al.* In this case, a copolymeric hydrogel containing protein derivatives of L-valine and L-histidine was developed.[43] Pilocarpine was loaded and released to the eye, taking advantage of the temperature and pH sensitivity of the hydrogel system. In a totally different application, the controlled delivery in this case of a food preservative, was engineered into a fruit packaging material composed of cellulose nanofiber-filled temperature- and pH-responsive hydrogel.[12] The release of the food preservative natamycin was accelerated under non-normal conditions like acid pH and high temperature, improving the effect of the preservative when needed.

Another common approach in the case of responsive hydrogels has been the improvement of their mechanical properties when the hydrogel toughness, elasticity, or durability for applications like artificial skin, joints, and tissue repair, among others, is desired without sacrificing responsiveness. Two general approaches have been used, either alone or combined with this goal: the preparation of semi-IPNs or fully interpenetrating polymer networks (IPNs) and the use of nanocomposite-filled hydrogels. As an example of the first case, Kalkan *et al.* reported the preparation of semi-IPNs starting with a poly(NIPAM-*co*-MAAc) hydrogel prepared in the presence of a PAM with high molecular weight ($M_w = 9.639 \times 10^5$ g/mol) inducing physical entanglements, increasing the apparent cross-linking density and enlarging the compressive elasticity resulting in an improvement of mechanical properties.[45] Huang *et al.* prepared a semi-IPN, but in this case, a PNIPAM hydrogel was prepared in the presence of a chitosan polymer (weight-average molecular weight from 100,000 to 300,000 g/mol), and additionally a 3-(trimethoxysilyl) propylmethacrylate-modified GO was used as a cross-linker.[46] The resulting thermo- and pH-responsive hydrogel showed extraordinary swelling/deswelling behavior and enhanced mechanical performances. A different chemistry was used by Li *et al.* when they reported the preparation of a poly(DMAEM-*co*-DEAM) hydrogel reinforced using Laponite as nanoclay.[40] A temperature- and pH-responsive hydrogel with mechanical strength as high as 2.219 MPa with compressive strain at 95%, was achieved. An even stronger hydrogel was reported by Zhiqiang *et al.* In this case, a double network hydrogel (doubly cross-linked IPN) was based on PNIPAM and PAAc filled with GO.[47] The compressive strength was as high as 3.5 MPa, while the deswelling velocity as a response to a temperature change was much faster than a conventional non-filled/non-IPN PNIPAM hydrogel. The deswelling took only 9 min, compared to 48 h for a 50% deswelling degree.

A different area of application of doubly responsive hydrogels is for actuators. Actuators can be used in chemomechanical valves, soft robotics, and also in shape memory materials. These materials possess a unique shape-change property in response to external stimuli. Hydrogels are usually isotropic materials, meaning that their response to temperature or other stimuli is the same in all directions. Therefore, a strategy for achieving hydrogel actuators is to create an internal mismatch of stresses, like softer vs stiffer, swelling vs less/non-swelling, helping to realize a controllable shape change for actuation. A general strategy is to prepare asymmetric/hybrid hydrogels generating filled/non-filled regions or regions using different polymers, where the use of layered structures is common. For example, Xiao et al. prepared bilayer hydrogels. One layer was PNIPAM as a temperature-sensitive layer and the second one was of poly(3-(1-(4-vinylbenzyl)-1H-imidazol-3-ium-3-yl)propane-1-sulfonate) (PVBIPS), a zwitterionic polymer responsive to salt concentration.[48] Both PNIPAM and PVBIPS layers exhibit a completely opposite swelling/shrinking behavior, where PNIPAM shrinks (swells), but PVBIPS swells (shrinks) in a salt solution (water) or at high (low) temperatures. This allows the fully reversible shape change of the bilayer hydrogel from rod to ring and back to rod. Using a different strategy, Zhu et al. combined layers of temperature-responsive PNIPAM hydrogels with pH-responsive PMAAc and intercalated layers of PNIPAM-grafted AuNPs and poly(NIPAM-co-DMAEM)-grafted gold nano-cages (AuNC).[49] The constructs were prepared by a spin coating technique over a silicon wafer. The use of AuNPs and AuNC leads to light-sensitive layers, so that the hydrogel construct can be heated by illuminating at specific wavelengths where the nanoparticles show a localized surface plasmon resonance (photothermal effect). Illuminating a surface of a hydrogel containing AuNPs can evolve into patterning of actuating effects. This approach was tested by Hawser et al. who demonstrated on-demand reconfigurable buckling of poly(NIPAM-co-AAc) hydrogel network films containing AuNPs by patterned photothermal deswelling.[50] Predictable, easily controllable, and reversible transformations from a single flat gel sheet (25 microns) to numerous different three-dimensional forms were shown when illuminating through a pattern. A different strategy was pursued by Choi et al. The multi-responsiveness of alginate-PNIPAM hydrogels was used in the shape memory of materials via temporary cross-linking.[51] Thus, the hydrogels were tested for reversible shape changes by increasing the temperature and adding metal ions that build additional cross-links; these cross-links were removed finally by adding an ethylenediaminetetraacetate (EDTA) solution, achieving reversibility in the shape changes. A simple but powerful approach for reversible biomimetic shape changes was reported by Kobayashi et al.[52] This group used different copolymer hydrogels from oligo (ethylene glycol) methyl ether methacrylate (OEGMA) of different side chain lengths and composition to achieve hydrogels with three VPTTs, also termed "response temperatures". These hydrogels were assembled as bilayers over a non-thermally responsive OEGMA 500 hydrogel and created by photopatterning in the form of grippers resembling petals of a flower of different VPTT. Figure 10.4 shows the schematic of the prepared grippers, pictures of the actual grippers when they are heated from left to right from 10 °C to 63 °C, and then cooling back to 10 °C, demonstrating the gradual and reversible shape changes of the actuating hydrogels.

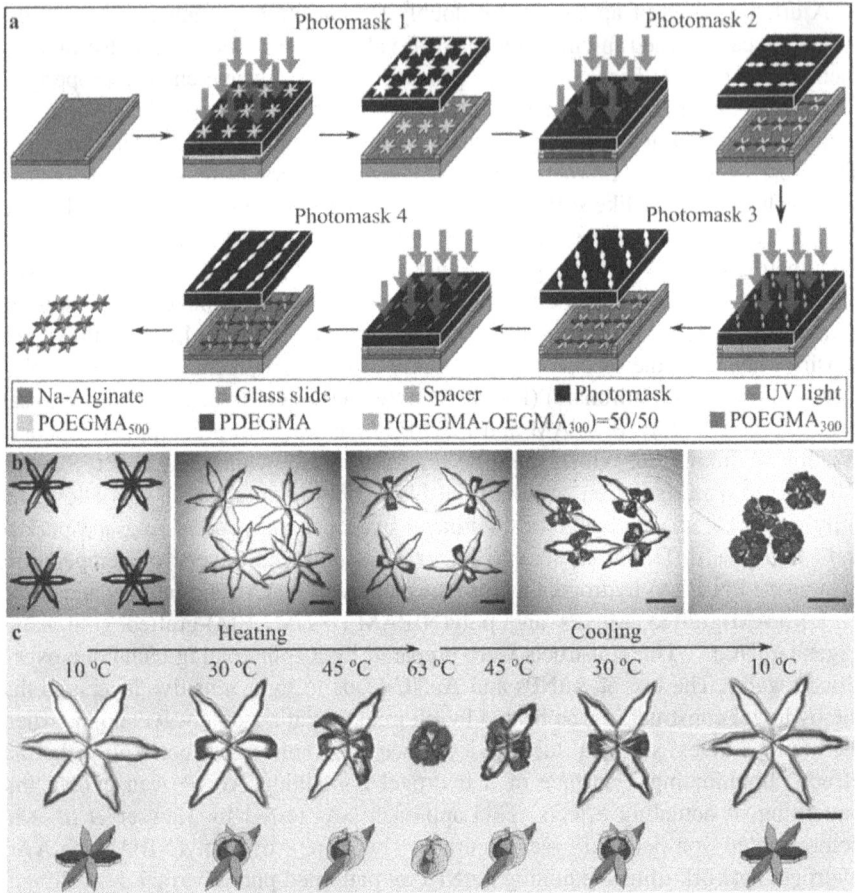

FIGURE 10.4 Schematics of process flow and four-state shape changes of hydrogel grippers. a) Schematic representation of the fabrication process for four-state thermally responsive grippers. Patterning required four photomasks that were used in sequential order to pattern the overall shape of the gripper with the passive layer of macromonomer OEGMA500 followed by pairs of fingers on top of the passive layer in spatial registry using macromonomers DEGMA, DEGMA-OEGMA300, and OEGMA300. b) Representative images of grippers. The images show the grippers on the glass slide, at State I, at State II, at State III, and at State IV from left to right, indicating the parallel nature of the fabrication and actuation process. c) Optical images and schematic representations of four-state shape changes of grippers during heating and cooling. The images from left to right are at 10, 30, 45, 63, 45, 30, and 10 °C. Scale bars represent 2 mm. "Reprinted from Reference[52]: Macromolecular Rapid Communications, Vol 39, Kobayashi K.; Oh S.H.; Yoon C.K.; Gracias D.H., Multitemperature Responsive Self-Folding Soft Biomimetic Structures, 1700692, Copyright (2018), with permission from WILEY-VCH."

A different application of responsive hydrogel is reported in smart windows. In this case, the turbidity associated with the VPTT has been exploited to decrease the transparency of windows exposed to sunshine. The response time and the grading of the turbidity have been an issue for research. Xin *et al*. used double network hydrogels based on PNIPAM, together with maleyl-gelatin to obtain smart windows.[53] The VPTT can be finely tuned since it shows a linear dependence on the concentration of Na^+, Mg^{2+}, and Al^{3+}. For example, VPTT decreased from 33 °C to 22 °C with Al^{3+} concentration increasing from 0 to 0.60 M. This opacity shift to lower temperatures follows the so-called Hofmeister series and is caused by the breaking of hydrogen bonds within the polymers in the hydrogel and water. The hydrogel is proposed as a temperature and metal ions-controlled smart window. Liu et al. reported a similar hydrogel based on PNIPAM, but in this case, with maleyl-chitosan and Al^{3+} as smart window material.[54] The authors claim that the hydrogel exhibited a smooth thermoresponse during heating and a novel recovery hysteresis during cooling. As a proof of concept, pictures of hydrogels at different temperatures by heating and by cooling are reported. By enabling additional gentle transition in translucency, it is expected that the hydrogel smart windows are able to retain a comfortable indoor temperature/lightness balance during an increase in outdoor temperature, in avoidance of sudden indoor darkness at an elevated temperature. Further, the authors claim that the response degree and response temperature can be adjusted by the contents of chitosan and Al^{3+}, respectively, so that the material can be customized personally.

A different kind of application with a bright future for responsive hydrogels is the development of wearable sensors or electronic skin. Smart wearable devices should essentially be flexible, attach to the skin properly, and possess a high degree of stretchability and conductivity. Di et al. reported a double-layer nanofilled hydrogel with excellent adhesiveness to a broad range of substrates including human skin with the help of synergistic multiple coordination bonds between clay, PNIPAM, and polydopamine nanoparticles (PDA-NPs).[55] The prepared hydrogel showed controllable near-infrared (NIR) responsive deformation after the incorporation of PDA-NPs as highly effective photothermal agents in the thermosensitive PNIPAM network. A bilayer hydrogel that bends was also prepared by selecting differently filled nanocomposite layers. Zhou et al. reported the preparation of thermoresponsive, magnetic, adhesive, and conductive nanocomposite hydrogels for wireless and non-contact flexible sensors.[56] Poly(NIPAM-co-AAm) hydrogel and Fe_3O_4 nanoparticles are designed as the thermoresponsive and magnetic component, respectively; poly(vinyl alcohol) (PVA) is introduced in a semi-interpenetrating way to enhance the adhesiveness and mechanical property; and potassium chloride (KCl) works as the conductive component. The nanocomposite hydrogels have ionic conductivity due to the presence of K^+ and Cl^- as charge carriers. By connecting a nanocomposite hydrogel strip as the wire into circuit, light-emitting diode (LED) shines accordingly, and the brightness of LED changes with stretching and releasing of the hydrogel strip. This means that such hydrogels have the potential to be used as strain sensors to convert mechanical deformation into electronic signal variation. The performances of the sensors are evaluated by recording the resistance over time when different strains or compression are applied. The relative resistance variation ($\Delta R/R_0$) are calculated according to $\Delta R/R_0 = (R_i-R_0)/R_0$, in which R_i and R_0 are the resistance at a certain strain and the original resistance, respectively. The response time of the sensor was 140 ms and about 2,000 sensing cycles were successfully conducted. The hydrogel sensors were

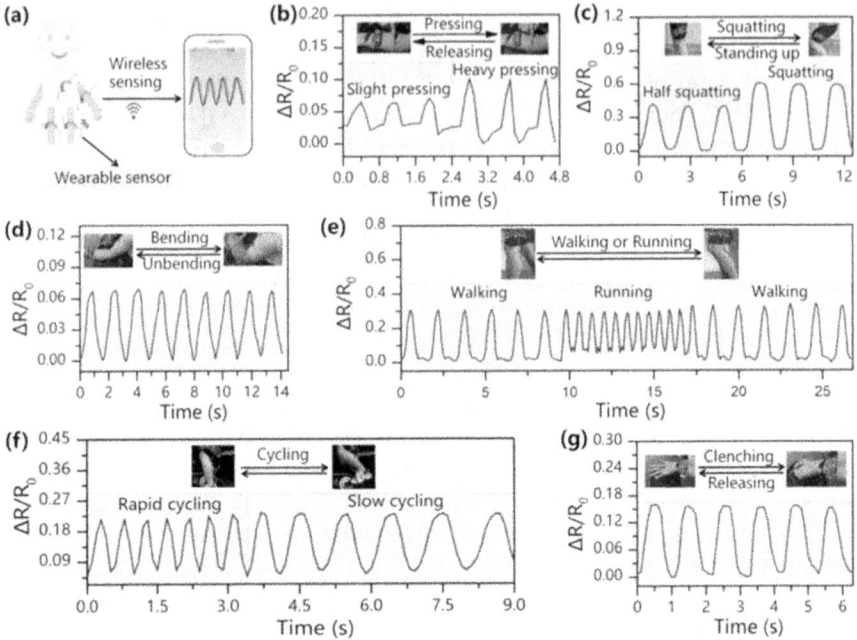

FIGURE 10.5 (a) Illustration of the wireless wearable hydrogel sensors for monitoring diverse motions and the displaying of sensing results on the screen of a cell phone. (b) Resistance response of the wireless wearable sensor to periodical slight and heavy finger pressing. (c) Half squatting and squatting. (d) Motion of bicipital muscle. (e) Walking and running. (f) Cycling. (g) Clenching. "Reprinted from Reference[56]: Colloids and Surfaces A: Physicochemical and Engineering Aspects, Vol 636, Zhou H.; Jin Z.; Gao Y.; Wu P.; Lai J.; Li A.; Jin X.; Liu H.; Chen W.; Wu Y.; Ma A., Thermoresponsive, magnetic, adhesive and conductive nanocomposite hydrogels for wireless and non-contact flexible sensors, 128113, Copyright (2022), with permission from Elsevier."

integrated with a Bluetooth transmitter and a Bluetooth receiver to construct wireless sensors. By this way, wireless wearable sensors were successfully fabricated and applied in human motion monitoring (Figure 10.5).

Finally, due to the presence of Fe_3O_4 nanoparticles, the nanocomposite hydrogels exhibit excellent deformation ability in a magnetic field. Based on the magnetic-field-induced deformation, non-contact sensors were fabricated. The resistance variation of a sensor when a magnetic field was periodically approached and withdrawn was recorded. Approaching the magnet induced an increase in resistance while withdrawing the magnet led to a decrease in resistance. When the magnet was held, $\Delta R/R_0$ remained constant. The resistance variation of the sensor was similar to that observed in the cases of directly applying mechanical strain. Furthermore, the hydrogel sensor also had the ability to respond in a noncontact way to cyclic magnetic fields applied with different frequencies. With this, the potential of hydrogel wearable and non-contact sensors was demonstrated.

10.7 CONCLUSIONS AND FUTURE PERSPECTIVES

Hydrogels with thermal responsiveness have been studied for a wide range of applications, mainly in the biomedical, pharmaceutical, biotechnological, and agricultural areas. In addition, each application requires hydrogels with specific properties. The physical-chemical, mechanical, and biological properties of these materials depend to a great extent on the nature of their components and on the possible combinations between different components in their structure. Hydrogels obtained from natural polymers are biocompatible with low toxicity; however, their chemical structures have limited capacity for further modification, in addition to being generally weak in mechanical properties, while the synthetic polymer networks are often strong and relatively easy to chemically modify. In this sense, simple hydrogels, semi-IPNs, full-IPNs, covalently cross-linked hydrogels, and composite materials have been obtained.

Thermal response (VPTT) represents a balance between attractive and repulsive intermolecular forces. LCST-based VPTT-type phase transition hydrogels became the main thermosensitive hydrogels, since there is a large number of polymeric materials with this behavior. Poli(N-isopropylacrylamide) (PNIPAM), poly(diethylene glycol methyl ether methacrylate) (PDEGMA), poly(N-vinylcaprolactam) (PNVCL), and poly(2-dimethylamino ethyl methacrylate) (PDMAEM) are the most widely used in the synthesis of thermosensitive hydrogels. On the other hand, although less numerous, reports of positive volume phase transition (UCST-based) hydrogels have shown properties that reflect similar applications and utility. The acrylamide (AAm) and acrylic acid (AAc); polysaccharides like carrageenans, starch, agarose, and gellan gum; or proteins like gelatin, collagen, and elastin-like polypeptides have been used to obtain hydrogels with UCST behavior. Combinations of polymers with a LCST and UCST response could be used to obtain hydrogels with double thermal response, too, although homopolymers such as PDMAEM present a dual response by itself (pH and temperature).

The study of thermosensitive hydrogels continues to be a topic of interest for future works, since the optimization of VPTT by the hydrophilic/hydrophobic balance in its chemical composition, monomer sequence, and microscopic order are all aspects that must be considered in addition to the improvement in its mechanical and biological properties. The introduction of a second responsivity to the hydrogels besides thermal, has been tested; the majority of reports deal with pH sensitivity as the second type of response. This is done by incorporating cationic, anionic, or hydrophobic functionalities into their internal structure. However, how to mix the benefits together of the two responsive entities is not a simple question to answer. In dual pH and thermo-responsive materials, the balance between intermolecular interactions, such as Van der Waals interactions, hydrogen bonds, hydrophobic interactions, and electrostatic forces, can modulate the response of the phase change.

Thermosensitive hydrogels with improved mechanical properties have been obtained by adding nanoparticles in their structure, generally in a small proportion (less than 10 wt%), obtaining thermosensitive nanocomposite hydrogels. Nanoceramics, carbon-based nanostructures, metallic nanomaterials, and nanosized polymeric materials are the most common materials used to reinforce polymeric networks.

We have arrived at the opportunity of different applications where a combination of the thermoresponsive hydrogels with a second or third material with other functions and engineering concepts leads to appealing soft actuators, sensors, shape-memory materials, and biomaterials with superior properties; the future of smart hydrogels is bright.

REFERENCES

[1] Jagur-Grodzinski J. Polymeric gels and hydrogels for biomedical and pharmaceutical applications. *Polymers for Advanced Technologies.* 2010; 21(1):27–47.

[2] Lanzalaco S, Mingot J, Torras J, Alemán C, Armelin E. Recent advances in poly(N-isopropylacrylamide) hydrogels and derivatives as promising materials for biomedical and engineering emerging applications. *Advanced Engineering Materials.* 2023; 25:2201303.

[3] Gonzalez-Urias A, Licea-Claverie A, Sañudo-Barajas JA, González-Ayón MA. NVCL-based hydrogels and composites for biomedical applications: Progress in the last ten years. *International Journal of Molecular Sciences.* 2022; 23(9):4722.

[4] Akimoto AM, Ueki T, Yoshida R. Thermoresponsive polymers. In: Kobayashi S, Müllen K, eds. *Encyclopedia of polymeric nanomaterials.* Springer, Berlin; 2014:1–5.

[5] Zhang Y, Huang Y. Rational design of smart hydrogels for biomedical applications. *Frontiers in Chemistry.* 2021; 8:615665.

[6] Jeong B, Kim SW, Bae YH. Thermosensitive sol–gel reversible hydrogels. *Advanced Drug Delivery Reviews.* 2012; 64:154–162.

[7] Seuring J, Agarwal S. Polymers with upper critical solution temperature in aqueous solution. *Macromolecular Rapid Communications.* 2012; 33(22):1898–1920.

[8] Gandhi A, Paul A, Sen SO, Sen KK. Studies on thermoresponsive polymers: Phase behaviour, drug delivery and biomedical applications. *Asian Journal of Pharmaceutical Sciences.* 2015; 10(2):99–107.

[9] Chatterjee S, Hui PC leung. Review of applications and future prospects of stimuli-responsive hydrogels based on thermo-responsive biopolymers in drug delivery systems. *Polymers.* 2021;13(13):2086.

[10] Teotia AK, Sami H, Kumar A. Thermo-responsive polymers. In: Zhang Z, ed. *Switchable and responsive surfaces and materials for biomedical applications.* Elsevier, Amsterdam; 2015:3–43.

[11] Lehmann M, Krause P, Miruchna V, von Klitzing R. Tailoring PNIPAM hydrogels for large temperature-triggered changes in mechanical properties. *Colloid and Polymer Science.* 2019; 297(4):633–640.

[12] Shaghaleh H, Hamoud YA, Xu X, Lui H, Wang S, Sheteiwy M, Dong F, Guo L, Qian Y, Li P, Zhang S. Thermo-/pH-responsive preservative delivery based on TEMPO cellulose nanofiber/cationic copolymer hydrogel film in fruit packaging. *International Journal of Biological Macromolecules.* 2021; 183:1911–1924.

[13] Dragan ES, Apopei Loghin DF, Cocarta AI, Doroftei M. Multi-stimuli-responsive semi-IPN cryogels with native and anionic potato starch entrapped in poly(N,N-dimethylaminoethyl methacrylate) matrix and their potential in drug delivery. *Reactive and Functional Polymers.* 2016;105:66–77.

[14] Guo H, Sanson N, Marcellan A, Hourdet D. Thermoresponsive toughening in LCST-type hydrogels: Comparison between semi-interpenetrated and grafted networks. *Macromolecules.* 2016; 49(24):9568–9577.

[15] Wang Q, Li S, Wang Z, Liu H, Li C. Preparation and characterization of a positive thermoresponsive hydrogel for drug loading and release. *Jornal of Applied Polymer Science.* 2009; 111(3):1417–1425.

[16] Eckert T, Abetz V. Polymethacrylamide—An underrated and easily accessible upper critical solution temperature polymer: Green synthesis via photoiniferter reversible addition–fragmentation chain transfer polymerization and analysis of solution behavior in water/ethanol mixtures. *Journal of Polymer Science.* 2020; 58(21):3050–3060.

[17] Alzari V, Ruiu A, Nuvoli D, Sanna R, Martinez JI, Appelhans D, Voit B, Zschoche S, Mariani A. Three component terpolymer and IPN hydrogels with response to stimuli. *Polymer.* 2014; 55(21):5305–5313.

[18] Ding Y, Yan Y, Peng Q, Wang B, Xing Y, Hua Z, Wang Z. Multiple stimuli-responsive cellulose hydrogels with tunable LCST and UCST as smart windows. *ACS Applied Polymer Materials.* 2020; 2(8):3259–3266.

[19] Higashi N, Sonoda R, Koga T. Thermo-responsive amino acid-based vinyl polymers showing widely tunable LCST/UCST behavior in water. *RSC Advances.* 2015; 5(83):67652–67657.

[20] Salgado-Rodríguez R, Licea-Claverie A, Zizumbo-López A, Arndt KF. Smart pH/temperature sensitive hydrogels with tailored transition temperature. *Journal of the Mexican Chemical Society.* 2013; 57(2):118–126.

[21] Taktak F. Rapid Deswelling of PDMAEMA hydrogel in response to pH and temperature changes and its application in controlled drug delivery. *AKU-Journal of Science and Engineering.* 2016; 16(1):68–75.

[22] París R, Quijada-Garrido I. Temperature- and pH-responsive behaviour of poly(2-(2-methoxyethoxy)ethyl methacrylate-co-N,N-dimethylaminoethyl methacrylate) hydrogels. *European Polymer Journal.* 2010; 46(11):2156–2163.

[23] Emileh A, Vasheghani-Farahani E, Imani M. Swelling behavior, mechanical properties and network parameters of pH- and temperature-sensitive hydrogels of poly((2-dimethyl amino) ethyl methacrylate-co-butyl methacrylate). *European Polymer Journal.* 2007; 43(5):1986–1995.

[24] Ilgin P, Ozay H, Ozay O. A new dual stimuli responsive hydrogel: Modeling approaches for the prediction of drug loading and release profile. *European Polymer Journal.* 2019; 113:244–253.

[25] Curcio M, Spizzirri UG, Cirillo G, Spataro T, Picci N, Iemma F. Tailoring flavonoids' antioxidant properties through covalent immobilization into dual stimuli responsive polymers. *International Journal of Polymeric Materials and Polymeric Biomaterials.* 2015; 64(11):587–596.

[26] Atta AM, Ahmed SA. Chemically cross-linked pH- and temperature-sensitive (*N*-isopropylacrylamide-co-1-vinyl-2-pyrrolidone) based on new cross-linker: I. swelling behavior. *Journal of Dispersion Science and Technology.* 2010; 31(11):1552–1560.

[27] Song F, Gong J, Tao Y, Cheng Y, Lu J, Wang H. A robust regenerated cellulose-based dual stimuli-responsive hydrogel as an intelligent switch for controlled drug delivery. *International Journal of Biological Macromolecules.* 2021; 176:448–458.

[28] Güngör A, Demir D, Bölgen N, Özdemir T, Genç R. Dual stimuli-responsive chitosan grafted poly(NIPAM-co-AAc)/poly(vinyl alcohol) hydrogels for drug delivery applications. *International Journal of Polymeric Materials and Polymeric Biomaterials.* 2021; 70(11):810–819.

[29] Wei QB, Luo YL, Fu F, Zhang YQ, Ma RX. Synthesis, characterization, and swelling kinetics of pH-responsive and temperature-responsive carboxymethyl chitosan/polyacrylamide hydrogels. *Journal of Applied Polymer Science.* 2013; 129(2):806–814.

[30] Huang H, Hou L, Zhu F, Li J, Xu M. Controllable thermal and pH responsive behavior of PEG based hydrogels and applications for dye adsorption and release. *RSC Advances.* 2018; 8(17):9334–9343.

[31] Guo H, Mussault C, Brûlet A, Marcellan A, Hourdet D, Sanson N. Thermoresponsive toughening in LCST-type hydrogels with opposite topology: From structure to fracture properties. *Macromolecules*. 2016; 49(11):4295–4306.

[32] Mellati A, Hasanzadeh E, Gholipourmalekabadi M, Enderami SE. Injectable nanocomposite hydrogels as an emerging platform for biomedical applications: A review. *Materials Science and Engineering: C.* 2021; 131:112489.

[33] Chatterjee S, Hui P, Kan C wai. Thermoresponsive hydrogels and their biomedical applications: Special insight into their applications in textile based transdermal therapy. *Polymers*. 2018; 10(5):480.

[34] Christiani T, Mys K, Dyer K, Kadlowec J, Iftode C, Vernengo AJ. Using embedded alginate microparticles to tune the properties of in situ forming poly(N-isopropylacrylamide)-graft-chondroitin sulfate bioadhesive hydrogels for replacement and repair of the nucleus pulposus of the intervertebral disc. *JOR Spine*. 2021; 4(3):e1161.

[35] Sreekanth S, Radhakrishnan A, Rauf AA, Kurup GM. Nanohydroxyapatite incorporated photocross-linked gelatin methacryloyl/poly(ethylene glycol)diacrylate hydrogel for bone tissue engineering. *Progress in Biomaterials*. 2021; 10 (1): 43–51.

[36] Strachota B, Strachota A, Horodecka S, Šlouf M, Dybal J. Monolithic nanocomposite hydrogels with fast dual T- and pH- stimuli responsiveness combined with high mechanical properties. *Journal of Materials Research and Technology*. 2021; 15:6079–6097.

[37] Céspedes-Valenzuela DN, Sánchez-Rentería S, Cifuentes J, Gómez SC, Serna JA, Rueda-Gensini L, Ostos C, Muñoz-Camargo C, Cruz JC. Novel photo- and thermo-responsive nanocomposite hydrogels based on functionalized rGO and modified SIS/chitosan polymers for localized treatment of malignant cutaneous melanoma. *Frontiers in Bioengineering and Biotechnology*. 2022; 10:947616.

[38] Park SY, Kang JH, Kim HS, Hwang JY, Shin US. Electrical and thermal stimulus-responsive nanocarbon-based 3D hydrogel sponge for switchable drug delivery. *Nanoscale*. 2022; 14(6):2367–2382.

[39] Han WJ, Lee JH, Lee JK, Choi HJ. Remote-controllable, tough, ultrastretchable, and magneto-sensitive nanocomposite hydrogels with homogeneous nanoparticle dispersion as biomedical actuators, and their tuned structure, properties, and performances. *Composites Part B: Engineering*. 2022; 236:109802.

[40] Li H, Wu R, Zhu J, Guo P, Ren W, Xu S, Wang J. pH/temperature double responsive behaviors and mechanical strength of laponite-cross-linked poly(DEA-co -DMAEMA) nanocomposite hydrogels. *Journal of Polymer Science Part B: Polymer Physics*. 2015; 53(12):876–884.

[41] Bardajee GR, Khamooshi N, Nasri S, Vancaeyzeele C. Multi-stimuli responsive nanogel/hydrogel nanocomposites based on κ-carrageenan for prolonged release of levodopa as model drug. *International Journal of Biological Macromolecules*. 2020; 153:180–189.

[42] Casolaro M, Casolaro I, Lamponi S. Stimuli-responsive hydrogels for controlled pilocarpine ocular delivery. *European Journal of Pharmaceutics and Biopharmaceutics*. 2012; 80(3):553–561.

[43] Wei W, Li J, Qi X, Zhong Y, Zuo G, Pan X, Su T, Zhang J, Dong W. Synthesis and characterization of a multi-sensitive polysaccharide hydrogel for drug delivery. *Carbohydrate Polymers*. 2017; 177:275–283.

[44] Chang Y, Yandi W, Chen WY, Shih YJ, Yang CC, Chang Y, Ling QD, Higuchi A. Tunable bioadhesive copolymer hydrogels of thermoresponsive poly(N-isopropyl acrylamide) containing zwitterionic polysulfobetaine. *Biomacromolecules*. 2010; 11(4):1101–1110.

[45] Kalkan B, Orakdogen N. Negatively charged poly(N-isopropyl acrylamide-co-methacrylic acid)/polyacrylamide semi-IPN hydrogels: Correlation between swelling and compressive elasticity. *Reactive and Functional Polymers*. 2022; 174:105245.

[46] Huang S, Shen J, Li N, Ye M. Dual pH- and temperature-responsive hydrogels with extraordinary swelling/deswelling behavior and enhanced mechanical performances. *Journal of Applied Polymer Science*. 2015; 132: 41530.

[47] Li Z, Shen J, Ma H, Lu X, Shi M, Li N, Ye M. Preparation and characterization of pH- and temperature-responsive nanocomposite double network hydrogels. *Materials Science and Engineering: C*. 2013; 33(4):1951–1957.

[48] Xiao S, Zhang M, He X, Huang L, Zhang Y, Ren B, Zhong M, Chang Y, Yang J, Zheng J. Dual salt- and thermoresponsive programmable bilayer hydrogel actuators with pseudo-interpenetrating double-network structures. *ACS Applied Materials & Interfaces*. 2018; 10(25):21642–21653.

[49] Zhu Z, Senses E, Akcora P, Sukhishvili SA. Programmable light-controlled shape changes in layered polymer nanocomposites. *ACS Nano*. 2012; 6(4):3152–3162.

[50] Hauser AW, Evans AA, Na JH, Hayward RC. Photothermally reprogrammable buckling of nanocomposite gel sheets. *Angewandte Chemie International Edition*. 2015; 54(18):5434–5437.

[51] Choi EJ, Ha S, Lee J, Premkumar T, Song C. UV-mediated synthesis of pNIPAM-cross-linked double-network alginate hydrogels: Enhanced mechanical and shape-memory properties by metal ions and temperature. *Polymer*. 2018; 149:206–212.

[52] Kobayashi K, Oh SH, Yoon C, Gracias DH. Multitemperature responsive self-folding soft biomimetic structures. *Macromolecular Rapid Communications*. 2018; 39(4):1700692.

[53] Xin F, Lu Q, Liu B, Yuan S, Zhang R, Wu Y, Yu Y. Metal-ion-mediated hydrogels with thermo-responsiveness for smart windows. *European Polymer Journal*. 2018; 99:65–71.

[54] Liu B, Liu J, Yu Y. Precise control over tunable translucency and hysteresis of thermo-responsive hydrogel for customized smart windows. *European Polymer Journal*. 2022; 162:110929.

[55] Di X, Kang Y, Li F, Yao R, Chen Q, Hang C, Xu Y, Wang Y, Sun P, Wu G. Poly (N-isopropylacrylamide)/polydopamine/clay nanocomposite hydrogels with stretchability, conductivity, and dual light- and thermo- responsive bending and adhesive properties. *Colloids and Surfaces B: Biointerfaces*. 2019; 177:149–159.

[56] Zhou H, Jin Z, Gao Y, Wu P, Lai J, Li S, Jin X, Liu H, Chen W, Wu Y, Ma A. Thermoresponsive, magnetic, adhesive and conductive nanocomposite hydrogels for wireless and non-contact flexible sensors. *Colloids and Surfaces A: Physicochemical and Engineering Aspects*. 2022; 636:128113.

11 Mechanical Properties of Multifunctional Hydrogels

Serap Sezen
Sabanci University, Istanbul, Turkey

Çiğdem Bilici
Istinye University, Istanbul, Turkey

Atefeh Zarepour
Saveetha University, Chennai, India

Arezoo Khosravi
Istanbul Okan University, Istanbul, Turkey

Ali Zarrabi
Istinye University, Istanbul, Turkey

Ebrahim Mostafavi
Stanford University, Stanford, California, USA

11.1 INTRODUCTION TO MECHANICAL PROPERTIES OF MULTIFUNCTIONAL HYDROGELS

Hydrogels are cross-linked polymeric networks with different mechanical features which have been used in several research areas. Several theoretical and experimental models are applied to determine the mechanical characteristics of these materials [1]. The mechanical features of hydrogels could be affected by the different preparation parameters. For instance, nano- and microparticles, as well as the fibers are added to the structure of these compounds to alter their mechanical properties to reach their ultimate structure. As smart materials, some hydrogels change their mechanical behavior as a response to external stimuli, so-called mechanoresponsive materials. Contrary to conventional hydrogels, mechano-responsive hydrogels show great attention for on-demand biomedical applications [2].

DOI: 10.1201/9781003340485-11

11.2 MECHANICAL MODELING OF HYDROGELS

11.2.1 RUBBER-LIKE ELASTICITY

Rubbers can be reversibly stretched hundreds of times and completely return to their original dimensions immediately after removing deformation. This unique property, defined as rubber elasticity, results from their three-dimensional (3D) cross-linked networks that allow rapid rearrangement of the flexible polymer chain segments under external stresses [3]. Hydrogels are like rubbers in their ability to respond by stretching under compression and returning to their previous shape depending on the applied force. This elastic behavior of hydrogels is generally expressed in the rubber-like elasticity theory, which was developed by Flory with inspiration from the second thermodynamic law. The retractive force is obtained from a decrease in entropy rather than a change in enthalpy. The stretching application causes changes in configurational entropy by transforming a random molecular network into an extended molecular network. Therefore, equilibrium thermodynamics can be used to determine how stress is related to changes in internal energy and entropy. The equation of state for rubber elasticity is derived from classical thermodynamics, as follows (Eq. 11.1):

$$f = \frac{\partial U}{\partial L_{T,V}} + T\frac{\partial f}{\partial T_{L,V}} \qquad (11.1)$$

where U, L, V, and T are the internal energy, length, volume, and temperature, respectively, and f is the retractive force in reaction to a tensile force. Stretching the polymers with normal rubber behavior leads to a change in the total internal energy because of the dimensional alteration between polymer chains. However, this internal energy change is negligible due to the loose network structures of the rubber-like polymers. Additionally, the alignment of the chains by stretch causes them to be in a better-organized form with lower amounts of entropy. Considering the ideal rubber conditions, a perfect differential mathematical relationship between the retractive force and entropy was determined as follows (Eq. 11.2) [4]:

$$\frac{\partial U}{\partial L_{T,V}} = 0, \text{ and } \frac{\partial f}{\partial T_{L,V}} = -\frac{\partial S}{\partial L_{T,V}} \qquad (11.2)$$

where U, L, V, T, and S are the internal energy, length, volume, temperature, and entropy, respectively, and f is the retractive force in reaction to a tensile force.

By combining equations 11.1 and 11.2, the retractive force is expressed by (Eq. 11.3):

$$f = -T\frac{\partial S}{\partial L_{T,V}} \qquad (11.3)$$

The tangent line is drawn in the way of extending to 0 K. For an ideal rubber, the value of $(\partial U/\partial L)_{T,V}$ is 0, and the entropic part of the tangent line passes through the origin. Under ideal conditions, the experimental line is straight, with the slope being proportional to $-(\partial S/\partial L)_{T,V}$ or $(\partial f/\partial T)_{L,V}$.

11.2.2 VISCOELASTICITY

Viscoelasticity is the ability of a material to have both viscous and elastic behavior when deformed. A hydrogel is a material consisting of a solid polymer network that can be stretched like an elastic material under deformation and can expand into its area, like a viscous fluid. In short, it has an intermediate scale of features between elastic solid and viscous liquid depending on experimental temperature and time. This ability is based on molecular relaxation; the viscoelastic deformation of a polymeric material changes the conformation and possibly the relative position of its macromolecules. When stress is removed, these chains tend to return partially to their initial state, but they require a given relaxation time for returning. To define viscoelasticity, a combination of Hooke (Eq. 11.4) and Newton (Eq. 11.5) equations for elastic and viscous components, respectively, is used [5].

$$\sigma = E\varepsilon, \quad \frac{\partial \sigma}{\partial t} = E\frac{\partial \varepsilon}{\partial t} \tag{11.4}$$

$$\sigma = \mu\frac{\partial \varepsilon}{\partial t} \tag{11.5}$$

where σ, ε, E, and μ are stress, strain, Young's modulus, and viscosity of the material, respectively.

As the polymer chain segments move in response to an applied mechanical tension, the stress or strain exhibits a time-dependent change. This movement triggers an internal response, causing time-dependent recovery when removing the applied tension. This can be considered viscoelastic behavior if the recovery is completed during long periods [1].

11.2.3 EQUILIBRIUM SWELLING THEORY

According to the Flory-Rehner theory [6], for neutral hydrogels, the thermodynamic mixing force and the elastic force of the 3D network are two opposing forces that determine the swelling behavior of the hydrogel. The Gibbs free energy equation (Eq. 11.6) can be used for this situation:

$$\Delta G_{total} = \Delta G_{mixing} + \Delta G_{elastic} \tag{11.6}$$

where ΔG_{total} is the total Gibbs free energy change in the hydrogel, $\Delta G_{mixture}$ is the Gibbs free energy change resulting from the mixing of the amorphous and unstressed network structure with the pure solvent, and $\Delta G_{elastic}$ is the free energy change associated with the conformation changes of the gel structure. The chemical potential of the solvent permeating into the hydrogel network is represented as follows (Eq. 11.7):

$$\mu_1 - \mu_{1,0} = \Delta\mu_{mixing} + \Delta\mu_{elastic} \tag{11.7}$$

μ_1 is the chemical potential of the solvent trapped inside the gel network and $\mu_{1,0}$ is the chemical potential of the pure solvent. At the equilibrium swelling, the difference between the chemical potentials of the solvent inside and outside of the gel network should be 0. Additionally, chemical potentials concerning elastic and mixing forces must be equal. The chemical potential of the mixture can be calculated from the heat and the entropy of the mixture.

11.3   EXPERIMENTAL METHODS FOR MECHANICAL CHARACTERISTICS

11.3.1   STRESS-STRAIN TESTS

The response of the materials to applied forces is defined by their mechanical properties. The forces acting on deformed objects are described by two terms: strain and stress. Stress is the force applied per unit cross-sectional area of the material, which is classified based on different factors. If the applied force is divided by the current cross-sectional area, it is known as true stress, and if the force is divided by the initial cross-sectional area, it is called nominal (engineering) stress. If the force applied to the material is perpendicular to the cross-section and causes dimensional changes (elongation or compression), the stress is defined as normal stress, while if it causes an angle change in the material, it is called shear stress.

Strain is the deformation or displacement of material caused by an applied force. The deformation types of elastic and viscoelastic materials and their strain definitions are seen in Table 11.1. Two critical deformations for elastic materials are elongation and compression. Change in length after deformation is positive for elongation, while it is negative for compression. Shear deformation (or angular deformation) is a kind of deformation for viscoelastic materials that occurs when opposing forces are applied tangentially in a parallel direction to the material's surface [7].

The relationship between stress and strain is acquired by the stress-strain diagram given in Figure 11.1a. The elastic property is quantified by the slope of the stress-strain curve, which is defined as Young's modulus (E) and shear modulus (G) for normal and shear deformations, respectively. These modulus values are calculated from the linear region of the diagram [3].

Toughness is the crucial ability of the material to absorb the energy required to fracture and plastically deform without fracturing. It is calculated from the whole area under the stress-strain curve [8] and provides both strength and ductility for the hydrogels. Ductile materials can sustain plastic deformation under applied stress, and absorb mechanical overload, but they do not have high modulus. On the other hand, brittle materials break suddenly under stress in the elastic region without reaching plastic deformation, although they are very strong. As seen in Figure 11.1b, the curves with green, blue, and yellow colors represent the toughness of brittle, tough, and ductile hydrogels, respectively [9].

E and G are connected with Poisson's ratio if the gel is isotropic, which means that the properties of the gel do not rely on the force direction. The Poisson's ratio

TABLE 11.1
The Main Deformation Types and the Strain Definitions of Elastic and Viscoelastic Materials

	Elastic	Viscoelastic
Deformation types	 **Elongation Compression**	 **Shear**
Strain	$\alpha = \dfrac{L}{L_o} = \dfrac{\Delta L \pm L_o}{L_o}$ $\varepsilon_{el.} = \dfrac{L - L_o}{L_o} \quad \varepsilon_{com.} = \dfrac{L_o - L}{L_o}$	$\gamma = tan\theta$ $\gamma = \dfrac{\Delta x}{l}$

where α, ε, and γ are the deformation ratio, normal strain, and shear strain, respectively. L_o and L are the initial and final lengths in the elongation and compression tests. Besides, $\varepsilon_{el.}$ and $\varepsilon_{com.}$ are the elongation and compression strains, respectively. Δx is the change in lateral direction and l is the sample length.

a)   b)

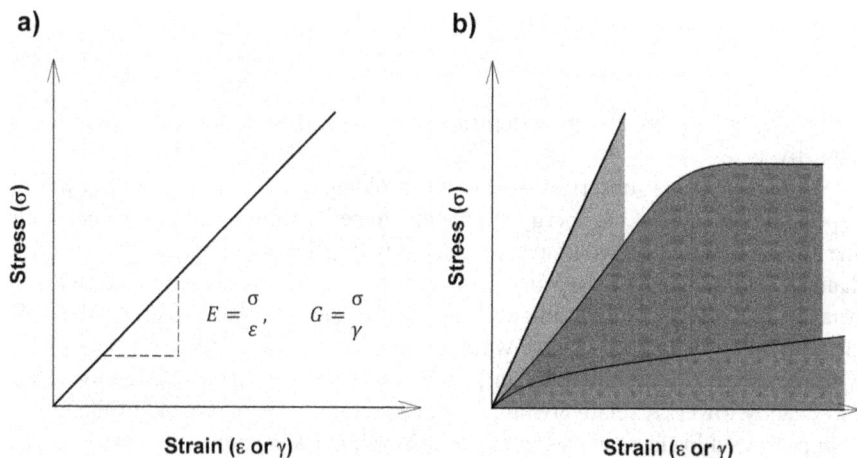

FIGURE 11.1   a) Stress-strain curve for elastic materials. b) Typical stress-strain curves for brittle (green), tough (blue), and ductile (yellow) hydrogels, indicating the areas under their respective curves.

(v) expresses a correlation between lateral and axial deformations of the material, which ranges between 0.0 and 0.5. If the volume of the gel does not change, v is 0.5 and E is equal to the threefold G.

11.3.2 CREEP AND STRESS RELAXATION

Creep is the measurement of time-dependent deformational change or strain under constant load. It represents the tolerance of hydrogel for deformation following the application of a constant static load. The ratio of applied shear stress to shear strain is defined as creep compliance. Creep compliance exhibits negligible time dependence when examined over a short or long period [1].

Stress relaxation is another time-dependent measurement that applies hydrogel specimens under constant strain and qualifies stress responded to by the specimen. It can be considered as the opposite of the creep test. When the specimen undergoes a certain deformation, a molecular chain relaxation occurs inside the gel network due to exponential stress dissipation. Such stress distribution deforms the specimen after the test period, and the remaining part that does not return to its full initial position is termed "residual strain". Returning action in hydrogels is time-dependent, and the time decay of stress is comparable to molecular relaxation [10].

11.3.3 CYCLIC DEFORMATION

The fatigue of hydrogels is characterized by testing them under cyclic loading-unloading deformation. When hydrogel specimens are subjected to cyclic deformation with maximum strength exceeding the anelastic limit of the hydrogel, their stress-strain curve displays a hysteresis loop, where the energy is dissipated because of internal structure friction [11]. Hysteresis is calculated from the area between the loading and unloading curves with the following (Eq.11.8) [12]:

$$U_{hys} = \int_0^{\varepsilon_{max}} \sigma d\varepsilon - \int_{\varepsilon_{max}}^0 \sigma d\varepsilon \qquad (11.8)$$

where U_{hys}, ε, ε_{max}, and σ are hysteresis energy, strain, maximum strain, and stress, respectively.

A shakedown is observed following prolonged cycles. The first cycle is responsible for most of the energy dissipation, material softening, and residual strain. The stress-stretch curve continues to change with each additional cycle, and these changes get smaller by increasing the cycle number. The stress-stretch curves, for instance, are essentially the same for the 8,000th and 10,000th cycles. In short, the stress-strain curve becomes stable when a tough hydrogel has reached a steady state. When compared to the initial loading cycle, the maximum stress in a steady state is significantly lower. A relatively high residual strain is seen at the end of the test.

In prolonged loading tests subjecting a hydrogel, a healing process can be applied to the hydrogel specimen between each loading-unloading cycle, such as elevated temperature, exposure to light, or immersion in a solvent. In these processes, the specimen is healed for some time after removing the applied load and then subjected to load again. The healing efficiency of a hydrogel is determined by subjecting the specimen to another period of prolonged load. Hydrogel healing is similar to thermoplastic healing, metal annealing, or ceramic sintering [13,14]. Some hydrogels with self-recovery properties can even heal without any other external stimuli in waiting time. These hydrogels can recover under many loading-unloading cycles due to their non-covalent and reversible bonds, such as hydrogen bonds [15], hydrophobic interaction [16], crystalline domains [17,18], and host–guest interaction [19]. They also have covalent bonds, providing elasticity, and the covalent bonding network excites their recovery properties. When the hydrogel deforms, these reversible bonds break, causing energy dissipation and toughening the material. The broken bonds can be reconstructed followed by unloading, which brings the material in self-recovery properties. Aself-recovery property can be dependent on the time scale of healing between loading and unloading. Here, a longer healing time leads to better recovery properties.

11.3.4 Fracture Processes

Fracture mechanics deal with crack propagation in materials. Once a crack is generated by a sharp tip, elastic forces in the material create a stress state that could show two types of fracture criterion: linear elastic fracture mechanics (LEFM), in which the length of the crack is larger than its tip (plastic zone), and elastic-plastic fracture mechanics (EPLM), in which the size of the plastic zone at the crack tip is high [20].

The fracture criterion of elastic solids is mainly defined by the LEFM, which has two mechanisms for mechanical load for crack growth, known as energetic and field approaches. In the energetic approach, the energy release rate (G) was considered the driving force. However, the field approach uses the singular stress field surrounding the crack tip. In LEFM, crack tip stress has a universal structure, and stress intensity factors are employed for fracture criteria, which are related to the J-integral, that is a path independent of the integral surrounding the crack tip. For non-linear deformation ahead of the crack tip, J-integral is expressed as (Eq. 11.9) [21]:

$$J = \int_C \left(W N_1 - \frac{\partial u_\alpha}{\partial X_{\text{linear viscoelastic (LVE) region}}} S_{\alpha\beta} N_\beta \right) d\xi \; \alpha, \beta = 1, 2, \quad (11.9)$$

Here, C is the smooth counter in the reference configuration enclosing the crack tip from lower to upper fracture face, W is the density of strain energy, u_α is the in-lane displacement field, $S_{\alpha\beta}$ is the first Piola-Kirchhoff stress, ξ is arc length, and N_β is the Cartesian components of the unit outward normal vector C. For pure elastics, J-integral is independent from the C and equal to the G.

11.3.5 DYNAMIC MECHANICAL ANALYSIS

Dynamic mechanical analysis (DMA) is a characterization technique to mechanically examine hydrogels depending on various conditions, such as time, temperature, frequency, and so on. The DMA concept is based on rheology, which examines the viscoelastic properties of the materials. The DMA measurement is mostly subjected to solid-like specimens and the analyzer has applied the force and deformation in two ways, axial and torsional, given in Figure 11.2.

The DMA is carried out under a certain sinusoidal stress or strain. The sinusoidal stress and strain curves for ideal elastic materials do not exhibit a phase shift ($\delta = 0$), while those for viscous materials indicate a phase shift angle of 90°. For viscoelastic materials, the phase shift angle is between 0° and 90°.

11.3.5.1 Amplitude Sweep

Amplitude sweep tests are carried out to find the linear viscoelastic (LVE) region of hydrogels. The LVE defines the range in which the test can be run without disturbing the specimen structure. Therefore, it should be the first measurement to be performed before starting other DMA tests, especially if the material is unknown. Temperature and frequency are fixed in the amplitude test, while the applied strain amplitude is variable at certain limits. In the LVE region, the elastic and viscous modulus exhibits a constant value called the plateau value [22].

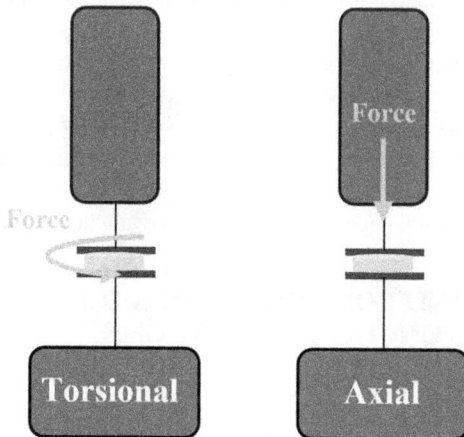

FIGURE 11.2 Torsional and axial deformations in DMA analysis.

11.3.5.2   Frequency Sweep

Frequency sweeps are generally performed to describe the time-dependent behavior of hydrogel specimens in the LVE range. High frequencies are utilized to imitate quick motion over short periods, while low frequencies are used to simulate slow motion over long periods. Frequencies can be expressed as either frequency (f) in Hz or angular frequency (ω) in rad/s. While the temperature and applied strain or stress are stable during the test, the frequency is varied [23].

A frequency sweep provides information about the mechanical characteristics of hydrogels, including stiffness and dynamic structures. Mechanically weak hydrogels exhibit frequency-dependent behavior in a frequency sweep, whereas the storage and loss moduli of strong hydrogels do not almost depend on frequency. Moreover, chemically cross-linked hydrogels are independent of frequency due to their high and strong cross-linking density. On the other hand, physically cross-linked hydrogels are highly frequency-dependent; their storage modulus is substantially higher than their loss modulus at the highest frequency. In addition, the presence of crossover in a frequency sweep indicates a dynamic and reversible cross-linking structure, while there is no such trend for chemically cross-linked hydrogels.

11.3.5.3   Temperature Sweep

Temperature sweeps are generally used to determine the transition temperatures of hydrogels. The glass transition temperature, sol-gel, and crystalline-amorphous can be given as examples of important transitions, depending on the temperature of hydrogels. These sweeps are also a method to search for the thermal stability of some hydrogels. In these hydrogels, the storage and loss modulus are stable until a certain temperature, while they start to have a negative slope after a certain temperature. This behavior is usually associated with changes in the chemical structure of hydrogels, such as degradation, decomposition, and de-cross-linking. All temperature sweeps are operated on the condition of constant frequency and strain (or stress) within the LVE region [24].

11.3.5.4   Time Sweep

A time sweep is frequently used for identifying gelation points as well as some structural properties such as degradation, recovery, and solvent evaporation over a certain time. In all time sweeps, the hydrogels' properties are investigated over time within the LVE region when temperature and frequency are kept constant. The gelation point is described as a crossover point of storage and loss moduli, at which a hydrogel passes from a liquid to a solid state. It provides information about the kinetics of gelation, including reaction speed and gelation times. In addition, measurements of changes in dynamic moduli depending on time help evaluate the hydrogel stability [25].

11.4   TUNING AND CONTROL OF THE MECHANICAL PROPERTIES OF MULTIFUNCTIONAL HYDROGELS

11.4.1   Multifunctional Hydrogel Network Types

Hydrogels are cross-linked polymeric networks that are obtained by the cross-linking of polymers or cross-linking polymerization of small monomers, generating

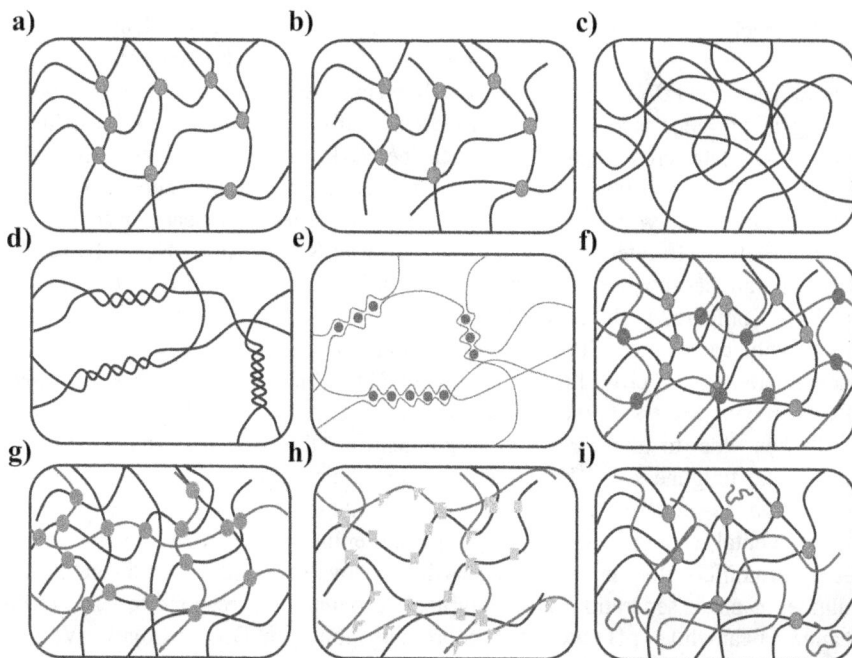

FIGURE 11.3 Hydrogel network configurations. a) Ideally cross-linked SN, b) non-ideally cross-linked SN, c) physical entanglement, d) helix configuration, and e) ionic cross-linking for egg-box model. Reproduced with permission from Reference [7], Copyright (2013), Taylor & Francis. f) double network model (interpenetrating network, IPN), g) dual network model, h) host-guest hydrogels, i) semi-IPN network. Reproduced with permission from Reference [26], Copyright (2017), under Creative Commons licenses, European cells & materials.

multiple network configurations (Figure 11.3). Cross-linking of one polymer generates single network (SN) hydrogels, which could be fully cross-linked or have free ends or polymeric loops. In some cases, hydrogels are formed by the entanglement of polymeric chains and cross-linking is achieved by hydrophobic interactions, as also observed in the helix configuration [7,26].

Hydrogels are mechanically weak, usually having elastic modulus in the kPa range and, cannot resist sub-MPa loading. The resistance to mechanical deformation has a straightforward relation to how easily a network can dissipate the energy. In brittle hydrogels, the energy dissipation is limited, thus causing the network to crack; meanwhile, in tough hydrogels, the energy dissipation along with fracture energy is high and the network can compensate for the applied force with less deformation [27]. This is due to the level of distribution of bonds through hydrogels. In tough hydrogels, dynamic non-covalent cross-linking (hydrophobic-hydrophobic interactions, dipole-dipole interactions, hydrogen bonds, and coordinate bonds) and chain entanglement overcome the issue of energy dissipation compared to the brittle hydrogels, which are dominated by straightforward covalent bonds [27,28].

The mechanical behavior of SN hydrogels could be further enhanced by the addition of other polymers or nanocomposites. Various tough hydrogels, including double network (DN), dual network, slide-ring hydrogels, and nanocomposite hydrogels, are generated. In DN hydrogels, two distinct polymers are cross-linked with a separate cross-linker, and each network is entangled with the other (Figure 11.3). In non-ideal DNs, also known as interpenetrating networks (IPNs), the ideal DN, along with free polymers and polymeric loops, appears. In semi-IPNs, a strong polyelectrolyte is dispersed into a neutral SN hydrogel. Here, the second polymer can be separated from the main polymer without any chemical bond breaking, compared to the IPNs [29]. In dual networks, polymers are directly cross-linked with each other. Supramolecular assemblies are formed because of host-guest interactions as cross-linkers (Figure 11.3) [7,26]. Compared to SN hydrogels, DN hydrogels possess extraordinary mechanical features. They have better energy dissipation due to the strong entanglement of the polymer chains and the mechanical toughness can be adjusted during their preparation [29]. Along with this, some as-prepared DN hydrogels alter their stiffness when exposed to environmental changes. Toughening and softening of these hydrogels are the results of non-covalent interactions, scission of covalent bonds, and change in their volume, or phase separation. For example, hydrogen bonds respond to temperature change; meanwhile, pH alters the coordinate bonds in DN hydrogels. Volume change associated with water content strongly influences the elastic and fracture behavior of hydrogels [28].

11.4.2 THE EFFECT OF GEL NETWORK COMPOSITIONS AND SWELLING BEHAVIOR ON MECHANICAL PROPERTIES

Hydrogels are composed of polymeric volume and water. There is a direct relationship between the water uptake and their toughness/rigidity. These features are strongly influenced by the type and amount of each component of hydrogel, including monomer, cross-linker, and polymer, and cross-linking conditions.

11.4.2.1 The Effect of Swelling on the Mechanical Behavior of Hydrogels

The rigidity and toughness, two characteristics of hydrogels, have a strong relationship with swelling. The water uptake of a hydrogel (Q) can be determined by (Eq. 11.10):

$$Q = \frac{1}{\phi} = \frac{V_{sw}}{V_{dry}} \tag{11.10}$$

where the ϕ is polymer volume fraction and V_{sw} and V_{dry} are the volumes of swollen and dry gels, respectively.

The toughness of hydrogels can be determined by crack propagation. In the experiment of compression tests, the crack propagation in the sample can be found at stress and strain values at the breaking point of the matrix. In this point, the interfacial crack energy $\Gamma(v)$ is expressed by (Eq. 11.11):

$$\Gamma(v) = \Gamma_0(1 + f(a_T v)) \tag{11.11}$$

where $f(a_T v)$ is the velocity-dependent crack factor, v is crack velocity, a_T is the superposition shift factor of time and temperature, and Γ_0 is threshold fracture energy to dismiss crack velocity. For Γ_0, the molecular theory assumes that the total energy of individual stretch bonds is lost when the main chain bond is broken or the number of single carbon-carbon bonds for a given N-chain. The fracture energy of hydrogels is strongly enhanced by the crack velocity, molecular friction, and chain entanglements by non-covalent bonds, which also increases the energy dissipation. To sum up, the rigidity and toughness of hydrogels are reduced as gel swells, which can be altered by the monomer type/composition, cross-link density, polymer type/concentration, and so on [28].

11.4.2.2 The Effect of Monomer Concentration and Composition on the Mechanical Behavior of Hydrogels

Cross-linking polymerization of monomers allows the control of hydrogel structure, compared to the cross-linking of polymers. Monomer type and concentration, as well as modified monomers, alter the mechanical properties of hydrogels. The mechanical stiffness increases with the enhanced monomer and cross-linker concentrations. The addition of hydrophobic co-monomers enhances loss factor tan δ (ratio of loss modulus to storage modulus), which is directly proportional to the toughness. This is because alkyl chains in the hydrogels create temporary junction zones, which minimize water exposure. During the local deformation, energy dissipation was dominated by the alkyl chains, therefore blocking the macroscale cracks in the hydrogels (Figure 11.4) [30].

11.4.2.3 The Effect of Cross-linker Density and Type on the Mechanical Behavior of Hydrogels

Cross-linker ratio, cross-linker type, or additional cross-link points in hydrogels alter the swelling, as well as the mechanical property. For polyacrylamide (PAAm) hydrogels, the increase in cross-linking time and cross-linker concentration enhanced the storage modulus, which is also associated with the reduced Q [31].

The type of cross-linker altered the mechanical property of chitosan hydrogels. Chitosan can be cross-linked with genipin or glutaraldehyde. Glutaraldehyde in a solution tends to be dimerized by itself compared to genipin. The degree of cross-linking was high in genipin-cross-linked chitosan hydrogels compared to genipin-cross-linked ones. The Q was reported to be high in glutaraldehyde-cross-linked hydrogels and maximum tensile strength was obtained with the genipin cross-linking [32].

11.4.2.4 The Effect of Cross-linking Temperature on the Mechanical Behavior of Hydrogels

Hydrogels with macroporous structures can be generated by porogen leaching, gas foaming, or cryogelation techniques. In cryotropic gelation, a gel precursor is placed in relatively lower temperatures, such as $-20\,^{\circ}C$ or $-40\,^{\circ}C$. In such a frozen system, ice crystals expel a polymer and cross-linker into unfrozen domains, and

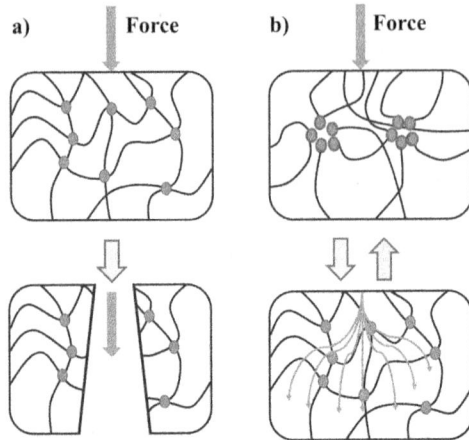

FIGURE 11.4 Schematic representation of a) crack energy localization in fully covalently cross-linked hydrogels and b) crack energy distribution in hydrophobically modified hydrogels. Reproduced with permission from Reference [30], Copyright (2009), Elsevier.

cross-linking reaction forms here. Compared to conventional hydrogels, cryogels exhibit superior mechanical properties, as well as macroporosity [33,34].

11.4.2.5   The Effect of Polymer Type and Content on the Mechanical Behavior of Hydrogels

Polymer properties, including charge, molecular weight, crystallinity, and the degree of functionalization, alter the swelling of hydrogels, as well as their toughness and rigidity. For example, gelatin methacrylate (GelMA) hydrogels with different stiffness were generated from the GelMA macromers with different methacrylation degrees. A GelMA hydrogel prepared from a higher methacrylation degree (91.7 ± 1.4%) possessed a high E value (29.9 ± 3.4 kPa) and lower Q (7.20 ± 0.12 g/g) [35].

For SN hydrogel, high molecular weight polymer usually results in high toughness. Sodium polyacrylate (PAAS) hydrogels with different molecular weights were prepared with laponite. Hydrogel with high molecular weight polymer was able to stretch up to 700% of its original shape; however, a hydrogel with low molecular weight broke just after the stretching. The interesting point here is that with certain clay concentrations, the increase in the percentage of PAAS in hydrogels reduced the elastic modulus, due to the disruption of homogeneity in the gels [36].

11.4.3   Design of Composite Hydrogels for Enhancing Mechanical Properties

A composite hydrogel can be generated by the addition of several materials, including nanoparticles, microparticles, and fibers. Once combined with a hydrogel, nanoparticles can either act as reinforcement agents or help the gelation occur via chemical or physical cross-linking [37].

For example, nano-Hydroxyapatite (nHAp) was incorporated into polyethyleneglycol (PEG) hydrogels to obtain an elastomeric network. E of PEG hydrogel was remarkably improved from 3.7 ± 0.6 kPa to 15.1 ± 1.3 kPa when 15 %wt. nHAp was used. Fracture stress, extensibility, ultimate strain, and compressive modulus values increased proportionally to the nHAp ratio. An increase of ten times in toughness was reported for PEG gels containing 15 %wt. nHAp, compared to 0 % wt. nHAp hydrogel [38].

The incorporation of synthetic silicate (clay) nanoparticles enhances the mechanical properties of hydrogels. Clays not only improve the rheology and rigidity of the hydrogels but also alter the swelling, cross-linking density, and injectability of the hydrogels, as well as provide sites for the attachment of other molecules [37].

The incorporation of metal and metal-oxide nanoparticles is another strategy to improve the mechanical properties of hydrogels, providing additional features including thermal and electrical conductivity. The mechanical feature of these composites can further be enhanced by the conjugation of metallic nanoparticles. For example, the addition of gold nanoparticles did not alter the mechanical behavior of hyaluronic acid/gelatin hydrogels. However, the cross-linking of thiol-functionalized gold nanoparticles with a polymeric matrix enhanced the mechanical properties [39].

As well as the inorganic nanoparticles, polymeric nanoparticles with various functionalities were embedded into hydrogel networks. Dendrimers and hyper-branched polymeric nanoparticles are used among all, whereby enabling the loading of drugs and providing mechanical stiffness. For example, poly-ethylene imine (PEI)/poly-acrylic acid (PAA) hydrogels were mechanically reinforced with *in situ* formed poly-diallyl dimethylammonium chloride (PDDA)/poly-sodium 4-styrene sulfonate (PSS) nanoparticles. *In situ* formed PDDA/PSS nanoparticles enhanced the toughness, tensile strength, E, and strain at break values. This is due to the flexible hydrogen bonds and electrostatic interactions between PEI and PAA chains and strong electrostatic interactions between PDDA and PSS in the nanoparticle system, which acted as an energy dissipation system when under the deformation of tough composite hydrogels [40].

Polymeric fibers also play a key role in matrix reinforcement and flexibility. Once the modified polycaprolactone (PCL) nanofibers were introduced into the PCL-*b*-PEG-*b*-PCL (PCEC) block copolymer, proper distribution of nanofibers into the matrix as well as the mechanical reinforcement was achieved. The compression modulus of nanocomposite hydrogels was increased when 2% wt. PCL nanofiber is added to PCEC. The rheological properties, storage, and loss modulus of PCEC were enhanced by the nanofiber concentration (from 0% to 2.5% nanofiber content) [41].

The elasticity and energy dissipation of nanocomposite hydrogels rely on reversible cross-links between nanoparticles and polymer chains. When cellulose nanofibrils are introduced into the PAAm matrix, reversible hydrogen bonding between the fibrils and PAAm chains occurs. The cross-linked PAAm and physical interactions enhanced toughness and mechanical strength. When the deformation was applied, the interface between fibrils and polymer was stiffening, and the

PAAm chains were desorbed from the fibril surface and became entangled during the resisting. Pulling out the fibrils activated a range of dissipation events, which ultimately enhanced the resistance to the crack propagation [42].

11.4.4 Mechanoresponsive Hydrogels with Different Biomedical Applications

Hydrogels have a broad range of applications in the biomedical field, due to their versatility including biocompatibility and biodegradation, tuneable micro-structure, a wide range of modifications, porosity, various mechanical features, and ability to respond to several stimuli. Therefore, hydrogels become the very first materials in biomedical applications including drug and gene delivery, diagnosis, wound healing, and most frequently tissue engineering (TE) when the mechanical features are the main criteria.

In general, tissue mimetic materials should be biocompatible. Enough mechanical properties, as well as mechanical stability, are required. The materials should support cellular growth and new tissue development and degrade within s controlled time with no toxic products. Porosity is another issue that the porous structure should allow the attachment and migration of the cells, as well as the diffusion of smaller molecules [43]. The physiological performance of hydrogels is strongly regulated by their mechanical properties. Since each tissue has a dynamic structure and experiences various mechanical forces, biomimetic constructs of tissues should also possess similar dynamic behavior [2]. Regarding the dynamic environment, mechano-responsive materials are able to alter their structure under force, pressure, or deformation, reaching the ultimate structure or returning to its original shape. Understanding the complexities of tissues switched the studies from conventional gels to mechano-responsive ones as smart materials [2].

The mechanism of mechano-responsiveness can be implemented by four mechanisms: strain-stiffening, shear-thinning, self-healing, and mechanochromic. The biomedical applications of these materials include artificial tissue, tissue scaffold, biosensor, 3D printing, drug delivery, and wound repair (Figure 11.5) [2].

11.4.4.1 Strain-Stiffening of Hydrogels

Once the external stress or strain is applied, tissues like muscle, skin, and blood vessels become stiffer to avoid large deformations. For example, networks comprising fibrin, collagen, vimentin, and actin exhibited improved shear stress when the strain was applied above the critical value. This is due to the presence of semi-flexible filaments in the self-organized matrix, in which filaments have comparable persistent and contour lengths. Under small strain, filaments were straightened, acting as cross-linkers. When strain is further enhanced, filaments are bent to show elasticity. By these two mechanisms, these matrices were able to stiffen under the strain. For some synthetic polymers, a strain-stiffening property is reduced due to polymeric loops, incomparable contour, and persistent length [2].

Strain-stiffening and self-healing of nanocomposite hydrogels were generated by dynamic cross-linking. Tannic acid–coated cellulose nanocrystals were incorporated into the PVA-borax network, yielding highly adhesive, self-healing, and toughness, as

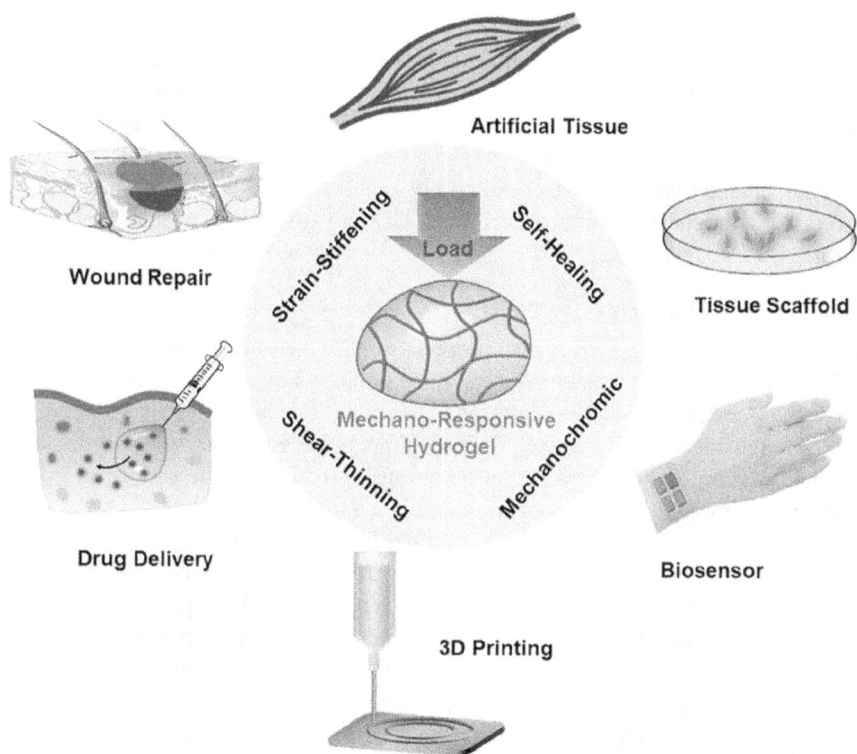

FIGURE 11.5 Mechano-responsive hydrogels show the response to several external mechanical stimuli which are used in various biomedical applications. Adapted with permission from Reference [2], Copyright (2020), American Chemical Society.

well as strain-stiffening properties [44]. Strain-stiffening hydrogels have mechanical stress of about 1 kPa, which is applicable in only soft tissues. By exactly mimicking the muscle training, a network was developed where the repetitive training of the network enhanced the mechanical strength up to MPa values. The self-growing network was prepared by the DN, one of which is rigid and brittle, and the other is soft and stretchable. When prepared DN was placed in a monomer bath, stretching caused the rigid network to crack and form mechano-radicals. Then, monomers are polymerized with formed radicals, yielding a new network that is mechanically stronger than the previous one. Inspired by the muscle training, a threefold increase in mechanical stiffness was reported after four-cycle training [45].

11.4.4.2 Shear-Thinning and Self-Healing of Hydrogels

These hydrogels possess supramolecular assembly, host-guest interactions, non-covalent bonds, and dynamic covalent bonds, which ultimately provide shear-thinning and self-healing ability. A shear-thinning property is the disruption of dynamic cross-linking points and decreases in viscosity under the shear, which provides injectability to hydrogels. Reversible bonds are also observed in self-healing hydrogels, whereby the hydrogel re-formed when the cut pieces were placed in contact with each other [2]. Injectable hyaluronic acid (HA) hydrogels were generated with a shear-thinning and self-healing ability for drug or cell delivery. HA chains were cross-linked with host-guest complex formation, where one chain is substituted with β-cyclodextrin (forming CD-HA) and the other is attached with the adamantane (named Ad-HA). Assembly and disassembly were observed upon a reversible host-guest interaction between β-CD and Ad, indicating self-healing and shear-thinning, respectively [46].

11.4.4.3 Mechanochromic Hydrogels

These materials change their optical properties under mechanical stimuli by mechanophore addition or switchable structural coloration. For example, the rhodamine derivative can act as a mechanophore, and change its color when force is applied, from yellow to red due to isomerization. Mechanophore particles or photonic crystals can be covalently attached to the hydrogels or physically added, so the change in optical properties in these hydrogels is observable upon exposure to force [2].

Except for the mechanical stimuli, thermal fluctuations and photo-induced color changes were also observed in those gels. A multi-responsive hydrogel was generated by the PAAm-co-poly (methyl acrylate) (PMA)/Spiropyran (SP) in which SP was used as a cross-linker and mechanophore that responds to the force, UV light, and heat. Upon exposure to UV, force, and heat, SP turns into merocyanine (MC), with a color change from yellow to purple. SP cross-linked hydrogels exhibited remarkable mechanical features with 1.45 MPa tensile strength at 570% fracture strain. The fracture energy of gels was found about 7,300 J m^{-2}, and they also exhibited a large hysteresis loop and dissipated energy. As the tensile strain increases, the color change is observed from resting state to purple due to the energy transfer from the outer layer to the inner SP molecules [47].

11.5 CONCLUSION

Hydrogels are polymeric network structures composed of different components that provide them with different mechanical properties. Understanding the mechanical properties of hydrogels is critical for their biomedical application. Accordingly, different mathematical models and methods have been introduced and some of them are summarized in this chapter. Besides, the structural network type, addition of functionalizing agents and nanomaterials, and cross-linking density could affect the type of bonds and energy distribution among the system, and in doing so altering the swelling ratio of hydrogel by affecting their mechanical features. It is revealed that increasing the amounts of hydrophobic monomers or cross-linking could enhance the stiffness of the hydrogels. Moreover, the preparation method and its related temperature could also change the structural and mechanical features of the hydrogels. On the other hand, due to the effects of different properties of polymeric backbone, including its charge, crystallinity, and molecular weight on the swelling behavior of the hydrogel, it is critical to consider them during the preparation. It is also revealed that the mechanical property could be affected by different microenvironments, a point that should be considered for the biomedical applications of hydrogels.

Moreover, this property was applied in some recent biomedical research for the fabrication of mechanoresponsive hydrogels via mechanisms like strain-stiffening, shear-thinning, self-healing, and mechanochromic. Application of these mechanisms led to fabrication of tunable hydrogels with features like self-healing ability, enhanced mechanical properties, and real-time sensing features.

REFERENCES

[1] K.S. Anseth, C.N. Bowman, L. Brannon-Peppas, Mechanical Properties of Hydrogels and Their Experimental Determination, Biomaterials, 1996, 17, 1647–1657.

[2] J. Chen, Q. Peng, X. Peng, L. Han, X. Wang, J. Wang, H. Zeng, Recent Advances in Mechano-Responsive Hydrogels for Biomedical Applications, ACS Applied Polymer Materials, 2020, 2(3), 1092–1107.

[3] L.H. Sperling, Introduction to Physical Polymer Science, 2006, John Wiley & Sons, New Jersey, USA.

[4] F.T. Wall, Chemical Thermodynamics: A Course of Study, 1974, W.H. Freeman & Co Ltd., San Francisco, USA.

[5] J.D. Ferry, Viscoelastic Properties of Polymers, 3rd edition, 1980, John Wiley & Sons, New York, USA.

[6] P.J. Flory, J. Rehner, Statistical Mechanics of Cross-Linked Polymer Networks II. Swelling, The Journal of Chemical Physics, 1943, 11(11), 521–526.

[7] M.L. Oyen, Mechanical Characterisation of Hydrogel Materials, International Materials Reviews, 2014, 59(1), 44–59.

[8] J. Li, H. Liu, C. Wang, G. Huang, A Facile Method to Fabricate Hybrid Hydrogels with Mechanical Toughness Using a Novel Multifunctional Cross-linker, RSC Advances, 2017, 7(56), 35311–35319.

[9] S. Ahmed, T. Nakajima, T. Kurokawa, M.A. Haque, J.P. Gong, Brittle–Ductile Transition of Double Network Hydrogels: Mechanical Balance of Two Networks as the Key Factor, Polymer (Guildf), 2014, 55(3), 914–923.

[10] S. Ganguly, P. Das, N.C. Das, Characterization tools and techniques of hydrogels, Yu Chen (Editor), Hydrogels Based Natural Polymers, 2020, Elsevier, Amsterdam, Netherlands.

[11] Y. Tanaka, R. Kuwabara, Y.H. Na, T. Kurokawa, J.P. Gong, Y. Osada, Determination of Fracture Energy of High Strength Double Network Hydrogels, Journal of Physical Chemistry B, 2005, 109(23), 11559–11562.

[12] A.G. Thomas, Rupture of Rubber. V. Cut Growth in Natural Rubber Vulcanizates, Journal of Polymer Science, 1958, 31(123), 467–480.

[13] R. Bai, J. Yang, Z. Suo, Fatigue of Hydrogels, European Journal of Mechanics- A/ Solids, 2019, 74, 337–370.

[14] R. Bai, Q. Yang, J. Tang, X.P. Morelle, J. Vlassak, Z. Suo, Fatigue Fracture of Tough Hydrogels, Extreme Mechanics Letters, 2017, 15, 91–96.

[15] X. Hu, M. Vatankhah-Varnoosfaderani, J. Zhou, Q. Li, S.S. Sheiko, Weak Hydrogen Bonding Enables Hard, Strong, Tough, and Elastic Hydrogels, Advanced Materials, 2015, 27(43), 6899–6905.

[16] M.A. Haque, T. Kurokawa, G. Kamita, J.P. Gong, Lamellar Bilayers as Reversible Sacrificial Bonds to Toughen Hydrogel: Hysteresis, Self-Recovery, Fatigue Resistance, and Crack Blunting, Macromolecules, 2011, 44(22), 8916–8924.

[17] C. Bilici, D. Karaarslan, S. Ide, O. Okay, Toughness Improvement and Anisotropy in Semicrystalline Physical Hydrogels, Polymer (Guildf), 2018, 151, 208–217.

[18] C. Bilici, S. Ide, O. Okay, Yielding Behavior of Tough Semicrystalline Hydrogels, Macromolecules, 2017, 50(9), 3647–3654.

[19] M. Nakahata, Y. Takashima, H. Yamaguchi, A. Harada, Redox-Responsive Self-Healing Materials Formed from Host-Guest Polymers, Nature Communications, 2011, 2(1), 1–6.

[20] H.R. Brown, A Model of the Fracture of Double Network Gels, Macromolecules, 2007, 40(10), 3815–3818.

[21] J.K. Knowles, E. Sternberg, Large Deformations Near a Tip of an Interface-Crack Between Two Neo-Hookean Sheets, Journal of Elasticity, 1983, 13(3), 257–293.

[22] L. Mendoza, W. Batchelor, R.F. Tabor, G. Garnier, Gelation Mechanism of Cellulose Nanofibre Gels: A Colloids and Interfacial Perspective, Journal of Colloid and Interface Science, 2018, 509, 39–46.

[23] R.G. Jones, E.S. Wilks, W.V. Metanomski, J. Kahovec, M. Hess, R. Stepto, T. Kitayama, Compendium of Polymer Terminology and Nomenclature, 2009, The Royal Society of Chemistry, Cambridge, UK.

[24] I. Antoniuk, D. Kaczmarek, A. Kardos, I. Varga, C. Amiel, Supramolecular Hydrogel Based on pNIPAm Microgels Connected via Host⁻Guest Interactions, Polymers, 2018, 10(6), 566.

[25] G. Stojkov, Z. Niyazov, F. Picchioni, R.K. Bose, Relationship Between Structure and Rheology of Hydrogels for Various Applications, Gels, 2021, 7(4), 255.

[26] S.L. Vega, M.Y. Kwon, J.A. Burdick, Recent Advances in Hydrogels for Cartilage Tissue Engineering, European Cells and Materials, 2017, 33, 59–75.

[27] C. Norioka, Y. Inamoto, C. Hajime, A. Kawamura, T. Miyata, A Universal Method to Easily Design Tough and Stretchable Hydrogels, NPG Asia Materials, 2021, 13(1), 34.

[28] X. Lin, X. Wang, L. Zeng, Z.L. Wu, H. Guo, D. Hourdet, Stimuli-Responsive Toughening of Hydrogels, Chemistry of Materials, 2021, 33(19), 7633–7656.

[29] Q. Chen, H. Chen, L. Zhu, J. Zheng, Fundamentals of Double Network Hydrogels, Journal of Materials Chemistry B, 2015, 3(18), 3654–3676.

[30] S. Abdurrahmanoglu, V. Can, O. Okay, Design of High-Toughness Polyacrylamide Hydrogels by Hydrophobic Modification, Polymer, 2009, 50(23), 5449–5455.

[31] R. Subramani, A. Izquierdo-Alvarez, P. Bhattacharya, M. Meerts, P. Moldenaers, H. Ramon, H. Van Oosterwyck, The Influence of Swelling on Elastic Properties of Polyacrylamide Hydrogels, Frontiers in Materials, 2020, 7, 212.

[32] Y. Liu, Z. Cai, L. Sheng, M. Ma, Q. Xu, Y. Jin, Structure-Property of Crosslinked Chitosan/Silica Composite Films Modified by Genipin and Glutaraldehyde under Alkaline Conditions, Carbohydrate Polymers, 2019, 215, 348–357.

[33] F. Ak, Z. Oztoprak, I. Karakutuk, O. Okay, Macroporous Silk Fibroin Cryogels, Biomacromolecules, 2013, 14(13), 719–727.

[34] S. Sezen, V.K. Thakur, M.M. Ozmen, Highly Effective Covalently Crosslinked Composite Alginate Cryogels for Cationic Dye Removal, Gels, 2021, 7(4), 178.

[35] X. Li, S. Chen, J. Li, X. Wang, J. Zhang, N. Kawazoe, G. Chen, 3D Culture of Chondrocytes in Gelatin Hydrogels with Different Stiffness, Polymers, 2016, 8(8), 269.

[36] H. Takeno, C. Sato, Effects of Molecular Mass of Polymer and Composition on the Compressive Properties of Hydrogels Composed of Laponite and Sodium Polyacrylate, Applied Clay Science, 2016, 123, 141–147.

[37] C. Tipa, M.T. Cidade, J.P. Borges, L.C. Costa, J.C. Silva, P.I.P. Soares, Clay-Based Nanocomposite Hydrogels for Biomedical Applications: A Review, Nanomaterials, 2022, 12(19), 3308.

[38] A.K. Gaharwar, S.A. Dammu, J.M. Canter, C.-J. Wu, G. Schmidt, Highly Extensible, Tough, and Elastomeric Nanocomposite Hydrogels from Poly(ethylene glycol) and Hydroxyapatite Nanoparticles, Biomacromolecules, 2011, 12(5), 1641–1650.

[39] A.K. Gaharwar, N.A. Peppas, A. Khademhosseini, Nanocomposite Hydrogels for Biomedical Applications, Biotechnology and Bioengineering, 2014, 111(3), 441–453.

[40] T. Yuan, X. Cui, X. Liu, X. Qu, J. Sun, Highly Tough, Stretchable, Self-Healing, and Recyclable Hydrogels Reinforced by in Situ-Formed Polyelectrolyte Complex Nanoparticles, Macromolecules, 2019, 52(8), 3141–3149.

[41] H. Zhong, Z. Li, T. Zhao, Y. Chen, Surface Modification of Nanofibers by Physical Adsorption of Fiber-Homologous Amphiphilic Copolymers and Nanofiber-Reinforced Hydrogels with Excellent Tissue Adhesion, ACS Biomaterials Science and Engineering, 2021, 7(10), 4828–4837.

[42] J. Yang, F. Xu, Synergistic Reinforcing Mechanisms in Cellulose Nanofibrils Composite Hydrogels: Interfacial Dynamics, Energy Dissipation, and Damage Resistance, Biomacromolecules, 2017, 18(8), 2623–2632.

[43] S. Utech, A.R. Boccaccini, A Review of Hydrogel-Based Composites for Biomedical Applications: Enhancement of Hydrogel Properties by Addition of Rigid Inorganic Fillers, Journal of Materials Science, 2016, 51, 271–310.

[44] C. Shao, L. Meng, M. Wang, C. Cui, B. Wang, C.R. Han, F. Xu, J. Yang, Mimicking Dynamic Adhesiveness and Strain-Stiffening Behavior of Biological Tissues in Tough and Self-Healable Cellulose Nanocomposite Hydrogels, ACS Applied Materials and Interfaces, 2019, 11(6), 5885–5895.

[45] T. Matsuda, R. Kawakami, R. Namba, T. Nakajima, J.P. Gong, Mechanoresponsive Self-Growing Hydrogels Inspired by Muscle Training, Science, 2019, 363(6426), 504–508.

[46] C. Loebel, C.B. Rodell, M.H. Chen, J.A. Burdick, Shear-Thinning and Self-Healing Hydrogels as Injectable Therapeutics and for 3D-Printing, Nature Protocols, 2017, 12(8), 1521–1541.

[47] H. Chen, F. Yang, Q. Chen, J. Zheng, A Novel Design of Multi-Mechanoresponsive and Mechanically Strong Hydrogels, Advanced Materials, 2017, 29(21), 1606900.

12 Hydrogels for Bioelectronics

Bushara Fatma, Nicholas G. Hallfors,
Nazmi B. Alsaafeen, and Charalampos Pitsalidis
Khalifa University of Science and Technology,
Abu Dhabi, UAE

12.1 INTRODUCTION

Machines that seamlessly interface with the human body are long-standing motif in science fiction, with major real-world implications. From the electrical transduction of neural signals to motor control, cardio-respiratory rhythm, and the ionic currents that dictate low-level cellular function, electricity plays a central role in the human body's functionality. With the growth of the modern electronics industry, researchers have naturally been drawn to the idea of interfacing modern electronics with biology giving rise to the field of bioelectronics. Many bioelectronic technologies have matured to a level of everyday use, e.g., pacemakers and cochlear implants, which have already transformed the way we approach healthcare [1,2]. Others have yet to merit widespread adoption, e.g., myoelectric prostheses and permanently implanted deep brain stimulation electrodes, though their time appears to be just on the horizon. Decades of research have brought the technology of wearable and implantable electronics to a point that they offer real value, in a form that is conducive to everyday use, both in and out of a clinical setting.

Still, major challenges remain in engineering the interface between the soft and flexible human tissue and electronics, which traditionally have been designed upon rigid silicon and metal. The human body is a complex network of solids, liquids, proteins and ions, all of which need to be addressed when designing a wearable sensor. Human skin has a highly variable resistance, from megaohms if dry, to several hundred ohms if wet, and even fewer if the skin is punctured [3]. Our bodies are always in motion, which is not a problem for tissue that is inherently flexible and elastic, but for a rigid electronic interface that requires planar contact, this is a critical flaw. For wearable devices to be considered for everyday or long-term use, they need to be not only effective in their function, but also comfortable and non-invasive. Biofouling, drying, and extended surface-surface transduction all need to be addressed at the material level when designing a functional bioelectronic interface, and many of these challenges can be overcome through the application of hydrogels. As such, hydrogels can be engineered to resist freezing or drying, to be self-healing and durable, and to meet specific criteria of sensitivity, compatibility,

DOI: 10.1201/9781003340485-12

and functionality for bioelectronics. In this chapter, we will discuss some of the materials currently being explored to design hydrogels for bioelectronics, as well as the fundamental principles that govern their effectiveness as an electrical/electro-chemical interface. Also, we will discuss the applications that are currently being explored for functional hydrogel bioelectronics, as well as the requirements for further advancing the development of this exciting class of electronic devices.

12.2 CONDUCTIVITY IN HYDROGELS

Both synthetic and natural hydrogels demonstrate desirable properties for skin-like hydrogel devices as they are soft, stretchable, conformable, biocompatible, adhesive, and exhibit anti-dehydration behavior. However, the key property for incorporating hydrogels in bioelectronic devices is the electrical conductivity, as it dictates the communication between skin/tissue and electronics. In their neutral form, natural hydrogels do not solely provide the electrical conductivity required for recording or stimulating purposes in bioelectronics. Therefore, they are supple-mented with other materials through synthetic chemical routes and/or configurations to achieve the desirable electrical properties.

In general, conductivity can be incorporated into a hydrogel system by following two different approaches: (i) *single-component conducting hydrogels,* which can be formed via self-assembly or self-polymerization of the conductive component (i.e., conducting polymer) and (ii) *two (or multi)-component systems,* with the conductive component being deposited either on top of the hydrogel, thus forming a bilayer structure, or introduced as a filler inside the hydrogel matrix. Based on the incorporated conductive component, the charge conduction mechanism can also be classified as electronic, ionic, or mixed. Electronic conductivity can be achieved by using conductive films/fillers such as metal nanoparticles/wires, carbon-based nanostructures (e.g., carbon nanotubes (CNTs), graphene) or 2D transition metal carbides/nitrides (MXenes ($Ti_3C_2T_x$)). Another approach suggests the use of conducting polymers, such as poly(3,4-ethylenedioxythiophene) (PEDOT), polypyrrole (PPy), and polyaniline (PANI). Ionic conductivity on the other hand can be introduced through the addition of ionic liquids and salts, which allow the motion of ionic charges throughout the hydrogel matrix. Finally, mixed conductors i.e., poly(3,4-ethylenedioxythiophene)-poly(styrenesulfonate) (PEDOT:PSS) represent an important family of materials for bioelectronic hydrogels, as they are known to promote ion-to-electron coupling for enhanced biosignal transduc-tion. The classification of conducting hydrogels, according to their composition and structures, is shown in Figure 12.1, and is further discussed in the following sections.

12.2.1 IONIC CONDUCTIVITY: POLYELECTROLYTES AND SALTS

Ionic conductive hydrogels can be developed by employing ionic species from polyelectrolytes and/or salts to hydrogels, which in turn can enable ion transport [4]. Polyelectrolyte hydrogels consist of repeating units bearing functional groups with fixed charges. As hydrogels are neutrally charged with respect to the fixed charges on the polymer backbone, there can be an equal amount of mobile opposite charges in the hydrogel that result in its ionic conductivity. Polyelectrolytes hydrogels can be

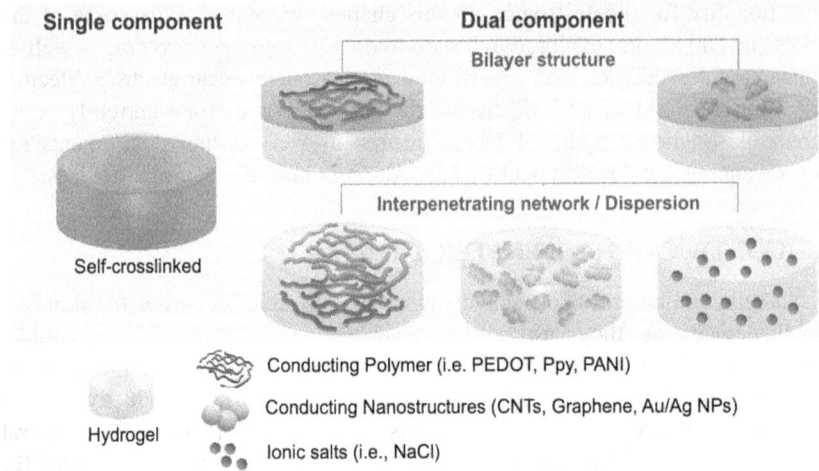

FIGURE 12.1 Schematic representation of the different approaches used to introduce conductivity in a hydrogel matrix.

classified, according to the type of mobile ions in the polymer complex, as polyanions, polycations, and zwitterionic, as shown in Figure 12.2a [4]. Examples of polyanion, polycation, and zwitterionic monomer are 2-acrylamido-2-methylpropane sulfonic acid, 2-(acryloyloxy) ethyltrimethyl ammonium chloride, and 2-methacryloyloxyethyl phosphorylcholine, respectively. The ionic conductivity of a hydrogel depends on various factors, including water uptake/retention, charge carrier density, and salt concentration. Water retention ability is an important factor when designing an ionic conductive hydrogel, since higher water content means more conducting pathways within the pore network of the gel and thus more beneficial ion transport (Figure 12.2b) [4,5]. To render the hydrogels insoluble and provide a 3D network structure, physical or chemical cross-linking is typically used. This often results in a decline of the hydrogel electrical properties, as it decrease the porosity and the water content [4]. To address this issue, Chen-Jung lee et al. replaced the typical cross-linker *N, N'*-methylenebisacrylamide with the more hydrophilic cross-linker tetraethylene glycol dimethactylate, and obtained a more hydrophilic poly(sulfobetaine methacrylate) hydrogel with high water uptake and increased porosity, and hence increased ionic conductivity [5]. Most water-based hydrogels tend to dry with time, which often affects their ionic conductivity. A common approach to prolong the water retention ability of ionic hydrogels is to use anti-freezing agents, such as methyl alcohol, glycerol, and ethylene glycol [6]. However, the addition of alcohols often increases the viscosity of the solution and decreases its ion mobility and the resulting ionic conductivity [6,7]. Indeed, Yongqi Yang et al. showed that increasing the glycerol content in a conductive organohydrogel decreases substantially the ionic conductivity (from 0.035 S/cm for 0% glycerol to ~ 0.00035 S/cm for 100% glycerol) (Figure 12.2c) [7].

Introducing highly hydrating salts (e.g., lithium chloride (LiCl), magnesium chloride ($MgCl_2$), potassium acetate) in hydrogels is known to improve water uptake and retention properties [6,8]. Indeed, Bai et al. showed that addition of a

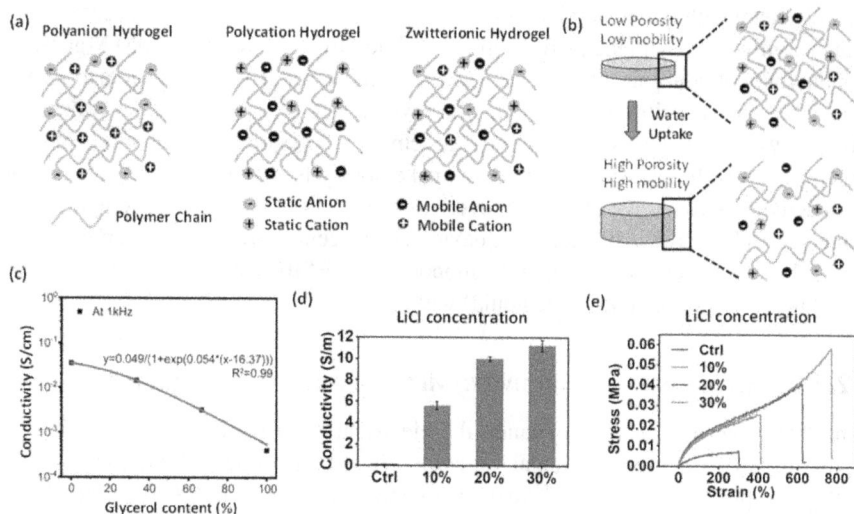

FIGURE 12.2 (a) Schematic of the structure of polyanion hydrogel, polycation hydrogel and zwitterionic hydrogel. (b) Increase in the water uptake ability of hydrogel increases the porosity and permits faster mobility of free ion throughout the hydrogel matrix leading to high ionic conductivity. (c) Linear decrease in the conductivity of hydrogel made of sodium methacrylate and [2-(methacryloyloxy)-ethyl]. trimethyl ammonium chloride with different glycerol concentrations at 1 kHz (conductivity decreases from the highest value of 0.035 S/cm for 0% glycerol to ~ 0.00035 S/cm for 100% glycerol). Adapted with permission from Reference [7]. Copyright (2019) American Chemical Society. (d) Conductivity and (e) stress-strain curves of poly(sulfobetaine-co-acrylic acid) hydrogel swelled in water (control), 10%, 20% and 30% of LiCl solutions. Adapted with permission from Reference [6]. Copyright (2021) Elsevier.

high amount of LiCl salt into a polyacrylamide (PAAm) hydrogel can retain 70% of the water as compared to the initial state, even when the surroundings have a rather low relative humidity of 10% [8]. The addition of salts into hydrogels not only enhances the water retention ability but also increases the conductivity of the hydrogel by incorporating free ions [5,6,8]. Sui et al. prepared an ionic conductive poly(sulfobetaine-co-acrylic acid) hydrogel by soaking it in highly hydrating salt solutions with different concentrations [6]. As shown in Figure 12.2d, the hydrogel soaked in 30% LiCl increased its conductivity to the value of ~11 S/m as compared to 10% LiCl (~5 S/m). Also, the resulting hydrogels maintained their water content after one week. However, in the same study, it was found that with increasing LiCl concentration, the rate of elongation is decreased (Figure 12.2e), which can be attributed to the swelling-driven weak interactions between polymer chains.

Another factor that determines the conductivity of hydrogels is the amount of static charge carrier density resulting from the number of charged functional groups in the polymer network [4,5]. Lee et al. studied the effect of a charged functional group of polyelectrolyte on the ionic conductivity of the polyelectrolyte hydrogel [5]. The zwitterionic polyelectrolyte, i.e., poly(carboxybetaine acrylamide) (PCBAA)-based hydrogel was found to have increased conductivity compared to non-ionic poly(ethylene glycol) methyl ether methacrylate (PEGMA) hydrogel when the same concentration of salts was used [5]. With the increase in salt addition from 2 mM to 100 mM $MgCl_2$, the conductivity of non-ionic hydrogel PEGMA was increased from ~0.2 mS/cm to ~2 mS/cm, which is very low compared to the zwitterionic hydrogel PCBAA where the conductivity was increased from ~0.4 mS/cm to ~7 mS/cm. Besides inorganic salt solvents, ionic liquids can be used to make ionogels that exhibit excellent ionic conductivity and hardly evaporate over time and, therefore, can be employed as conducting electrodes in bioelectronic devices. In a recent study, Wang et al. prepared an ionogel of N,N-dimethylacrylamide monomer in $EMIMBF_4$ (1-ethyl-3-methylimi dazolium tetrafluoroborate, ionic liquid) with a conductivity of 8.7 mS/cm [9].

12.2.2 ELECTRONIC CONDUCTIVITY: METAL/CARBON NANOSTRUCTURES

An alternative approach to enhance the electrical properties of a non-conductive hydrogel is to incorporate metal or carbon-based nanofillers. Examples include metal nanoparticles (e.g., gold (Au)), nanowires (e.g., silver (Ag), CNTs), graphene, or MXenes, which can be dispersed into hydrogel networks to enhance electrical conductivity.

Metals exhibit excellent electrical conductivity and represent the main component of modern bioelectronic interfaces. More specifically, noble metals are commonly adopted as the conducting component in hydrogels to avoid degradation in wet/physiological conditions. Au nanoparticles represent one of the most commonly used conducting components for hydrogels. In a study from Baei et al., a thermosensitive conductive hydrogel based on chitosan and Au nanoparticles has been demonstrated. The conductivity and the gelation of the resulting hydrogel were tailored by adjusting the concentration of Au nanoparticles. Culturing mesenchymal stem cells within these hydrogels revealed enhanced cardiomyogenic differentiation even in the absence of electrical stimulation [10]. A platinum (Pt) nanoparticle is another candidate structure for use in conductive hydrogels since they offered additional capabilities for biomedical research (i.e., biocatalytic activity). Other nanoparticles have been also used investigated, such as Ag, zinc peroxide (ZnO_2), and titanium dioxide (TiO_2).

Carbon-based conductors such as CNTs not only can greatly improve the electrical properties of non-conducting hydrogels, but they can also contribute to their mechanical stability. Xia Sun et al. modified CNTs by oxidizing and subsequently functionalizing them with gelatin via hydrogen bonding between the carboxyl groups present on the oxidized CNT and the hydroxyl/carboxyl groups of gelatin. This increased its dispersibility in water as well as its interaction with the PAAm-based hydrogel [11]. The conductivity of the CNT-reinforced nanocomposite hydrogel was found to be 0.067 S/m, with a stretchability of 700%, tensile strength of 0.71 MPa, and durability of over 300 cycles. Although chemical modification improves the dispersibility of CNTs, it was found to compromise their conductivity. In addition, increasing the CNTs' content to compromise for the conductivity loss has been shown to cause biocompatibility issues. Therefore, stability and toxicity of the dual component hydrogel with nanofiller remains an issue, notably when used for *in vivo* bioelectronics applications.

Hydrogels based on graphene or graphene derivatives have also been used for biomedical applications. In a recent study, hydrogel-graphene nanocomposite fibers have been developed with the aim to develop a 3D conducting scaffold for tissue engineering. Remarkably, these hydrogel fibers showed excellent electrochemical properties and cytocompatibility [12]. In another study, an ion-conducting hydrogel based on polyvinyl alcohol (PVA) and graphene oxide was demonstrated. The resulting composite exhibited excellent mechanical properties, with a tensile stress of up to 65 kPa and high ionic conductivity of up to 3.38 S/m [13].

As mentioned before, the dispersibility of nanofillers is a major issue causing inhomogeneity in the electrical and mechanical properties of the hydrogels. To address this issue, a bilayer structure can be developed as a separate layer of conducting nanostructure(s) on top of a non-conducting hydrogel. This approach gives the flexibility to take control over the mechanical and electrical properties of the different components without compromising their characteristics. In a work from Ahn et al., an Ag nanowire-based microelectrode array was fabricated on top of a PAAm-based hydrogel, which resulted in high electrical conductivity, flexibility, and stability in a wet condition [14]. Similarly, carbon-based electrodes have also been deposited on top of non-conducting hydrogels, even though they exhibit limited resistance to mechanical deformations (i.e., elongation).

12.2.3   ELECTRONIC CONDUCTIVITY: CONDUCTING POLYMERS

Conducting polymers are commonly used as an electrically conducting component to introduce conductivity to non-conducting hydrogel. Different approaches have been adapted for the use of conducting polymers to make hydrogels with a conducting component; single-component hydrogel where the complete hydrogel is made of conducting polymer, dual component with conducting polymer on top of hydrogel, and the third approach is to uniformly disperse the conducting polymer in the non-conducting hydrogel matrix. Wang et al. prepared a double network conductive hydrogel formed by a hydrogen bond between poly(acrylamide-co-hydroxylethyl methyl acrylate) (P-(AAm-*co*-HEMA)) and interpenetrating network of conducting polymer PANI, reaching a conductivity of ~8 S/m with 200% strain and high toughness [15]. The hydrogel showed a linear increase in the resistance change ratio upon high strains (up to 300%) without significant alterations in its mechanical properties upon various loading and unloading cycles. Single-component hydrogels can be synthesized using pure conducting polymers, avoiding the use of other components, such as a non-conducting hydrogel matrix and/or nanofillers.

12.2.4   HYBRID CONDUCTIVITY

Mixed, ionic-electronic hydrogels can be made with more than one conductive component, taking advantage of both electronic and ionic transport. While many conducting polymers exhibit such a binary conducting character, PEDOT:PSS has been the material of choice for the development of various bioelectronic devices. This is no exception for conductive hydrogels, with PEDOT:PSS being one of the most commonly used materials due to its high conductivity, stability, and tailorability. PEDOT:PSS hydrogels can be directly developed by adding positive ionic species (i.e., Ca^{2+}, $Fe^{2+}/3+$, Mg^{2+}), ionic liquids or acids (i.e., sulfuric acid (H_2SO_4)), which can electrostatically interact with the negatively charge PSS⁻. While this strategy ensures sufficient cross-linking and conductivity, the mechanical properties of the resulting hydrogels are rather poor. In a study by Lu et al., PEDOT:PSS hydrogels were developed by adding volatile dimethyl sulfoxide (DMSO) into aqueous PEDOT:PSS solutions followed by controlled annealing to dry the structure and later rehydrate it to yield pure PEDOT:PSS hydrogels [16]. The amount of DMSO and the annealing were controlled to tailor the mechanical and electrical properties. The electrical conductivity of the fabricated hydrogel was found to be 20 S/cm in PBS (phosphate buffered saline) and ~ 40 S/cm in DI (deionized) water with ~2 MPa young modulus and over 35% stretchability in a wet environment. Moreover, the conductivity of the PEDOT:PSS hydrogel remained the same at different strains, ranging from 0 to 20% in both DI water and a PBS solution. In a different approach, Ren et al. added a positively charged conducting polymer PPy through in-situ polymerization to the negatively charged PEDOT:PSS, resulting in a highly conductive hybrid PPy-PEDOT:PSS hydrogel (867 S/m) with excellent biocompatibility for bioelectronics applications [17]. It has been shown that the mechanical stability and strength of PEDOT:PSS hydrogels can be improved by introducing PEDOT:PSS into a non-conducting hydrogel, thus forming an interpenetrating network or a bilayer structure. For bioelectronics and tissue engineering applications, such integration can be done in biocompatible synthetic (e.g., gelatin methacryloyl (GelMA)) or natural (e.g., collagen) hydrogels.

12.3   HYDROGEL MATERIALS FOR BIOELECTRONIC APPLICATIONS

12.3.1   Natural Hydrogels

12.3.1.1   Gelatin

Gelatin is a protein-derived compound obtained from the hydrolysis of animal collagen. It has various functional groups like amine (R-NH₂), alcohol (R-OH), and carboxyl (R-COOH) groups, which could be physically or chemically cross-linked to other polymers through physical interactions or certain cross-linking agents such as genipin and glutaraldehyde, respectively [18–20]. It is biocompatible, bio-degradable, abundant, renewable, transparent, thermo-reversible, cost-effective, and cyto/bio-compatible, distinguishing it from other biopolymers. Gelatin is an FDA (Food and Drug Administration)–approved material used as a backbone in natural hydrogels. Conductive gelatin hydrogels are typically developed by supplementing the gelatin hydrogel matrix with a conductive material. Some gelatin-based hydrogels tend to have a porous structure that aids in the movement of charge carriers introduced through carbon-based conductive fillers, metallic nanoparticles, and conductive polymers. Gelatin could be modified to enhance its properties, interaction with other materials, and processability. For instance, Gelatin metha-cryloyl (GelMA) hydrogels promote cellular adhesion and proliferation, two properties that are fundamental for tissue engineering and therapeutics delivery. Furthermore, GelMA is more likely to develop dynamic contacts with other macromolecules, such as chitosan. This can be done by broadening the number of potential bonding interactions, including hydrophobic due to the nonpolar methyl group ($R-CH_3$) replacing the amine group, as well as through hydrogen, and electrostatic bonding interactions [21]. A photo-initiator-infused GelMA can be photo-cross-linked by UV (ultraviolet) radiation into the desired shape through soft-lithography. Gelatin-based hydrogels are used in a variety of bioelectronics applications, such as strain and movement sensing, as well as in biopotential signals conduction. In addition, gelatin-based conductive hydrogels have also been used for physiological recordings or stimulation applications. For example, they can be developed as electrodes for long-term electrocardiography (ECG) [18]. They exhibit good mechanical properties over long-term operation, minimal skin irritation, good recording performance, and better skin conformability compared to conventional Ag/AgCl (silver/silver chloride) electrodes. Gelatin electrodes have also been used for muscle stimulation [22]. In another application, a robust gelatin-based hydrogel augmented with tannic acid, demonstrated self-powering potential and was used for strain sensing and movement tracking in wearable devices [23]. The fabricated device was stretchable, conformable, highly ionically conductive, and environmentally friendly.

12.3.1.2 Chitosan

Chitosan is a linear polysaccharide with rigid N-acetylglucosamine and D-glucosamine repeating units derived from the deacetylation of the second most ubiquitous low-cost polysaccharide on earth (chitin). Like gelatin, chitosan contains a plethora of reactive groups useful in cross-linking. Chitosan is a very lucrative biomaterial due to its fascinating properties, including being biocompatible, biodegradable, bio-adhesive, ionically conductive, and able to form intermolecular hydrogen bonds and polyelectrolyte complexes [24]. Chitosan could also be modified as carboxymethylated chitosan to improve its solubitility in water and its overall properties and processability [24]. Kang et al. created a chitosan/tannic acid hydrogel with excellent fatigue resistance, mechanical and adhesion properties through a double network configuration [25]. The various hydrogel interactions, such as hydrophobic interactions and metal coordination, was found to promote its replicable adhesiveness on various materials, including metals and porcine skin. Also, it was observed that peeling off the hydrogel doesn't leave residues, thanks to the large catechol groups of tannic acid. These features can be highly beneficial in a variety of applications, including motion tracking and electrophysiological monitoring. Interestingly, it also showed the capacity to act as a dual sensor for temperature and strain deformation. This duality paves the way for multifunctional flexible hydrogel wearable sensors. Hu et al. developed a hyaluronic acid and chitosan conductive hydrogel with good mechanical properties, self-healing, anti-dehydration, and anti-freezing properties [26]. It is worth noting that the addition of potassium chloride and glycerol significantly contributed to the anti-freezing properties. These properties make this hydrogel a potential candidate for wearable strain-sensing applications, such as human movement monitoring and electronic skin.

12.3.1.3   Cellulose

Cellulose is a linear polysaccharide with linear repeating units of glucose. Cellulose is the most abundant polysaccharide, as it is the key structural component in plants and microorganisms like bacteria and algae. In addition to being biodegradable and biocompatible, cellulose is affordable, has high chemical stability, and good mechanical properties with tensile strength and Young's Modulus reaching mega-pascals (MPa) and giga-pascals (GPa) range, depending on the source and processing. Furthermore, cellulose has hydroxyl functional groups (-OH), which aid in developing sacrificial hydrogen bonds necessary for developing hydrogels. Also, it is water-insoluble and, like chitosan, it can be modified to enhance its properties and water solubility. In addition it can be modified to obtain cellulose nanocrystals, nanofibrils, and carboxymethylated cellulose (CMC). Cellulose can be cross-linked to form a homopolymeric hydrogel or introduced in a composite hydrogel. Hu et al., for instance demonstrated that cellulose-based hydrogels can be developed via chemical crosslinking in a NaCl solution. The resulting hydrogels were found to be transparent, freeze tolerant, conductive as well as mechanically responsive and stable [27]. Furthermore, a similar hydrogel was developed by Cui et al.; however, the cellulose used was extracted from okra rather than cotton and consequently had denser cross-links and stronger tensile strength [28]. In another approach a composite hydrogel composed of CMC, chitosan (Ch), and polydopamine was developed [29]. The injectability and self-healing behavior were primarily due to the dynamic Schiff base or imide bonds, while the other aspects were orchestrated by the amount of CMC-dialdehyde-PDA. This hydrogel was evaluated in different bioelectronics applications such as electrophysiological recordings (ECG) and body movement energy, and harvesting as triboelectric nanogenerators (TENGs). In addition this hydrogel was found to be also suitable for tissue regeneration applications, as confirmed through *in-vivo* and *in-vitro* studies.

12.3.1.4   Alginate

Sodium alginate or alginate is a linear copolymeric polysaccharide with several block arrangements of (1–4)-linked β-D-mannuronic acid (M), and α-L-guluronic acid(G). Alginate is derived from brown seaweed. It is biocompatible, biodegradable, hydrophilic, affordable, customizable, and has tunable physical properties by modifying the (M/G) content ratio. For example, alginate with a high G block proportion elicits more ionic cross-linking favorable for rigid hydrogels. In contrast, alginate with a high M block proportion makes it softer, which is favorable for soft hydrogels. Moreover, alginate has an abundance of hydroxyl and carboxyl groups in its polymer chain, which increases the potential to interact with biopolymers. This property makes it appealing for chemical modifications to suit the desired properties. Using an aqueous solution of sodium alginate, alginate hydrogels can easily be developed through ionic cross-linking between its chains and divalent ions; for example, calcium cation Ca^{2+} coordination with the G-block in an alginate chain [30].

Alginate can be also cross-linked along with a conducting polymer to create a conductive hydrogel. For example, Yang et al. created a PPy/Alginate conductive hydrogel by physically cross-linking sodium alginate with Ca^{2+} and chemically cross-linking that to PPy [31]. The resulting hydrogels were in vitro cultured with human mesenchymal stem cells, exhibiting improved cell adhesion and growth, as well as enhanced expression of neural differentiation markers. Moreover, long term (eight weeks) in vivo implantation of the PPy/Alginate hydrogels revealed a moderate immune response, highlighting their potenial for use in tissue engineering applications. In another work by Zhu et al., composite conductive hydrogels were fabricated using oxidized alginate dialdehyde-gelatin and MXene. The resulting hydrogels exhibited self-healing properties and good biocompatibility characteristics [32].

12.3.2 Synthetic Hydrogels

12.3.2.1 Polyacrylamide (PAAm)

PAAm is a synthetic polymer fabricated via polymerization of acrylamide. It has gained its place among the most preferred materials for developing tough hydrogels. PAAm is stable, hygroscopic, non-resorbable, non-immunogenic, and non-toxic while simultaneously possessing many desirable properties such as bio-compatibility, viscoelasticity, hydrophilicity, and cohesiveness. It is commonly used in conjunction with other synthetic and natural polymers, including PVA and alginate. PAAm is not inherently conductive; however, it can be integrated with conductive materials for use in bioelectronic applications such as electrophysiolo-gical recordings and human motion detection. For example, Shekh et al. created an elastic composite hydrogel for motion detection composed of PAAm, MXene, Fe (III) ions, oxidized alginate, and chitosan [33]. The composite hydrogel demon-strated fast-self recovery, high fracture strain, toughness, high sensitivity, elasticity, adherence to skin, plastic, and glass. In another work by Carvalho et al., a PAAm-based hydrogel was developed as a proof of concept for neuromuscular electro-stimulation of the forearm, and for wearable devices that monitor the brain, heart, and facial expressions [34].

12.3.2.2 Polyvinyl Alcohol (PVA)

PVA is a linear synthetic polymer commonly synthesized via hydrolysis of polyvinyl acetate and found to be highly hydrophilic, biocompatible, biodegradable, thermo-stable, transparent, adhesive, and viscoelastic. PVA chains possess an abundance of hydroxyl side groups, eliciting physical and chemical interactions with several compounds. PVA has been used for a wide range of applications in the biomedical field, namely wearable sensors for healthcare monitoring and wound dressings. For example, Wang et al. blended PVA with CMC and PEDOT:PSS to develop a conductive hydrogel with high response time, low contact impedance, fatigue resistance, conformability, flexibility, elasticity, and biocompatibility [35]. The resulting hydrogels demonstrated excellent electrophysiological monitoring capabili-ties with high quality ECG signals. PVA has been also used in wearable bioelectronics for strain sensing and e-skin applications, resulting in a low cost, stretchable, strain sensor, as demonstrated by Zhang et al. [36].

12.3.2.3 Polyethylene Glycol (PEG)

PEG is a linear synthetic polymer produced by the polymerization of ethylene glycol. PEG is biocompatible and hydrophilic. PEG has been used for various biomedical applications such as healthcare monitoring and tissue engineering. In a work by Cai et al., PEG-grafted/chitin nanocrystals-enhanced hydrogel was designed as a strain sensor for wearable bioelectronic applications [37]. It featured stretchability, freeze-tolerance, high sensitivity, and ionic conduction. Also, Xu et al. created a linear peptide-PEG/PEDOT:PSS conductive hydrogel for biosensors, nerve grafts, and neuroprosthetics [38]. It featured injectability, electron conduction, self-healing, biocompatibility, and cell adhesion, promoting the proliferation of 3D cell cultures. For more details about the discussed hydrogels' synthesis and properties, refer to the summary table (Table 12.1).

12.4 APPLICATIONS OF HYDROGELS IN BIOELECTRONICS: SENSING, DIAGNOSTIC, AND THERAPEUTICS

12.4.1 ON-SKIN BIOELECTRONICS

12.4.1.1 Epidermal Bioelectronics

Epidermal bioelectronics, pioneered by J. Rogers, are devices with stiffness, elastic moduli, and thickness matched to the epidermis, leading to a seamless contact and good adhesion [39]. Unlike typical wearables, epidermal bioelectronics adapt to the skin in a way that is mechanically "invisible" to the wearer. The non-invasive nature of the tissue-electronic contact is beneficial for electrical stimulation. The most common clinical applications of epidermal bioelectronics include transcutaneous electrical nerve stimulation (TENS), which has been widely used for pain relief. At the tissue-electrode interface, the communication is bidirectional, allowing the recording of signals using non-invasive epidermal electrodes. Most common clinical example of epidermal bioelectronic recording is electroencephalography (EEG), where the epidermal electrode can record the brain activity. Other widely used examples of diagnostics include ECG and electromyography (EMG) where the electrode can record the cardio-muscular activities. The working mechanism of bioelectronic stimulation and recording is the same but acts in the reverse manner. However, in both cases, the tissue-electrode interaction is based on biochemical interactions at the interface. Being epidermal, the electrode can experience bending, stretching and compression affecting the performance and reliability of the device. Therefore, the electrode should possess the properties of biological entities having softness, high water content (> 70%), and the ability to undergo cyclic mechanical deformation during biomechanical movement (skin, muscle, etc. can undergo 30% strain). Commercially available epidermal bioelectronic for ECG, EMG, EEG, and TENS make use of electrodes with hydrogels for stimulation and recording, owing to their hydrating nature, conformability to the skin, and ionic conductivity [3].

Toward this direction, Li et al. developed a multifunctional epidermal sensor based on MXene, polylactic acid, and amorphous calcium carbonate (ACC)–based hydrogel for physiological monitoring. The sensor could detect the finger and elbow bending angle along with swallowing and blood pulse when attached to the throat and the radial artery of the wrist, respectively (Figure 12.3a-d)[40]. Additionally, the composite hydrogel was used to monitor electrophysiological signals (EMG, ECG)

TABLE 12.1

Commonly Used Natural and Synthetic Conducting Hydrogels

Hydrogel Constituents	Synthesis Method	Tensile Strength (kPa)	Elongation at Break (%)	Conductivity (S/m)	Special Features	Ref.
PAAm/MXene/oxidized alginate/chitosan/Fe(III)	Physical interactions and chemical (MBAA)	910	820	1.31	High sensitivity, fast self-recovery, adhesive	[33]
PAAm/glycerol	Chemical (MBAm)	~62	600	–	Non-drying, adhesive, transparent, low cost, high SNR	[34]
PVA/PEDOT:PSS	Chemical (Glutaraldehyde)	70	239	0.36	Strain sensitive	[36]
PVA/CMC/PEDOT:PSS	Physical interactions (Freeze thawing)	85	73.8	75.4	Adhesive, elastic, porous, fatigue resistant	[35]
PEG/chitin nanocrystals	Physical (H-Bonds)	2000	435	0.01	Anti-freezing, high stretchability, and strain sensitivity	[37]
Gelatin/PEDOT:PSS	Chemical (Genipin)	72	235	0.04	Adhesive, biodegradable	[18]
Gelatin/tannic acid	Physical interactions	50	1600	–	Rapid self-healing, adhesive, strain sensitive, self-powered	[23]
Chitosan/tannic acid	Physical interactions	85	320	0.0093	Rapid response and recovery, fatigue resistant, versatile, adhesive	[25]
Chitosan-Hyaluronic acid	Physical interactions and chemical (Schiff base reaction)	46	200	0.0638	Anti-drying, anti-freezing, strain sensitive self-healing	[26]
Cellulose/NaCl solution	Chemical (epichlorohydrin)	5200	235	4.03	Anti-freezing, transparent, rapid response	[27]

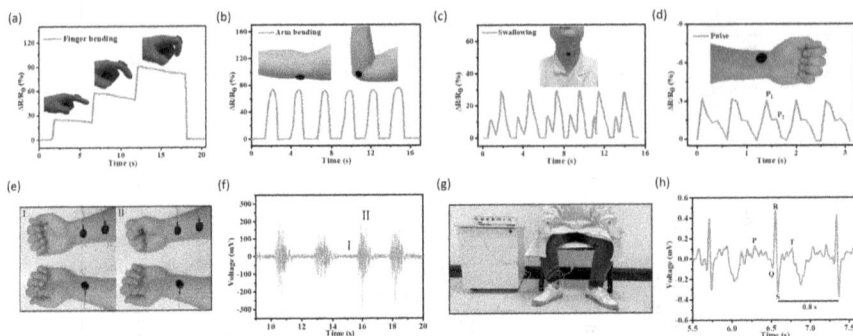

FIGURE 12.3 MXene, polylactic acid, and ACC-based hydrogel based epidermal sensors for biomechanical motion monitoring by recording the resistance change caused by strain due to (a) finger movement, (b) arm movement, (c) swallowing, and (d) blood pulse and the image showing the corresponding position of the sensor is shown in the insets. (e) Schematic showing the use of MXene/PAA-ACC-based hydrogels as EMG electrodes and (f) the corresponding EMG signals (I) before and (II) after making a fist. (g) Schematic of the setup made of MXene/Polylactic acid-ACC based hydrogels for ECG detection and (h) the corresponding ECG signals. Here, (A) is a working electrode, (B) is a reference electrode, and (C) is a ground electrode. Adapted with permission from Reference [40]. Copyright (2021) American Chemical Society.

(Figure 12.3e-h) with a reported signal-to-noise ratio higher than Ag/AgCl electrodes and commercial electrodes, owing to the better mechanical and electrical interfacial contact between the hydrogel electrode and the skin tissue.

12.4.1.2 Wound Patch Devices

Chronic wounds are a major problem in diabetic patients, often resulting in additional complications, including infection and chronic inflammation. Novel treatments for diabetic wounds are being explored to address the mismatch between the soft and elastic skin surface, and wound treatment devices that seek to protect the wound site from infection and induce healing before additional complications arise. Methods such as controlled release of drugs or electrical stimulation have both shown promise in promoting the regeneration of damaged tissue while reducing the likelihood of infection, and hydrogels are showing promise as an effective foundation for these technologies. In addition to providing a favorable mechanical interface for the injection of electric current, the dense fiber network of a hydrogel matrix can be modified with weakly bound drug particles via physical interaction forces, to then be activated and released via various stimuli including pH, temperature, or electrical stimulation. Many standard hydrogel matrix materials are poor electrical conductors and have limited capacity to electrically actuate the release of bound drugs without functional modifications. Functional modifications such as embedded nanoparticles or deep eutectic solvents have been developed to maintain the structural and mechanical advantages of various hydrogels while improving electrical and chemical functionality.

Deep eutectic solvents are a promising method of improving the electrical functionality of various hydrogels via the chemical alteration of the matrix surface to effectively bind drugs, which can then be released via low-voltage electrical signals sent from on-device batteries. Deep eutectic solvents can even operate as drug-free antibacterial agents due to the strong electrical interaction with bacterial cell walls. Guo et al. produced a deep eutectic-hydrogel with surface-bound fullerenes to produce an inherently antibacterial, battery-powered drug release system utilizing the electrically favorable properties of carbon nanomaterials within a biologically advantageous agarose gel [41]. Voltage-current tests of bare agarose gel showed no conductivity up to 10 volts, while the addition of deep eutectic solvents alone caused a dramatic rise in conductivity (28.1 mS/cm), and even further conductivity with the addition of fullerenes (31.0 mS/cm). In addition to improving the mechanical properties of the unmodified agarose gel, the authors found that the improved electrical properties allowed this fullerene modified gels to demonstrate voltage-modified behavior, showing a fivefold increase in the amount of drug delivered to tissue under a battery-powered 3-V load.

The benefits of conductive and functional hydrogels have further been investigated in wound dressings that actively promote wound healing. Diabetic patients regularly suffer from chronic wounds, which are subject to slow healing and risk of infection. Many studies have demonstrated the benefits of electrical stimulation on the proliferation of fibroblasts for accelerated wound healing; however, the application of long-term rigid electrodes remains highly impractical. As demonstrated by Liu et al., the Konjac glucomannan hydrogel shows promise as a wound dressing, owing to its wettability and mechanical similarity to skin, allowing wound site protection and promoting healing [42]. Functionalized with a polymerized ionic liquid (PIL), this hydrogel further exhibits passive antibacterial activity, as well as allowing for electrically stimulated fibroblast proliferation and a multimodal wound healing advantage. Electrical stimulation of the PIL hydrogel between 3 and 12 volts at 25–125 Hz resulted in significantly increased fibroblast migration and proliferation in-vitro, while the passive antibacterial activity of the PIL-hydrogel functional electrode was demonstrated against S. aureus and E. coli at nearly 100%. A comparable approach from Guan et al. utilized a composite hydrogel composed of PVA, cellulose nanocrystals, Ag nanoparticles, and PDA, dubbed PACPH resulting in excellent electrical conductivity, good skin-hydrogel mechanical interaction, and high carrying capacity for drugs [43]. The PACPH hydrogel displayed good flexibility, and an adhesion strength of 17.2–25.95 kPa to materials such as glass, polymethyl methacrylate (PMMA), and various metals. Antibacterial assays of S. aureus and E. coli showed significant bacterial inhibition in the presence of PACPH, owing to the natural antibacterial behavior of both Ag nanoparticles and PDA. Their platform was additionally shown to enhance M2 macrophage polarization, enhancing the immunoregulation-mediated wound repair process and resulting in increased cell proliferation and reduced inflammation in-vitro. Immunochemical analysis of the PACPH-treated cells showed significantly increased expression of anti-inflammatory cytokines CD86 and CD206, and reduction of the pro-inflammatory cytokine TNF-α.

The application of electrical stimulation via an electronic hydrogel device has been shown to have numerous advantages; therefore, an effective power delivery system is needed to fully realize the potential of a wearable electronic device. Hardwired systems offer uninterrupted power, but limit mobility and are typically limited to use in a clinical or confined setting. Battery power has been implemented in wearable hydrogel bioelectronics, offering mobility at the cost of power and life span as determined by the choice of battery. Alternatively, a self-powered system could potentially address both the mobility limitation of wired power, as well as the life span limitation of a battery power source. Sharma et al. describe a piezoelectric driven TENG (PTENG) powering a hydrogel electrode for a wearable wound dressing as a treatment for diabetic wounds [44]. A polyvinylidene difluoride (PVDF) membrane was produced by electrospinning and sandwiched between two layers of conductive carbonized PDA (CPDA) hydrogel to produce the TENG. The device was capable of producing up to 60 nA of current under short circuit, and 60 mV of voltage open circuit under a load pressure of 15 kPa from 1–5 Hz. PTENG wound dressings showed no biocompatibility issues, and significantly increased endothelial biomarker CD-31 expression, indicating enhanced blood vessel formation along with notably faster wound closure and reformation of hair follicles. Due to their multimodal antibacterial activity, wound site protection and enhanced wound healing, these bioelectronic hydrogels are showing significant potential in the treatment of diabetic injury.

12.4.1.3   Biochemical Sensor

Many of the intrinsic properties of hydrogels e.g., electrical conductivity, mechanical flexibility, and good skin-hydrogel interfacing make them attractive not only for electrical transduction, but for chemical transduction and sensing as well. The sponge-like tendency of hydrogels to swell as they absorb fluid allows them to collect various analytes of medical importance directly from the skin in blood or sweat. In the context of a skin surface biochemical sensor that uses a hydrogel interface to analyze sweat, we need to first consider how best to collect sufficient sweat for analysis. At rest, people typically don't produce significant volumes of sweat, so some approaches involve simply having the subjects exercise to induce sweating through exertion [45,46]. One approach involves the attachment of a simple agarose hydrogel patch to the skin, which is worn for a period of physical exertion (Figure 12.4a) [46]. The patch is subsequently removed, and placed onto a secondary hydrogel sensor stack composed of a glucose oxidase-chitosan functional hydrogel layer, on top of a Prussian blue-PEDOT hydrogel electrode. This approach is advantageous in requiring only a very low-profile passive gel patch be worn, which encourages regular use. However, this design still relies on passive sweat collection, making it impractical for short periods, and potentially uncomfortable or inconvenient if it requires strenuous activity.

Incorporation of more sophisticated technologies to collect sweat for analysis can improve the passive collection of sweat from the skin surface without the need for physical exertion. A series of microfluidic channels, for example, could be employed to direct multiple small surface flows of sweat to a central location, as was devised by Xu et al. in their uric acid detector [47]. In this example, a soft and flexible polydimethylsiloxane (PDMS) device composed of multiple orbiting microfluidic channels collect sweat from a larger area of the skin, allowing ample sweat collection under ambient rest conditions (Figure 12.4b). Utilizing the passive capillary activity of microfluidics, their device channels sweat toward a central reservoir where it would then be absorbed into a functional hydrogel made from PEDOT-PSS. The large surface area of the porous PEDOT hydrogel improves the detection limit and sensitivity of the sensor, which then measures the amperometric response of the collected sweat to determine uric acid content. Alternatively, an active sweat solution could further enhance the available material for analysis. Electrodes that puncture the skin to directly measure fluid offer superior signal quality, but are highly intrusive and uncomfortable, limiting their widespread adoption in wearable bioelectronics. Active yet nonintrusive fluid collection is a challenge, particularly in the form of a skin surface sensor. Hakala et al. chose a novel approach involving magnetohydrodynamics to noninvasively extract skin sweat by electroosmotic flow [48]. A gelatin methacryloyl hydrogel with two implanted Ag/AgCl electrodes was placed on the subject's skin, and an applied magnetic field drove fluid up into the gel, where it was absorbed for later analysis (Figure 12.4c).

Once the issue of fluid collection has been addressed, hydrogels provide an excellent platform for the absorption and storage of body fluids for analysis, and various effective methods have been employed for that purpose. Colorimetric and pH-sensitive dyes have been incorporated into hydrogel matrices for the detection of a number of biologically relevant chemicals but lack the broader range of an electronic sensor. To achieve biochemical sensing within a hydrogel electronic device, a number of methods exist, including enzyme-activated functional electrodes that interact with the target molecules, and amperometric sensors that measure chemicals by changes in the impedance of the hydrogel. Enzyme functionalized sensors for the detection of sweat glucose or lactate utilize glucose oxidase (GOx) or lactate deoxidase (LOx), respectively, to elicit a voltage response. Nagamine et al. developed an electrochemical sensor composed of Prussian blue carbon ink, modified with lactate deoxidase and built on a PEN (poly(ethylene 2,6-naphthalate) substrate [49]. An Ag/AgCl reference electrode was patterned nearby, and the electrode pair was enclosed in an agarose hydrogel (Figure 12.4d). The hydrogel absorbed sweat (obtained from donors) and activated the LOx sensor. The electrodes were shorted through a 100 kohm resistor, and then subjects would place their finger on the gel for 30 seconds and then remove it. Voltage across the resistor was measured continuously for 300 seconds to measure the level of lactate in the gel. The device showed good response to the presence of lactate; however, the authors noted that its accuracy was limited by the diffusion of sweat into the gel, thus diluting the sweat into the gel and giving a reading different from the true lactate concentration. Lin et al. developed a conductive composite of PEDOT and Prussian blue (PB-PEDOT) fabricated on top of a printed carbon electrode [46]. A glucose oxidase-chitosan gel was fixed on top of the PB-PEDOT layer to begin the electrochemical reaction of the glucose sensor. An agarose hydrogel patch was placed on the subject's skin for 15 minutes, after which it was transferred to the

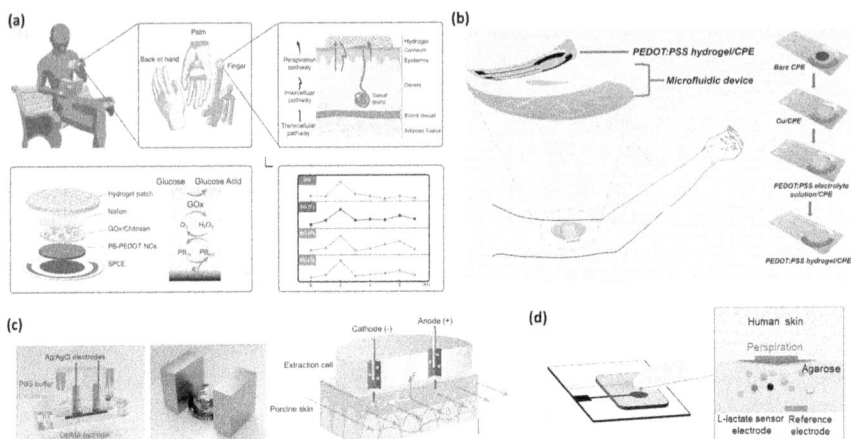

FIGURE 12.4 (a) A hydrogel composite electrode for a wearable glucose sensor. Adapted with permission from Reference [46]. Copyright (2022) Elsevier. (b) An array of PDMS microfluidic channels for the passive collection of sweat in a wearable uric acid sensor. Adapted with permission from Reference [47]. Copyright (2021) Elsevier. (c) Magnetically driven electroosmotic hydrogel sensor for improved sweat collection. Adapted with permission from [48]. Copyright (2021) Nature. (d) Hydrogel sweat sensor for touch-based measurement of sweat lactose. Adapted with permission from Reference [49]. Copyright (2019) Nature.

hydrogel composite electrode, resulting in a signal proportional to the level of glucose in the sweat.

Amperometric sensors measure the contents of relevant chemicals within the absorbed electrolyte by supplying a constant voltage and measuring the resulting changes in current as the hydrogel swells with sweat. These types of devices can be specifically calibrated to measure the analyte of choice e.g., uric acid or glucose. An SB-MB (SBMA/MBPA) zwitterionic hydrogel was produced by Guo et al. to construct a biocompatible, antibacterial patch for the sensing of sweat glucose, while simultaneously functioning as a strain sensor [50]. The uptake of glucose into the hydrogel matrix and resulting osmotic shifts cause ions to transfer between the absorbed fluid and the SB-MB matrix, causing a change in conductivity proportional to the glucose concentration. A microfluidics-couple hydrogel sensor from Xu et al. similarly used the variable impedance of hydrogels to measure uric acid in sweat [47]. Uric acid in sweat can be an indicator of gout and hyperuricemia. PDMS microfluidic channels are placed on the skin, which use capillary action to direct surface sweat towards a central reservoir. In their device, a PEDOT hydrogel on a copper functional electrode placed at the center of the wearable device was used to measure the concentration of uric acid in sweat by amperometric response. The large and conductive surface area of the PEDOT:PSS hydrogel contributed to the sensitivity of the electrochemical sensor, while the microfluidic structure allowed efficient collection of sweat from the skin surface.

12.4.2   Implantable Devices and Tissue Engineering

12.4.2.1   Neurological Signal Monitoring

Electrical interfacing with living neurons e.g., brain tissue, central nerves, or peripheral nerves has made significant progress since scientists first discovered the role of electricity in living tissue. The complex network of cellular electric circuitry that composes our nervous system controls nearly every high-level function of our body and is an ideal candidate for treatment of disorders ranging from epilepsy and heart arrythmia, to next-generation prosthetics and artificial organs. There are significant inherent challenges to interfacing an electrical device, which typically operates on an electron current, to a biological electric device that almost exclusively operates on ionic currents through a liquid medium. Neurons are complex, delicate structures that are highly sensitive to their microenvironment and possess a limited ability to heal. Extended interaction between neural tissue and foreign devices is known to cause inflammation, glial scarring, and significant biofouling, leading to rapid degradation of the electrode functionality. Therefore, any electronic device designed to interface directly with neurons, in addition to enabling the effective transduction of electrical signals, needs to address the mechanical mismatch of the device and cellular surface and any potential toxicity to the nerve tissue. The tissue-like mechanical properties of hydrogels, as well as their adaptability through the choice of matrix material and infiltrant, make them good candidates for this application.

Park et al. developed a multifunctional probe composed of multiple thermally drawn fibers made from cyclic olefin copolymer (COC) clad polycarbonate optical fiber, poly(etherimide) (PEI) coated tin microwires for electrical interfacing, and PEI microfluidic tubes for delivery of drugs [51]. A PAAm-alginate hydrogel serves as a matrix and surface agent for the fibers, binding them into a bundle for group insertion into brain tissue. The brain-like mechanical properties of the PAAm-alginate as well as its biocompatibility and chemical stability make it ideal candidate for neural interfacing. In addition to being highly biocompatible, thus reducing the immune-response-driven degradation of the electrode, hydrogels benefit from having a dynamic elastic modulus spanning three orders of magnitude, depending upon the level of hydration of the matrix. During insertion of the probe bundle, the dehydrated gel is stiff enough to facilitate entry to the tissue; then, upon implantation, it can be fully hydrated, increasing pliability and providing a highly compatible mechanical match to the surrounding tissue. These properties allowed the device to provide long-term multimodal recording and actuation while fully implanted.

Oribe et al. developed a completely organic electrode array composed of PEDOT-modified carbon fabric encapsulated within a PVA hydrogel for long-term, continuous brain monitoring [52]. The system is biocompatible and contains no metallic elements, making it compatible with magnetic resonance imaging (MRI) and brain scanning in patients requiring treatment for epilepsy or other related brain conditions. The stretchable, flexible, and gas permeable hydrogel conforms to the brain surface, allowing good signal transduction, without hindering biological function during long-term implantation. This device could potentially be used in ECoG (electrocorticography)-MRI studies for advanced analysis of brain function disorders with real-time electrical monitoring.

Traditional Pt electrodes used for deep brain stimulation are size limited, as using an electrode that is too small to create unsafe charge densities, and reduce the charge injection limit of the device. Since the electrode size is a key factor in multiple ways, including the safety of insertion, the specificity of targeted implantation, and the mechanical conformity to the surrounding tissue, Adams et al. sought to apply a conductive hydrogel coating to address the charge density limitation, allowing smaller and safer implantable electrodes [53]. They developed a low swelling mixture of PVA, encapsulating a PEDOT-coated Pt electrode bundle. Comparison of this hydrogel-coated electrode to an uncoated one showed significantly lower voltage transient impedance, and higher charge injection limit. This improved electrochemical stability allows for a more miniaturized electrode design for better neural resolution, safer implantation, and better overall potential effectiveness for deep brain stimulation and recording.

12.4.2.2   Tissue Engineering

The development of biocompatible 3D scaffolds/hydrogels that can mimic the properties of different tissues in the human body has been driven by the need for transplants, as well by the need for more reliable cell-based models for fundamental studies and drug testing [54,55]. As such, conducting hydrogels have been heavily explored in the field of biomaterials to mimic the biological and electrical properties of tissues in the human body. Novel, "smart scaffolds", designed to be responsive to different stimuli such as physical (mechanical, temperature, light, electric, and magnetic), chemical (pH, glucose, and biological), have been recently developed [56,57]. In this direction, hydrogel-based scaffolds that can be electroactive while exhibiting the desired biocompatibility, can be used for electrical stimulation of cell cultures, notably electroactive, upon application of an external electric signal.

In myocardial tissue engineering, for example, the hydrogel should be both electrically conducting (as the heart muscles have electrically conductive Purkinje fibers with DC conductivity of 0.1 S/m) [58], but also exhibit mechanical properties and microstructure similar to that of the heart tissue. In a recent work, Shin et al. designed functional cardiac patches [58], where the cultured cardiac cells (neonatal rat cadiomyocytes) onto a CNT/gelatin derivative (GelMA) hydrogel. The CNTs alongside their high electrical conductivity and mechanical strength formed a uniform nanofibrous network, which led to improved cell adhesion, distribution, and cell-cell coupling. The CNT-reinforced GelMA hydrogel showed stability and synchronous beating in response to the electric field stimulation (up to 3 Hz) even after one day of culture with an average beating rate of 69.8 ± 19.1 BPM. Moreover, the addition of CNTs in GelMA reduced the excitation thresholds by 85%.

In neural tissue engineering, electrically conducting hydrogels are also especially promising for the dual role of biological signal recording and stimulation of living tissue. Zhang et al. were the first to develop a 3D-printable conductive hydrogel that can be photo-cross-linked without compromising its electrical conductivity. The conducting hydrogel was based on PEDOT:PSS and was printed using a table-top stereolithography 3D printer. The resulting 3D-printed hydrogel provided excellent support to dorsal root ganglion (DRG) cells and showed enhanced neuronal differentiation following electrical stimulation [59].

12.5 CONCLUSIONS

In this chapter, we have discussed the growing role of hydrogels in bioelectronics, including some common materials, designs, and applications. This family of devices shows promise in applications ranging from wearable health monitors and chemical sensors to implantable devices for wound healing and deep brain stimulation. Years of research on bioelectronics technology have made it clear that the interface between biological tissue and electronic devices remains a major issue. Hydrogel bioelectronics offer promising alternatives to classical metal-silicon structures, utilizing mechanical properties that closely match to the skin's surface, while also offering ionic mobility and hydration mechanics that mimic living tissue. We have discussed a number of popular materials used in hydrogel electronics, ranging from animal proteins to synthetic polymers and nanostructured materials. While these materials share some common characteristics allowing them to absorb and contain water, facilitate ionic currents, and mechanically match with living tissue, they each offer unique advantages that require dedicated research to fully explore. Functional hydrogels made from porous networks of PEDOT:PSS or MXene offer enormous reactive surface areas for enzyme-mediated electrochemical sensors, as well as the potential for semiconductor behavior in bioelectric transistors.

The electrical conductivity of a nonmetallic hydrogel has typically depended upon the ionic currents facilitated by the absorbed electrolyte. Ionic currents are intrinsically less conductive than metallic conductors of free electrons; therefore, conducting hydrogels of metallic or conducting polymers have been developed to address the functional limitations of nonconducting polymer matrices. In some cases, composite materials have been used, utilizing the conductive nature of carbon or metallic fibers dispersed within a structurally advantageous hydrogel. However, these composite structures may be limited by the interactions between materials, which can hinder the electrical properties, reducing the overall effectiveness of the device. Alternate designs have been proposed, e.g., spring-like conducting layers seek to match the mechanical flexibility of the underlying hydrogel while contributing its own electrical advantages. Despite the steady and observable progress that continues to be made in the advancement of hydrogel bioelectronics, a number of fundamental challenges have yet to be addressed in our desire to fully couple biology with electronics. The interface between the rigid structures of traditional electronics and soft biology presents a challenge in fully exploring the potential of bioelectronics. A better understanding of how the many materials available to us behave at the skin-electrode interface, and how they can best be combined to maximize efficiency while minimizing limitations will require a consistent and dedicated approach to this growing field of research.

REFERENCES

[1] F.V. Tjong, V.Y. Reddy, Permanent leadless cardiac pacemaker therapy: A comprehensive review, Circulation, 135 (2017) 1458–1470.

[2] F.G. Zeng, S. Rebscher, W. Harrison, X. Sun, H. Feng, Cochlear implants: System design, integration, and evaluation, IEEE Reviews in Biomedical Engineering, 1 (2008) 115–142.

[3] H. Yuk, B. Lu, X. Zhao, Hydrogel bioelectronics, Chemical Society Reviews, 48 (2019) 1642–1667.

[4] H. Dechiraju, M. Jia, L. Luo, M. Rolandi, Ion-conducting hydrogels and their applications in bioelectronics, Advanced Sustainable Systems, 6 (2022) 2100173.

[5] C.-J. Lee, H. Wu, Y. Hu, M. Young, H. Wang, D. Lynch, F. Xu, H. Cong, G. Cheng, Ionic conductivity of polyelectrolyte hydrogels, ACS Applied Materials & Interfaces, 10 (2018) 5845–5852.

[6] X. Sui, H. Guo, H. Cai, Q. Li, C. Wen, X. Zhang, X. Wang, J. Yang, L. Zhang, Ionic conductive hydrogels with long-lasting antifreezing, water retention and self-regeneration abilities, Chemical Engineering Journal, 419 (2021) 129478.

[7] Y. Yang, L. Guan, X. Li, Z. Gao, X. Ren, G. Gao, Conductive organohydrogels with ultrastretchability, antifreezing, self-healing, and adhesive properties for motion detection and signal transmission, ACS Applied Materials & Interfaces, 11 (2019) 3428–3437.

[8] Y. Bai, B. Chen, X. Feng, J. Zhou, H. Wang, Z. Suo, Transparent hydrogel with enhanced water retention capacity by introducing highly hydratable salt, Applied Physics Letters, 105 (2014) 151903.

[9] C. Yin, X. Liu, J. Wei, R. Tan, J. Zhou, M. Ouyang, H. Wang, S.J. Cooper, B. Wu, C. George, Q. Wang, "All-in-gel" design for supercapacitors towards solid-state energy devices with thermal and mechanical compliance, Journal of Materials Chemistry A, 7 (2019) 8826–8831.

[10] P. Baei, S. Jalili-Firoozinezhad, S. Rajabi-Zeleti, M. Tafazzoli-Shadpour, H. Baharvand, N. Aghdami, Electrically conductive gold nanoparticle-chitosan thermosensitive hydrogels for cardiac tissue engineering, Materials Science and Engineering: C, 63 (2016) 131–141.

[11] X. Sun, Z. Qin, L. Ye, H. Zhang, Q. Yu, X. Wu, J. Li, F. Yao, Carbon nanotubes reinforced hydrogel as flexible strain sensor with high stretchability and mechanically toughness, Chemical Engineering Journal, 382 (2020) 122832.

[12] S. Talebian, M. Mehrali, R. Raad, F. Safaei, J. Xi, Z. Liu, J. Foroughi, Electrically conducting hydrogel graphene nanocomposite biofibers for biomedical applications, Frontiers in Chemistry, 8 (2020).

[13] J. Wei, R. Wang, F. Pan, Z. Fu, Polyvinyl alcohol/graphene oxide conductive hydrogels via the synergy of freezing and salting out for strain sensors, Sensors, 22(8) (2022)3015.

[14] Y. Ahn, H. Lee, D. Lee, Y. Lee, Highly conductive and flexible silver nanowire-based microelectrodes on biocompatible hydrogel, ACS Applied Materials & Interfaces, 6 (2014) 18401–18407.

[15] Z. Wang, J. Chen, Y. Cong, H. Zhang, T. Xu, L. Nie, J. Fu, Ultrastretchable strain sensors and arrays with high sensitivity and linearity based on super tough conductive hydrogels, Chemistry of Materials, 30 (2018) 8062–8069.

[16] B. Lu, H. Yuk, S. Lin, N. Jian, K. Qu, J. Xu, X. Zhao, Pure pedot:Pss hydrogels, Nature Communications, 10 (2019) 1043.

[17] X. Ren, M. Yang, T. Yang, C. Xu, Y. Ye, X. Wu, X. Zheng, B. Wang, Y. Wan, Z. Luo, Highly conductive ppy–pedot:Pss hybrid hydrogel with superior biocompatibility for bioelectronics application, ACS Applied Materials & Interfaces, 13 (2021) 25374–25382.

[18] Y. Lee, S.-G. Yim, G.W. Lee, S. Kim, H.S. Kim, D.Y. Hwang, B.-S. An, J.H. Lee, S. Seo, S.Y. Yang, Self-adherent biodegradable gelatin-based hydrogel electrodes for electrocardiography monitoring, Sensors, 20 (20) (2020)5737.

[19] B. Ying, X. Liu, Skin-like hydrogel devices for wearable sensing, soft robotics and beyond, iScience, 24 (2021) 103174.

[20] P. Jaipan, A. Nguyen, R.J. Narayan, Gelatin-based hydrogels for biomedical applications, MRS Communications, 7 (2017) 416–426.

[21] F. Mushtaq, Z.A. Raza, S.R. Batool, M. Zahid, O.C. Onder, A. Rafique, M.A. Nazeer, Preparation, properties, and applications of gelatin-based hydrogels (ghs) in the environmental, technological, and biomedical sectors, International Journal of Biological Macromolecules, 218 (2022) 601–633.

[22] F. Zhang, K. Qu, X. Li, C. Liu, L.S. Ortiz, K. Wu, X. Wang, N. Huang, Gelatin-based hydrogels combined with electrical stimulation to modulate neonatal rat cardiomyocyte beating and promote maturation, Bio-Design and Manufacturing, 4 (2021) 100–110.

[23] J. Wang, F. Tang, Y. Wang, Q. Lu, S. Liu, L. Li, Self-healing and highly stretchable gelatin hydrogel for self-powered strain sensor, ACS Applied Materials & Interfaces, 12 (2020) 1558–1566.

[24] V. Zargar, M. Asghari, A. Dashti, A review on chitin and chitosan polymers: Structure, chemistry, solubility, derivatives, and applications, ChemBioEng Reviews, 2 (2015) 204–226.

[25] B. Kang, X. Yan, Z. Zhao, S. Song, Dual-sensing, stretchable, fatigue-resistant, adhesive, and conductive hydrogels used as flexible sensors for human motion monitoring, Langmuir, 38 (2022) 7013–7023.

[26] Y. Hu, N. Liu, K. Chen, M. Liu, F. Wang, P. Liu, Y. Zhang, T. Zhang, X. Xiao, Resilient and self-healing hyaluronic acid/chitosan hydrogel with ion conductivity, low water loss, and freeze-tolerance for flexible and wearable strain sensor, Frontiers in Bioengineering and Biotechnology, 10 (2022) 837750.

[27] Y. Hu, M. Zhang, C. Qin, X. Qian, L. Zhang, J. Zhou, A. Lu, Transparent, conductive cellulose hydrogel for flexible sensor and triboelectric nanogenerator at subzero temperature, Carbohydrate Polymers, 265 (2021) 118078.

[28] X. Cui, J.J.L. Lee, W.N. Chen, Eco-friendly and biodegradable cellulose hydrogels produced from low cost okara: Towards non-toxic flexible electronics, Scientific Reports, 9 (2019) 18166.

[29] V. Panwar, A. Babu, A. Sharma, J. Thomas, V. Chopra, P. Malik, S. Rajput, M. Mittal, R. Guha, N. Chattopadhyay, D. Mandal, D. Ghosh, Tunable, conductive, self-healing, adhesive and injectable hydrogels for bioelectronics and tissue regeneration applications, Journal of Materials Chemistry B, 9 (2021) 6260–6270.

[30] Z. Bao, C. Xian, Q. Yuan, G. Liu, J. Wu, Natural polymer-based hydrogels with enhanced mechanical performances: Preparation, structure, and property, Advanced Healthcare Materials, 8 (2019) 1900670.

[31] S. Yang, L. Jang, S. Kim, J. Yang, K. Yang, S.W. Cho, J.Y. Lee, Polypyrrole/alginate hybrid hydrogels: Electrically conductive and soft biomaterials for human mesenchymal stem cell culture and potential neural tissue engineering applications, Macromolecular Bioscience, 16 (2016) 1653–1661.

[32] H. Zhu, W. Dai, L. Wang, C. Yao, C. Wang, B. Gu, D. Li, J. He, Electroactive oxidized alginate/gelatin/MXene (Ti3C2Tx) composite hydrogel with improved biocompatibility and self-healing property, *Polymers*, 14 (18))(2022)3908.

[33] M.I. Shekh, G. Zhu, W. Xiong, W. Wu, F.J. Stadler, D. Patel, C. Zhu, Dynamically bonded, tough, and conductive mxene@oxidized sodium alginate: Chitosan based multi-networked elastomeric hydrogels for physical motion detection, International Journal of Biological Macromolecules, 224 (2023) 604–620.

[34] F.M. Carvalho, P. Lopes, M. Carneiro, A. Serra, J. Coelho, A.T. de Almeida, M. Tavakoli, Nondrying, sticky hydrogels for the next generation of high-resolution conformable bioelectronics, ACS Applied Electronic Materials, 2 (2020) 3390–3401.

[35] Y. Wang, Z. Qu, W. Wang, D. Yu, Pva/cmc/pedot:Pss mixture hydrogels with high response and low impedance electronic signals for ecg monitoring, Colloids and Surfaces. B, Biointerfaces, 208 (2021) 112088.

[36] Y.-F. Zhang, M.-M. Guo, Y. Zhang, C.Y. Tang, C. Jiang, Y. Dong, W.-C. Law, F.-P. Du, Flexible, stretchable and conductive pva/pedot:Pss composite hydrogels prepared by sipn strategy, Polymer Testing, 81 (2020) 106213.

[37] J. Cai, Y. He, Y. Zhou, H. Yu, B. Luo, M. Liu, Polyethylene glycol grafted chitin nanocrystals enhanced, stretchable, freezing-tolerant ionic conductive organohydrogel for strain sensors, Composites Part A: Applied Science and Manufacturing, 155 (2022) 106813.

[38] Y. Xu, X. Yang, A.K. Thomas, P.A. Patsis, T. Kurth, M. Kräter, K. Eckert, M. Bornhäuser, Y. Zhang, Noncovalently assembled electroconductive hydrogel, ACS Applied Materials & Interfaces, 10 (2018) 14418–14425.

[39] D.-H. Kim, N. Lu, R. Ma, Y.-S. Kim, R.-H. Kim, S. Wang, J. Wu, S.M. Won, H. Tao, A. Islam, K.J. Yu, T.-i. Kim, R. Chowdhury, M. Ying, L. Xu, M. Li, H.-J. Chung, H. Keum, M. McCormick, P. Liu, Y.-W. Zhang, F.G. Omenetto, Y. Huang, T. Coleman, J.A. Rogers, Epidermal electronics, Science, 333 (2011) 838–843.

[40] X. Li, L. He, Y. Li, M. Chao, M. Li, P. Wan, L. Zhang, Healable, degradable, and conductive mxene nanocomposite hydrogel for multifunctional epidermal sensors, ACS Nano, 15 (2021) 7765–7773.

[41] Y. Guo, Y. Wang, H. Chen, W. Jiang, C. Zhu, S. Toufouki, S. Yao, A new deep eutectic solvent-agarose gel with hydroxylated fullerene as electrical "switch" system for drug release, Carbohydrate Polymers, 296 (2022) 119939.

[42] P. Liu, K. Jin, W. Wong, Y. Wang, T. Liang, M. He, H. Li, C. Lu, X. Tang, Y. Zong, C. Li, Ionic liquid functionalized non-releasing antibacterial hydrogel dressing coupled with electrical stimulation for the promotion of diabetic wound healing, Chemical Engineering Journal, 415 (2021) 129025.

[43] L. Guan, X. Ou, Z. Wang, X. Li, Y. Feng, X. Yang, W. Qu, B. Yang, Q. Lin, Electrical stimulation-based conductive hydrogel for immunoregulation, neuroregeneration and rapid angiogenesis in diabetic wound repair, Science China Materials, (2022).

[44] A. Sharma, V. Panwar, B. Mondal, D. Prasher, M.K. Bera, J. Thomas, A. Kumar, N. Kamboj, D. Mandal, D. Ghosh, Electrical stimulation induced by a piezo-driven triboelectric nanogenerator and electroactive hydrogel composite, accelerate wound repair, Nano Energy, 99 (2022) 107419.

[45] L. Wang, T. Xu, X. He, X. Zhang, Flexible, self-healable, adhesive and wearable hydrogel patch for colorimetric sweat detection, Journal of Materials Chemistry C, 9 (2021) 14938–14945.

[46] P.-H. Lin, S.-C. Sheu, C.-W. Chen, S.-C. Huang, B.-R. Li, Wearable hydrogel patch with noninvasive, electrochemical glucose sensor for natural sweat detection, Talanta, 241 (2022) 123187.

[47] Z. Xu, J. Song, B. Liu, S. Lv, F. Gao, X. Luo, P. Wang, A conducting polymer pedot:Pss hydrogel based wearable sensor for accurate uric acid detection in human sweat, Sensors and Actuators B: Chemical, 348 (2021) 130674.

[48] T.A. Hakala, A. García Pérez, M. Wardale, I.A. Ruuth, R.T. Vänskä, T.A. Nurminen, E. Kemp, Z.A. Boeva, J.-M. Alakoskela, K. Pettersson-Fernholm, E. Hæggström, J. Bobacka, Sampling of fluid through skin with magneto-hydrodynamics for noninvasive glucose monitoring, Scientific Reports, 11 (2021) 7609.

[49] K. Nagamine, T. Mano, A. Nomura, Y. Ichimura, R. Izawa, H. Furusawa, H. Matsui, D. Kumaki, S. Tokito, Noninvasive sweat-lactate biosensor emplsoying a hydrogel-based touch pad, Scientific Reports, 9 (2019) 10102.

[50] H. Guo, M. Bai, C. Wen, M. Liu, S. Tian, S. Xu, X. Liu, Y. Ma, P. Chen, Q. Li, X. Zhang, J. Yang, L. Zhang, A zwitterionic-aromatic motif-based ionic skin for highly biocompatible and glucose-responsive sensor, Journal of Colloid and Interface Science, 600 (2021) 561–571.

[51] S. Park, H. Yuk, R. Zhao, Y.S. Yim, E.W. Woldeghebriel, J. Kang, A. Canales, Y. Fink, G.B. Choi, X. Zhao, P. Anikeeva, Adaptive and multifunctional hydrogel hybrid probes for long-term sensing and modulation of neural activity, Nature Communications, 12 (2021) 3435.

[52] S. Oribe, S. Yoshida, S. Kusama, S.-i. Osawa, A. Nakagawa, M. Iwasaki, T. Tominaga, M. Nishizawa, Hydrogel-based organic subdural electrode with high conformability to brain surface, Scientific Reports, 9 (2019) 13379.

[53] T. Hyakumura, U. Aregueta-Robles, W. Duan, J. Villalobos, W.K. Adams, L. Poole-Warren, J.B. Fallon, Improving deep brain stimulation electrode performance in vivo through use of conductive hydrogel coatings, Frontiers in Neuroscience, 15 (2021).

[54] C. Pitsalidis, D. van Niekerk, C.-M. Moysidou, A.J. Boys, A. Withers, R. Vallet, R.M. Owens, Organic electronic transmembrane device for hosting and monitoring 3d cell cultures, Science Advances, 8 eabo4761.

[55] K.Y. Lee, D.J. Mooney, Hydrogels for tissue engineering, Chemical Reviews, 101 (2001) 1869–1880.

[56] S. Mantha, S. Pillai, P. Khayambashi, A. Upadhyay, Y. Zhang, O. Tao, H.M. Pham, S.D. Tran, Smart hydrogels in tissue engineering and regenerative medicine, Materials (Basel, Switzerland), 12 (2019).

[57] J.F. Mano, Stimuli-responsive polymeric systems for biomedical applications, Advanced Engineering Materials, 10 (2008) 515–527.

[58] S.R. Shin, S.M. Jung, M. Zalabany, K. Kim, P. Zorlutuna, S.b. Kim, M. Nikkhah, M. Khabiry, M. Azize, J. Kong, K.-t. Wan, T. Palacios, M.R. Dokmeci, H. Bae, X. Tang, A. Khademhosseini, Carbon-nanotube-embedded hydrogel sheets for engineering cardiac constructs and bioactuators, ACS Nano, 7 (2013) 2369–2380.

[59] D.N. Heo, S.-J. Lee, R. Timsina, X. Qiu, N.J. Castro, L.G. Zhang, Development of 3d printable conductive hydrogel with crystallized pedot:Pss for neural tissue engineering, Materials Science and Engineering: C, 99 (2019) 582–590.

13 Hydrogels for Physical and (Bio)Chemical Sensors

Mojdeh Mirshafiei and Saleheh Shahmoradi
University of Tehran, Tehran, Iran

Amin Janghorbani
Semnan University, Semnan, Iran

Fatemeh Yazdian
University of Tehran, Tehran, Iran

Iman Zare
Sina Medical Biochemistry Technologies Co. Ltd.,
Shiraz, Iran

Ebrahim Mostafavi
Stanford University, Stanford, California, USA

13.1 AN INTRODUCTION TO DESIGNING SENSORS AND BIOSENSORS

13.1.1 Sensors: Definition and Classification

The term "sensor" is described as a device or module that detects and responds to environmental signals and stimuli. The first definition of the sensor is derived from the Latin word "sentire", which conceivably arises from the human senses (1). Sensors are classified into various categories according to the physical quantity (substance) or analyte to be measured. Based on signal detection, sensors can be divided into the two most fundamental and widely opted classes: physical and chemical sensors. Physical sensors provide information about the system's physical characteristics by measuring a physical quantity and converting it into a user-identifiable signal. Figure 13.1 shows a schematic diagram of various processes occurring in sensors. These sensors have been widely used in the biomedical field, especially for developing more accurate and compact sensors. The International Union of Pure and Applied Chemistry (IUPAC) defines chemical sensors as "instruments that convert chemical information from a concentration of a sample component to its complete composition into a useful

DOI: 10.1201/9781003340485-13

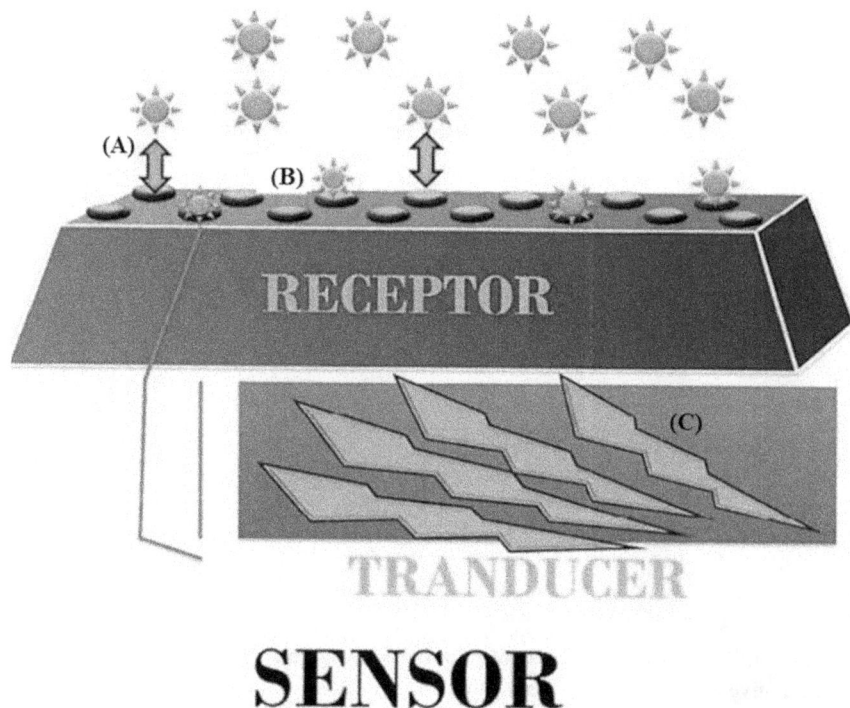

FIGURE 13.1 Schematic diagram of various processes occurring in sensors: (A) Analytes are attracted to receptor sites; (B) the electrical signal is generated through the chemical interaction between the analyte; (C) the electrical signal is transmitted to the processor by a transducer. Reproduced with permission from (1). Copyright (2017) Hilaris.

analytical information signal". These sensors monitor the concentration or activity of relevant chemical species in the gas or liquid phase in various fields, from the environment to food and drug analysis and clinical diagnostic purposes (2,3).

13.1.2 BIOSENSORS

It is worth mentioning that physical and chemical sensing are almost always closely linked. The intersection of physical and chemical sensing techniques can be considered the class of biosensors. In principle, biosensors are integrated receptor-transducer-based devices that could provide selective quantitative or semi-quantitative analytical information by integrating a biological recognition element with an electronic component to generate a measurable signal. Physiological or chemical changes or the presence of various chemical or biological materials in the environment are detected, recorded, and transmitted with the electronic component (3,4). The ideal biosensor should possess certain characteristics including selectivity, sensitivity, linearity, response time, reproducibility, high resolution, repeatability, and stability (5). A biosensor system consists of three fundamental components: (i) biomolecules/ bioagents with selective recognition mechanisms, (ii) converters, and (iii) electrons that

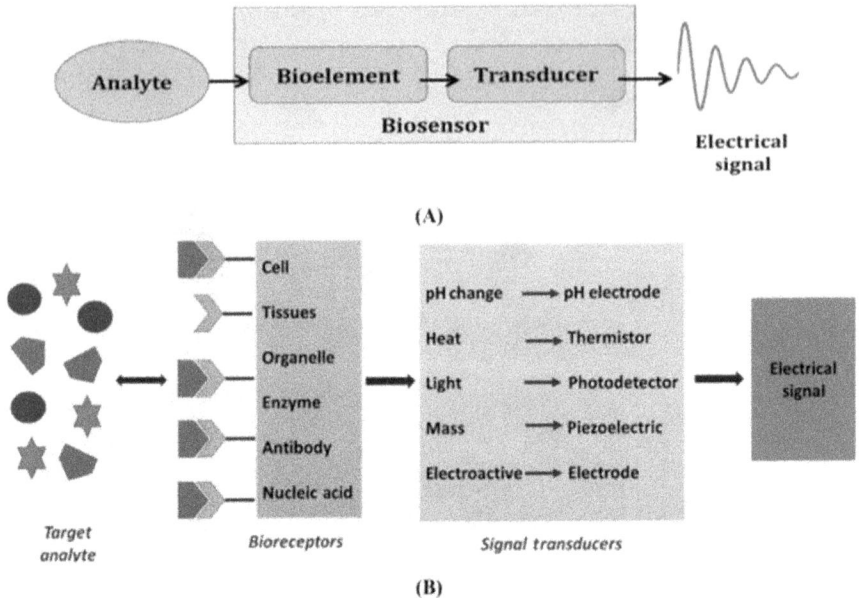

FIGURE 13.2 Schematic representations of biosensors: (A) working principle, (B) illustrative example. Reproduced with permission from (6). Copyright (2022) Elsevier.

can convert physicochemical signals resulting from the interaction of bioagents with the substance under investigation into electronic signals (Figure 13.2). The sensitive biological agents that interact with the target analyte in a highly selective but reversible manner are the key functions of these elements (6).

13.1.3 HYDROGEL-BASED SENSORS

Hydrogel-based sensors rely on the unique properties of hydrogels, such as high water content, high compliance, stimulus responsiveness, and high permeability to various molecules. The hydrophilic nature of hydrogels with high water absorption capacity and extracellular matrix (ECM)–like porous structure renders them intrinsically biocompatible and biodegradable. Therefore, hydrogels have been increasingly proposed in various applications (7). The progress of sensor technology has enabled the direct application of hydrogels as a base material to form a hydrogel-based biosensor. Their open, porous structure and unique large internal surfaces allow interaction and diffusion of analytes at the molecular level through the hydrogel matrix. Their increased cross-linking density, however, may limit or even prevent the diffusion of larger molecules, such as proteins (7,8).

In addition, various modifications of hydrogels have been proposed for analyte detection (9). For example, the use of functional monomers that have a favorable interaction with the target analyte enables molecular recognition (10). Molecular imprinting techniques can be employed to better regulate the incorporation of these functional groups and enhance the affinity for the analyte (11). To achieve recognition and confer response properties, biomolecules with a known affinity for an analyte can be conjugated into the polymer (12). Additionally, these systems can be designed to respond to a wide variety of analytes, from toxic heavy metal ions (such as mercury (Hg^{2+}), uranyl (UO_2^{2+}), and lead (Pb^{2+})), to macromolecules and cells (13,14). Moreover, other features have made hydrogels popular for sensing applications, including tunable viscoelasticity yet durability, stimuli-responsiveness, antifouling properties, and incorporating bioreceptors for detecting biochemical or biological interactions into their highly porous structure through a variety of well-known synthesis methods (8).

Alternatively, hydrogels can serve as an inert matrix for active or responsive elements, such as free ions, nanomaterials, biomolecules, and living cells. Due to the porous hydrogel structures and their unique large internal surfaces, the hydrogel matrix entraps or chemically anchors these active elements. The matrix is permeable to chemical inputs, allowing interactions between the active elements and these input signals. This method employs hydrogel-immobilized receptors to host biomolecular recognition components and identifies specific biological-system events. During sensing processes, passive hydrogel matrices typically remain constant in volume and phase, unlike stimuli-responsive hydrogels. For the design of this type of biosensor, immobilization of receptors, surface bonding strategies, and the prevention of nonspecific protein adsorption to the hydrogel surface are of primary importance.

Furthermore, desirable properties and characteristics can be achieved by designing and optimizing passive hydrogel matrices, such as porosity, biocompatibility, conductivity, and mechanical strength (15,16). A high-performance sensor is typically characterized by its short response time, high sensitivity, low detection limit, low limit of detectable input signals, and reversibility (no memory effect). As many hydrogel-based sensors rely on the diffusion of input signals and water within the hydrogel, the response time is prolonged due to the restriction of chemical diffusion throughout the hydrogel sensors. Therefore, to address this issue, according to diffusion kinetics ($\tau \sim L2/D$), increasing the diffusion coefficient of hydrogel (D) or decreasing the typical dimension (L) could reduce the diffusion time (τ) significantly (17,18).

13.1.4 Polymer Hydrogel-Based (Bio)Sensors

Numerous polymers have been employed in synthesizing hydrogel-based sensors with molecular chains cross-linked chemically through covalent bonds or physically through non-covalent interactions to form 3D network structures. For improving mechanical properties or facilitating unique sensing capabilities such as stimuli-responsiveness, chemical detection, conductivity, and self-healing, various polymers are used either as a base material or as a chemical, biological, or physical source in hydrogels. Various synthetic polymers such as poly(acrylamide) (PAM), poly(vinyl alcohol) (PVA), poly-(vinylpyrrolidone) (PVP), poly(vinyl imidazole) (PVI), poly(ethylene glycol) (PEG), etc. have been widely utilized in sensing applications. As sensors and actuators, these synthetic polymers possess desirable mechanical strength, flexibility, selective chemical reactivity, and controllable molecular structure (9). PVA hydrogels, for example, due to their biocompatibility, flexibility, unique mechanical properties, and biodegradation performance are widely utilized in sensing applications. Relying on its abundant hydroxyl groups, the possibility of chemical modifications on PVA polymer chains is provided; also, cross-linking of PVA can be achieved by either chemical cross-linking or physical cross-linking. PEG hydrogels also have the potential to be used in sensing applications conferred by excellent biocompatibility and anti-fouling properties. It is well known that PAM hydrogels are widely used in biosensors due to their superior biocompatibility and well-defined processes for fabricating robust platforms with optimal mechanical and physiological properties. PAM hydrogels have also been reported to have been immobilized with peptides, proteins, enzymes, or antibodies for molecular recognition (19). Moreover, natural polymers, including polysaccharides and polypeptide origins, such as cellulose, starch, chitin, chitosan, alginate, agarose, dextran, collagen, protein, and their derivatives, have also been employed for the preparation of hydrogel-based sensors; because these biopolymer-based hydrogels possess abundant functional groups and distinctive physicochemical properties, as well as unique characteristics such as biocompatibility, bioadhesion, high protein affinity, environmental responsiveness, swelling behaviors, and ease of surface modification, which makes them suitable and more attractive than synthetic polymer for biosensing applications (9,19).

Furthermore, some electroconductive polymers have also received considerable attention in fabricating hydrogel-based sensors. Electroconductive hydrogels are combinations of hydrated structures with electronic functionality with unique properties. Since a porous hydrogel provides a large surface area with greater diffusivity, conductive polymers facilitate electron transport across the interface in such hydrogels. Their flexibility and processability in the conduction of electrons and the functionalities conferred through chemical modifications gain prominence in sensing technologies. The most prevalent conductive polymers employed in conductive hydrogel fabrication are polyaniline, polypyrrole, and poly (ethylenedioxy thiophene) (19,20).

13.1.5 DESIGN AND PRINCIPAL

With a particular focus on designing hydrogel-based sensors and biosensors, the crucial parameters that should be considered during the designing and engineering of biosensors with high performance are (i) immobilization/fabrication of biomolecules in its native conformation, (ii) high accessibility of receptors to the targets, and (iii) specific adsorption of the analyte to the support. Indeed, the quality of sensing and the sensitivity level mainly depends on the accessibility and activity of the immobilized sensing molecules. So, using hydrogels to preserve enzymes and other biomolecules is an excellent solution to maintain their activity and functionality (21). In addition, an important criterion in designing a biosensor with high selectivity and sensitivity is the receptor that has to be immobilized in its native configuration to be capable of effectually interacting with the analyte. At the same time, the support material should be resistant to non-specific absorption (2). Most importantly, the properties of hydrogels may easily be tailored to form an ideal and versatile platform. Hydrogel-based sensors detect target molecules by measuring volume changes between analytes and sensing elements in this context. This volumetric change due to recognition creates a new sensing system as an alternative to classical biosensors based on electrochemical sensing. Thus, various advantages of hydrogels offer specific intriguing characteristics that suit a vast pool of sensing applications that comprise point-of-care testing, home diagnostics, environmental monitoring, etc. (22).

13.2 IMMOBILIZATION TECHNIQUES FOR FUNCTIONALIZATION OF HYDROGELS

To design the biorecognition portion of the biosensor, a reliable strategy for the immobilization of biological elements on the surface of the transducer is necessary. Selecting the appropriate immobilization technique is one of the most important and critical aspects of sensor design and preparation (8). Hydrogels with a 3D porous structure provide an enhanced surface area and more binding than traditional rigid supports such as glass, silicon, and plastic, as well as having a homogeneous liquid phase environment rather than a heterogeneous liquid-solid interface, which could improve the efficiency of immobilization and biochemical reactions (23). Moreover, a hydrogel exhibits excellent performance under physiological conditions, making it possible to detect biological phenomena *in vivo* with minimal impact on biological activity (24). The choice of an appropriate immobilization technique to minimize undesired interactions relies on the selected bio-detection element, transducer, physicochemical environment, and the characteristics of the analyte, as well as the availability of reactive groups and binding types. As illustrated in Figure 13.3, bioreceptors could be immobilized on hydrogel surfaces in two general techniques: physical (reversible) and chemical (irreversible) or some combination of these techniques. The bound bioreceptor could not be detached without destroying the hydrogel microstructure or bioreceptor activity in an irreversible immobilization technique (3).

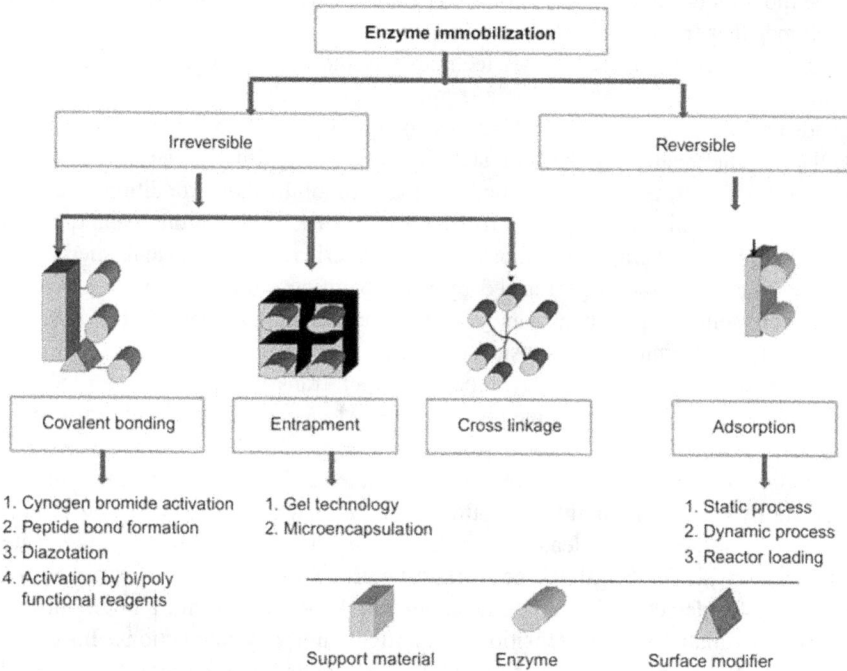

FIGURE 13.3 The classification of immobilization techniques. Reproduced with permission from (25). Copyright (2019) Elsevier.

13.2.1 Physical or Reversible Immobilization

The physical or reversible technique is based on attaching enzymes to the transducer surface without creating chemical bonds. Physical immobilizations comprise (i) physical adsorption and (ii) physical entrapment. The physical adsorption technique immobilizes a biorecognition element on the outer surface of an inert solid material via weak attractive forces such as hydrogen bonds, ionic bonds, electrostatic forces, and van der Waals forces. Simplicity, affordability, non-degradation of the activity of biological receptors, and no need to modify biological elements or create a matrix are the advantages of this technique. Disadvantages of this technique include immobilization's sensitivity to changes in temperature, ionic strength, weak interactions, and poor operational and storage stability (3).

The physical entrapment technique immobilizes biorecognition elements via covalent or non-covalent bonds within the 3D hydrogel matrices. The entrapment immobilization procedure could arise through two paths: The enzyme is mixed with a monomer solution, and then the monomer solution is polymerized by a chemical reaction or by changing the experimental conditions. Electropolymerization, sol-gel technique, and microencapsulation are common techniques used in this process. Electropolymerization is a plain procedure in which a current or potential is applied to an aqueous solution or electrolyte containing monomer molecules and biomolecules. On the surface of the electrode in the electrolyte, the reduction or oxidation of the monomers occurs, generating active radical species that bind together to form a polymer that traps biorecognition elements.

On the other hand, the sol-gel technique is the most commonly used method retained at low temperatures for enzyme trapping. Biorecognition elements are encapsulated by forming the 3D porous matrix with time and temperature. Easy synthesis, chemical, and thermal stability, and the ability to encapsulate high concentrations of biomolecules under mild immobilization conditions, are the advantages of this technique. Ultimately, a microencapsulation technique and biorecognition elements are enclosed into a spherical semipermeable membrane. Ideally, this technique is performed by (a) separating enzyme microdroplets in a water-immiscible liquid phase and (b) polymerizing monomers at the interface between the two immiscible substances (3,26).

For example, Wei et al. developed a direct transverse relaxation time (T_2) biosensing strategy using alkaline phosphatase (ALP)–mediated sol-gel transition in hydrogels for studying food-borne pathogens. ALP catalyzes reactions to generate an acidic environment to convert the sol-state alginate solution to hydrogel. Alginate hydrogel can significantly alter the transverse relaxation of water protons in aqueous solutions and lead to a reduction in the T_2 value by trapping water molecules inside its "egg-box" network structure. This T_2 biosensor exhibits high sensitivity for detecting 50 CFU/mL *Salmonella enteritidis* within 2 hours through enzyme-modulated sol-gel transition and antigen-antibody interactions. Instead of the magnetic probes used in conventional methods, this biosensing approach directly modifies the water molecules, providing a simple, fresh, and sensitive platform for pathogen detection (Figure 13.4) (27).

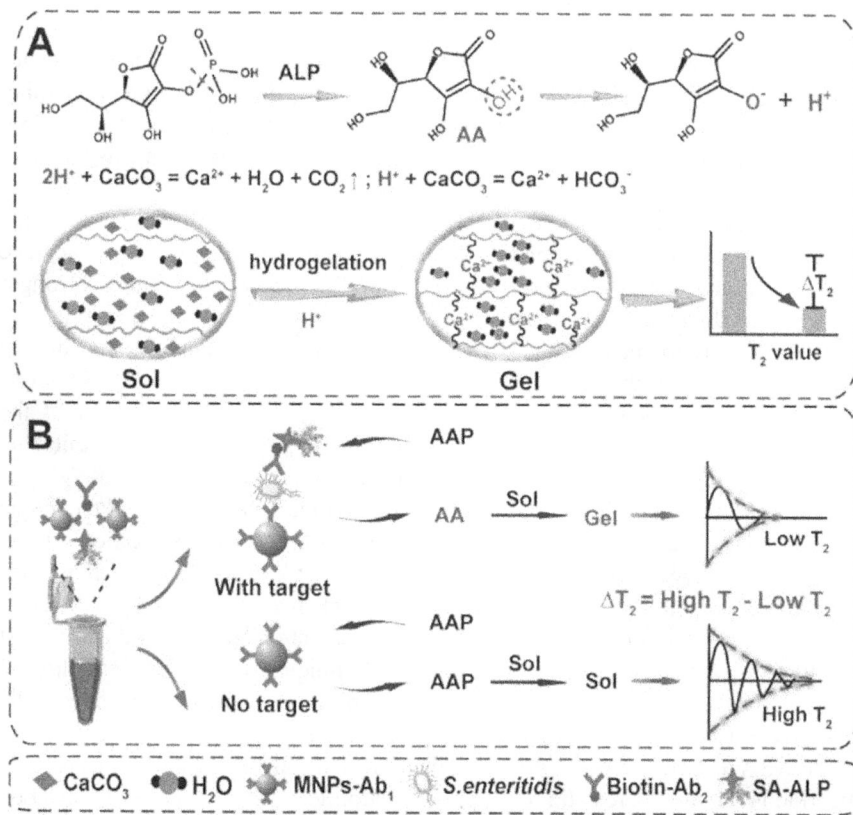

FIGURE 13.4 Schematic of a direct T_2 biosensing strategy by the enzyme-mediated sol-gel transition of hydrogels for S. enteritidis immunoassays. (A) ALP-catalyzed reactions that regulate hydrogelation of sodium alginate with an "egg box" network structure. (B) Schematic of a gelation-based T_2 biosensor for the detection of *S. enteritidis*, the presence of which forms MNPss-Ab1-antigen-Ab2-biotin-streptavidin-ALP complexes and encodes sol-gel. Reproduced with permission from (27). Copyright (2021) American Chemical Society.

13.2.2 CHEMICAL OR IRREVERSIBLE IMMOBILIZATION

In chemical or irreversible techniques, strong chemical bonds are formed between the functional groups of the biorecognition element and the transducer surface. Chemical immobilization comprises covalent binding and cross-linking. During covalent binding, the biorecognition element is firmly bonded to either the electrode/transducer surface or the membrane's inert matrix. The immobilization process via the membrane matrix involves the synthesis of functional polymer and covalent immobilization; the binding mechanism relies on interactions between biorecognition elements and protein functional groups and reactive groups on the transducer/membrane matrix surface. The formation of a strong bond between the biorecognition element and the matrix, strong resistance to environmental changes, and low biorecognition element leakage are the main advantages of this technique. However, this method has some disadvantages, including the use of harsh chemicals causes the inability to regenerate the deployed matrix after use (3,26). On the other side, the cross-linking mechanism works by creating intermolecular covalent cross-links between biorecognition elements or between biorecognition elements and inert proteins. This technique is carried out with the assistance of multifunctional reagents, which act as a linker to bind enzyme molecules in 3D cross-linked aggregates to the transducer surface. The optimal cross-linking conditions enable shorter response time, stronger binding, and higher catalytic activities. Stronger chemical binding, less biorecognition element leakage, and the possibility of adjusting the environment for the biological identification element by using suitable stabilizing agents are the main advantages of the cross-linking technique. However, the formation of covalent cross-links between protein molecules instead of the matrix and protein and the partial denaturation of protein structure limits the application of cross-linking immobilization (3,26).

To recapitulate aspects of immobilization techniques, the entrapment technique is appropriate for hydrogels prepared by the solution polymerization, albeit bioreceptor leakage is unpleasant. The entrapment technique is preferred when the average intermolecular spacing of the hydrogel (hydrogel porosity) is smaller than the size of the bioreceptor. Covalent immobilization techniques are the most widely used among various immobilization techniques due to the stability of covalent bonds formed between hydrogels and bioreceptors. However, due to the minimization of the release of bioreceptors during the process, to avoid blockage of the active site through the possible formation of covalent linkages, considerable attention is required. Cross-linking immobilization technique is primarily based on the ability of cross-linkers to react with bioreceptors and hydrogel surfaces. Cross-linkers with functional groups greater than two in this immobilization technique are widely used (8).

Physical adsorption and affinity ligand binding are simple immobilization techniques that bind bioreceptors to hydrogel surfaces via hydrogen bonds, van der Waals reactions, or salt linkage. Bioreceptor functionality is preserved with this technique. However, the immobilization process might be reversed by environmental conditions. Weak reproducibility, weak adhesion, and the random orientation of bioreceptors attached to the hydrogel surface are some circumscriptions of this technique. Chelation, known as metal binding, is an ideal immobilization technique when using bioderived hydrogels in biosensor construction. Precipitation of metal salts or hydroxides results in their binding via nucleophilic coordination using heating or neutralization processes. The mode of cleavage and bond formation and the environment and conditions of the active site are the factors that affect the unoccupied binding sites for interaction with bioreceptors. The heterogeneity of bioreceptor adsorption and metal ions' leakage reduces the chelation process's reproducibility. However, the use of chelating ligands improved this immobilization technique. A unique physical immobilization technique is establishing disulfide bonds between hydrogels and bioreceptors. Stable covalent bonds formed tend to be broken by reaction with appropriate agents or when modulation of reactivity occurs with changes in the ionic (8).

Another significant aspect of biosensor fabrication performed after immobilization is bioreceptor stabilization, which refers to storage or manipulation stability. Stabilization refers to the ability of a bioreceptor to retain its activity or conformation. The most well-known technique of imparting stability to bioreceptors is using polyelectrolytes to enhance the stability of certain enzymes and antibodies. Sugar components of polyelectrolytes modify the bioreceptor environment by replacing free water. The protective hydration shell helps the protein maintain its structure and activity. Several strategies that combine polymers and polyelectrolytes have a synergistic effect on bioreceptor stabilization. To stabilize the microorganisms within the biosensor, techniques such as lyophilization, vacuum drying, continuous culture, and encapsulation have been proposed (8,28).

13.3 TRANSDUCING STRATEGIES

Hydrogel-based sensors can be classified according to the target analyte and the transducing strategy. The nature of the materials, fabrication techniques, and design of a sensor matrix has a significant impact on the ultimate detection capability of the sensor. Also, it influences the selection of analyte and transducing strategy. Transducers process energy and convert biometric events into measurable signals. Based on the transducing strategy, (bio)sensors can be classified into electrochemical, optical, micromechanical, and stimuli-responsive (29).

13.3.1 ELECTROCHEMICAL SENSORS

In an electrochemical transducer, a certain sensing element reacts with a target analyte to make a sensing signal. A sensor is a special kind of transducer that changes the signal into a measurable electrical signal proportionate to the target analyte concentration; hence, it can be used in qualitative or quantitative analysis (30,31). This type of sensor contains two parts: a diagnosis system (a solidified sensing element) and a conversion system (a transducer).

To improve the analytical performance, which is the sensitivity of the electrochemical sensor, nanomaterials can be used as electrodes or assisting matrices. For signal enhancement, these requirements are needed: assisting electro-catalytic property, outstanding electron movement capability, and great biocompatibility with capture biomolecules. Figure 13.5 shows the analytical principle for electrochemical biosensors based on carbon and non-carbon nanomaterials (32).

In a study, a thin hydrogel micropatch (THMP) was made as an interface for sweat sampling and a medium for electrochemical sensing. In order to evaluate the THMP performance, caffeine (as a xenobiotic (33) and due to its therapeutic potential to treat airway obstruction and apnea of prematurity (34)) and lactate (lactate-targeting sensors have been widely used for sweat analysis (35,36)) were chosen as two target molecules. The hydrogel interface also makes the diffusion of analytes easier from sweat to the reservoir compared to the open-air interface. Moreover, the hydrogel interface sampled nearly three times as many analytes as the dry pad interface (37).

Modification of electrodes in electrochemical sensors is a way to increase their efficiency. Hydrogel, as a soft material, consists of cross-linking hydrophilic

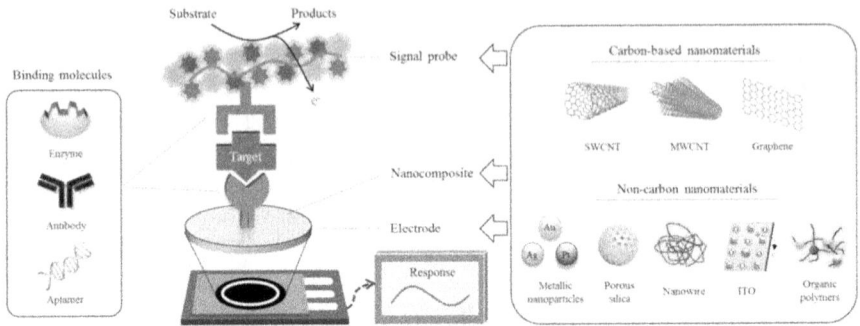

FIGURE 13.5 The analytical principle for electrochemical biosensors based on carbon and non-carbon nanomaterials. Reproduced with permission from (32). Copyright (2020) BioMed Central.

monomers containing ionic groups with covalent, hydrogen, or ligand bonds. Due to the ionic groups in these hydrogels, they have high electrical conductivity (38,39). Also, they have a considerable porous structure and good flexibility (40,41). To detect promyelocytic leukemia/retinoic acid receptor alpha (PML/RARa) fusion gene in acute promyelocytic leukemia (APL), Liu et al. used aldehyde-agarose hydrogel films. The results were obtained by changing the diol group of the hydrogel to an aldehyde group by a reaction with NaIO$_4$ and using methylene blue as an electrochemical indicator (42). To immobilize the enzyme periplasmatic aldehyde oxidoreductase to detect benzaldehyde electrochemically, research was conducted on the covalent and noncovalent biomolecules entrapment into a biocompatible hydrogel based on PEG and hyperbranched polyglycerol. Using a hydrogel maintained the enzyme stable for up to four days (43).

Hydrogels, as a modifier, have a 3D structure and can make a microenvironment that limits the analyte, which is a great platform to immobilize bioactive molecules and keep their activity (44); but they are poorly conductive. Using nanomaterials in composition with hydrogels enhances their electrical conductivity, which is favorable in hydrogel-based electrochemical sensors (38,45). Hence, a large specific surface area and more active sites on the surface can successfully speed up the electron transfer rate and raise the electrochemical reaction rate (46). In a study, various metal nanoparticles (MNPs) like Ni, Au, Pt, Pd, and Cu were deposited onto graphene hydrogel (GHG) to investigate their effect on electrocatalytic activity and sensing. The results showed that fabricated gold nanoparticles (AuNPs)/GHG composite had good sensing and electrocatalytic activity. On the other hand, since MNP–GHG composites have both large surface area and high electric conductivity of the graphene with the MNPs' high electrocatalysis activity, is suitable for electrochemical sensors (47).

13.3.2 OPTICAL

Optical sensors work based on the intensity change monitoring in light beams or their phase changes while reacting with the physical systems. This type of sensor is classified as an interferometric or intensity sensor. Other optical techniques exist based on light scattering, spectral transmission changes, Rayleigh and Raman scattering, radiative losses, etc. Optical sensors have remarkable benefits in their characteristics, including electrical passiveness, freedom from electromagnetic interference, greater sensitivity, and wide dynamic range. A sensor network or array sends and receives optical signals over long distances. Since in optical sensors, it is not necessary to create a conversion between electronics and photonics at each sensing site, the flexibility and the costs will increase and decrease, respectively (48,49).

Hydrogel-based photonic crystals (PCH) are an example of the hydrogel application in optical sensors benefiting from low price, high sensitivity, and portability. In (50), a PCH was designed to detect various biomolecules with different sizes and charges using an antibody/antigen interaction.

In another study, molecular imprinting and photonic hydrogels were applied to produce a photonic sensor for fast and label-free detection of vanillin. In this study, a non-covalent, self-assembly approach was applied in the fabrication of this photonic hydrogel (51).

Hydrogels are also applied in hydrophilic colorimetric gas sensors to deal with humidity interference. In this study, a hydrogel matrix is applied for loading sensing probes, regulating the humidity of the sensing environment based on its water absorption capability (52). Application of the hydrogel in this colorimetric sensing system leads to improvement of its reliability and accuracy.

Recently, bacterial cellulose-based re-swellable hydrogel fabricated through a facile approach was applied as a colorimetric sensor to measure sweat pH and glucose with a wide, linear sensing region and low sample volume requirements (53).

On the other hand, hydrogels were also applied to fabricate an easy-to-use, eco-friendly, compatible unaided-eye detection bromide colorimetric detection of the system. In this colorimetric system, Ag nano-prisms aggregating or transforming in a solution problem was resolved by immobilization of Ag nano-prisms in a biocompatible hydrogel (54).

In an interesting study, an optical-based system to detect humidity was fabricated. Researchers encapsulated fluorescent flavylium salts in a poly(22-hydroxyethyl methacrylate) (PHEMA) network matrix. The hydrogel swelling and deswelling in different humidities were measured by the flavylium salt's fluorescence decompose times (55).

13.3.3 MICROELECTROMECHANICAL SYSTEMS

A microelectromechanical system (MEMS) is a process technology applied to produce integrated and small devices with a combination of mechanical and electrical elements. They can not only sense, control, and actuate on the micro-scale but also can affect the macro scale. Generally, MEMSs have mechanical microstructures, microsensors, microactuators, and microelectronics as the principal parts. The MEMS functionality is based on the mechanical movements of their elements. The main idea is that the changes in the particular environment parameters lead to changes in the mechanical properties of the micromechanical transducer, which can be measured electronically, optically, etc. (56).

Several studies proposed using hydrogels as components in MEMS. Stimuli-responsive hydrogels are enormously paid attention to as active materials to be applied in MEMS (57,58). Controlling microvalves is an essential process in MEMS functioning. Several studies applied stimuli-responsive hydrogels as controllable microvalves (59–61) and pumps (62). Several environmental parameters, such as magnetic field, temperature, frequency, and light, were applied to control these microvalves in these studies. In addition, the sensitivity of hydrogels to volume changes makes them good candidates for a self-regulation microvalve (57). In addition, responsive hydrogels can be applied as sensors in MEMS. The swelling behavior of hydrogels in response to the pH, osmotic pressure, temperature, and concentration of an analyte transduce chemical energy to mechanical energy, making them applicable as *in vivo* sensors (63). Also, electroactive hydrogels have been successfully applied in MEMS-based drug delivery systems.

Moreover, in a MEMS-based biosensor, hydrogels were used as a sensor for measuring the concentration of pH and glucose based on their swelling property causing the deformation of a diaphragm of a pressure transducer (64).

13.3.4 STIMULI-RESPONSIVE SENSORS

Generally, based on the function of hydrogels in sensors, hydrogel-based sensors are divided into two categories: (i) Sensors that are stimuli-responsive hydrogels that can change their volume and/or phase in response to external stimuli; and (ii) sensors that are passive hydrogels matrices that host responsive substances and response to environmental stimuli (Figure 13.6).

Environmental influences detected by hydrogel sensors include physical stimuli (*e.g.,* temperature, mechanical stress, light, ionic strength, electric field, magnetic field), chemical stimuli (*e.g.,* pH, ions), and biological stimuli (*e.g.,* glucose, enzyme, antigen) (Figure 13.7).

In the presence of these chemical, physical, or biochemical stimuli, changes in the molecular level of polymer chains and networks lead to changes in the micro/macroscopic volume, and/or hydrogels are subjected to phase transition by sensor molecules; subsequently, the sensors convert the signals into measurable outputs.

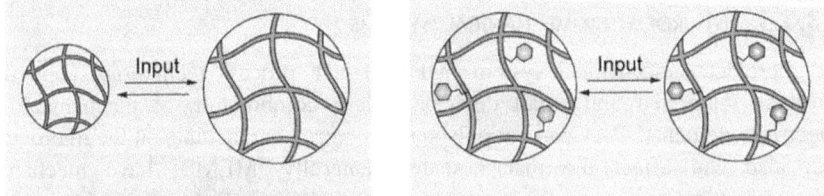

FIGURE 13.6 Schematic of hydrogel sensors based on (A) stimulus-responsive hydrogels as sensors that can display volume and/or phase changes according to environmental cues; (B) passive hydrogel matrix to accommodate responsive elements (i.e., free ions, nanoparticles, biomolecules, living cells). Reproduced with permission from (15). Copyright (2020) Elsevier.

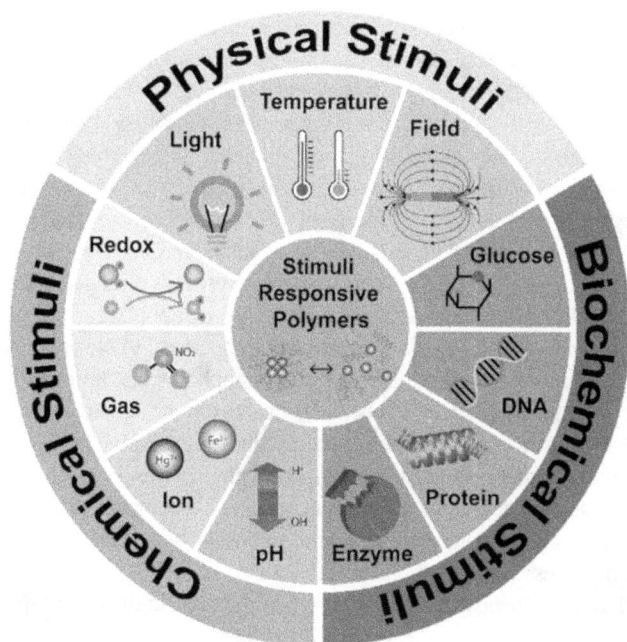

FIGURE 13.7 Classification of stimuli-responsive platforms for polymers, including physical, chemical, and biochemical stimuli. Reproduced with permission from (19). Copyright (2020) American Chemical Society.

Therefore, the stimulus plays an essential role in the response mechanism of the hydrogel (15). On the other hand, hydrogel sensors are highly hydrated, so their stimuli-responsive polymers, cross-linkers, and incorporated entities must be compatible with and workable in water and aqueous environments (65). The response rate of hydrogels depends on their composition, shape, and size; thus, several proposed techniques can be used to increase the response, such as reducing their size, decreasing their cross-linking density, and increasing their ionic group content and pore size. Hydrogel sensors produce output signals in the form of geometrical, mechanical, optical, electrical, and biological responses (15).

As a result of their intrinsic stimuli-responsive nature, many hydrogel strain sensors are reversible and capable of multiple sensing. However, hydrogel sensors lose their original sensing ability when external stimuli damage their irreversible bonds and cross-linkers; on the other hand, they can also lose their original sensing capability when exposed to external stimuli. Depending on the type of stimulation the sensors are intended to detect, polymer networks can sometimes be functionalized with different stimuli-responsive pendent groups and additives even before gelation, resulting in hydrogels that can be used directly as sensor platforms without any additional substrates or matrixes (66). In addition, using chemical synthesis and gelation methods, it is also possible to copolymerize or sequentially polymerize various stimuli-responsive monomers and cross-linkers in hydrogel sensors to achieve multiple stimuli-responsive properties (19).

Various types of stimuli-responsive hydrogels were reported according to the diversity of responsivities. The most common stimuli-responsive polymers for sensors include thermo-responsive (such as poly(N-isopropylacrylamide) (PNIPAM) and poly(vinyl methyl ether) (PVME) hydrogels), pH-responsive (such as carboxylated poly(2-hydroxyethyl methacrylate–ethylene glycol dimethacrylate) (poly(HEMA–EDMA)) hydrogels), light/force-sensitive (such as poly(acrylamide (AAm)-co-methyl acrylate (MA))/spiropyran hydrogels) (66). Thermoresponsive polymers are widely used and well known due to their large changes in solubility, volume, and compatibility with ambient temperature, which can be adjusted over a wide temperature range. Electroresponsive polymers can modify their physicochemical properties by applying an electrical stimulus such as rheological, optical, and/or mechanical properties. Photoresponsive or light-responsive polymers, on the other hand, have chromophores that absorb light at specific wavelengths and undergo specific chemical reactions or structural changes. These polymers have received much attention, mainly due to their ability to position- and/or intensity-dependent remote. Moreover, polymers with enzyme-responsive moieties have also been synthesized that change their physicochemical properties upon adding specific enzymes. Despite their popularity, enzyme-responsive sensors suffer from complex immobilization processes, poor stability, and high costs (67).

Table 13.1 shows various composites of hydrogel used in different types of sensors.

13.4 CONCLUSION AND PERSPECTIVES

Remarkable efforts have been made to integrate diverse hydrogels into modified electrodes for sensing applications; therefore, hydrogel-based (bio)sensors have witnessed explosive development in various fields in the past decades (88). Due to chemical diversity, tunability of various physicochemical properties, intrinsic properties such as swellability, and high sensitivity toward external stimuli, considerable attention has been focused on hydrogels in (bio)sensing applications (89). Indeed, the 3D swellability and permeability of hydrogels conduct electrical signals, and also negligible interactions with different swelling media, and the ability to optimally immobilize various biomolecules will exhibit their significant role in novel (bio)sensor platforms (90). Portability, sensitivity, selectivity, stability, easy fabrication, surface modifications, and simultaneous sensing performance are major points offered by the combination of hydrogels in (bio)sensor platforms. However, despite the great progress achieved in hydrogel-based sensors with excellent properties, some challenging issues have remained. Conventional hydrogel-based sensors typically have limited mechanical strength and are prone to permanent failure. As a result of this intrinsic mechanical weakness, hydrogel-based sensors are often susceptible to damage under continuous and responsive actions, leading to unstable sensitivity and specificity. Also, the limitation of mechanical strength and dynamic cues within hydrogels restricts their sensory capabilities (91–93). To address this issue, an alternative design strategy to improve the performance of hydrogel-based sensors is to use hard hydrogels (such as nanocomposite hydrogels and dual-network hydrogels) as structural platforms for integrating various sensor components. To address this issue, an alternative design strategy to improve the performance of hydrogel-based sensors is to use tough hydrogels (such as nanocomposite hydrogels and dual-network hydrogels) as structural

TABLE 13.1

Hydrogel Composites Used in Different Types of Sensors/Biosensors

Types of Sensors/ Biosensors	Composite	Results	Refs.
Electrochemical	Chitosan/dipeptide nanofibrous hydrogel and immobilized DNA probes with OTA aptamer	The sensor was used in the food industry to detect Ochratoxin A in white wine.	(68)
	Supramolecular biopolymer chitosan and zinc ions hydrogels	The zinc redox changes caused hydroxyl radical or hydrogen peroxide determination.	(69)
	Hydrogel conjugated aptamer	The composite was used for progesterone level monitoring in blood samples.	(70)
	Polymer polypyrrole hydrogel	Having excellent biocompatibility, suitable for bioelectronics applications.	(71)
	Platinum nanoparticles decorated graphite/gelatin hydrogel (PtNPs– GR/GLN) composite	The PtNPs–GR/GLN modified electrode was successfully used to determine H_2O_2 in real biological samples like human serum and saliva for feasible applications.	(72)
	Dopamine polymer doped polypyrrole (PDA-PPy) hydrogel modified screen-printed carbon electrode (SPCE) (noted as PDA-PPy/SPCE)	PDA-PPy/SPCE showed great sensing performance towards Pb(II) with low detection limit and high sensitivity.	(73)
	Collagen hydrogel filled into the interspace of the 3D graphene foam/Pt NPs/poly(3,4-ethylene dioxythiophene) sensor	Successful real-time monitoring of reactive oxygen species released from microglia in the collagen matrix.	(74)
	Peptide hydrogels loaded with ciprofloxacin and AuNPs	A typical neurotransmitter dopamine detection and improvements in the implantable sensors' construction without infection.	(75)
	Silver nanoparticles doped chitosan hydrogel film	A trichloroacetic acid amperometric sensor was fabricated with a low detection limit and high sensitivity.	(76)
Optical	Hydrogel films with 2D colloidal arrays attached on both surfaces	Having anti-curling performance and improved optical diffraction intensity was useful in improving the practical applications of visual and quantitative detection.	(77)
	A hydrogel-coated optical fiber surface plasmon resonance	Having notable repeatability and good stability. Because of the wide measuring range and simple fabrication process, the sensor is suitable for pH measurements in 1–12.	(78)

(Continued)

TABLE 13.1 *(Continued)*

Hydrogel Composites Used in Different Types of Sensors/Biosensors

Types of Sensors/ Biosensors	Composite	Results	Refs.
	AuNPs and glucose-responsive hydrogels composition	The reversible and reusable sensor detected physiological glucose levels with high linearity and negligible hysteresis.	(79)
	Chitosan-based carbon dots rooted in agarose hydrogel film	The sensing platform was the most profitable for use as an on-site operational, portable, cheap colorimetric-optical heavy metal ion detector with separation ability.	(80)
	Hydrogel-based optical sensor with polymer microarrays	The probe was evaluated by pH measurement in the ovine lung and distinguished tumorous and normal tissue, suggesting the potential for the rapid and accurate observation of tissue pH changes.	(81)
	A hydrogel bioink functionalized by adding luminescent optical sensor nanoparticles	Online imaging of O_2 was conducted to monitor cell growth in printed constructs.	(82)
	3-phenylboronic acid and a tertiary amine, dimethyl aminopropyl acrylamide, into a hydrogel matrix	The sensor measured physiological glucose levels in blood with a minimal impact from interfering molecules.	(83)
	A 2D Au nanosphere array attached to a polyacrylic acid (PAA) hydrogel film	The composite film showed visually diffraction color and stronger diffraction intensity.	(84)
MEMS	Polyvinyl alcohol and polyacrylic acid (PVA/PAA)	Measuring PH with long-term reproduction and relatively good sensitivity.	(64)
	Poly (2-hydroxyethyl methacrylate) (pHEMA)	Measuring calcium nitrate tetrahydrate exhibited the greatest sensitivity in the concentration range of 0–0.5 M.	(85)
	A crosslinked poly(methacrylic acid) network containing high amounts of poly(ethylene glycol) dimethacrylate	Measuring pH with an extraordinary maximum sensitivity of 1 nm/5 × $10^{-5}\Delta$pH.	(86)
	Hyaluronic acid hydrogel	Measuring the flow, the hydrogel-capped sensor showed a much higher output voltage than the naked hair cell sensor.	(87)

platforms for integrating various sensor components. In addition, the incorporation of various nanomaterials or biomolecules into polymer matrices through suitable immobilization techniques can improve properties of the hydrogel (94).

On the other hand, incorporating self-healing properties into hydrogel-based sensors will prevent structural damage, ensure reusability, increase stability, and restore some of the sensor's functionality. However, the design and fabrication of self-healing hydrogel-based sensors that combine features of rapid self-healing capabilities, high mechanical efficiency, and easy detection of response signals need to be addressed. Especially under harsh conditions, retaining the long-term self-healing of hydrogel-based sensors is another crucial consideration (95,96).

Another research direction is to employ hydrogel-based sensors for biological applications. In particular, similarities between hydrogels and biological tissues in terms of mechanical, chemical, and biological properties make hydrogels ideal interfaces for integrating engineering and biological systems where novel innovations are widely implemented to design and fabricate novel hydrogels with desired properties towards specific biosensing applications. Exploring the development of the biosensor, which is sensitive detection of very low amounts of analyte in a complex biological matrix, will be highly promising for specific applications. Meanwhile, the biocompatibility, flexibility, and stability of hydrogels for biosensing applications need to be addressed. However, hydrogel-based biosensors still have a long way to go before being used in commercialized health management systems due to some limitations, including lifetime, storage, and transducers for rapid quantitative analysis. Developing superior electro-chemically active sites, establishing sensitivity and selectivity by optimizing properties, modification, and functionalization of hydrogels by using appropriate functional organic or biomolecules, and integration of *in situ* electrochemical and physiological techniques are some considerations to develop better biosensors (97).

In parallel, the development of multifunctional hydrogel-based sensors that utilize multiple functions (sensing, optical/electrical response, actuation) should also be pursued. Computational modeling and machine learning technologies, along with polymer chemistry, synthesis, and gelation, should be combined to better understand the relationship between components, structures, and performance in hydrogels. Also, the structurally based design of hydrogel sensors and their interactions with target components and exogenous stimuli should be improved. Multiscale simulations coupled with experimental analyses of polymer chemistry, synthesis, and gelation technique should be performed at the atomic, molecular, and macroscopic levels to illuminate the performance, deformation, and recovery of hydrogels in terms of materials and component structures, enable improved design, and integrated system to perform a series of complex actions to enhance hydrogel-based sensor responses (98,99).

Apart from conventional hydrogel-based sensor fabrication techniques, printing techniques are an attractive alternative for rapid electrode prototyping and fabrication of portable yet advanced electrodes for sensing applications. With the printing ability of bioinspired hydrogels on solid transducers, sophisticated sensing devices can be developed. This will greatly contribute to the fabrication of the next generation of advanced, flexible, and wearable bioelectronics with superior sensor performance. Moreover, hydrogel sensors integrated into microfluidic or lab-on-a-chip systems will facilitate the development of miniaturized and portable sensing devices. However, ensuring the type of constituent material, biocompatibility, accuracy, and stability of the sensor is very important in developing hydrogel sensing devices. Although the future of hydrogel-based (bio)sensors is bright and promising, it requires a determined, cost-effective, and multidisciplinary approach to bring these systems from the development laboratory to the market.

REFERENCES

[1] Ali J, Najeeb J, Ali MA, Aslam MF, Raza A. Biosensors: Their fundamentals, designs, types and most recent impactful applications: a review. J Biosens Bioelectron. 2017;8(1):1–9.

[2] Buenger D, Topuz F, Groll J. Hydrogels in sensing applications. Prog Polym Sci. 2012;37(12):1678–1719.

[3] Naresh V, Lee N. A review on biosensors and recent development of nanostructured materials-enabled biosensors. Sensors. 2021;21(4):1109.

[4] Jose JM, Jose JV, Vijaykumar Mahamuni C. Multi-biosensor based wireless body area networks (WBAN) For critical health monitoring of patients in mental health care centers: An interdisciplinary study. Int J Res Eng Sci Manag. 2020;3.

[5] Ahmad R, Tripathy N, Ahn M-S, Bhat KS, Mahmoudi T, Wang Y, et al. Highly efficient non-enzymatic glucose sensor based on CuO modified vertically-grown ZnO nanorods on electrode. Sci Rep. 2017;7(1):1–10.

[6] Karunakaran R, Keskin M. Biosensors: components, mechanisms, and applications. In: Analytical Techniques in Biosciences. Elsevier; 2022. pp. 179–190.

[7] Bae J, Park J, Kim S, Cho H, Kim HJ, Park S, et al. Tailored hydrogels for biosensor applications. J Ind Eng Chem. 2020;89:1–12.

[8] Tavakoli J, Tang Y. Hydrogel based sensors for biomedical applications: An updated review. Polymers (Basel). 2017;9(8):364.

[9] Dhanjai, SA, Kalambate PK, Mugo SM, Kamau P, Chen J, et al. Polymer hydrogel interfaces in electrochemical sensing strategies: A review. TrAC Trends Anal Chem [Internet]. 2019;118:488–501. Available from: https://www.sciencedirect.com/science/article/pii/S0165993619301645

[10] Zhou Q, Xu Z, Liu Z. Molecularly imprinting–Aptamer techniques and their applications in molecular recognition. Biosensors. 2022;12(8):576.

[11] Nawaz N, Abu Bakar NK, Muhammad Ekramul Mahmud HN, Jamaludin NS. Molecularly imprinted polymers-based DNA biosensors. Anal Biochem. 2021 Oct;630:114328.

[12] Park R, Jeon S, Jeong J, Park SY, Han DW, Hong SW. Recent advances of point-of-care devices integrated with molecularly imprinted polymers-based biosensors: From biomolecule sensing design to intraoral fluid testing. Biosensors. 2022;12(3):136.

[13] Divya MS, Srivastava VR, Chandra P. Nanobioengineered sensing technologies based on cellulose matrices for detection of small molecules, macromolecules, and cells. Biosensors. 2021;11(6):168.

[14] Culver HR, Clegg JR, Peppas NA. Analyte-responsive hydrogels: intelligent materials for biosensing and drug delivery. Acc Chem Res [Internet]. 2017 Feb 21;50(2):170–178. Available from: 10.1021/acs.accounts.6b00533

[15] Liu X, Liu J, Lin S, Zhao X. Hydrogel machines. Mater Today. 2020;36:102–124.

[16] Erol O, Pantula A, Liu W, Gracias DH. Transformer hydrogels: A review. Adv Mater Technol. 2019;4(4):1900043.

[17] Hong W, Zhao X, Zhou J, Suo Z. A theory of coupled diffusion and large deformation in polymeric gels. J Mech Phys Solids. 2008;56(5):1779–1793.

[18] Tanaka T, Fillmore DJ. Kinetics of swelling of gels. J Chem Phys. 1979;70(3):1214–1218.

[19] Sun X, Agate S, Salem KS, Lucia L, Pal L. Hydrogel-based sensor networks: Compositions, properties, and applications—A review. ACS Appl Bio Mater. 2020;4(1):140–162.

[20] Wang L, Xu T, Zhang X. Multifunctional conductive hydrogel-based flexible wearable sensors. TrAC Trends Anal Chem. 2021;134:116130.

[21] Wadhera T, Kakkar D, Wadhwa G, Raj B. Recent advances and progress in development of the field effect transistor biosensor: A review. J Electron Mater. 2019;48(12):7635–7646.

[22] Pinelli F, Magagnin L, Rossi F. Progress in hydrogels for sensing applications: A review. Mater Today Chem. 2020;17:100317.

[23] Tang J, Xiao P. Polymerizing immobilization of acrylamide-modified nucleic acids and its application. Biosens Bioelectron. 2009;24(7):1817–1824.

[24] Völlmecke K, Afroz R, Bierbach S, Brenker LJ, Frücht S, Glass A, et al. Hydrogel-based biosensors. Gels. 2022;8(12):768.

[25] Sneha HP, Beulah KC, Murthy PS. Chapter 37 – Enzyme immobilization methods and applications in the food industry. In: Kuddus MBT-E in FB, editor. Academic Press; 2019. pp. 645–658. Available from: https://www.sciencedirect.com/science/article/pii/B9780128132807000372

[26] Sassolas A, Blum LJ, Leca-Bouvier BD. Immobilization strategies to develop enzymatic biosensors. Biotechnol Adv. 2012;30(3):489–511.

[27] Wei L, Wang Z, Feng C, Xianyu Y, Chen Y. Direct transverse relaxation time biosensing strategy for detecting foodborne pathogens through enzyme-mediated sol-gel transition of hydrogels. Anal Chem [Internet]. 2021 May 4;93(17):6613–6619. Available from: 10.1021/acs.analchem.0c03968

[28] Srivastava A, Ranjan K, Katyayni N, Tripathi S. Microbial Biosynthesis of Nanoparticles: A Review. Think India J. 2019;22(37):546–595.

[29] Purohit B, Vernekar PR, Shetti NP, Chandra P. Biosensor nanoengineering: Design, operation, and implementation for biomolecular analysis. Sensors Int. 2020;1:100040.

[30] Wu W, Zhou Q, Zheng Y, Fu L, Zhu J, Karimi-Maleh H. An electrochemical fingerprint approach for direct soy sauce authentic identification using a glassy carbon electrode. Int J Electrochem Sci. 2020;15:10093–10103.

[31] Fu L, Liu Z, Ge J, Guo M, Zhang H, Chen F, et al. (001)plan manipulation of α-Fe2O3 nanostructures for enhanced electrochemical Cr(VI)sensing. J Electroanal Chem. 2019;841:142–147.

[32] Cho I-H, Kim DH, Park S. Electrochemical biosensors: Perspective on functional nanomaterials for on-site analysis. Biomater Res. 2020;24(1):1–12.

[33] Dobson NR, Liu X, Rhein LM, Darnall RA, Corwin MJ, McEntire BL, et al. Salivary caffeine concentrations are comparable to plasma concentrations in preterm infants receiving extended caffeine therapy. Br J Clin Pharmacol. 2016;82(3):754–761.

[34] Schmidt B, Roberts RS, Davis P, Doyle LW, Barrington KJ, Ohlsson A, et al. Caffeine therapy for apnea of prematurity. N Engl J Med. 2006;354(20):2112–2121.

[35] Jia W, Bandodkar AJ, Valdés-Ramírez G, Windmiller JR, Yang Z, Ramírez J, et al. Electrochemical tattoo biosensors for real-time noninvasive lactate monitoring in human perspiration. Anal Chem. 2013;85(14):6553–6560.

[36] Khodagholy D, Curto VF, Fraser KJ, Gurfinkel M, Byrne R, Diamond D, et al. Organic electrochemical transistor incorporating an ionogel as a solid state electrolyte for lactate sensing. J Mater Chem. 2012;22(10):4440–4443.

[37] Lin S, Wang B, Zhao Y, Shih R, Cheng X, Yu W, et al. Natural Perspiration Sampling and in Situ Electrochemical Analysis with Hydrogel Micropatches for

User-Identifiable and Wireless Chemo/Biosensing. ACS Sensors [Internet]. 2020 Jan 24;5(1):93–102. Available from: 10.1021/acssensors.9b01727

[38] Fu L, Yu A, Lai G. Conductive hydrogel-based electrochemical sensor: A soft platform for capturing analyte. Chemosensors. 2021;9.

[39] Herrmann A, Haag R, Schedler U. Hydrogels and their role in biosensing applications. Adv Healthc Mater. 2021;10(11):2100062.

[40] Chimene D, Kaunas R, Gaharwar AK. Hydrogel Bioink Reinforcement for Additive Manufacturing: A Focused Review of Emerging Strategies. Adv Mater. 2020;32(1):1902026.

[41] Ahmed EM. Hydrogel: Preparation, characterization, and applications: A review. J Adv Res. 2015;6(2):105–121.

[42] Liu A, Chen X, Wang K, Wei N, Sun Z, Lin X, et al. Electrochemical DNA biosensor based on aldehyde-agarose hydrogel modified glassy carbon electrode for detection of PML/RARa fusion gene. Sensors Actuators, B Chem. 2011;160(1):1458–1463.

[43] Dey P, Adamovski M, Friebe S, Badalyan A, Mutihac R-C, Paulus F, et al. Dendritic polyglycerol–poly (ethylene glycol)-based polymer networks for biosensing application. ACS Appl Mater Interfaces. 2014;6(12):8937–8941.

[44] Rong Q, Lei W, Liu M. Conductive Hydrogels as Smart Materials for Flexible Electronic Devices. Chem – A Eur J. 2018;24(64):16930–16943.

[45] Power AC, Gorey B, Chandra S, Chapman J. Carbon nanomaterials and their application to electrochemical sensors: a review. Nanotechnol Rev. 2018;7(1):19–41.

[46] Putzbach W, Ronkainen NJ. Immobilization techniques in the fabrication of nanomaterial-based electrochemical biosensors: A review. Sensors. 2013;13(4):4811–4840.

[47] Du C, Yao Z, Chen Y, Bai H, Li L. Synthesis of metal nanoparticle@graphene hydrogel composites by substrate-enhanced electroless deposition and their application in electrochemical sensors. RSC Adv [Internet]. 2014;4(18):9133–9138. Available from: 10.1039/C3RA47950A

[48] Wallace PA, Yang Y, Campbell M. Application of sol-gel processes for fiber optic chemical sensor development. In: Tenth International Conference on Optical Fibre Sensors. SPIE; 1994. pp. 465–467.

[49] Kersey AD. A review of recent developments in fiber optic sensor technology. Opt fiber Technol. 1996;2(3):291–317.

[50] Qin J, Li X, Cao L, Du S, Wang W, Yao SQ. Competition-based universal photonic crystal biosensors by using antibody–antigen interaction. J Am Chem Soc [Internet]. 2020 Jan 8;142(1):417–423. Available from: 10.1021/jacs.9b11116

[51] Peng H, Wang S, Zhang Z, Xiong H, Li J, Chen L, et al. Molecularly imprinted photonic hydrogels as colorimetric sensors for rapid and label-free detection of vanillin. J Agric Food Chem [Internet]. 2012 Feb 29;60(8):1921–1928. Available from: 10.1021/jf204736p

[52] Yu J, Tsow F, Mora SJ, Tipparaju VV, Xian X. Hydrogel-incorporated colorimetric sensors with high humidity tolerance for environmental gases sensing. Sensors Actuators B Chem [Internet]. 2021;345:130404. Available from: https://www.sciencedirect.com/science/article/pii/S0925400521009722

[53] Siripongpreda T, Somchob B, Rodthongkum N, Hoven VP. Bacterial cellulose-based re-swellable hydrogel: Facile preparation and its potential application as colorimetric sensor of sweat pH and glucose. Carbohydr Polym [Internet]. 2021;256:117506. Available from: https://www.sciencedirect.com/science/article/pii/S0144861720316799

[54] Kim SH, Woo H-C, Kim MH. Solid-phase colorimetric sensing probe for bromide based on a tough hydrogel embedded with silver nanoprisms. Anal Chim Acta [Internet]. 2020;1131:80–89. Available from: https://www.sciencedirect.com/science/article/pii/S0003267020307674

[55] Galindo F, Lima JC, Luis SV, Melo MJ, Parola AJ, Pina F. Water/humidity and ammonia sensor, based on a polymer hydrogel matrix containing a fluorescent flavylium compound. J Mater Chem [Internet]. 2005;15(27–28):2840–2847. Available from: 10.1039/B500512D

[56] Datskos PG, Lavrik NV, Sepaniak MJ. Micromechanical sensors. In: Introduction to Nanoscale Science and Technology. Springer; 2004. pp. 417–439.

[57] Beck A, Obst F, Gruner D, Voigt A, Mehner PJ, Gruenzner S, et al. Fundamentals of hydrogel-based valves and chemofluidic transistors for lab-on-a-chip technology: A tutorial review. Adv Mater Technol. 2022;2200417.

[58] Saunders JR, Abu-Salih S, Khaleque T, Hanula S, Moussa W. Modeling theories of intelligent hydrogel polymers. J Comput Theor Nanosci. 2008;5(10):1942–1960.

[59] Li H, Chen J, Lam KY. Multiphysical modeling and meshless simulation of electric-sensitive hydrogels. J Polym Sci Part B Polym Phys. 2004;42(8):1514–1531.

[60] Eddington DT, Beebe DJ. Flow control with hydrogels. Adv Drug Deliv Rev. 2004;56(2):199–210.

[61] Tokarev I, Minko S. Stimuli-responsive hydrogel thin films. Soft Matter. 2009;5(3):511–524.

[62] Baldi A, Gu Y, Loftness PE, Siegel RA, Ziaie B. A hydrogel-actuated smart microvalve with a porous diffusion barrier back-plate for active flow control. In: Technical Digest MEMS 2002 IEEE International Conference Fifteenth IEEE International Conference on Micro Electro Mechanical Systems (Cat No 02CH37266). IEEE; 2002. pp. 105–108.

[63] Guenther M, Gerlach G. Hydrogels for chemical sensors. In: Hydrogel sensors and actuators. Springer; 2009. pp. 165–195.

[64] Thong TQ, Guenther M, Gerlach G. Development of hydrogel-based MEMS piezoresistive sensors for detection of solution pH and glucose concentration. Vietnam J Mech. 2012;34(4):281–288.

[65] Zhang D, Ren B, Zhang Y, Xu L, Huang Q, He Y, et al. From design to applications of stimuli-responsive hydrogel strain sensors. J Mater Chem B. 2020;8(16):3171–3191.

[66] Lavrador P, Esteves MR, Gaspar VM, Mano JF. Stimuli-responsive nanocomposite hydrogels for biomedical applications. Adv Funct Mater. 2021;31(8):2005941.

[67] Hu L, Zhang Q, Li X, Serpe MJ. Stimuli-responsive polymers for sensing and actuation. Mater Horizons. 2019;6(9):1774–1793.

[68] Li X, Falcone N, Hossain MN, Kraatz H-B, Chen X, Huang H. Development of a novel label-free impedimetric electrochemical sensor based on hydrogel/chitosan for the detection of ochratoxin A. Talanta [Internet]. 2021;226:122183. Available from: https://www.sciencedirect.com/science/article/pii/S0039914021001041

[69] Fu L, Wang A, Lyu F, Lai G, Yu J, Lin C-T, et al. A solid-state electrochemical sensing platform based on a supramolecular hydrogel. Sensors Actuators B Chem [Internet]. 2018;262:326–333. Available from: https://www.sciencedirect.com/science/article/pii/S0925400518302995

[70] Velayudham J, Magudeeswaran V, Paramasivam SS, Karruppaya G, Manickam P. Hydrogel-aptamer nanocomposite based electrochemical sensor for the detection of progesterone. Mater Lett [Internet]. 2021;305:130801. Available from: https://www.sciencedirect.com/science/article/pii/S0167577X21014981

[71] Yang M, Ren X, Yang T, Xu C, Ye Y, Sun Z, et al. Polypyrrole/sulfonated multi-walled carbon nanotubes conductive hydrogel for electrochemical sensing of living cells. Chem Eng J [Internet]. 2021;418:129483. Available from: https://www.sciencedirect.com/science/article/pii/S1385894721010706

[72] Thirumalraj B, Sakthivel R, Chen S-M, Rajkumar C, Yu L, Kubendhiran S. A reliable electrochemical sensor for determination of H2O2 in biological samples using platinum nanoparticles supported graphite/gelatin hydrogel. Microchem J

[Internet]. 2019;146:673–678. Available from: https://www.sciencedirect.com/science/article/pii/S0026265X18312827

[73] Zhong J, Zhao H, Cheng Y, Feng T, Lan M, Zuo S. A high-performance electrochemical sensor for the determination of Pb(II) based on conductive dopamine polymer doped polypyrrole hydrogel. J Electroanal Chem [Internet]. 2021;902:115815. Available from: https://www.sciencedirect.com/science/article/pii/S1572665721008419

[74] Hu X-B, Qin Y, Fan W-T, Liu Y-L, Huang W-H. A three-dimensional electrochemical biosensor integrated with hydrogel enables real-time monitoring of cells under their in vivo-like microenvironment. Anal Chem [Internet]. 2021 Jun 8;93(22):7917–7924. Available from: 10.1021/acs.analchem.1c00621

[75] Wang W, Han R, Tang K, Zhao S, Ding C, Luo X. Biocompatible peptide hydrogels with excellent antibacterial and catalytic properties for electrochemical sensing application. Anal Chim Acta [Internet]. 2021;1154:338295. Available from: https://www.sciencedirect.com/science/article/pii/S0003267021001215

[76] Liu B, Deng Y, Hu X, Gao Z, Sun C. Electrochemical sensing of trichloroacetic acid based on silver nanoparticles doped chitosan hydrogel film prepared with controllable electrodeposition. Electrochim Acta [Internet]. 2012;76:410–415. Available from: https://www.sciencedirect.com/science/article/pii/S0013468612008250

[77] Men D, Zhang H, Hang L, Liu D, Li X, Cai W, et al. Optical sensor based on hydrogel films with 2D colloidal arrays attached on both the surfaces: anti-curling performance and enhanced optical diffraction intensity. J Mater Chem C [Internet]. 2015;3(15):3659–3665. Available from: 10.1039/C5TC00174A

[78] Zhao Y, Lei M, Liu S-X, Zhao Q. Smart hydrogel-based optical fiber SPR sensor for pH measurements. Sensors Actuators B Chem [Internet]. 2018;261:226–232. Available from: https://www.sciencedirect.com/science/article/pii/S0925400518301266

[79] Guo J, Zhou B, Du Z, Yang C, Kong L, Xu L. Soft and plasmonic hydrogel optical probe for glucose monitoring. Nanophotonics. 2021;10(13):3549–3558. Available from: 10.1515/nanoph-2021-0360

[80] Gogoi N, Barooah M, Majumdar G, Chowdhury D. Carbon dots rooted agarose hydrogel hybrid platform for optical detection and separation of heavy metal ions. ACS Appl Mater Interfaces [Internet]. 2015 Feb 11;7(5):3058–3067. Available from: 10.1021/am506558d

[81] Gong J, Tanner MG, Venkateswaran S, Stone JM, Zhang Y, Bradley M. A hydrogel-based optical fibre fluorescent pH sensor for observing lung tumor tissue acidity. Anal Chim Acta [Internet]. 2020;1134:136–143. Available from: https://www.sciencedirect.com/science/article/pii/S0003267020308011

[82] Trampe E, Koren K, Akkineni AR, Senwitz C, Krujatz F, Lode A, et al. Functionalized bioink with optical sensor nanoparticles for O2 imaging in 3D-bioprinted constructs. Adv Funct Mater [Internet]. 2018 Nov 1;28(45):1804411. Available from: 10.1002/adfm.201804411

[83] Tierney S, Falch BMH, Hjelme DR, Stokke BT. Determination of glucose levels using a functionalized hydrogel–optical fiber biosensor: Toward continuous monitoring of blood glucose in vivo. Anal Chem [Internet]. 2009 May 1;81(9):3630–3636. Available from: 10.1021/ac900019k

[84] Men D, Zhou F, Hang L, Li X, Duan G, Cai W, et al. A functional hydrogel film attached with a 2D Au nanosphere array and its ultrahigh optical diffraction intensity as a visualized sensor. J Mater Chem C [Internet]. 2016;4(11):2117–2122. Available from: 10.1039/C5TC04281J

[85] Strong ZA, Wang AW, McConaghy CF. Hydrogel-actuated capacitive transducer for wireless biosensors. Biomed Microdevices. 2002;4(2):97–103.

[86] Hilt JZ, Gupta AK, Bashir R, Peppas NA. Ultrasensitive biomems sensors based on microcantilevers patterned with environmentally responsive hydrogels. Biomed Microdevices. 2003;5(3):177–184.

[87] Prakash Kottapalli AG, Bora M, Kanhere E, Asadnia M, Miao J, Triantafyllou MS. Cupula-inspired hyaluronic acid-based hydrogel encapsulation to form biomimetic MEMS flow sensors. Sensors (14248220). 2017;17(8).

[88] Hasan S, Kouzani AZ, Adams S, Long J, Mahmud MAP. Recent progress in hydrogel-based sensors and energy harvesters. Sensors Actuators A Phys [Internet]. 2022;335:113382. Available from: https://www.sciencedirect.com/science/article/pii/S0924424722000206

[89] Bustamante-Torres M, Romero-Fierro D, Arcentales-Vera B, Palomino K, Magaña H, Bucio E. Hydrogels classification according to the physical or chemical interactions and as stimuli-sensitive materials. Gels. 2021;7(4):182.

[90] Shafique H, de Vries J, Strauss J, Khorrami Jahromi A, Siavash Moakhar R, Mahshid S. Advances in the translation of electrochemical hydrogel-based sensors. Adv Healthc Mater [Internet]. 2023 Jan 1;12(1):2201501. Available from: 10.1002/adhm.202201501

[91] Ghorbanizamani F, Moulahoum H, Guler Celik E, Timur S. Ionic liquid-hydrogel hybrid material for enhanced electron transfer and sensitivity towards electrochemical detection of methamphetamine. J Mol Liq [Internet]. 2022;361:119627. Available from: https://www.sciencedirect.com/science/article/pii/S0167732222011655

[92] Baretta R, Raucci A, Cinti S, Frasconi M. Porous hydrogel scaffolds integrating Prussian Blue nanoparticles: A versatile strategy for electrochemical (bio)sensing. Sensors Actuators B Chem [Internet]. 2023;376:132985. Available from: https://www.sciencedirect.com/science/article/pii/S0925400522016288

[93] Juska VB, Pemble ME. A critical review of electrochemical glucose sensing: Evolution of biosensor platforms based on advanced nanosystems. Sensors (Switzerland). 2020;20(21):1–28.

[94] Kailasa SK, Joshi DJ, Kateshiya MR, Koduru JR, Malek NI. Review on the biomedical and sensing applications of nanomaterial-incorporated hydrogels. Mater Today Chem [Internet]. 2022;23:100746. Available from: https://www.sciencedirect.com/science/article/pii/S2468519421003268

[95] Yao X, Zhang S, Qian L, Wei N, Nica V, Coseri S, et al. Super stretchable, self-healing, adhesive ionic conductive hydrogels based on tailor-made ionic liquid for high-performance strain sensors. Adv Funct Mater [Internet]. 2022 Aug 1;32(33):2204565. Available from: 10.1002/adfm.202204565

[96] Jing X, Mi HY, Lin YJ, Enriquez E, Peng XF, Turng LS. Highly stretchable and biocompatible strain sensors based on mussel-inspired super-adhesive self-healing hydrogels for human motion monitoring. ACS Appl Mater Interfaces [Internet]. 2018 Jun 20;10(24):20897–20909. Available from: 10.1021/acsami.8b06475

[97] Yuk H, Lu B, Zhao X. Hydrogel bioelectronics. Chem Soc Rev [Internet]. 2019;48(6):1642–1667. Available from: 10.1039/C8CS00595H

[98] Moin A, Zhou A, Rahimi A, Menon A, Benatti S, Alexandrov G, et al. A wearable biosensing system with in-sensor adaptive machine learning for hand gesture recognition. Nat Electron [Internet]. 2021;4(1):54–63. Available from: 10.1038/s41928-020-00510-8

[99] Singh A, Sharma A, Ahmed A, Sundramoorthy AK, Furukawa H, Arya S, et al. Recent advances in electrochemical biosensors: Applications, challenges, and future scope. Biosensors. 2021;11(9):336.

14 Hydrogels for Biomedical Applications

A. C. Liaudat, Virginia Capella, Martin F. Broglia,
Maria A. Molina, Cesar A. Barbero, Pablo Bosch,
Nancy Rodríguez, and Claudia R. Rivarola
National University of Rio Cuarto, Rio Cuarto, Argentina

14.1 HYDROGELS FOR BIOMEDICAL AND TECHNOLOGICAL APPLICATIONS

Biomedical researchs can be focused on different thematic areas: immunology, molecular biology, cell biology, molecular pharmacology, etc. The biomedical area searches mainly for the development of new drugs and techniques to help treat diseases through the basic understanding of the different pathologies, such as infectious, immune response, neurodegenerative diseases, cancer, etc. It also includes researchs about new therapeutic targets, therapeutic strategies, pharmaceuticals, and biopharmaceutical products.

Currently, there is great interest in the biomedical applications of advanced materials based on biocompatible polymers and hydrogels due to their water content and compatibility with living tissues. These materials are used for prostheses, contact lenses, biosensors, and even in tissue reconstruction.[1] Certainly some hydrogels have the capacity to absorb, retain, and release substances in a controlled manner depending on the physicochemical conditions of the environment, so these are applied in pharmacology as a controlled drug release system.[2]

The selection of the material to apply in biomedicine must meet certain physicochemical, mechanical, and biologic conditions. If one of them is not present, the material will not be useful, so an exhaustive study of the material properties should be previously carried on. Basically, the study implies an interdisciplinary work that encompasses from material chemical to cell biology. This implies starting to analyze the composition/origin, properties, and morphology of the material, at the same time that *in vitro* biological studies are carried out, such as the biocompatibility, cytotoxicity, and adhesion between the living system and the material (scaffold), as well as the immunologic response, among others. Figures 14.1A and B summarize some of the considerations to take into account when selecting a material, according to the field of biomedical application.

In addition, the scaffold material of medical devices should be evaluated and tested by means of the ISO 10993 international norm, before being applied. The characterization of material should be considered according to: i) the biological system that it will

DOI: 10.1201/9781003340485-14

A)

Selection of natural, synthetic, or combined hydrogels

Polymer, copolymer, semi/interpenetrated, nanocomposite...

Define the physicochemical, mechanical and biological properties

Hydrophilicity, rigidity, elasticity...

Design and stablish a synthetic method

Film, 3D form, rugosity, porosity...

B)

Cell morphology, junction point, tissue development

Cell adhesion, and proliferation

Cell line, microorganism, multicellular system

Living system

Tissue irritation, rejection, adaptation...

Immunologic responsive

Biologic assays

Biocompatibility, cyto and genotoxicity

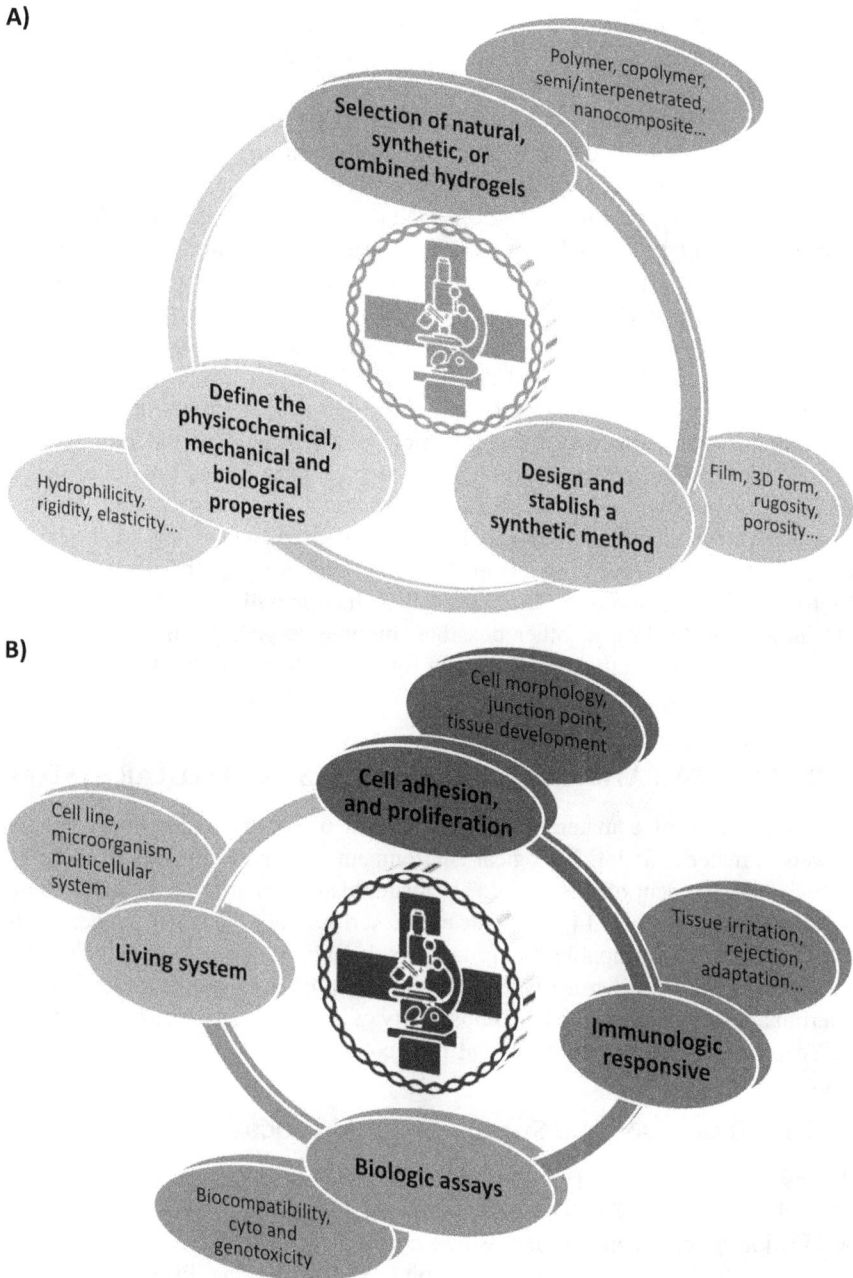

FIGURE 14.1 Relevant information from the point of view of advanced materials (A) and cell biology (B) to apply the biocompatible materials based on hydrogels in biomedicine.

be in contact with, such as skin, blood, tissue/bone/dentin, mucosal membrane, etc., and ii) the adverse biological effect that it could be causing, meaning levels of cytotoxicity, genotoxicity, sensitization, irritation, among others.

The development of *in vitro* cell tissues to be implanted or replaced for damaged tissue or wound closure can be possible through the cell seeding over 2D or inside 3D structures of the hydrogels.[3] The cell tissue would be implanted as that or together with the scaffold. In the latter case, the material could be fully or only partially biodegradable, since it represents a minimum percentage of tissue implanted. It has been demonstrated that the cell adaptation is not predictable since each living system interacts differently with each scaffold material, and vice versa. For that, another consideration to take into account is the capacity of cell adhesion and proliferation of the cell when they are exposed to the polymeric material. It is known that the cell adaptation in contact with the scaffold material depends on the favorable characteristic of the biointerfacial zone defined between them. The interactions present in the cell-material system will be subject to the extracellular membrane characteristics and physico-chemical, mechanical, and morphological characteristics of the surface. The shape, size, porosity, rugosity, elasticity of material, among others, can influence the foreign body response.[4] Then, the cell behavior will depend on the signals and indicators, leading to other possible immune responses. In this way, most studies are focused on the signals shown for macrophages through *in vitro* and *in vivo* assays.

14.2   BIOCOMPATIBILITY OF HYDROGELS IN CELLULAR SYSTEMS

Biocompatibility of a material is an expression of the benignity of the relation between a material and its biological environment.[5] In order to verify it, aspects of biological assessment of new or "old" biomaterials used in new applications should be considered.[6] Figure 14.2 graphically describes the sequence suggested for determining the biocompatibility of materials through *in vitro* assays, which should be validated with *in vivo* assays. Details of the protocol applied for each of the determinations to analyze the biocompatibility of hydrogels are described, including the adaptation of several conventional protocols.

14.2.1   PURIFICATION AND STERILIZATION OF HYDROGELS

Hydrogels are generally synthesized under non-sterile conditions and must be purified and sterilized before being exposed to *in vitro/in vivo* systems. The purification process consists of washing cycles with ultrapure water followed by washing with physiologic solution (phosphate buffered saline, PBS) depending on the application. There are several sterilization processes that can be carried out: exposure to ultraviolet radiation, ethylene oxide, use of chemical disinfectants (ethylene oxide, ethanol), or use of carbon dioxide gas.[7] However, ultraviolet radiation is not recommended for hydrogels because chemical modifications have been observed after application. One of the simplest processes is washing the material with 70% v/v ethanol followed by an incubation period with an

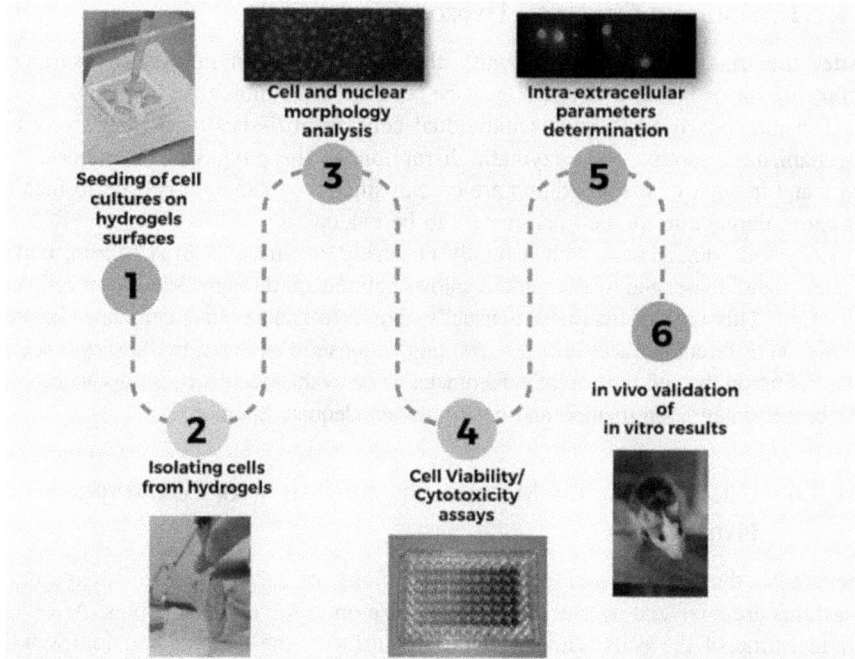

FIGURE 14.2 Steps suggested to take into account for the biocompatibility analysis of hydrogels.

antimicrobial/antimycotic agent.[8] The steps of sterilization protocol are the following:

1. Wash one time with 70% v/v ethanol for 45 minutes.
2. Wash three times with sterile 1X PBS.
3. Incubate hydrogels with antimicrobial-antimycotic solution for 2 hours.
4. Wash three times with sterile 1X PBS.
5. Store at 4°C in 1X PBS until used.

14.2.2 SEEDING OF CELL CULTURES ON HYDROGELS SURFACES

The physicochemical structure of a hydrogel determines the cell-material interaction and therefore the possibility of using this scaffold for culture in exposure, on the surface or inside the hydrogel. Regardless of the type of surface on which the cells are to be seeded, it is recommended to maintain the number of cells-unit surface ratio, even when the cells are seeded on the hydrogels (around 100,000 cells/cm^2).

The cell culture in exposition consists of adding the hydrogel to cells dispersed in the culture medium. The seeding on the surface requires hydrogel swelling in the culture medium and subsequent seeding of cells onto the surface. Seeding inside the hydrogel could be carried out by encapsulation with a hydrogel scaffold or seeding on one defined region of hydrogel and growing the cells inside the 3D structure.

14.2.3   ISOLATING CELLS FROM HYDROGELS

After the treatment (or cell growth), the cells grown on polystyrene surfaces (control), on or inside hydrogels must be recovered for biological analysis.

The methods used to isolate individual cells from tissues or organs are often mechanical disruption and enzymatic disruption. In these types of techniques, the time and manner of disintegration are crucial since they can alter cell performance or cause damage to the cell parameters to be evaluated.

Enzymatic degradation with naturally occurring enzymes, such as trypsin, collagenase, nattokinase, and hyaluronidase, allows cell release through cleavage of cell-cell junctions. This type of treatment is usually chosen to isolate intact cells after having grown on different surfaces and then cell populations are analyzed by flow cytometry. Depending on the cell type and the parameter to be evaluated, it is necessary to choose the best disintegration method and determine its adequate duration.

14.2.4   CELL AND NUCLEAR MORPHOLOGY ANALYSIS AFTER CONTACTING HYDROGELS

Generally, the nuclear and cellular morphology of cells attached to hydrogel scaffolds are analyzed by electron, optical, or atomic force microscopies. After the contact time of the cells with the hydrogel surfaces, the treatment to follow will depend on the type of microscopy to be used.

For scanning electron microscopy (SEM), cells should be fixed with 2.5% glutaraldehyde in PBS and dehydrated with increasing concentrations of ethanol (60, 70, 80, 90, 100%). A thin film of conductive metal must be deposited on the solid materials obtained by dehydration to be analyzed by a scanning electron microscope.[9] For atomic force microscopy, the cell-hydrogel sample follows the same processing as that for SEM, except the metallic sprayed.

To perform optical microscopy, the cells must be stained. Acridine orange and Hoechst 33258 staining techniques are used to assess cytoplasmic and nuclear morphology, respectively, by inverted optical, fluorescence, or confocal microscopies.

The acridine orange and Hoechst 33258 staining protocol is the following:

1. A film composed of cells adhered to the hydrogel are fixed with 4% formaldehyde/PBS and then immersed in 1 μg mL^{-1} acridine orange solution for 15 minutes to analyze cytoplasmic morphology or in 20 μL mL^{-1} Hoechst 33258/PBS solution for nuclear tension.
2. Subsequently, the sample is washed with a PBS solution to be observed by microscopy. Acridine orange fluorescence intensity is measured at excitation and emission wavelengths of 520 and 650 nm and for Hoechst 33258 at 345 and 446 nm.[8]

14.2.5   CELL VIABILITY/CYTOTOXICITY ASSAYS

Numerous techniques exist to assess cell viability/cytotoxicity of biological systems that have grown on hydrogel surfaces or come into contact with the materials.[10] Cell cytotoxicity assay is commonly used to evaluate the effect of certain compounds or materials on cell functionality. The 3-(4,5-dimethylthiazol-2-yl)-2,5-diphenyltetrazolium bromide (MTT) assay is used as an indicator of cell viability. This colorimetric assay measures cellular metabolic activity based on the reduction of tetrazolium salt to formazan crystals. Viable cells contain NAD(P)H-dependent oxidoreductase enzymes that reduce MTT to formazan. Insoluble formazan crystals are dissolved using a solubilizing solution and the colored resulting solution is quantified by measuring the absorbance at 540 nm with a multiwall spectrophotometer.

The applied protocol for MTT assay is the following:

1. Seed the necessary number of cells according to the available surface and incubate during a defined treatment time.
2. Remove the hydrogels and wash with 1X PBS.
3. Then, add the MTT solution in DMEM-10% FBS (0.5 mg/mL) and incubate for 3 hours at 37°C.
4. Remove the supernatant and replace it with 100 μL of DMSO.
5. Measure the absorbance at 540 nm.

In another way, the live/cytotoxic/dead cell stain is one of the most recent and simplest assays that assess membrane integrity and esterase activity and can be determined by fluorescence microscopy, flow cytometry, or microplate readers.

LIVE/DEAD® protocol is the following:

1. Add 100–150 μL of the combined LIVE/DEAD® assay reagents (2 μM calcein AM and 4 μM EthD-1) on the surface with attached cells.
2. Incubate in a covered dish for 30–45 minutes at room temperature.
3. Analyze cells under the fluorescence microscope. Calcein is excited at 485 nm, whereas EthD-1 is at 530 nm. Then, the fluorescence emissions intensity is acquired for calcein at 530 nm and for EthD-1 at 645 nm.

14.2.6   Intra- and Extracellular Parameters Determination

There are numerous intra- and extracellular parameters that can be evaluated in cells that have been in contact with hydrogels. The most commonly studied processes are cell death, nuclear DNA fragmentation, mitochondrial membrane potential, cell cycle, gene expression, and protein translation.

Since the cell cycle directs the proliferation and the correct functioning of cells adhered to the scaffold, it is of utmost importance to verify material bio-compatibility in a biological medium. The applied protocol to study the cell cycle is the following:

1. After contact with hydrogels, harvest cells by trypsinization.
2. Centrifuge for 5 minutes at 2,500 rpm.
3. Fix overnight with 100% ethanol at −20°C.
4. Incubate with 5 μg mL^{-1} PI/PBS (propidium iodide in PBS solution) and 0.015 U mL^{-1} RNase in PBS during 20 minutes at room temperature in the dark.

5. Analyze fluorescence intensity of PI in a flow cytometer with excitation and emission wavelengths setting at 484 and 500 nm, respectively.

For certain applications, it is necessary to analyze the distribution of actin filaments in cells that are growing on or inside hydrogels. Phalloidin is a highly selective bicyclic peptide used for staining actin filaments.[11]

The applied protocol for phalloidin staining is the following:

1. Rinse cell-hydrogel sample 3X for 5 minutes with PBS.
2. Fix gels by incubation in 4% formaldehyde for 30 minutes at room temperature.
3. Rinse constructs 3X with a blocking solution containing 3% (w/v) BSA (bovine serum albumin protein) and 0.5% (w/v) Tween in PBS.
4. Permeabilize membrane with 0.25% (w/v) Triton X in the blocking solution for 20 minutes.
5. Rinse constructs 3X with blocking solution.
6. Stain for F-actin with 0.66 μg mL^{-1} FITC (isothiocyanate de fluoresceine) or rhodamine phalloidin in blocking solution for 2 hours at room temperature.
7. Rinse 3X with a blocking solution.
8. Stain cellular nuclei using 0.4 μL mL^{-1} DAPI (4 ',6-diamidino-2-fenilindol) in PBS for 20 minutes.
9. Rinse 3X with PBS.
10. Visualize cells by confocal microscopy.

14.3   CELL/HYDROGEL BIOINTERFACE

The biointerface is defined as the zone of contact between the scaffold surface and the living system. The cell adhesion, proliferation, and differentiation could be different according to interactions present in the interface. The same cell line can adopt a tapered, shrined, or agglomerated shape in contact with different materials.[12] If the cells tend to quickly adapt over scaffold, there is greater probability of tissue growth but if the cells tend to take a shrined or agglomerated shape, the hydrogel could be used as scaffold of cell 3D systems, which allows mimicking early the environment conditions of a living system. Another possibility to develop 3D cell growth is using a 3D hydrogel scaffold[13] or microencapsulation.[14]

Physicochemical, mechanical, and morphological properties of biocompatible hydrogels can be regulated by synthetic methods in order to simulate the characteristics of an extracellular matrix (ECM). A cell scaffold can be constituted by synthetic, natural, or both kinds of polymers combined. In this way, it is possible to provide a platform of biocompatible scaffold materials that mimic the native cellular environment. For that, it is needed to know the behavior of living systems in contact with each scaffold material.

14.3.1 BIOINTERFACIAL WETTABILITY AND CHEMICAL COMPOSITION OF THE SCAFFOLD

One of the superficial properties that defines cell adhesion and adaptation is wettability. This can be tailored by including polymeric materials with polar non charged (e.g., -OH) or ionic functional groups, to achieve the best biointerfacial conditions for cells to adapt to the scaffold.[12] Wettability of some materials called "smart polymers" considerably change with the environment conditions (temperature, ionic force, solvent polarity, etc.).[12] These responsive materials are allowed to control the cell attach/detach condition through environment changes, demonstrating that the cell adhesion process depends on the biointerfacial characteristics. There are thermosensitive scaffolds that exhibit a collapse phenomenon, for instance at human body temperature (37°C), exposing hydrophilic functional groups, particularly ionic groups resulting from electrostatic repulsions. Consequently, this altered surface arrangement modifies the scaffold's superficial wettability. Generally, the wettability changes should be analyzed by measuring contact angles in water at room temperature and cell culture conditions, 37°C, and culture medium prior to cell seeding. Thus, the biologic behaviors of cell-hydrogel systems (such as adhesion, proliferation, nuclear, and cytoplasmatic morphology) can be related to the biointerfacial properties.

Poly-N-isopropylacrylamide (PNIPAM) is one of the most used polymeric materials in the biomedical field due to its biocompatibility, thermosensibility, and soft texture, among other properties. It is possible to modify the superficial characteristics of a PNIPAM hydrogel, which will be in contact with a living system, by synthetic methods (e.g., copolymerization) or interpenetration with different polymers. Previous research has demonstrated that PNIPAM hydrogels can be modified by the addition of neutral and ionic comonomers such as N-acryloyl-tris-(hydroxymethyl)aminomethane (THMA), 2-acrylamido-2-methylpropanesulfonic acid (AMPS), or (3-acrylamidopropyl)trimethyl ammonium chloride solution (APTA). Thus, changes in the wettability, phase transition temperature, and swelling capacity are observed for the newly obtained materials. Table 14.1 shows information about cell lines in contact with materials based on modified PNIPAM hydrogels. In some cases, the cell goes adapting the scaffold surface with the culture days, taking a tapered typical morphology, but in other cases, the surface is rejected by the cell. It was also demonstrated that bovine fetal fibroblast (BFF) cells produce collagen to improve the adhesion on the surface, which allows the development of BFF tissue inside the 3D structure of PNIPAM.[13] At the same time, cell proliferation inside the scaffold and elastic modulus changes of the whole can be observed during cell growth. HEK296 cells form agglomerates like 3D organelles during the first five culture days, time enough to apply an in-vitro therapeutic treatment. These cell organelle systems allow the study of *in vitro* therapy treatments in order to select the more suitable to be applied as a possible *in vivo* treatment.

Therefore, the biologic behavior of different cell lines on scaffolds is a function of the wettability, surface characteristics, and exposition or adaptation time. It has also been demonstrated that most cells adhere better to moderately hydrophilic surfaces and dislodge from extremely hydrophilic surfaces.[17] Even though a surface

TABLE 14.1

Biointerfacial Behavior of Cellular Lines in Contact with Hydrogels based on PNIPAM and Copolymers, at Culture Conditions (37°C/DMEM Medium)

Living System	Scaffold	Adhesión/Morphology
BFF	Polystyrene	Typical flat morphology is adopted.[15]
	PNIPAM	Typical flat morphology is adopted.[12,15]
	PAAm	Clump shape with tendency for adhesion at 2 culture days. After 5 culture days, cells tend to adopt the typical flat shape with high cell confluence.[12]
	PNIPAM-co-2%AMPS	Clump shape at 24 h culture with tendency for adhesion until 5 culture days with high cell confluence.[15]
	PNIPAM-co-10%AMPS	Clump shape with tendency for adhesion until 5 culture days and with high cell confluence is observed.[15]
	PNIPAM-co-10%THMA	Typical flat shape is adopted.[15]
	PNIPAM-co-20%THMA	Clump shape with tendency for adhesion and adopt a typical flat morphology (until 15 culture days).[12]
	PNIPAM-co-10%APTA	Typical flat shape is adopted.[12]
3T3-L1 A549	PNIPAM	Typical flat shape is adopted. High cell confluence after 5 culture days.[16]
HEK293	PNIPAM	These cells have stronger cell-cell cohesion and like-cluster growth (agglomerate cells) until 5 culture days. After, these adapt, adhere, and proliferate. High cell confluence after 5 culture days.[8, 16]
	PNIPAM-co-20%THMA	Typical epithelial and flattened morphology is adopted at 5 culture days. High cell confluence.[8]
	PNIPAM-co-3%APTA	Typical epithelial and flattened morphology is adopted. Rounded cells are also observed after 5 culture days perhaps high cell confluence.[8]
	PNIPAM-co-10%APTA	Tendency to agglomerate. Cells reject surface with higher wettability. Nuclear morphology changes are observed.[8]
	PNIPAM-co-2%AMPS	Typical flat shape is adopted. Rounded cells are also observed after 5 culture days perhaps high cell confluence.[8]
	PNIPAM-co-10%AMPS	Tendency to agglomerate. Cells reject surface with higher wettability. Nuclear morphology changes are observed.[8]

with ionic charges increases the wettability of the interface, the high number of charges leads to low cell adhesion.

Hence, there is a situation of compromise between the chemical composition and the biocompatibility of scaffold material.[9] Noteworthy, these studies must be parallelly complemented with immune responsiveness of each cell-material system.

Cells: Bovine fetal fibroblasts (BFFs), murine preadipose cells (3T3-L1), human embryonic kidney cells (HEK293), human carcinoma-derived cells (A549).

Monomers: N-isopropylacrylamide (NIPAM), acrylamide (AAm), 2-acrylamido-2-methylpropanesulfonic acid (AMPS), N-acryloyl-tris-(hydroxymethyl)aminomethane (THMA), 3-acrylamidopropyl)trimethyl ammonium chloride solution (APTA).

14.4   *IN VIVO* ASSESSMENT OF HYDROGELS' BIOCOMPATIBILITY

In vivo biocompatibility of foreign material to a living system can be defined from a healthcare perspective based on the following criteria: capacity to cause adverse immunological rejection in the host; to avoid carcinogenic reactions; to be biostable or durable time while fulfilling the function it performs; and in case of being biodegradable, that the resulting chemical products are eliminated by the organs involved in detoxification without causing systemic or local toxicity.[18] The tissue microenvironment is specific to each organ and is essential for its anatomic-physiological development; hence, the biocompatibility for a particular tissue varies according to the nature, chemical composition, and three-dimensional structure of each material. An exhaustive analysis of the tissue biological response to the chosen hydrogel is extremely important for the design of a biomaterial with potential biomedical applications. Hydrogels as scaffolds for cell growth and differentiation represent a relevant advance in the field of tissue engineering due to the possibility of reducing the wait time for a transplant, and thus minimizing the rates of mortality and the risk of immune rejection. Tissues and organs delineated by these biomedical strategies could avoid complications associated with allogeneic transplants, allowing faster recovery and improving the patient's quality of life.[19]

Organs and tissues have a great capacity to remodel and regenerate damaged areas caused by serious injuries or trauma. Tissue regeneration refers not only to anatomical but also functional recomposition, with the new tissue being very similar to the original and avoiding repair processes with too many collagen deposits and other components of the ECM.[20] Nevertheless, not all organs have a similar proliferative capacity due to several reasons, such as the complexity of the organ, the size of the caused damage, or the age of the organism. Thus, the medicine should decide to carry out transplants or prostheses. In the tissue engineering field, cell and developmental biology, medical sciences, and materials chemistry are combined according to the therapeutic approach from replacement to the stimulation of local tissue regeneration. Tissue engineering seeks to provide structural integrity and biochemical signals through the use of biomaterials, with dimensional characteristics equal to the desired tissue microenvironment.

The events of cell migration and differentiation that lead to tissue development and its various functions are accompanied by the continuous and dynamic remodeling of the ECM.[21] A key concept in this field of study is the proper selection and design of a biomaterial able to function as an optimal 3D support system, where the deposit of ECM, the free circulation of nutrients and wastes, and cell interconnection. In this way, the desired cell proliferation and distribution during the lifetime of the material are favored. Several techniques adopt a bottom-up approach, being 3D bioprinting a novel and promising technology capable of creating high-precision geometric constructions, delimiting the size and arrangement of the pores. In some cases, bioprinting includes the cells and growth factors within the matrix during the printing process, replicating the architecture of native tissues. Implants created with this technology hold great promise, but challenges remain for the development of organs with optimal physiology. On the other hand, currently the most used tissue engineering technique employs a top-down approach in which cells are harvested and seeded onto (or inside) a prefabricated scaffold. The appropriate *in vitro* conditions are established and the cells are exposed to biochemical signals and mechanical stimulations to achieve the formation of the desired tissue. This technique presents great advantages, such as generating scaffolds with good mechanical properties and correct migration within the matrix and homogeneous distribution, but it is difficult to achieve experimentally.[22]

Moreover, the macroscopic shape of the scaffold could be determined by a mold made by a well-established 3D printing (e.g., filament forming molding, FFM) technique. The use of *in vivo* biomaterials implies the study of the interactions with the host's immune system as a relevant variable that determines the efficacy of its application. The interface of the biomaterial, the recognition by cells, and interstitial fluid surrounding the insertion site determine the responses that can be triggered, such as inflammation, wound healing, and immunological reaction/immunotoxicity. A well-established sequence of reactions may occur after surgery: wound, hematoma, acute inflammation, chronic inflammation, and granular tissue. Regarding the inflammatory reaction, it can be acute with consequent tissue regeneration or chronic with evident immunological rejection, depending on the implantation area and the elapsed time. After implantation of biocompatible material, a minimal inflammatory response should be obtained after two to three weeks, with the formation of fibrotic tissue surrounding the implant, without the development of progressive fibrosis or adjacent inflammation, which is a typical result compared to surgical trauma. Few macrophages may be present in the area, with no indication of a local or systemic adverse response. If the hydrogel and/or degradation products are pro-inflammatory, the host's immune system will respond by triggering a foreign body reaction, an ancestor response to implant rejection, where monocytes and macrophages form multinucleated giant cells in an attempt to phagocytose the implant and release defensive oxidizing agents. For this reason, it is important to recover the implants after the passage of time and evaluate them through histological assays, observing the presence or absence of a fibrous capsule, the neovascularization of the area, and the stability or degradation of the material. Figure 14.3 represents the strategies that can be carried out for the *in vivo* biocompatibility analysis of biomaterials. Thus, it is possible to verify whether the material is suitable for biomedical applications.

Macrophages are cells belonging to the immune system with a key behavior in immune rejection. Depending on the microenvironment, macrophages have the

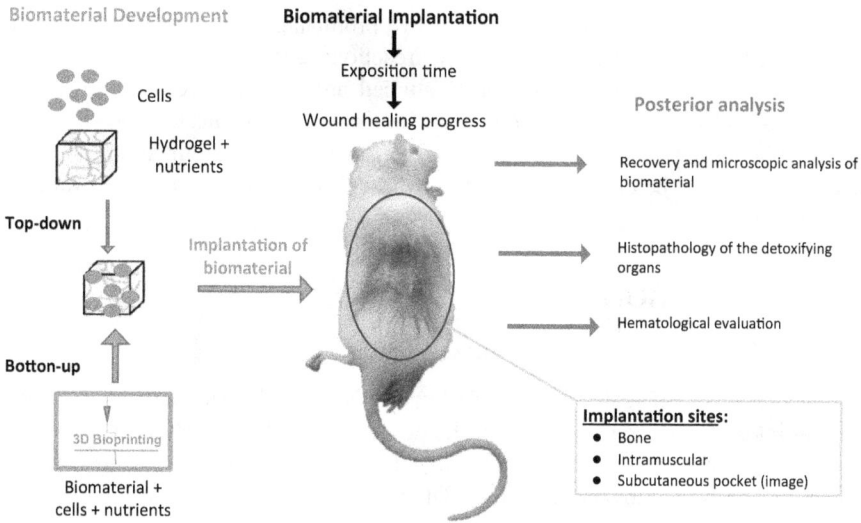

FIGURE 14.3 Strategies for the *in vivo* biocompatibility analysis of implanted materials before biomedical application.

ability to polarize into classic activated macrophages (M1), which exhibit pro-inflammatory properties that contribute to the destruction of foreign agents, and alternatively in activated anti-inflammatory macrophages (M2), which are involved in wound healing and in tissue regeneration. Hence, macrophages are associated with immune response processes and are polarized based on their interaction with surface receptors, growth factors, cytokines, and other substances present in the tissue microenvironment.

Polymeric matrices could trigger the release of soluble agents and signal molecules from cells adjacent to the implantation and from immune cells that ultimately control the wound healing process. One habitual approach to reduce inserted tissue rejection is the use of the patient's own stem cells, and differentiate them with soluble factors inserted into the material. When considering biomaterials and the possibility of immunological rejection, biotransformation or metabolization reactions and the enzymatic processes triggered for their neutralization and/or elimination must be taken into account. Hepatic and renal tissues are organs with metabolic and excretory functionality, in which systemic toxicity can be generated by implant constituents that can be released.[23] In addition, hemocompatibility tests evaluate systemic effects in the body, in particular on the blood and/or blood components. The substances obtained from the degraded materials could travel through this fluid until they reach metabolic or detoxifying organs, and blood cells could be potential targets for cell damage. For this reason, the analysis of the organ anatomy, the quantification of enzymes and products indicative of hepatocellular injury/repair and renal functional alteration, and the determinations of blood parameters become relevant in the biocompatibility study of materials. In the case of rejection, the degree of toxicity and the consequences on animal homeostasis must be determined.

In conclusion, hydrogel biomaterials have promising potential to be applied in the biomedical field. The host immune reactions and the presence/absence of cytotoxic effects need to be previously studied and characterized to ensure their success as scaffolds. Due to the complexity of tissues and organs and the variability of existing hydrogels, a general theory of the functioning of biomaterials and their acceptance cannot be developed, and the hydrogel-tissue system must be evaluated individually.

14.5 APPLICATIONS

14.5.1 NANOMEDICINE APPLICATIONS

Among the different scales and morphologies of hydrogels that can be synthetized, special interest has been dedicated to the development of nanogels. These are high molecular weight cross-linked polymers, yielding dispersible particles with sizes in the range between 20 and 200 nm. There are mainly two approaches for the obtention of nanogels: i) cross-linking of polymeric precursors to form nanoparticles and ii) synthesis of a nanogel network by heterogeneous polymerization of monomers onto nuclei.[24] Table 14.2 summarizes the different cross-linking strategies of polymeric precursors and several ways to obtain nanogels in heterogeneous systems.

Nanogels are highly used in biomedical applications including tissue engineering, living cell encapsulation, and controlled drug delivery.[33]

Particularly stimuli-sensitive nanogels based on "smart hydrogels" are polymeric macro architectures consisting of a cross-linked, three-dimensional network that can respond to local environmental conditions. They can swell, shrink, or degrade in response to external stimuli such as temperature, pH, ionic strength, and electrical or magnetic fields. This capability has been explored since the last decade to develop drug delivery systems for different biomedical applications, including cancer therapy, infectious diseases, vaccines, etc.[34]

There are several mechanisms to trigger the release of the active principles from smart nanogels, including pH-responsive mechanism, thermosensitive and volume

---

TABLE 14.2

Synthetic Procedures of Nanogels in Heterogeneous Systems

Synthesis Methods	Procedures
Crosslinking of polymeric precursors	Click chemistry[25]
	Disulfide-based cross-linking[26]
	Photo-induced cross-linking[27]
	Physical cross-linking[28]
Heterogeneous polymerization of monomers	Miniemulsion[29]
	Microemulsion[30]
	Precipitation polymerization[31,32]

transition mechanism, and degradation of disulfide linkages, among others. pH-responsive nanogels have attracted great attention due to the relevance of pH in the gastrointestinal route in oral drug delivery applications.[35] In this mechanism, pH-responsive nanogels can protect their cargo through the acidic stomach (pH 2) and release it at the neutral intestine (pH 7.4); most of the examples involve carboxyl-containing precursors. The characteristic temperature that produces changes in the volume of thermoresponsive nanogels is known as volume phase transition temperature (VPTT) and is used to trigger the release of therapeutic cargo at the action site. This release mechanism is extensively used in infected tissues (i.e., tumor cells), since the higher metabolic rate in tumor cells increases their temperature. A nanogel that releases its cargo when it reaches a VPTT will deliver a drug (e.g., an antitumoral chemical) close to the infected cell, but not to normal.[36] Thermoresponsive nanogels are used in transdermal drug delivery, taking advantage of the gradient temperature of the different layers of the skin.[37] Nanogels that present disulfide cross-linking can degrade in a reducing environment. This mechanism is mostly used for intracellular drug delivery, where the degradation of the nanogels and release of the therapeutic agent is caused by the presence of reductive glutathione (GSH) inside cells.[38]

More interestingly, the particulate nature of nanogels allows further functionalizing at different sites, creating multifunctional materials, where the carrier is transformed from a simple vehicle to actively participate in the desired therapy. Thus, introducing the "combine-synergistic therapy" idea or even the "theranostic" approach.[39,40] Nanotheranostic is defined as a combination of therapeutic and diagnostic tools or imaging into one system. In this regard, nanocomposites based on nanogels and nanomaterials are being explored.

While great advances have been made in the last years on nanogel research for biomedical applications, few developments have been tested in clinical studies. There is an urgent need for relevant clinical data and to overcome unsolved issues related to the effects that could be caused in cell metabolism, pharmacodynamics, and pharmacokinetics. These responses will allow the transition from clinical trial to clinical application.

14.5.2 Mammalian Sperm Selection by Attachment to Hydrogel Surfaces

Assisted reproductive techniques (ARTs) are increasingly used to achieve successful reproduction in humans and animals. Therefore, there is a need to develop more precise and efficient techniques to select gametes for ART. In particular, males ejaculate a heterogeneous population of spermatozoa regarding maturation, morphological, and functional features. Furthermore, many ART procedures rely on the use of refrigerated or frozen spermatozoa, which include additives called extenders. These must be removed before performing the procedure. *In vivo*, the female reproductive tract acts as an efficient "filter" for spermatozoa, allowing it to arrive at the site of fertilization for those cells with higher fertilizing ability, which maximizes successful reproduction. However, during ART, female *in vivo* selection mechanisms are obviated in different degrees; therefore, there is real concern about the possibility of using spermatozoa with suboptimal fertilization and embryo development potential. Experimental data indicate that normal sperm morphology is not necessarily associated with DNA integrity,[41] which raises concerns about potential transmission of DNA alterations to the next generations. This is especially true during the intra-cytoplasmic sperm injection (ICSI), in which a single spermatozoon is mechanically introduced into the mature oocyte, obviating all natural sperm selection processes.

Sperm cell preparation techniques routinely used for ART are density gradient centrifugation (DGC) or swim-up. These techniques are based on sedimentation or sperm migration, respectively, to separate the best cells. Both methods are efficient in selecting discrete numbers of morphologically normal cells with a high degree of nuclear maturity. However, the development of more refined techniques that permit the identification of the best sperm at the single-cell level is needed. Different sperm selection protocols for ART are being developed to improve conventional selection techniques.[42]

Polymeric synthetic materials based on hydrogels with the capacity of absorbing vast amounts of water are being intensively investigated to characterize their physicochemical properties and potential on-field applications in regenerative medicine. PNIPAM-based hydrogels are being used for various *in vitro* cell culture applications. The innocuity of these materials for different types of mammalian cells has been also demonstrated.[12, 15, 16] In addition, the softness of hydrogels, plus their hydrophilic character, closely resembles the extracellular matrix of living tissues, which facilitate the best biointerfacial connection. It was demonstrated that mammalian cells can grow attached to the surface of microporous PNIPAM hydrogels, and even colonize the internal portions of macroporous hydrogels.[13] These results sparked interest in investigating the possible interaction between hydrogel surfaces and frozen-thawed bull spermatozoa. Preliminary experiments showed that a subpopulation of bull sperm attached to the hydrogel surface with adhered cells had better quality parameters was hypothesized. Initially, bull sperm were cultured with different hydrogels, i.e., PNIPAM, PNIPAM-co-20%HMA, PNIPAM-co-10%AMPS, or PNIPAM-co-10%APTA. Depending on the hydrogel type, the percentage of sperm attached to the hydrogel surface with regard to the total initial number of sperm put in coculture with the hydrogel ranged from 15% to 40%. Higher percentages were determined in non-ionic hydrogels, such as PNIPAM and PNIPAM-co-20%HMA, in which almost 40% of sperm were attached after 15 minutes.[43] Sperm attachment to cationic PNIPAM-co-10%APTA reached 30%. Much lower percentages of sperm attachment were observed (<15%) on anionic PNIPAM-co-10%AMPS.[41] These results suggest that surface charges could be a factor involved in sperm-hydrogel attachment. Based on the known negative Z potential of the sperm surface, electrostatic attraction to positively charged surfaces may explain, at least in part, sperm adhesion to hydrogels. Furthermore, sperm attachment to PNIPAM-co-20%HMA might be mediated by hydrogen bonding interactions between sperm glycocalyx and hydrophilic groups present on the hydrogel surface.

In order to study the population of sperm that are attached to the surface, and in the future use these sperm in ART, it is convenient that the cells detach from the polymeric surface. To this end, an experiment to achieve this objective was designed. Initially, treatments that had been successful in releasing sperm bound to oviductal epithelial cells *in vitro* were tested. For instance, experimental evidence indicates that removing bivalent cations from a culture medium induces detachment of bull sperm from oviductal cell monolayers.[44] Similarly, a significant number of spermatozoa were released from oviductal cells when sperm capacitation was induced chemically by heparin.[42] The same conditions used in these reports were re-created in our sperm-hydrogel setting; however, despite testing different treatments, sperm release from the hydrogel was negligible.

Besides direct interaction of sperm with hydrogel components, hydrogels could act as support of macromolecules that favor the sperm attachment. One such species is hyaluronic acid, a polymer broadly distributed in animal tissues and implicated in many reproductive processes.[45] Sperm-hyaluronic acid interaction has been reported and a method to select mature spermatozoa based on cell attachment to polystyrene-immobilized hyaluronic acid has been developed (PICSI® Dish for Sperm Selection, CooperSurgical Fertility Solutions, Denmark). This method can be used to identify mature spermatozoa (i.e., those that express HA receptors) from a human semen sample for use in ICSI.[46]

Thus, a surface based on PNIPAM-co-20%HMA semi-interpenetrated with HA of low molecular weight was generated.[41,47] Gravimetric experiments confirmed that the HA were semi-interpenetrated into the hydrogel network and the contact angle assay evidenced that HA polymers were exposed on the hydrogel surface, due to a reduction of contact angle by 10° compared with the same hydrogel without HA.

Motile bull sperm attached to HA semi-interpenetrated hydrogels and when treated with hyaluronidase, almost 47% of attached sperm cells are released from the hydrogel. It is plausible that the enzymatic action of hyaluronidase on HA is responsible for the higher percentage of released sperm from the semi-interpenetrated hydrogel. The biological characteristics of the attached/released sperm fraction were also investigated. Released bull spermatozoa had 70% progressive motility and scored 4 for vigor.[41] The higher motility observed in sperm released from HA semi-interpenetrated hydrogels could be due to hyaluronic fragments adhered on sperm heads, which in turn can trigger intracellular pathways, leading to enhanced motility. Furthermore, sperm released from semi-interpenetrated hydrogels had a significantly higher percentage of viable cells than those in the unbound fractions and raw sperm sample. Therefore, the cell viability is conserved when sperm adhered to hydrogel surfaces. This result also confirms the biocompatibility of synthetic materials based on PNIPAM.

A similar selection process mediated by sperm attachment to hydrogel surfaces has been demonstrated in porcine[48] and equine species.[49] Overall, attached/released sperm fraction had acceptable functional characteristics for use in ART. In this way, these synthetic and natural combined materials can be proposed for active sperm selection as part of ART programs.

14.6 CONCLUSIONS

Today, biocompatible hydrogels are being extensively used in several biomedical applications. Noteworthy is the selection of biomaterial implies interdisciplinary hard work. From the point of view of the material chemical, there are several known strategies to design materials with 2D, 3D, or nano structures, and to adapt them with the desired properties for a well-developed tissue. From the cell biology point of view, it is important to achieve the adaptation and develop tissue, and at the same time, an acceptable immunology response to the material is expected. It is very important to take into account that each cell line can respond differently to the same biocompatible material, so the cell-material behavior is not predictable.

Therefore, this is a research area that is booming and continually developing in order to contribute positively to nano and biomedicine to address key challenges in healthcare.

NOTES

1 Memic A., Colombani T., Eggermont L.J., Rezaeeyazdi M., Steingold J., Rogers Z.J., Navare K.J., Mohammed H.S, Bencherif, S.A. Latest advances in cryogel technology for biomedical applications. Advanced Therapeuthics 2, 2019, 1800114.
2 Nazir F., Tabish T.A, Tariq F., Iftikhar S., Wasim R., Shahnaz G. Stimuli-sensitive drug delivery systems for site-specific antibiotic release. Drug Discovery Today, 27, 2022, 1698–1705.
3 Bello A.B., Kim D., Kim D., Park H., Lee S.H. Engineering and functionalization of gelatin. Biomaterials: From cell culture to medical applications. Tissue Engineering Part B: Reviews, 26, 2020, 164–180.
4 Zhang B., Su Y., Zhou J., Zheng Y., Zhu D. Toward a better regeneration through implant-mediated immunomodulation: Harnessing the immune responses. Advanced Science 8, 2021, 2100446.
5 Naahidi S., Jafari M., Logan M., Wang Y., Yuan Y., Bae H., Dixon B., Chen P. Biocompatibility of hydrogel-based scaffolds for tissue engineering applications. Biotechnology Advances 35, 2017, 530-544.
6 Anderson J. Future challenges in the in vitro and in vivo evaluation of biomaterial biocompatibility. Regenerative Biomaterials, 3, 2016, 73–77.
7 Karajanagi S.S, Yoganathan R., Mammucari R., Park H., Cox J., Zeitels S.M., Langer R., Foster N.R. Application of a dense gas technique for sterilizing soft biomaterials. Biotechnology and Bioengineering, 108, 2011, 1716–1725.
8 Capella V., Liaudat A.C, Broglia M.F., Barbero C.A., Bosch P., Rodríguez N., Rivarola C.R. Biointerfacial analysis of HEK293 cells in contact with hydrogels based on Poly-N-isopropylacrylamide and copolymers. Advanced Materials Interfaces, 9, 2022, 2201162.
9 Gilarska A., Lewandowska-Łańcucka J., Horak W., Nowakowska M. Collagen/chitosan/hyaluronic acid - based injectable hydrogels for tissue engineering applications - Design, physicochemical and biological characterization. Colloids and Surfaces B: Biointerfaces, 170, 2018, 152–162.
10 Dominijanni A.J., Devarasetty M., Forsythe S.D., Votanopoulos K.I., Soker S. Cell viability assays in three-dimensional hydrogels: A comparative study of accuracy. Tissue Engineering Part C: Methods, 27, 2021, 401–410.
11 Khetan S., Burdick J. Cellular encapsulation in 3D hydrogels for tissue engineering. Journal of Visualized Experiments, 32, 2009, e1590.

12 Rivero R., Alustiza F., Capella V., Liaudat C., Rodríguez N., Bosch P., Barbero C., Rivarola C. Physicochemical properties of ionic and non-ionic biocompatible hydrogels in water and cell culture conditions: Relation with type of morphologies of bovine fetal fibroblasts in contact with the surfaces. Colloids and Surfaces B: Biointerfaces 158, 2017, 488–497.

13 Rivero R.E., Capella V., Liaudat A.C., Bosch P., Barbero C.A., Rodríguez N., Rivarola C.R. Mechanical and physicochemical behavior of a 3D hydrogel scaffold during cell growth and proliferation. RSC Advances, 10, 2020, 5827–5837.

14 Lopez-Mendez T.B., Santos-Vizcaino E., Pedraz J.L., Orive G., Hernández R.M. Cell microencapsulation technologies for sustained drug delivery: Latest advances in efficacy and biosafety. Journal of Controlled Release, 335, 2021, 619–636.

15 Rivero R.E., Alustiza F., Rodríguez N., Bosch P., Miras M.C., Rivarola C.R., Barbero C.A. Effect of functional groups on physicochemical and mechanical behavior of biocompatible macroporous hydrogels. Reactive and Functional Polymers 97, 2015, 77–85.

16 Capella V., Rivero R.E., Liaudat A.C., Ibarra L.E., Roma D.A., Alustiza F., Mañas F., Barbero C.A., Bosch P., Rivarola C.R., Rodríguez N. Cytotoxicity and bioadhesive properties of poly-N-isopropylacrylamide hydrogel. Heliyon 5, 2019, e01474.

17 Gao L., Gan H., Meng Z., Gu R., Wu Z., Zhang L., Zhu X., Sun W., Li J., Zheng Y. Effects of genipin cross-linking of chitosan hydrogels on cellular adhesion and viability. Colloids Surfaces B: Biointerfaces 117, 2014, 398–405.

18 Naahidi S., Jafari M., Logan M., Wang Y., Yuan Y., Bae H., Dixon B., Chen P. Biocompatibility of hydrogel-based scaffolds for tissue engineering applications. Biotechnology Advances, 35, 2017, 530–544.

19 Bhat S., Kumar A. Biomaterials and bioengineering tomorrow's healthcare. Biomatter, 3, 2013, e24717.

20 Colazo J.M., Evans B.C., Farinas A.F., Al-Kassis S., Duvall C.L., Thayer W.P. Applied bioengineering in tissue reconstruction, replacement, and regeneration. Tissue Engineering Part B: Reviews, 25, 2019, 259–290.

21 Hussey G.S., Dziki J.L., Badylak S.F. Extracellular matrix-based materials for regenerative medicine. Nature Reviews, 3, 2018, 159–173.

22 Ebrahimi M., Editors: Inamuddin, Abdullah M. Asiri, Ali Mohammad, Applications of Nanocomposite Materials in Orthopedics: Biomimetic Principle for Development of Nanocomposite Biomaterials in Tissue Engineering, 2019. Swaston, UK, Woodhead Publishing.

23 Meyer D.J. Harvey J.W. Veterinary Laboratory Medicine: Interpretation & Diagnosis. 3rd ed., 2004, United States Saunders.

24 (a) Hamzah Y.B., Hashim S., Abd Rahman W.A.W. Synthesis of polymeric nano/microgels: A review. Journal of Polymer Research. 24, 2017, 134–143 (b) Mauri E., Giannitelli S.M., Trombetta M., Rainer A. synthesis of nanogels: Current trends and future outlook. Gels 7, 2021, 36.

25 (a) Wang J., Wang X., Yan G., Fu S., Tang R. pH-sensitive nanogels with ortho ester linkages prepared via thiol-ene click chemistry for efficient intracellular drug release. Journal of Colloid and Interfaces Science, 508, 2017, 282–290. (b) Zhang Y., Andrén O.C.J., Nordström R., Fan Y., Malmsten M., Mongkhontreerat S., Malkoch M. Off-stoichiometric thiol-ene chemistry to dendritic nanogel therapeutics. Advanced Functional Materials, 29, 2019, 1806693. (c) Zhang Y., Ding J., Li M., Chen X., Xiao C., Zhuang X., Huang Y., Chen X. One-step "click chemistry"-synthesized cross-linked prodrug nanogel for highly selective intracellular drug delivery and upregulated antitumor efficacy. ACS Applied Materials and Interfaces, 8, 2016, 10673–10682.

26 (a) Zhang X., Malhotra S., Molina M., Haag R. Micro- and nanogels with labile crosslinks – from synthesis to biomedical applications. Chemical Society Reviews, 44,

2015, 1948–1973 (b) Zhou W., Yang G., Ni X., Diao S., Xie C. Recent advances in crosslinked nanogel for multimodal imaging and cancer therapy. Polymers 12, 2020, 1902.

27 (a) Chen W., Hou Y., Tu Z., Gao L., Haag R. pH-degradable PVA-based nanogels via photo-crosslinking of thermo-preinduced nanoaggregates for controlled drug delivery. Journal of Controlled Release 259, 2017, 160–167. (b) Kim J., Gauvin R., Yoon H.J., Kim J.-H., Kwon S.-M., Park H.J., Baek S.H., Cha J.M., Bae H. Skin penetration-inducing gelatin methacryloyl nanogels for transdermal macromolecule delivery. Macromolecular Research, 24, 2016, 1115–1125.

28 Moshe H., Davizon Y., Menaker Raskin M., Sosnik A. Novel poly(vinyl alcohol)-based amphiphilic nanogels by non-covalent boric acid crosslinking of polymeric micelles. Biomaterials Science, 5, 2017, 2295–2309.

29 Sarika P.R., Anil Kumar P.R., Raj D.R., James N.R. Nanogels based on alginic aldehyde and gelatin by inverse miniemulsion technique: Synthesis and characterization. Carbohydrate Polymers, 19, 2015, 118–125.

30 McAllister K., Sazani P., Adam M., Cho M.J., Rubinstein M., Samulski R.J., DeSimone J.M. Polymeric nanogels produced via inverse microemulsion polymerization as potential gene and antisense delivery agents. Journal of American Chemical Society, 124, 2002, 15198–15207.

31 Pérez M.LS., Funes J.A., Flores C., Ibarra K.E, Forrellad M.A., Taboga O., Cariddi L.N., Salinas F., Ortega H.H., Alustiza F., Molina M. Development and biological evaluation of pNIPAM-based nanogels as vaccine carriers. International Journal of Pharmaceutics, 630, 2022, 122435.

32 Theune L.E., Buchmann J., Wedepohl S., Molina M., Laufer J., Calderón M. NIR- and thermo-responsive semi-interpenetrated polypyrrole nanogels for imaging guided combinational photothermal and chemotherapy. Journal of Controlled Release, 311–312, 2019, 147–161.

33 Kaewruethai T., Laomeephol C., Pan Y., Luckanagul J.A. Multifunctional polymeric nanogels for biomedical applications. Gels. 7, 2021, 228.

34 Preman N.K., Jain S., Johnson R.P. "Smart" polymer nanogels as pharmaceutical carriers: A versatile platform for programmed delivery and diagnostics. ACS Omega 6, 2021, 5075–5090.

35 Zhao Q., Zhang S., Wu F., Li D., Zhang X., Chen W., Xing B. Rational design of nanogels for overcoming the biological barriers in various administration routes. Angewandte Chemie International Edition, 60, 2021, 14760–14778.

36 Ghaeini-Hesaroeiye S., Bagtash H.R., Boddohi S., Vasheghani-Farahani E., Jabbari J. Thermoresponsive nanogels based on different polymeric moieties for biomedical applications. Gels, 6, 2020, 20.

37 Witting M., Molina M., Obst K., Plank R., Eckl K.M, Hennies H.C., Calderón M., Frieß W., Hedtrich S. Thermosensitive dendritic polyglycerol-based nanogels for cutaneous delivery of biomacromolecules. Nanomedicine: NBM 11, 2015, 1179–1187.

38 Sousa-Herves A., Wedepohl S., Calderón M. One-pot synthesis of doxorubicin-loaded multiresponsive nanogels based on hyperbranched polyglycerol. Chemical Communications, 51, 2015, 5264–5267.

39 Chambre L., Degirmenci A., Sanyal R., Sanyal A. Multi-functional nanogels as theranostic platforms: Exploiting reversible and nonreversible linkages for targeting, imaging, and drug delivery. Bioconjugate Chemistry, 29, 2018, 1885–1896.

40 Vijayan V.M., Vasudevan PN, Thomas, V. Polymeric nanogels for theranostic applications: A mini-review. Current Nanoscience, 16, 2020, 392–398.

41 Avendaño C., Oehninger S. DNA fragmentation in morphologically normal spermatozoa: How much should we be concerned in the ICSI era? Journal of Andrology, 16, 2020, 392–398.

42 Ortega N., Bosch P. Editor S. Friedler. In Vitro Fertilization – Innovative Clinical and Laboratory Aspects, Methods for Sperm Selection for In Vitro Fertilization, 2012, InTechOpen, pp. 71–76.

43 Blois D.A., Liaudat A.C., Capella V., Morilla G., Rivero R.E., Broglia M.F., Barbero C.A., Rodríguez N., Bosch P., Rivarola C.R. Interaction between hyaluronic acid semi-interpenetrated hydrogel with bull spermatozoa: Studies of sperm attachment–Release and sperm quality. Advanced Materials Interfaces, 8, 2021, 2101155.

44 Bosch P., de Avila J.M., Ellington J.E., Wright Jr. R.W. Heparin and Ca^{2+} free medium can enhance release of bull sperm attached to oviductal epithelial cell monolayers. Theriogenology, 56, 2000, 247–260.

45 Bergqvist A.S., Yokoo M., Heldin P., Frendin J., Sato E., Rodríguez-Martínez H. Hyaluronan and its binding proteins in the epithelium and intraluminal fluid of the bovine oviduct. Zygote, 13, 2005, 207–218.

46 Erberelli R.F., Salgado R.M., Pereira D.H., Wolff P. Hyaluronan-binding system for sperm selection enhances pregnancy rates in ICSI cycles associated with male factor infertility. JBRA Assist Reproduction, 21, 2017, 2–6.

47 Liaudat A.C., Blois D., Capella V., Morilla G., Rivero R., Barbero C., Rodríguez N., Rivarola C., Bosch P. Short communication: Bull sperm selection by attachment to hyaluronic acid semi-interpenetrated hydrogels. Reproduction in Domestic Animals, 57, 2022, 228–232.

48 Morilla G., Liaudat A.C., Blois D., Capella V., Rivarola C., Barbero C., Pablo B., Rodríguez N. Development of poly-n-isopropylacrylamide surfaces for the selection of swine sperm. Corpus Journal of Dairy and Veterinary Science, 2, 2021, 1028.

49 Ebel F., Liaudat A.C., Blois D., Rodríguez N., Rivarola C., Bosch P. Adhesion of stallion spermatozoa on poly(acrylamide) based hydrogels surfaces: Effect of hydrogel ionic charge. Biocell, 45, 2021.

15 Hydrogels for Drug Delivery

Shuo Wang
Nankai University, Tianjin, China

Xiaobao Chen
Scindy Pharmaceutical Co. Ltd., Suzhou, China

Meng Meng
Nankai University, Tianjin, China

Wei Wang
University of Bergen, Bergen, Norway

15.1 INTRODUCTION

The effective delivery and controlled release of drugs over an extended period of time are essential for optimal therapeutic outcomes. To this end, the use of hydrogel delivery systems has emerged as a promising approach. These systems possess adjustable physicochemical properties and controllable degradability, which provide a favorable environment for the sustained and controlled release of both small and large drug molecules.[1]

Hydrogels can be composed of either polymers (polymeric hydrogels) or small molecules (supramolecular hydrogels), both of which feature a unique three-dimensional network structure.[2] This network structure enables hydrogels to hold a significant amount of water and expand without dissolving in the swelling state. Hydrogels can be classified using various criteria, such as the network bonding, the response to external stimuli, and the synthetic materials used. Hydrogels have the ability to undergo chemical cross-linking via covalent bonds between polymer materials, physical cross-linking via non-covalent interactions, or a combination of both. Water adsorption in hydrogels is due to capillary, osmotic, and hydration forces, which are balanced by the cross-linked polymer chains during swelling. The extent of swelling equilibrium in hydrogels depends on the magnitude of these opposite effects, which in turn determines important properties of hydrogels such as their external mechanical strength and drug delivery efficiency. These properties are influenced not only by the swelling degree of hydrogels but also by the chemical properties and morphology of the polymer materials used to synthesize them.

Hydrogels are used as drug delivery systems due to their high water content, which makes their properties similar to those of biological tissues. Additionally, the high

DOI: 10.1201/9781003340485-15

water content and variable shape of hydrogels make them an effective fit for targeted sites without causing inflammation in the body. Compared to other delivery systems, drug delivery hydrogels have good biocompatibility. Stimuli-responsive and controlled release systems are designed based on different bonding principles, which facilitate efficient delivery and sustained release of loaded drugs, thereby reducing the number of doses, and allowing for low dose by preventing drug degradation.

In this chapter, we aim to provide a comprehensive overview of polymeric hydrogel-based drug delivery systems. Specifically, we summarize the general classification and gelation mechanism of hydrogels, highlighting their physical and chemical properties that enable their use as drug delivery systems. We explore the relationship between different stimuli-responsive hydrogel drug delivery systems. Moreover, we discuss the challenges associated with the translation of different hydrogel systems from the laboratory to clinical settings. Through an analysis of existing hydrogel drug delivery systems, we provide insights into the obstacles that must be overcome to achieve successful clinical translation. Finally, we will assess the future prospects of hydrogel applications and provide guidance for the rational design of hydrogel drug delivery systems.

15.2 CLASSIFICATION OF HYDROGELS

Drug delivery hydrogels come in a range of sizes, from centimeters to nanometers, and can be tailored to possess varying water content and structural designs, resulting in distinct drug delivery functions and efficiencies. The hydrogel's macroscopic design determines how it will be introduced into the body. Hydrogels can be formed into any size or shape, and their micropores can influence their overall physical properties, such as deformability, while still allowing for drug delivery. These hydrogels contain an open space, the mesh size of which is determined by the cross-linked network. Crucially, the mesh size determines how the drug will diffuse within the hydrogel network.

Hydrogels are engineered with molecular and atomic scale chemical interactions between the drug and polymer, enabling drug loading. The polymer materials used to construct hydrogels can possess multiple sites for binding and interacting with drugs, and these sites can be pre-designed to achieve optimal drug-polymer binding. The mechanism for controlling drug release can also be rationally designed at the atomic level. Because the molecule's internal design is independent of the hydrogel's macroscopic characteristics, drug loading and controlled release at the molecular and atomic scales can be largely independent. This independence allows for a more rational, modular design of drug delivery hydrogels, allowing them to be built into multi-functional gel platforms that can meet various delivery requirements. The multi-scale properties of hydrogels are especially important for the controlled release of drugs and their ability to target disease sites. Drug delivery hydrogels can be broadly categorized into macroscopic hydrogels, in situ injectable hydrogels, shear-thinning hydrogels, microgels, and nanogels based on their macroscopic properties.

15.2.1 MACROSCOPIC HYDROGELS

Macroscopic hydrogels are hydrogels that are significantly larger than millimeters in size. These hydrogels are typically administered through two methods: surgical implantation or transdermal delivery for skin adhesion. Currently, hydrogels used for drug delivery in clinical settings are often achieved through surgical implantation at specific sites such as the skin, intestinal epithelium, and mucosa. Despite their large size and impenetrable appearance, drugs loaded in these hydrogels can be delivered to the targeted site via passive diffusion. In recent years, transdermal delivery of hydrogels has seen various applications, including the development of wound dressings and hydrogels for delivering drugs to superficial tumors.

15.2.2 In Situ Injectable Hydrogels

When dealing with deep injuries, surgical implantation of hydrogels can be burdensome for patients. As a result, researchers have shifted their focus toward studying injectable hydrogels. In situ injectable hydrogels are a type of hydrogel that can be injected into the body in liquid form and undergo a sol-gel transition inside the body, allowing for slow drug release throughout the body. These hydrogels utilize a variety of gelation mechanisms.

Compared to gel dressings or solid skin substitutes, in situ injectable hydrogels offer many unique advantages in the treatment of wound infections and other applications.[3] This type of gel system can be easily adapted to irregularly shaped skin defects, and the hydrogels can be mixed with therapeutic bioactive factors, small molecule drugs, or stem cells before injection. Once formed, the gel can act as a controlled release carrier to protect cells and active factors from harmful environmental damage, promote cell proliferation, reduce inflammation, and decrease scar formation.

One example of an injectable antibacterial hydrogel is a tetrahydropyrimidine cationic antibacterial polymer that was designed by researchers (as shown in Figure 15.1).[4] The

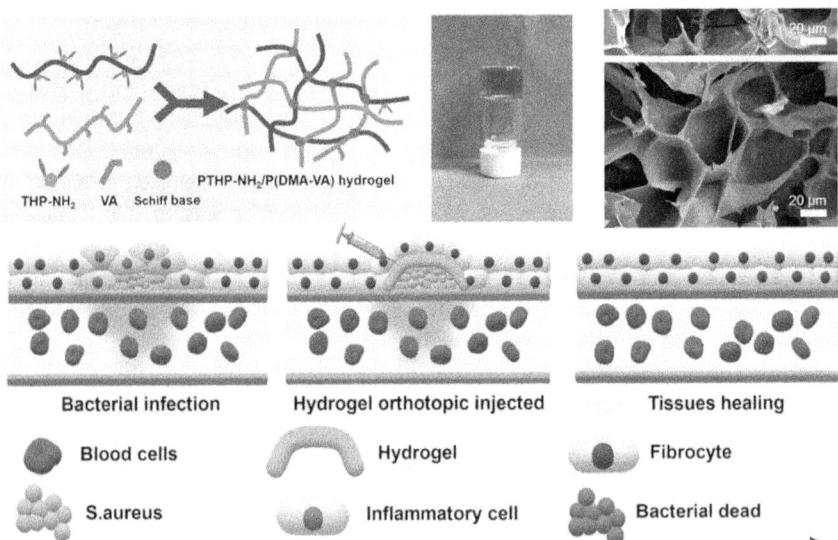

FIGURE 15.1 Schematic illustration of the synthetic method for PTHP-NH₂/P(DMA-VA) hydrogel. Adapted with permission from Reference [4], Copyright (2020), American Chemical Society.

polymer served as the skeleton of the hydrogel through a Schiff base reaction and cross-linking with polyvinyl acrylate. Finally, an injectable, self-healing, and highly efficient antibacterial hydrogel with good biocompatibility was formed. In this study, the tetrahydropyrimidine structure demonstrated strong inhibition against microorganisms, and the bacteriostatic properties were not lost after small molecules were incorporated into the polymer. The injectable hydrogel was prepared by mixing it with polyaldehyde polymer and exhibited notable bactericidal properties against various microorganisms, especially *Staphylococcus aureus*.

15.2.3   SHEAR THINNING HYDROGELS

In addition to injectable hydrogels that undergo sol-gel transformation in vivo, certain hydrogels can also form in vitro to achieve injectable purposes by applying shear stress. This process involves applying shear stress after gelation, causing the hydrogel to flow like a fluid, and then return to its original gelatinous state after injection. The shear thinning behavior of hydrogels is due to their reversible physical cross-linking. Unlike chemical cross-linking, this type of hydrogel is typically achieved through reversible cross-linking via electrostatic interaction, hydrophobic interaction, hydrogen bonds, and other mechanisms, resulting in superior drug release properties compared to chemical hydrogels.[5]

However, most shear-thinning hydrogels developed to date are prone to permanent damage caused by irreversible bond breakage, leading to poor shear-thinning injectability. Generally, the shear-thinning efficiency of hydrogels is negatively correlated with their strength. The high mechanical strength is caused by irreversible cross-linking between polymer chains, thus limiting the shear-thinning ability of the hydrogels. To achieve excellent shear-thinning injectability, hydrogels must contain dynamic covalent bonds or reversible non-covalent interactions, which may result in poor mechanical strength of the hydrogels themselves. Therefore, developing hydrogels with both high mechanical strength and good shear-thinning injection capacity remains a challenge in the field. Hydrogels with high mechanical strength and injectable properties have attracted wide attention in biomedical and tissue engineering. However, designing hydrogels with these two properties is challenging because they are usually inversely correlated.

To address this challenge, researchers have successfully prepared high-strength shear-thinning injectable supramolecular hydrogels by constructing a multi-hydrogen bond system.[6] A hydrogel was constructed by the self-assembly of a monomer nucleoside gelator (2-FA) in a phosphate buffer saline. Its storage modulus reaches 1 MPa, the concentration is 5.0% by weight, and it is composed of ultra-low molecular weight (MW 300). In addition, the hydrogel can complete the sol-gel state transformation within seconds after injection at 37°C, achieving good injectability. The poly-hydrogen bond system is based on the synergistic interaction between double NH_2 groups, water molecules, and 2'-F, which enables a 2-FA hydrogel to ensure good biocompatibility and at the same time have a high drug load to achieve good antibacterial activity. When applied on animals, 2-FA hydrogels exhibit faster degradation and induce less osteoclast activity and inflammatory infiltration, leading to more complete bone healing, compared to

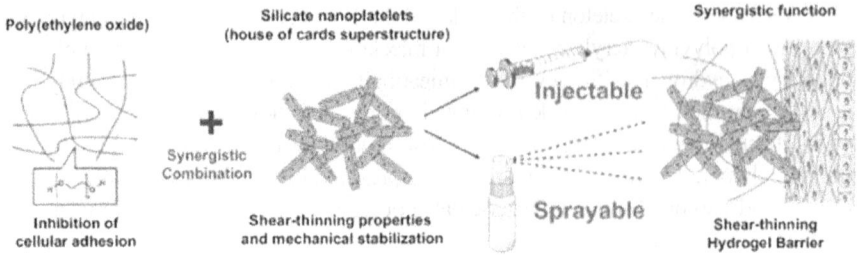

FIGURE 15.2 A schematic illustration of the synthetic designs for injectable and sprayable hydrogels. Adapted with permission from Reference [7], Copyright (2021), Springer Nature.

natural healing and commercial bleeding agent gelatin sponges. This study provides ideas for proposing multifunctional, high-strength, injectable supramolecular hydrogels for various biomedical engineering applications.

In a separate instance, researchers have created hydrogel barriers that are both injectable and sprayable and have the unique property of being shear-thinning. The hydrogels are composed of silicate nanoplatelets and poly(ethylene oxide), which prevent the formation of adhesions postsurgery through their exceptional physical and mechanical properties (as depicted in Figure 15.2).[7] These hydrogels are in liquid form under stress, but return to their original viscoelastic solid state once the stress is removed, making them ideal for injection and spray applications. In vitro studies have demonstrated that these hydrogels significantly reduce the adhesion of fibroblasts and macrophages while maintaining low cytotoxicity. Additionally, in vivo studies on rat models with peritoneal injuries have shown that these hydrogels are more effective than commercially available antiadhesion products, resulting in lower adhesion index scores or no adhesion formation. This suggests that the hydrogels are highly effective in preventing the formation of postoperative adhesions. Moreover, immunohistochemical analysis has revealed a decrease in immune cell infiltration, indicating that the hydrogels are biocompatible.

15.2.4 MICROGELS AND NANOGELS

In the realm of minimally invasive surgery, nanogels offer greater promise than their macro and injectable hydrogel counterparts. Their reduced dimensions, typically under a micron in size, facilitate both injection and drug loading, as well as enhance penetration of tissue barriers. With size and deformability playing critical roles in drug half-life and biodistribution, nanogels offer an effective means of targeted drug delivery. Transportation and adhesion are also influenced by the size of hydrogels introduced into blood vessels, airways, or the gastrointestinal tract. Microgels, with diameters less than 5 μm, are generally reserved for oral or pulmonary administration, whereas nanogels between 10–100 nm are better suited for systemic administration, due to their ability to leave blood vessels and extravasate into tissues. Clearance rates are similarly dependent on size, with hydrogels under 10 nm being cleared by kidney filtration, and those between 0.5–10 μm being phagocytized by macrophages. When designing micro- and nanogels, consideration must also be given to gel morphology, surface charge, and variability.

Nanogels with positive zeta potential and high aspect ratios, for instance, have been observed to take up rodlike shapes more quickly, offering yet another level of customization in the field of targeted drug delivery.

In light of its vast surface area, the nanogel is a highly efficacious medium for intravenous injection, making it widely utilized in the realm of biomedicine. This is especially true when treating tumors and brain ailments. The nanoscale size permits the nanogel to bind with targeting ligands, resulting in the enhancement of permeability to retain effect (EPR), thereby precisely targeting tumors. This also ensures intracellular drug delivery through endocytosis and the crossing of the blood-brain barrier (BBB). Locally injected nanogels demonstrate superior permeability and increased responsiveness to stimulation, as they can inhibit the degradation of the loaded drug. Moreover, some inorganic materials with poor colloidal stability and low water solubility may be loaded into nanogels for in vivo cancer diagnosis, imaging, and treatment.

Gene therapy represents a promising approach to the treatment of various diseases, such as cancer, hemophilia, and viral infections. Nanogels are highly advantageous in the delivery of nucleotide-based drugs, such as plasmid DNA, mRNA, siRNA, or other nucleic acid drugs. The nanogel delivery system improves the uptake of DNA/RNA by cells and prolongs the circulation time in the body, which ensures the uptake of more drugs than with free DNA/RNA. Additionally, nanogels are capable of achieving effective tumor targeting due to the enhanced permeability and retention effect, whereby the tumor site vascular system enhances the accumulation of nanoparticles, while ineffective lymphatic drainage limits the clearance of nanoparticles. Some researchers have designed a drug delivery system comprising self-assembled aggregations of small interfering RNA (siRNA), epigallocatechin gallate (EGCG), and protamine sulfate as "invisible" drug carriers, hyaluronic acid (HA) shells, and penetrating peptides of tumor homing cells as target ligands for synergistic delivery of therapeutic genes and anticancer drugs.[8] This self-assembled nanogel consisting of siRNA and EGCG represents a promising approach to finding a "stealth" carrier and provides an effective solution by integrating high-dose therapeutic genes and drugs into a single biodegradable carrier.

15.3 MECHANISM OF GELATION

15.3.1 Chemical Hydrogels

In materials science, hydrogels prepared by means of chemical cross-linking are typically achieved through covalent bonds, such as those formed by Michael addition, Schiff base reaction, or click reaction. The creation of hydrogels through chemical cross-linking generally involves the introduction of small cross-linked molecules, polymer-polymer conjugation, photoinitiated reactions, or enzymatic reactions. While physical cross-linking exhibits reversible properties, the bonds formed through chemical cross-linking are irreversible, yielding a heightened degree of stability in these materials by virtue of the covalent bonds established between polymers. This added stability also confers upon chemically cross-linked hydrogels the capacity to resist dissolution in surrounding fluids, thus preventing sudden drug release. As a result, the production of chemically cross-linked hydrogels facilitates greater control over factors such as gelation time, drug release, and action site selection, ultimately endowing these materials with more reliable in vitro and in vivo properties as compared to their physically cross-linked counterparts.[9]

For chemically cross-linked hydrogels, polymer chains interconnect gradually within a reaction system, resulting in the formation of a giant molecule of "infinite" size known as gelation. Thus, the foremost consideration when producing chemically cross-linked hydrogels is to devise a cross-linking mechanism that can react in opposition to the precursor and remains effective in water-based media. In clinical applications, the characteristics of hydrogels and drugs necessitate strict compliance. Therefore, controlling gel stability by means of chemical cross-linking renders it more suitable for preparing drug-delivery hydrogels with favorable prospects for clinical application.

To form intermolecular cross-links in a progressively growing gel system, the average functional degree of the reactants must exceed 2, i.e., at least one component ought to have a functional degree equal to or greater than 3, while the remainder should have a functional degree equal to or greater than 2. This can be accomplished either by cross-linking polymer chains with multiple (i.e., ≥3) overhanging reaction segments, or by using multi-arm polymer chains with double, or sometimes multi-functional, cross-linking agents that contain functional groups at each chain end, producing side chain link networks or end-link networks, respectively (Figure 15.3).[10] The former type of network is prevalent in hydrogels made from long linear polymers, such as natural chitosan-based drug delivery hydrogels. Polymer chains of this nature typically possess numerous reaction sites, resulting in more adequate cross-linking and smaller internal cross-network sizes. For the latter type of network, such as polyethylene glycol (PEG)–based drug delivery hydrogels prepared from long chains, the gel mesh is larger, with higher integrity and uniformity. Although there are disparities between the two hydrogels, each one boasts distinct advantages and divergent clinical applications. Chitosan hydrogel, for instance, exhibits a more stable structure, while PEG hydrogel possesses a higher drug-loading efficiency.

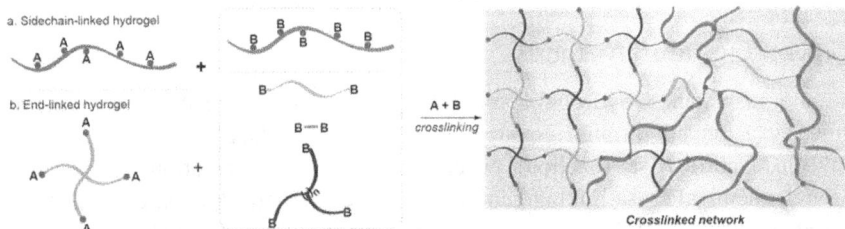

FIGURE 15.3   Schematic illustration depicting the fabrication of chemically cross-linked hydrogels using stepwise cross linking chemistries. (a) The hydrogel is chemically cross-linked by a polymer chain that contains multiple (≥3) side chain reactive groups. (b) Alternatively, the hydrogel can be formed by linking multiarmed polymer chains that have reactive groups at each chain end with difunctional or multifunctional cross-linkers. These stepwise cross-linking processes lead to the construction of the hydrogel network. Adapted with permission from Reference [10], Copyright (2021), Wiley.

While the preparation of most hydrogels requires dissolution or preparation in organic solvents, their cross-linking is typically achieved in aqueous solutions. This is especially crucial for tissue adhesives and wound dressings, which are integral in clinical settings. As such, the cross-linking reactions of hydrogels must be kinetically or thermodynamically compatible with an aqueous environment, which requires hydrogels to possess water expansion properties. Furthermore, drug delivery hydrogels ought to demonstrate good solubility and stability in an aqueous solution. One approach to stabilizing the properties of hydrogels is to introduce small molecular compound modifications with differing properties.

The selection of appropriate cross-linked chemicals and precursors, or the adoption of complex manufacturing techniques, is another critical aspect of designing chemically cross-linked drug delivery hydrogel systems. Through this manipulation of the structure and properties of hydrogels, ideal drug-loaded hydrogels can be manufactured. Notably, the relationship between the structure and properties of hydrogels is often pivotal. While such strategies have been extensively researched and hold immense importance in basic research, they are often inadequate to support the clinical translation of drug delivery hydrogels. Therefore, the rational design of hydrogels for clinical translation calls for further investigation.

15.3.1.1 Biorthogonal Chemical Reactions

Biological orthogonal reactions, a class of reactions developed by Bertozzi and her colleagues for biomacromolecular conjugating, are based on non-interference with the biological environment.[11] These reactions have gained widespread use in chemical biology due to their many advantages. In particular, they can occur under mild reaction conditions and have good biocompatibility. As a result, biological orthogonal chemistry is commonly employed in the manufacture of hydrogels, especially for cell encapsulation applications. The development of biological orthogonal reactions for the preparation of hydrogels for drug delivery has been rapid in recent years, and here we focus on reactions that have successfully been employed in the production of hydrogels.

The Staudinger ligation, which was the first biological orthogonal reaction developed, utilizes two functional groups, azide and triarylphosphine, to form an amide bond (Figure 15.4).[12] This reaction is characterized by high chemical selectivity for the amide bond in complex biological environments, making it suitable for many chemical biology studies. However, its reaction speed is slow and, therefore, it is not ideal for the preparation of drug delivery hydrogels. In fact, an incubation time of 20–24 hours is often required to ensure complete polymerization of the polymers used in the preparation of hydrogels employing this reaction.

As this field has developed, many studies have been undertaken to optimize this reaction. For instance, Stabler et al. applied this cross-linking reaction to cross-link elastin-like protein (ELP) and detailed gel kinetics studies and demonstrated that hydrogels could form in about an hour.[13] This reaction can be employed in conjunction with other cross-linking mechanisms, such as ion cross-linking, to rapidly form stable gel materials. Combining this reaction with other cross-linking mechanisms is a mainstream approach that can achieve a reduction in gel formation time while retaining its advantages.

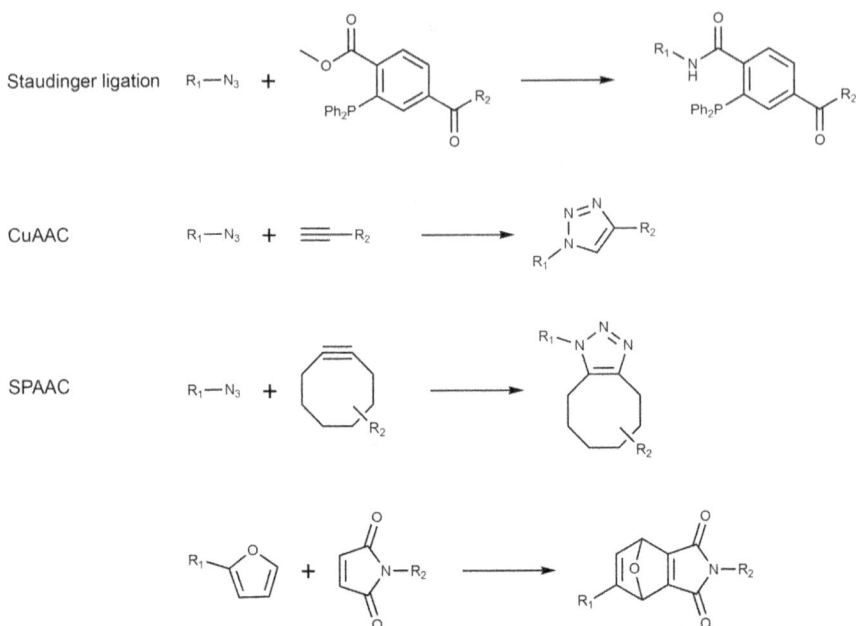

FIGURE 15.4 Reaction routes of some bioorthogonal chemical reactions.

Another example is that of Cu(I) catalyzed azide-alkyne cycloaddition, commonly referred to as CuAAC. This process constitutes a modified Huisgen 1, 3-dipole cycloaddition reaction (Figure 15.4). CuAAC is employed extensively in both drug development and polymer preparation. Additionally, there are some applications in the synthesis of hydrogels, particularly in the study of hydrogel structure. However, this reaction requires the assistance of Cu ions as a catalyst, which is undoubtedly its primary shortcoming. Recently, some researchers have devised a solution to this dilemma with the advent of strain-promoted azide-alkyne cycloaddition (SPAAC). By utilizing a more reactive strain-promoted cyclooctyne in the absence of any metal catalyst (Figure 15.4), azides can be effectively reacted with great efficacy.[14] Numerous catalysts have been thoroughly examined, and their efficacy optimized for use in vivo. This reaction boasts excellent biocompatibility and chemical selectivity, leading to its burgeoning application in the design of biocompatible hydrogels for drug delivery.

15.3.1.2 Non-Biorthogonal Reaction

While biological orthogonal reactions hold a significant place in the realm of drug delivery hydrogels, it is not without notice that non-biorthogonal reactions also warrant recognition. Among them, reactions involving amines, mercaptans, and 1, 2-amino-mercaptans (Figure 15.5) hold importance in the development of hydrogels. The functional group of amines, in particular, is ubiquitously distributed within proteins and peptides, rendering them crucial in reactions that involve the aforementioned macromolecules. Such chemical reactions have found relevance in

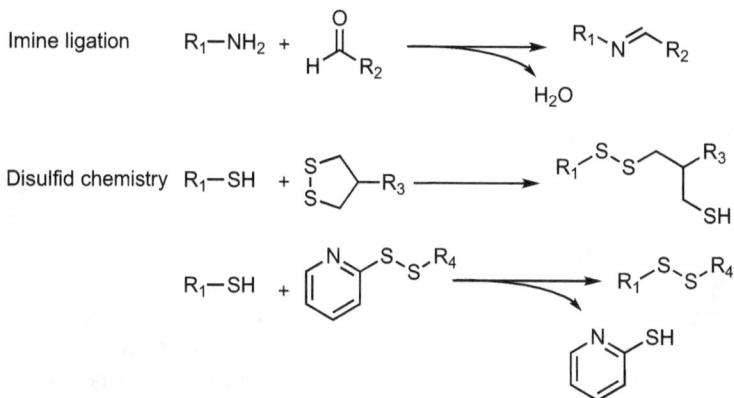

FIGURE 15.5 The synthetic routes of imine ligation and disulfide chemistry.

the manufacture of polypeptide hydrogels and have been utilized in the creation of various drugs approved by the Food and Drug Administration (FDA).

Imine bonding, which encompasses Schiff base chemistry, refers to the reaction of carbonyl compounds such as aldehydes and ketones with primary amines, resulting in dynamic covalent imine bonds. Likewise, the hydrazone and oxime reactions constitute two other conventional Schiff base reactions, known as carbonyl click chemistry. These reactions are uncomplicated, efficient, chemically specific, and have a benign by-product: water. Primary amines, being a rather common compound, have been the focus of research in protein hydrogels. By cross-linking additional amines on proteins or peptides with other compounds containing aldehydes, many drug delivery hydrogels such as bovine albumin (BSA)–based surgical adhesives have been produced.

Among the various reactions that take place in click chemistry, the involvement of mercaptan stands out as a prominent functional group for cross-linking reactions. Mercaptan's ready availability from cysteine and its ability to react with multiple functional groups makes it a highly sought-after element in this domain. When applied to hydrogels, mercaptan chemistry primarily drives the selective protein mercaptan conjugation and the hydrogel's high adaptability to cross-linking.

Disulfide bond chemistry, which plays a crucial role in regulating the folding and stability of many proteins, is also widely employed in the coupling of proteins. This coupling method finds extensive use in the delivery of small molecules, nucleic acids, and protein drugs. Disulfide cross-linking chemistry proves particularly advantageous for hydrogels that exhibit a stimulus response. For example, bovine serum albumin (BSA) was prepared as a pH- and redox-sensitive hydrogel using simple disulfide bond cutting, and the gel was then stimulated in the pH range of 4.5–8.1.[15] This cross-linking principle typically applies to polypeptide compounds or proteins containing mercaptan, and the straightforward reaction conditions make them suitable for hydrogel preparation.

However, it must be noted that such reactions often require longer reaction times under physiological conditions and are limited in the process of hydrogel preparation. This limitation hinders their application in hydrogel manufacturing.

To overcome these limitations, these reactions can be combined with other cross-linking methods, as detailed in the following section.

15.3.2 Physical Hydrogels

Physical cross-linked hydrogels utilize weak and reversible supramolecular forces, including hydrogen bonds, ionic bonds, hydrophobic interactions, physical entanglement, and electrostatic interactions, to form their network. Notably, physically cross-linked hydrogels are self-healing due to the dispersion of tensile stress through the breakage of physical interactions during hydrogel stretching, followed by reversible self-repair once the tensile stress is removed.[16] In contrast, covalently cross-linked hydrogels are vulnerable to irreversible and permanent damage when subjected to significant mechanical deformation, leading to a decline in their overall mechanical properties. In this context, substituting covalent cross-linking with physical cross-linking provides an opportunity to regulate cross-linking mechanics and enables stimuli-response, self-healing, and self-recovery attributes. Moreover, the cross-linking structure of physically cross-linked hydrogels can adjust to environmental conditions, and since no cross-linking agent is required, the preparation process for physical hydrogels is uncomplicated and safe. Thus, hydrogels have found wide application in biomedicine, particularly as stimulus-response hydrogels that facilitate the efficient delivery of small molecule drugs. Further details on this will be expounded upon below.

15.3.2.1 Physical Hydrogels Based on Hydrogen Bonding

Hydrogen bonding, a ubiquitous and fundamental form of interaction in nature and living organisms, serves as a key component in the synthesis of hydrogels. The strength, directivity, and affinity of this interaction make it a suitable choice for such applications. The highly controllable nature of hydrogen bonding allows for the manipulation of physical properties of the gel through the adjustment of reaction conditions and material ratios. The stability of the hydrogel can also be regulated by the number of hydrogen bonds, despite the weak nature of the force between individual bonds and the limited bonding energy of approximately 1–5 kcal/mol.

In certain molecules, the formation of double, triple, or quadruple hydrogen bonds between two complementary hydrogen bond groups can occur. For example, adenine and thymine form double hydrogen bonds, while guanine and cytosine form triple hydrogen bonds in the DNA molecule of biological systems. By introducing special functional groups that can form hydrogen bonds into polymer chains, such as carboxyl group, amide bond, ureacymidone, and other functional groups, the self-healing gel can be prepared. Upon damage to the hydrogel, complementary functional groups at the damaged area can form hydrogen bonds, reconnecting the broken surface and enabling the material to repair itself.

15.3.2.2 Physical Hydrogels Based on Hydrophobic Interaction

Hydrophobic interaction, exemplifying a quintessential non-covalent interaction, assumes an indispensable function in biological systems. Notably, hydrophobicity plays a vital role in preserving the structural integrity of biomembranes, containing a phospholipid bilayer, wherein the long fatty chains, intricately oriented with hydrophilic molecular groups, display remarkable water-repelling characteristics. In recent years, researchers, taking cues from biological systems, have been increasingly exploring the hydrophobic effect to enhance the mechanical properties of hydrogels.

15.3.2.3 Physical Hydrogels based on Ionic Bonds

Ion cross-linked hydrogels are formed through the process of ionic bonding between polymer side chains, including metal-ion coordination bonds and complex salts of weak acids and bases. The former is achieved through the coordination between metal ions and carboxyl groups, while the latter involves electrostatic interaction. An exemplary instance is the entirely physically cross-linked poly(acrylic acid)-cellulose-iron (PAA-CNF-Fe^{3+}) hydrogel, synthesized with hydrogen bonding and double coordination serving as the cross-linking modes. The mechanical properties of the PAA-CNF-Fe^{3+} gel can be modulated by regulating the concentration of Fe(III). This gel possesses remarkable mechanical attributes, including high breaking strength (1.37 MPa), elongation at break (1803%), rapid self-recovery (95.7% recovery rate within 60 min), and excellent self-healing efficiency (94.2% for 48 h continuous repair at 25°C).[17] Hydrogen bonds, which function as a skeleton for maintaining the integrity of the primary structure, tend to break before coordination bonds. The surviving coordination bonds with dynamic characteristics can also be utilized as sacrifice bonds, which dissipate additional energy after the hydrogen bond breaks. This phenomenon not only improves the elasticity of the gels but also maximizes the contribution of sacrifice bonds to energy dissipation. Moreover, non-covalent interactions synergistically function as dynamic but highly stable associations, resulting in an effective self-healing efficiency of over 90%.

15.4 PHARMACEUTICAL STIMULI-RESPONSIVE HYDROGELS

Stimuli-responsive hydrogels possess the remarkable ability to release drugs by undergoing volume changes brought about by alterations in the environment. These alterations may be triggered by various stimuli, including pH, temperature, light, electricity, or enzyme. In vivo, this responsive property finds particular utility since the physiological condition of the disease site is often subject to changes such as pH reduction, temperature elevation, and changes in various physiological indicators. Consequently, stimuli-responsive hydrogels loaded with drugs facilitate the precise release of therapeutic agents specifically at the disease site, thereby reducing drug toxicity to other parts of the body, diminishing the necessary dosage, and producing more efficacious therapeutic outcomes.

Notwithstanding the potential of stimuli-responsive hydrogels for drug delivery, it has been difficult to control the rapid release of drugs, inefficiently load hydrophobic drugs, and achieve long-lasting on-demand drug delivery. To overcome these challenges, the incorporation of nanomaterial hydrogels has been adopted for stimuli-responsive drug delivery applications. The addition of nanoparticles possessing diverse properties addresses the challenges of hydrogel drug release, while simultaneously producing an auxiliary effect on the structure and stability of hydrogels. Furthermore, the incorporation of nanomedicine permits the loading and release of drugs possessing different physicochemical properties, consequently raising the drug loading rate of hydrogels.

15.4.1 Injectable pH-Responsive Hydrogels

Injectable pH-responsive hydrogels are designed to respond to changes in pH, resulting in the controlled release of drugs. These hydrogels typically contain ionizable groups, such as acidic or alkaline groups, which respond to changes in the surrounding pH environment.[18] At low pH values, the degree of dissociation of acidic functional groups decreases while that of basic functional groups increases. Conversely, in high-pH environments, the dissociation degree of acidic functional groups increases while that of basic functional groups decreases. These changes can affect the cross-linking of the gel network, resulting in modifications to the swelling degree of the hydrogel.

Hydrogels with carboxylic groups absorb less water and have lower swelling capacity in acidic environments, which can reduce the drug release rate. In contrast, under alkaline conditions, the volume of hydrogels with carboxylic groups increases due to repulsive forces between ionized carboxyl groups, leading to faster drug release. For hydrogels containing polymers with alkaline groups, the opposite is true, with hydrogel volume decreasing in alkaline conditions and expanding in acidic environments. Several injectable pH-responsive hydrogels have been reported, including those loaded with model molecules such as Hoechst 33342 (Hst) in porous nanohydrogels (Figure 15.6). These drug-loaded nanomaterials were

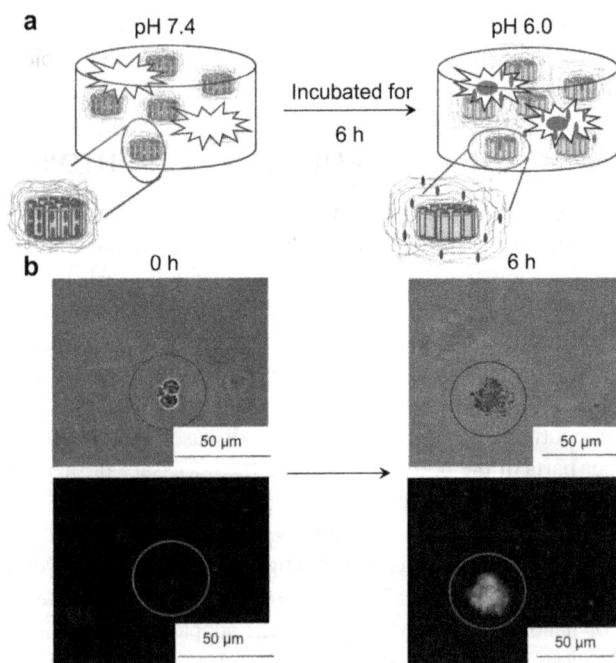

FIGURE 15.6 (a) Schematic illustration of the release mechanism of drugs from a pH-responsive hydrogel. (b) Microscopic images of hydrogel before and after the fluorescence dye are release from the hydrogel after adjusting pH. Adapted with permission from Reference [19], Copyright (2021), Wiley.

then coated with polymers of opposite charges and embedded in an alginate saline gel system. Controlled release of the model molecules was achieved by changing the pH from 7.4 to 6.0 for 6 hours.[19]

Polymers that contain a substantial number of ionizable acidic or alkaline groups are the key components of pH-responsive hydrogels. These polymers enable proton exchange in response to changes in pH values, causing hydrogels to dissociate and associate in external environments and exhibit different swelling characteristics. However, pH-responsive hydrogels face constraints due to hydrogen ion diffusion. Polyacrylic acid (PAAc) is a frequently used acidic polymer that can undergo deprotonation in an acidic environment and protonation in a neutral or alkaline environment. Similarly, poly(4-vinylpyridine) is a widely used basic polymer. In hydrogel preparation, when the pH drops below pKa, the polymer's ionization boosts the electrostatic repulsive force among polymers, resulting in hydrogel instability. For example, polyethylene oxide-block-poly(N,N-diethylaminoethyl methacrylate) (PEO-PDEAMA) diblock copolymers dissolve in acidic environments, but the hydrophilic blocks (PDEAMA, pKa = 7.3) become insoluble in water due to protonation. pH-responsive hydrogels have been constructed using cross-linked particles with high porosity and elasticity. Additional research has demonstrated that altering the initial pH of the prerequisite solution also affects swelling and gelation kinetics, resulting in hydrogels with improved mechanical properties. pH-responsive hydrogels are primarily used to respond to pH changes in pathological tissue (e.g., local tissue inflammation, infection, and cancer). However, predicting lesion site pH accurately is difficult in clinical practice due to many factors, such as metabolic status, disease type, and treatment history. This uncertainty presents a significant challenge for preparing pH-responsive hydrogels.

In numerous studies, Schiff bases have been employed in the synthesis of pH-responsive injectable stimuli-responsive hydrogels. Schiff bases arise from the reaction between aldehydes and nucleophilic amines and exhibit significant pH sensitivity in acidic environments, resulting in hydrogel transformation. Consequently, Schiff base hydrogels can regulate the release of required drugs in response to microenvironmental changes, ensuring the targeted delivery and release of drugs. For instance, a recent investigation developed a pH-responsive hydrogel composed of xanthan gum-PEG with Schiff base. The researchers cross-linked xanthan gum to PEG using dynamic covalent hydrazone bonds. This hydrogel is ideal for delivering drugs to tissues with acidic microenvironments (e.g., tumor tissues) while having little impact on untargeted healthy tissues.[20]

15.4.2 TEMPERATURE-RESPONSIVE HYDROGELS

Temperature-sensitive hydrogels are a type of material that undergoes volume changes in response to temperature fluctuations. These changes occur due to alterations in hydrophobic interactions and hydrogen bonding between polymer chains. The solubility of polymers in water increases as temperature rises, leading to an increase in volume for temperature-responsive hydrogels at elevated temperatures and a decrease in volume at lower temperatures. However, hydrogels composed of polymer compounds with lower critical solution temperatures

(LCST) experience a decrease in water solubility above their LCST, causing thermosensitive hydrogels to shrink, with the volume change being negatively correlated with temperature.

Hydrogels used for drug delivery often contain hydrophobic groups. When the temperature rises above a certain threshold, hydrophobic interactions between polymer chains become stronger than hydrogen bond interactions, causing the hydrogels to contract to minimize hydrophobic surface area in contact with water. Conversely, below a certain temperature, the hydrogen bond interactions between the hydrophilic units become dominant, leading to hydrogel expansion. By adjusting the proportion of hydrophilic and hydrophobic groups in the polymer, the critical temperature can be modified.

Temperature-responsive injectable pre-gels should be liquid or semi-solid at room temperature and form a hydrogel after injection into the body. The gelation and stimulation response of hydrogels facilitate on-demand drug release. Temperature-responsive polymers are a primary constituent in creating temperature-responsive hydrogels, such as poly(N-isopropylacrylamide) (PNIPAAM). PNIPAAM is composed of both hydrophilic and hydrophobic groups, with a LCST of 32–34°C. When the temperature surpasses the LCST, the hydrogen bonds between the amide group and water molecules weaken, causing an increase in hydrophobic interaction between the groups. This leads to hydrogel contraction and the rapid discharge of the encapsulated drug. Nevertheless, the non-biodegradable nature of the polymer restricts its applicability. Various strategies have been developed to address this concern, including copolymerization with other polymers, such as PEG, to improve hydrophilicity, as well as copolymerization with polycaprolactone (PCL) or polylactic acid (PLA) to enhance biodegradability.[21]

Another commonly used method for creating temperature-responsive hydrogels is with Poloxamer, which forms a gel at 37°C. The tetrandrine thermosensitive gel (TTG) was generated using Poloxam 188 and 407, and it can remain in the nasal cavity for up to 4 hours. This prolonged retention significantly improved pathological changes in the hippocampus, prefrontal cortex, and amygdala of mice modeled for post-traumatic stress disorder (PTSD).

Poly(N-vinyl caprolactam) (PVCL) is another type of temperature-responsive polymers with a phase transition temperature in the range of 30–40°C. These hydrogels possess several excellent properties, such as strong coordination ability, good biocompatibility, and non-toxic derivatives after degradation. Compared to PNIPAAM, PVCL exhibits similar temperature responsiveness but superior biocompatibility. As a result, it has numerous biomedical and non-biomedical applications, including drug delivery. For instance, researchers synthesized a stimuli-responsive PVCL nanogel via aqueous precipitation polymerization. The resulting PVCL nanogel particles were temperature- and pH-responsive. In reducing medium conditions, the PVCL nanogel with N,N'-methylene bisacrylamide as the cross-linking agent remained stable, while the PVCL nanogel with N,N-bis(acryloyl) cystamine as the cross-linking agent degraded due to the disulfide bond breaking. This PVCL nanogel exhibits low cytotoxicity and good biocompatibility, making it an ideal drug carrier.

15.4.3 PHOTON-RESPONSIVE HYDROGELS

Photon-responsive drug delivery hydrogels are created by incorporating photo-sensitive functional groups into a hydrogel network. These functional groups can sense changes in the external environment under light, causing the hydrogel to expand or contract. This, in turn, facilitates the controlled and on-demand release of drugs. Exposure to light induces photoionization, generating ions that disrupt the osmotic balance of the gel medium. This disturbance results in water molecules and ions moving into or out of the gel network to compensate for the osmotic imbalance, leading to a phase transition of the gel. The photosensitive hydrogel polymers consist of a skeleton and a photosensitive part that capture light signals and convert them into chemical signals through photoreactions such as isomerization, cleavage, or dimerization.

One of the most commonly used light-switching molecules is azobenzene, and it can be used as a cross-linking agent. Ultraviolet irradiation can de-cross-link azobenzene, leading to a solid-liquid phase transition. For example, a photon-sensitive supramolecular hydrogel was constructed by linking urea-pyrimidones with tetrafluoroazobenzene. Under visible light irradiation, reversible isomerization was achieved between the two isomers. In another example, a photon-responsive hydrogel was constructed based on the photo-controlled host-guest supramolecular interaction of azo-phenyl group with α-cyclodextrin. The swelling rate of the photosensitive hydrogel can be finely regulated by adjusting the content of the azobenzene group in the polymer structure. The cis-trans isomerization of azobenzene can induce a volume change of about 60%, and the photon-stimulated deformation of the thin hydrogel sheet can be completed within tens of seconds, demonstrating excellent photon response.[22]

Spiropyrane is an organic photochromic compound that finds widespread use in various fields, including photoelectric devices, ultra-high-density information storage, molecular logic switches, ion recognition, molecular self-assembly, controlled drug release, and super resolution imaging. Additionally, spiropyrane can function as a magnetic control switch, a fluorescence control switch, a fluid channel switch, a biomolecular activity or detection switch, and a switch of material surface characteristics. One of the most significant characteristics of spiropyrane is its ability to rapidly and reversibly transform between closed loop (SP) and open loop (MC) states under multiple external stimuli, such as light, heat, solvent, acid-base, electricity, and force. This transformation causes significant changes in spiropyrane properties, such as coloration, hydrophilicity, fluorescence, and aggregation morphology.

Researchers have introduced orthogonal wavelength cross-linking of hydrogel networks using two highly redshifted chromophores: acryl pyrene (AP) and styrenyl pyridine [2,3-b] pyrazine (SPP). The wavelengths of activiation for these two chromophores are at 410–490 nm and 400–550 nm, respectively. By using a spiropyrane-based photon-producer with appropriate absorption wavelengths, the team was able to limit the activation wavelength of the SPP portion to the green-light region with the wavelength of activation at 520–550 nm, thereby achieving wavelength-orthogonal activation of the AP group. This wavelength-orthogonal photochemical system has been successfully applied to the design of hydrogels, and the mechanical properties of the hydrogel can be independently adjusted by green or blue light.[23]

In addition to the above-mentioned groups, nitrobenzyl ether groups are also used for the preparation of light-responsive hydrogels. Nitrobenzyl-based compounds have several advantages, such as simple synthesis, easily modifiable structures, high photochemical quantum yields, and rich labeling groups. Thanks to the non-invasiveness and superior spatiotemporal resolution of light-triggered technology, photolabile compounds have found extensive applications in organic chemistry, biochemistry, materials science, and drug delivery hydrogel preparation. However, the application of these light-responsive hydrogels is limited by the tissue penetration depth of light, making it challenging to achieve deep tissue applications. Therefore, future research should focus on the development of photoresponsive structures that can respond to different wavelengths.

15.4.4 ENZYME-RESPONSIVE HYDROGELS

Protease overexpression has been strongly linked to various diseases. Consequently, in addition to conventional pH and photon-responsive hydrogels, protease-responsive hydrogels exhibit significant potential for localized and controlled drug delivery in the treatment of these diseases. For instance, an injectable enzyme-responsive hydrogel was developed for controlled drug release.[24] The hydrogel was comprised of micelles that formed through a chemical reaction between cysteine and thioester groups, namely N-(2-hydroxypropyl)methacrylamide-cysteine (HPMA-Cys) or N-(2-hydroxypropyl)methacrylamide-ethylthioglycolate succinic acid (HPMA-ETSA). Hydrogels created through cross-linking of these micelle systems can effectively respond to changes in enzyme activity in the environment to achieve controlled drug release.

In recent years, glucose-responsive drug delivery hydrogels have garnered attention in the field of insulin delivery for the treatment of diabetes. Upon detecting glucose, insulin is released from the glucose-responsive hydrogel and enters the receptor environment. The response group of glucose-sensitive hydrogel comprises glucose-sensitive compounds, such as phenylboric acid. For instance, a multifunctional hydrogel that responds to pH and glucose to release metformin was reported.[25] In an environment of weak acid, the Schiff base structure dissociates, and the increased level of glucose at the wound site can competitively combine with phenylboric acid, resulting in the dissociation of the phenylboric acid structure and catechol to achieve the pH/glucose response and metformin release.

15.4.5 ELECTRO-RESPONSIVE HYDROGELS

In electro-responsive hydrogels, the movement of ions within and outside of the gel is directed by an electric field, resulting in a reversible macroscopic phase transition. Electro-responsive drug delivery hydrogels typically incorporate polymer compounds containing electrolytes and other groups. Upon application of an electric field, these hydrogels absorb water and expand, thereby releasing the drug in a controllable manner based on the time and intensity of the electric field.

To control drug release in these hydrogels, changes in internal osmotic pressure and pH are often induced via an electric field. Recently, researchers have developed injectable electro-responsive nanohydrogels using conductive polymer nanoparticles. These nanohydrogels can be synthesized using oxidized alginate cross-linking gelatin reinforced by electroactive tetraaniline-graft-oxidized alginate nanoparticles. The content of oxidized alginate in the nanohydrogel network is proportional to the number of tetraaniline units, which can be stimulated by electrical fields. The mechanical and electrical conductivity of the nanohydrogel increase with higher content of oxidized alginate.[26]

15.5   REGULATION ON HYDROGELS

Hydrogels are gaining traction in clinical applications, with their use expanding from wound dressings to contact lenses, cell therapy, and other biomedical needs. These water-containing materials have good biocompatibility and are becoming more prevalent in our lives. The hydrogel industry is also benefiting from the development of polymer materials. However, hydrogels still face regulatory issues that need to be addressed.

Although guidelines and regulations governing the regulatory process of hydrogels in drug delivery are regularly updated, different regulatory agencies exhibit significant variations. The regulatory approach to hydrogels by the FDA is illustrated in Figure 15.7. As per this regulatory approach, the classification of hydrogels determines their approval for drug use. Hydrogels prepared solely by chemical bonding and those created by both chemical bonding and intermolecular forces cannot be approved as drugs but instead are classified as medical devices. Additionally, hydrogels that are prepared by administering drugs, biological products, or cells, are subject to complex processes such as new drug application (NDA) and biological licensing application (BLA) before being approved as drug products.

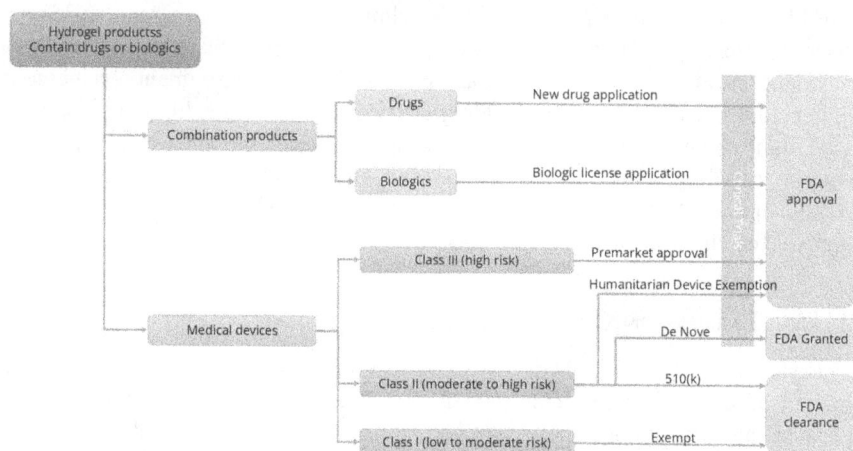

FIGURE 15.7   FDA Regulatory pathways for diverse types of hydrogel products. Adapted with permission from Reference [10], Copyright (2021), Wiley.

15.5.1 DRUG PRODUCTS

New drugs, biologics, and drug delivery hydrogels require NDA or BLA to enter clinical use. Hydrogel products have increasingly used this pathway over the past years. Most hydrogel products on the market are hydrates, such as wound dressings. Hydrogel systems have good biocompatibility, adjustable network structure, excellent stimulation response, and drug control release ability, which promote their clinical transformation. To date, over 200 hydrogel products have been approved for clinical use in various diseases.

Hydrogels are mostly classified as Class II medical devices and approved for production. Compared to Class III medical devices, their supervision is lighter, and their application is less strict. Drug delivery hydrogels in Class II medical devices focus on effectiveness and safety. They are commonly composed of long chains of polymers approved by the FDA, which reduces their manufacturing costs. However, their biosafety cannot be degraded, which is the basis and most important point for clinical approval. Polymeric drug delivery hydrogels based on different functional PEG polymers have been approved by the FDA for use in clinical wound dressings or the treatment of arthritis.

Macroscopic hydrogels, such as subcutaneous fillers and tissue adhesives for knee joints, are usually administered by surgical implantation. They play a tissue and bonding role, fill tissue gaps, or glue postoperative folds. Hydrogels can also block the contact between the wound and the external environment after surgery, reducing inflammation and infection. These hydrogels have a high requirement in clinical application, including good physical and chemical properties, fluidity for intraoperative injection, and biological safety. Most surgical implantation hydrogels are classified as Class III medical devices with strict management. They need to have good injectable properties and can be stably stored in physiological buffers for various clinical scenarios.

In addition to biosafety, the physical appearance of hydrogels is also important. Thin-film hydrogel dressings used in the clinic are often transparent, making it easier to observe the wound. The dressings need to closely adhere to the surface of the wound, maintain wound exudate, and provide a moist environment conducive to wound closure and promote the shedding of necrotic tissue. The dressings also need to be tightly bonded with the uneven wound surface to reduce the chance of bacterial breeding, prevent wound infection, accelerate neovascularization, and promote epithelial cell growth. These advantages have led to hydrogels' increasing approval for clinical use in recent years.

15.5.2 BARRIERS ON CLINICAL TRANSLATION

In recent years, while more hydrogels have received clinical approval, only a small number of drug delivery hydrogels have successfully navigated the rigorous approval and regulatory process. To ensure successful clinical translation, researchers must consider many factors, including biosafety, cost-effectiveness of raw materials, ability to prepare the gel on a large scale, and adequate preclinical studies of drugs. However, even with perfect functions, new hydrogels prepared in laboratories may not be suitable for clinical use due to issues such as high raw material costs, extended gelation times, and inconsistent material batches.

To address these challenges, current research is focused on the development of low-cost, high-efficacy hydrogels that can be accepted by a larger number of patients. However, achieving clinical translation for new hydrogels requires large-scale production to meet the needs, which poses a significant challenge. Additionally, new bio-hydrogels developed using new polymers, polypeptides, or proteins as raw materials may face difficulties in maintaining material consistency and verifying the interaction of new materials in a short time, which can affect the biosafety control of hydrogels.

Despite the wide availability of materials and gel chemicals, hydrogels are relatively new as a soft material, and their highly complex properties pose significant challenges for scientists to fully understand and characterize the structure-property relationship. For example, determining the mesh size of hydrogels, which controls various properties such as mechanical strength, swelling rate, loading capacity, and molecular permeability, remains an ongoing challenge.[27] Moreover, characterizing the degree of cross-linking and the accurate interaction between molecules is still unclear for physically and chemically cross-linked hydrogels. Finally, understanding the interaction between hydrogels and surrounding normal tissues or diseased sites, and conducting immunology-response assays to assess the impact of hydrogels on tissues after injection, also presents significant challenges.[28]

To further develop novel hydrogels, it is essential to rely on the power of analytical chemistry and biomedical analysis to develop tools to better understand hydrogels. In conclusion, while hydrogels have shown great promise in drug delivery and tissue engineering, further development of novel hydrogels requires careful consideration of numerous factors and continued efforts to address significant challenges in their characterization, production, and clinical translation.

15.6 CONCLUSION AND FUTURE PERSPECTIVES

Drug delivery hydrogels have made significant progress because of the advancement of polymer synthesis. In addition, the field of drug delivery is rapidly evolving with the development of "smart" medicines, and hydrogels are no exception. Despite these advancements, a deeper understanding of hydrogel properties and functions is still necessary. It is particularly crucial to manipulate the properties and functions of drug delivery hydrogels at the molecular and atomic levels to develop cross-linking chemistry that can promote the evolution of hydrogels to meet clinical needs.

The ultimate goal of drug delivery hydrogels is to make medication convenient for patients, and current development trends focus mainly on injectable hydrogel systems. Several stimuli-responsive hydrogels use a minimally invasive approach to treatment, which is highly attractive. To achieve the transition of injectable hydrogels from their solution state outside the body to their gel state after injection, different functional groups are necessary for cross-linking. These groups can also provide targeted delivery and on-demand release functionality after hydrogels are administered, thereby expanding the clinical application of drug delivery hydrogels.

Based on the above summary, we can suggest future developments for drug delivery hydrogels. With the improvement of hydrogel functionality, single-mode drug loading will gradually develop into the loading and controlled release of more and different types of drugs in one hydrogel. These advantages make hydrogels highly beneficial for clinical applications. As the field advances, we anticipate regulatory changes that will approve more "smart" hydrogels for clinical use. This is also the goal that researchers and relevant departments need to work together for in the coming decades.

REFERENCES

[1] J. Li and D. J. Mooney, Designing Hydrogels for Controlled Drug Delivery, *Nat. Rev. Mater.*, 2016, **1**, 1–18.

[2] H. Wu, J. Zheng, A. Kjøniksen, W. Wang, Y. Zhang and J. Ma, Metallogels: Availability, Applicability, and Advanceability, *Adv. Mater.*, 2019, **31**, 1806204.

[3] S. Gu, Y. Lu, Y. Wang, W. Lu and W. Wang, Low Molecular Weight Hydrogel for Wound Healing, *Pharmaceutics*, 2023, **15**, 1119.

[4] Y. Tian, L. Pang, R. Zhang, T. Xu, S. Wang, B. Yu, L. Gao, H. Cong and Y. Shen, Poly-tetrahydropyrimidine Antibacterial Hydrogel with Injectability and Self-Healing Ability for Curing the Purulent Subcutaneous Infection, *ACS Appl. Mater. Interfaces*, 2020, **12**, 50236–50247.

[5] H. Wu, Y. Lei, X. Song, Y. Tan, Z. Sun, Y. Zhang, A. L. Kjøniksen, W. Wang and J. Ma, Real Time Rheological Study of First Network Effects on the In Situ Polymerized Semi-Interpenetrating Hydrogels, *Colloids Surf. A: Physicochem. Eng. Asp.*, 2019, **575**, 111–117.

[6] Z. Wang, Y. Zhang, Y. Yin, J. Liu, P. Li, Y. Zhao, D. Bai, H. Zhao, X. Han and Q. Chen, High-Strength and Injectable Supramolecular Hydrogel Self-Assembled by Monomeric Nucleoside for Tooth-Extraction Wound Healing, *Adv. Mater.*, 2022, **34**, 2108300.

[7] G. U. Ruiz-Esparza, X. Wang, X. Zhang, S. Jimenez-Vazquez, L. Diaz-Gomez, A.-M. Lavoie, S. Afewerki, A. A. Fuentes-Baldemar, R. Parra-Saldivar, N. Jiang, N. Annabi, B. Saleh, A. K. Yetisen, A. Sheikhi, T. H. Jozefiak, S. R. Shin, N. Dong and A. Khademhosseini, Nanoengineered Shear-Thinning Hydrogel Barrier for Preventing Postoperative Abdominal Adhesions, *Nano-Micro Lett.*, 2021, **13**, 212.

[8] J. Ding, T. Liang, Q. Min, L. Jiang and J.-J. Zhu, "Stealth and Fully-Laden" Drug Carriers: Self-Assembled Nanogels Encapsulated with Epigallocatechin Gallate and siRNA for Drug-Resistant Breast Cancer Therapy, *ACS Appl. Mater. Interfaces*, 2018, **10**, 9938–9948.

[9] W. Wang and S. A. Sande, Kinetics of Re-equilibrium of Oppositely Charged Hydrogel–Surfactant System and Its Application in Controlled Release, *Langmuir*, 2013, **29**, 6697–6705.

[10] Y. Gao, K. Peng and S. Mitragotri, Covalently Crosslinked Hydrogels via Step-Growth Reactions: Crosslinking Chemistries, Polymers, and Clinical Impact, *Adv. Mater.*, 2021, **33**, 2006362.

[11] H. C. Kolb, M. G. Finn and K. B. Sharpless, Click Chemistry: Diverse Chemical Function from a Few Good Reactions, *Angew. Chem. Int. Ed.*, 2001, **40**, 2004–2021.

[12] E. Saxon and C. R. Bertozzi, Cell Surface Engineering by a Modified Staudinger Reaction, *Science*, 2000, **287**, 2007–2010.

[13] C. M. Madl, L. M. Katz and S. C. Heilshorn, Bio-Orthogonally Crosslinked, Engineered Protein Hydrogels with Tunable Mechanics and Biochemistry for Cell Encapsulation, *Adv. Funct. Mater.*, 2016, **26**, 3612–3620.

[14] N. J. Agard, J. A. Prescher and C. R. Bertozzi, A Strain-Promoted [3 + 2] Azide–Alkyne Cycloaddition for Covalent Modification of Biomolecules in Living Systems, *J. Am. Chem. Soc.*, 2004, **126**, 15046–15047.

[15] S. T. K. Raja, T. Thiruselvi, A. B. Mandal and A. Gnanamani, pH and Redox Sensitive Albumin Hydrogel: A Self-derived Biomaterial, *Sci. Rep.*, 2015, **5**, 15977.

[16] Z. Wang, W. Li, X. Yang, J. Cao, Y. Tu, R. Wu and W. Wang, Highly Stretchable and Compressible Shape Memory Hydrogels Based on Polyurethane Network and Supramolecular Interaction, *Mater. Today Commun.*, 2018, **17**, 246–251.

[17] C. Shao, H. Chang, M. Wang, F. Xu and J. Yang, High-Strength, Tough, and Self-Healing Nanocomposite Physical Hydrogels Based on the Synergistic Effects of Dynamic Hydrogen Bond and Dual Coordination Bonds, *ACS Appl. Mater. Interfaces*, 2017, **9**, 28305–28318.

[18] H. Xie, M. Asad Ayoubi, W. Lu, J. Wang, J. Huang and W. Wang, A Unique Thermo-Induced Gel-to-Gel Transition in a pH-Sensitive Small-Molecule Hydrogel, *Sci. Rep.*, 2017, **7**, 8459.

[19] B. Ergün, L. De Cola, H.-J. Galla and N. S. Kehr, Surface-Mediated Stimuli Responsive Delivery of Organic Molecules from Porous Carriers to Adhered Cells, *Adv. Healthc. Mater.*, 2016, **5**, 1588–1592.

[20] P. K. Sharma, S. Taneja and Y. Singh, Hydrazone-Linkage-Based Self-Healing and Injectable Xanthan–Poly(ethylene glycol) Hydrogels for Controlled Drug Release and 3D Cell Culture, *ACS Appl. Mater. Interfaces*, 2018, **10**, 30936–30945.

[21] V. Pertici, C. Pin-Barre, C. Rivera, C. Pellegrino, J. Laurin, D. Gigmes and T. Trimaille, Degradable and Injectable Hydrogel for Drug Delivery in Soft Tissues, *Biomacromolecules*, 2019, **20**, 149–163.

[22] A. S. Kuenstler, M. Lahikainen, H. Zhou, W. Xu, A. Priimagi and R. C. Hayward, Reconfiguring Gaussian Curvature of Hydrogel Sheets with Photoswitchable Host–Guest Interactions, *ACS Macro Lett.*, 2020, **9**, 1172–1177.

[23] V. X. Truong, J. Bachmann, A. Unterreiner, J. P. Blinco and C. Barner-Kowollik, Wavelength-Orthogonal Stiffening of Hydrogel Networks with Visible Light, *Angew. Chem. Int. Ed.*, DOI: 10.1002/anie.202113076.

[24] M. Najafi, H. Asadi, J. van den Dikkenberg, M. J. van Steenbergen, M. H. A. M. Fens, W. E. Hennink and T. Vermonden, Conversion of an Injectable MMP-Degradable Hydrogel into Core-Cross-Linked Micelles, *Biomacromolecules*, 2020, **21**, 1739–1751.

[25] A. Kikuchi, K. Suzuki, O. Okabayashi, H. Hoshino, K. Kataoka, Y. Sakurai and T. Okano, Glucose-Sensing Electrode Coated with Polymer Complex Gel Containing Phenylboronic Acid, *Anal. Chem.*, 1996, **68**, 823–828.

[26] Q. Wang, Q. Wang and W. Teng, Injectable, Degradable, Electroactive Nanocomposite Hydrogels Containing Conductive Polymer Nanoparticles for Biomedical Applications, *IJN*, 2016, 131.

[27] M. A. Wisniewska, J. G. Seland and W. Wang, Determining the Scaling of Gel Mesh Size with Changing Cross-Linker Concentration Using Dynamic Swelling, Rheometry, and PGSE NMR Spectroscopy, *J. Appl. Polym. Sci.*, 2018, **135**, 46695.

[28] M. A. Ning Li, H. Ayoubi, J. W. Chen and W. Wang, Co-hydrogelation of Dendritic Surfactant and Amino Acids in Their Common Naturally-occurring Forms: A Study of Morphology and Mechanisms, *Colloid J.*, 2019, **81**, 253–260.

16 Hydrogels for Anti-Pathogen Applications

Yingnan Liu and Su Li
Shenzhen Institute of Advanced Technology Shenzhen,
China and Paris Lodron University of Salzburg, Salzburg,
Austria

Guofang Zhang
Shenzhen Institute of Advanced Technology, Shenzhen, China

Benjamin Punz and Martin Himly
Paris Lodron University of Salzburg, Salzburg, Austria

Yang Li
Shenzhen Institute of Advanced Technology, Shenzhen, China

16.1 INTRODUCTION

Infectious diseases caused by microorganisms, *e.g.*, bacterium, fungus, virus, and parasite, can lead to severe health problems, which attribute to millions of deaths annually. For instance, *Pseudomonas aeruginosa* is a genus of gram-negative aerobic bacteria found in the human intestine and skin. Although it does not present pathogenic in most cases, it can cause disease or even death when the individual's immunity is in a compromised state. Systemic *P. aeruginosa* infection can lead to sepsis, pneumonia, urinary tract infection, and meningitis. In addition, virus infection can cause enormous health and economic threats to society. Taking the prominent coronavirus disease that started in 2019 (COVID-19) as an example, induced by severe acute respiratory syndrome coronavirus-2 (SARS-CoV-2) has caused a global pandemic of an infectious disease lasting for years and is unlikely to be ever completely eradicated. As of November 29th, 2022, the WHO (World Health Organization) has reported 641,915,931 diagnosed COVID-19 cases worldwide, with more than 6,622,760 deaths, resulting in a mortality rate of more than 1.03% (WHO, December 4, 2022). At present, the most common therapeutics for infectious diseases is the administration of anti-microbial drugs; however, the existing anti-microbial drugs are incapable of pathogen inhibition thoroughly. On this occasion, the application of anti-microbial drugs combined with photodynamic and photothermal therapy has been developed in clinical treatment; nevertheless, many challenges remain to be solved. For instance, the current treatment with broad-spectrum could also result in heavy side effects, *e.g.*, toxicity. This is because

DOI: 10.1201/9781003340485-16

of the cytotoxicity induced by these drugs towards normal cells or tissues indistinguishably. Furthermore, there is an increasing issue of drug resistance due to the abuse of common drugs, which limits their further application. Taking all these challenges into consideration, many efforts need to be undertaken for the development of strategies for solving, preventing, and treating pathogen infection, while avoiding the previously mentioned disadvantages.

The rapid development of biomaterials draws great attention in preventive medicine, including diagnosis and treatment for pathogen infection. An appropriate biomaterial used to treat infectious diseases is required to meet the following needs: (1) excellent biocompatibility to the normal cells and the surrounding tissues; (2) selective binding to pathogens; (3) appropriate and tunable mechanical strength to fulfill the practical application; and (4) promoted repair to the damaged tissue. Among the various kinds of biomaterials, *e.g.,* electrospun nanofibers, inorganic nanoparticles, polymers, dendrimers, and liposomes, hydrogels have evolved as highly promising biomaterials to treat infectious diseases due to their satisfying biocompatibility, appropriate mechanical strength, and three-dimensional porous structures. The applicability of hydrogels has improved from single to multiple functions, including physical coverage and drug delivery. Because hydrogels become a promising material platform in disease treatment, herein, as shown in Figure 16.1, we will emphasize the functions and effects of hydrogels, specifically on therapy for infectious diseases. An introduction from the basic principles of hydrogels to the medical applications as well as their applications for pathogen treatment will help us to better understand the importance of alternative approaches to treat infectious diseases and gain enough attention for the next generation of anti-pathogen hydrogel development. Importantly, the current progress of hydrogels for infectious disease therapy will also pave the way for its translational approaches.

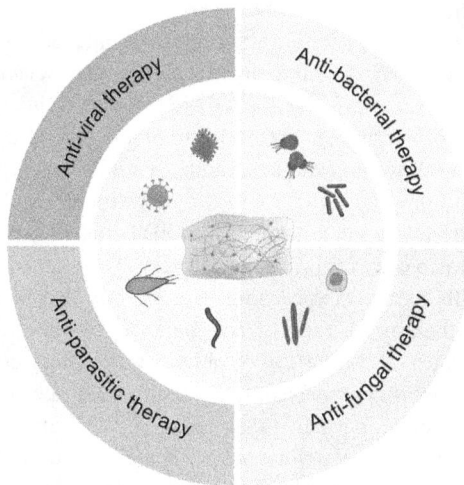

FIGURE 16.1 Hydrogels used in the therapy of infectious diseases, including anti-bacterial, anti-fungal, anti-viral and anti-parasitic treatments.

16.2 HYDROGELS FOR ANTI-BACTERIAL THERAPY

Antibiotics are the most commonly used anti-bacterial strategy. In the past few decades, the abuse of antibiotics has resulted in increasing severity of bacterial resistance and even "superbugs". In addition, antibiotics can lead to severe side effects, such as immediate-type hypersensitivity reactions. To explore innovatively anti-bacterial therapy, like antibiotic-free strategies, is urgently needed. Fortunately, the anti-bacterial hydrogels catch the attention.

Hydrogels are an appropriate platform for drug delivery and local release. The positively charged molecules, polyethyleneimine (PEI) and tobramycin (Tob), interacted with biocompatible oxidized carboxymethylcellulose (OCMC) via a Schiff base reaction to construct the OCMC-Tob/PEI cationic hydrogel, which showed injectable, self-healing, and good mechanical properties [1]. The biocompatible positive charged OCMC-Tob/PEI hydrogel could not only be used to eliminate a variety of negatively charged risk molecules (*e.g.,* cell-free DNA, LPS, and TNF-α) by electrostatic interaction to modulate inflammation, but also can respond to low pH values for releasing Tob as a broad-spectrum antimicrobial agent at the infectious site. This approach, which combines the broad-spectrum and long-term anti-bacterial strategy with cationic hydrogels has been found to accelerate the healing of *P. aeruginosa*–infected diabetic wounds through a dual function way of anti-inflammation and anti-bacteria. Thus, hydrogels provide a useful platform for a sustained release of anti-bacterial drugs and functional molecules to treat bacterial infections. Hydrogels could be used for photodynamic therapy (PDT), a promising method to tackle antibiotic resistance. Not like antibiotics, PDT kills the bacteria with photosensitizers that can generate reactive oxygen species (ROS) to destroy the cell walls or even the DNA of the bacteria. Unfortunately, most of the photosensitizers are water-insoluble and have aggregation-caused quenching (ACQ) effects, limiting their interaction with the bacterial cell wall, as well as reducing the ROS yields and the effectiveness of photodynamic therapy against bacteria. While a water-soluble near-infrared fluorescence-emitting rho-damine photosensitizer (CS-2I) with an aggregation-induced emission (AIE) effect has been reported recently, which can effectively and selectively identify and eliminate drug-resistant gram-positive bacteria [2]. Notably, due to the AIE property, CS-2I can combine with Carbomer 940 to form an anti-microbial hydrogel dressing (CS-2I@gel), which showed excellent anti-microbial and pro-wound healing properties *in vitro* and *in vivo*. This provides a new strategy and could be used as a blueprint for the next-generation anti-microbial PDT materials.

Hydrogels can be also used for probiotic delivery. To be noted, the ROS generated by PDT not only kills the disease-causing bacteria but also inhibits the growth of beneficial bacteria, thereby disrupting the balance of the microbial ecosystem surrounding the wounds. To solve this problem, researchers developed a creative hydrogel scaffold containing live bacteria, which could accelerate the healing of infected wounds by spreading beneficial bacteria with anti-bacterial substances (Figure 16.2). The probiotic, *Lactobacillus royale*, could inhibit the growth of pathogenic bacteria by lowering the local pH and producing reuterin, an anti-microbial agent, during metabolism [3]. The *Lactobacillus royale* was encapsulated in hydrogel microspheres and formed into a hydrogel dressing *in situ* in the wound by

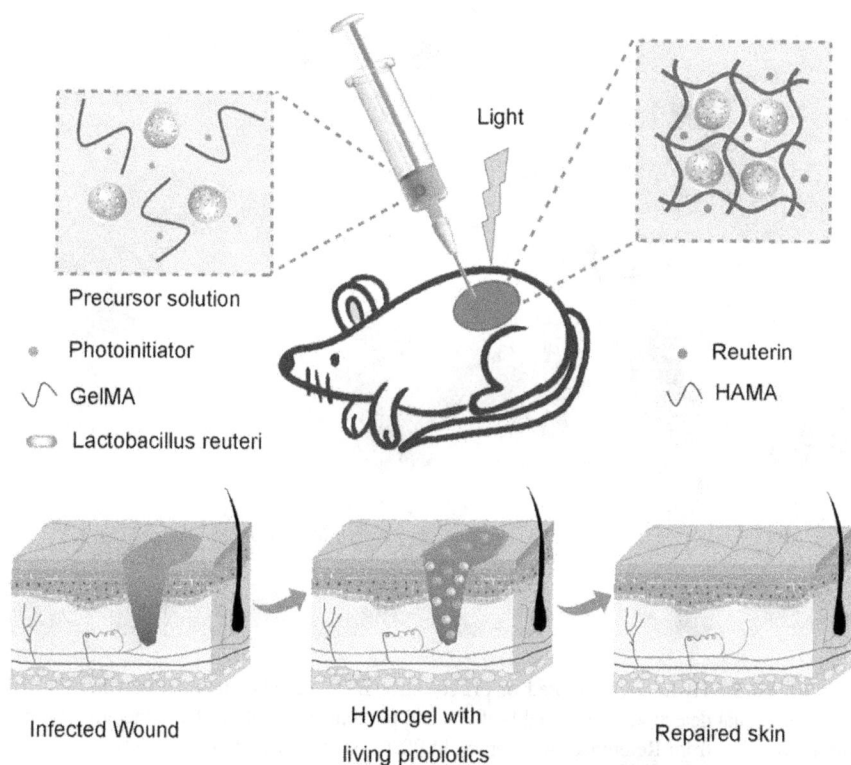

FIGURE 16.2 The hydrogel containing microspheres encapsulated live probiotics for killing harmful bacteria by releasing reuterin from probiotics. Adapted with permission from Reference [3]. Copyright 2021, Wiley.

photo-cross-linking. This hydrogel scaffold protects the beneficial bacteria from the immune system, in parallel to preventing the beneficial bacteria from escaping into the local environment, averting a potential threat. Both *in vitro* and *in vivo* experiments have shown that this hydrogel has excellent resistance to harmful bacteria by releasing reuterin from probiotics, as well as an anti-inflammatory capacity for promoting closure of infected wounds and regeneration of new tissue.

Hydrogels could be used as a theranostic platform as well. Theranostics have attracted extensive attention to anti-bacterial therapy, whereas the rapid diagnosis of bacterial infection is difficult. Numerous studies are mostly focused on the detection of planktonic bacteria and are limited to practical applications. In addition, even when a bacterial infection is detected, the subsequent treatment normally cannot be applied immediately and the effectiveness of the treatment cannot be assessed timely either. Integrating rapid diagnosis and effective treatment is urgently needed. Recently, a smart hydrogel integrating *in situ* visual diagnosis of bacterial infection and photothermal therapy has been developed (Figure 16.3). In this study, the Bromothymol blue (BTB), as a pH-sensitive diagnostic reagent, poly-{2,5-thiophen-co-[3,6-di(thiophen-2-yl)-2,5-bis(N,N,N-trime-thylhexan-1-aminiu-m)pyrrolo[3,4-c]

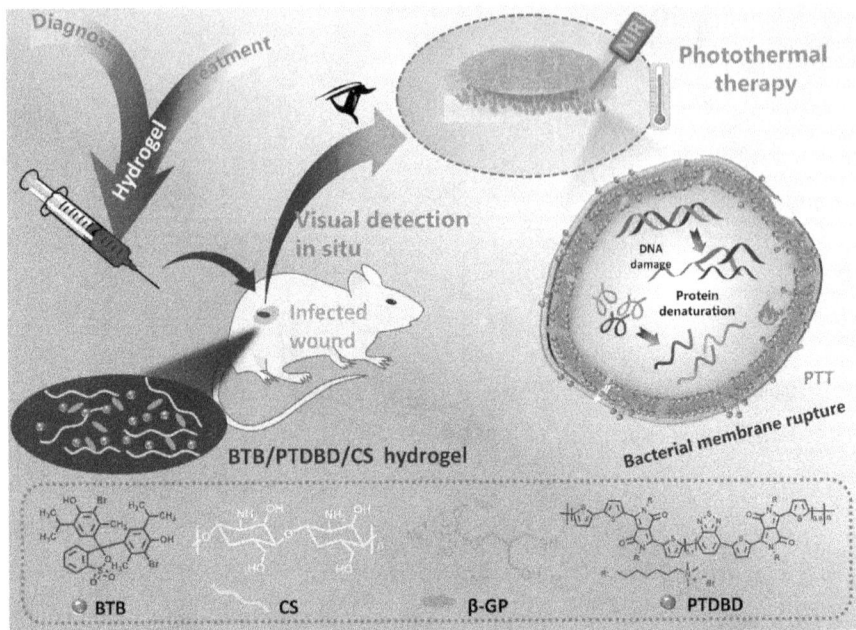

FIGURE 16.3 Hydrogel composed of pH-sensitive BTB and photothermal polymer PTDBD for *in situ* visual detection of bacterial biofilms and photothermal anti-bacterial therapy. Adapted with permission from Reference [4]. Copyright 2020, American Chemical Society.

pyrrole-1,4(2 H,5 H)-dione]-bromide-co-4,7-(2,1,3-benzothia-diazole)} termed PTDBD for photothermal conversion was loaded into a chitosan (CS)-based hydrogel by self-assembly. The obtained BTB/PTDBD/CS hydrogel enables visual detection of biofilms and naked-eye diagnosis of infected wounds because of the color change of BTB in the acid environment after bacterial infection [4]. After rapid diagnosis, the hydrogel immediately treated the infected area with local thermal therapy under near-infrared laser (808 nm) irradiation. The hydrogel showed good biocompatibility on cells *in vitro* and major organs in mice. It opens a new path for the development of an intelligent and simple diagnostic and therapeutic platform for bacterial infection. In another study, a fluorescent anti-bacterial hydrogel with excellent optical properties and good biocompatibility was developed using cationic carbon dots (CDs) with intrinsic anti-bacterial properties, together with pectin and acrylic acid as cross-linking agents [5]. The variations in the surrounding environment due to bacterial growth caused hydrogen bond breakage to release CDs from the hydrogel. In this way, the fluorescent anti-bacterial hydrogel obtained a long-term anti-bacterial activity. The hydrogels combine anti-bacterial function and fluorescence sensing, and thus open a wide range of promising biological applications.

Most hydrogels contain metal or inorganic nanoparticles that exhibit poor biocompatibility, and even the cross-linking reagents can also bring cytotoxicity. On this occasion, natural products have been applied for anti-bacteria. Inspired by the capability of mussels to adhere strongly on organic or inorganic substrates, the filament

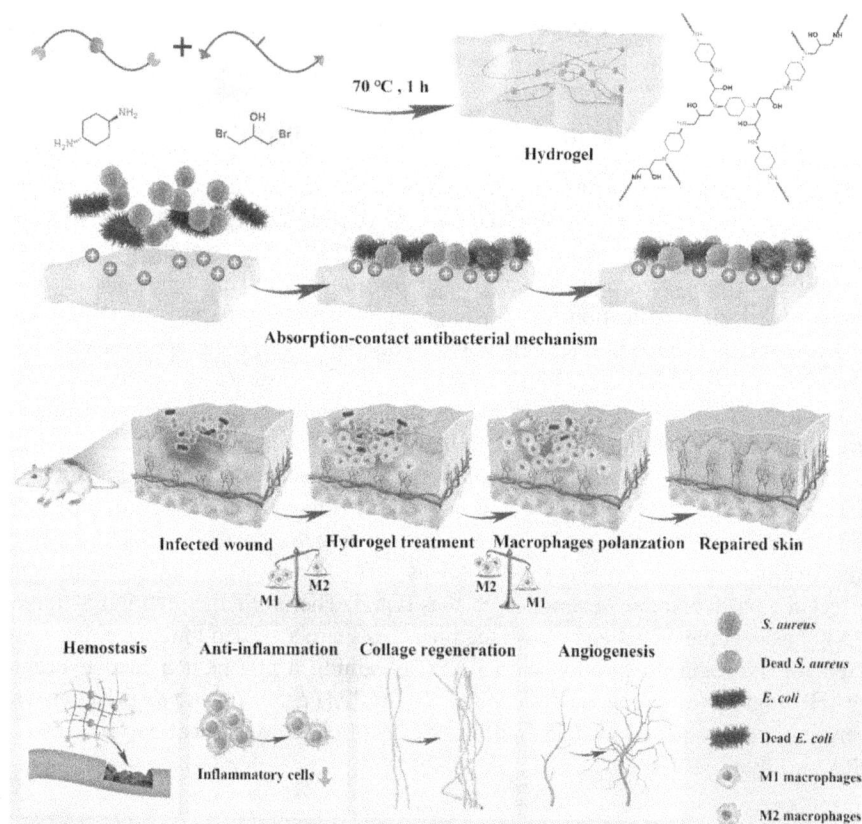

FIGURE 16.4 A multifunctional cationic polymeric dendritic hydrogel for promoting wound healing in bacterial infection by electrostatically adsorbing and killing bacteria, promoting hemostasis, anti-inflammation, collaged regeneration, and angiogenesis. Adapted with permission from Reference [7]. Copyright 2022, American Chemical Society.

protein of mussel foot and mussel foot protein-5 (Mfp-5) has been designed for a novel adhesive hydrogel [6]. To be noted, Mfp-5-derived peptide shows anti-bacterial property and exhibits inhibitory activity against drug-resistant gram-positive bacteria. In another study, a new efficient synthetic strategy was developed for the preparation of nontoxic cationic hydrogel poly(2-hydroxypropylene imine cyclohexene imine) (PHCI) by cross-linking *trans*-1,4-cyclohexanediamine with 1,3-dibromo-2-propanol without the use of toxic cross-linkers into the condensation reaction [7]. The prepared PHCI hydrogel had an intrinsically anti-microbial capacity that electrostatically adsorbed and killed *Staphylococcus aureus* and *Escherichia coli* (Figure 16.4). Notably, *in vivo* experiments with normal or diabetic rat models showed that PHCI hydrogels could rapidly stop bleeding, effectively kill bacteria, promote the polarization of macrophages from a pro-inflammatory M1 to an M2 phenotype, and accelerate collagen deposition and blood vessel formation, resulting in rapid wound healing. Overall, this work presents an effective anti-microbial and nontoxic hydrogel that may provide an easy and effective approach to clinical wound management.

16.3 HYDROGELS FOR ANTI-FUNGAL THERAPY

The infected fungi can proliferate at the site of the lesion such as skin, tissues, or organs, and lead to enteritis, pneumonia, or even mortality, which makes fungal infections vastly different from bacterial infections. The wetting property of hydrogels makes them a breeding ground for fungal colonialization and greatly limits hydrogels' application for anti-fungi. A widely adopted anti-fungal strategy of hydrogels is to incorporate anti-fungal drugs. However, there are relatively few effective anti-fungal agents. Furthermore, the drug resistance to commonly used anti-fungal agents, for instance, azoles and echinocandins, is increasing, creating an urgent need for new anti-fungal agents.

Previous research showed that a solution rich in reactive free radicals, known as plasma activated water (PAW), could be generated through the interaction of low-temperature plasma with an aqueous solution, which showed a good microbial inactivation capacity [8]. However, it was difficult to sustain its anti-microbial effect at a particular site because PAW was liquid. Therefore, the immobilization of PAW has become a challenge for researchers. Researchers tried to replace the water used for hydrogel preparation with PAW, obtaining a plasma-activated hydrogel (PAH) (Figure 16.5). The anti-fungal activity of PAH against *Candida albicans* was verified, which showed a better inhibitory effect than conventional polyacrylamide (PAAm) hydrogel.

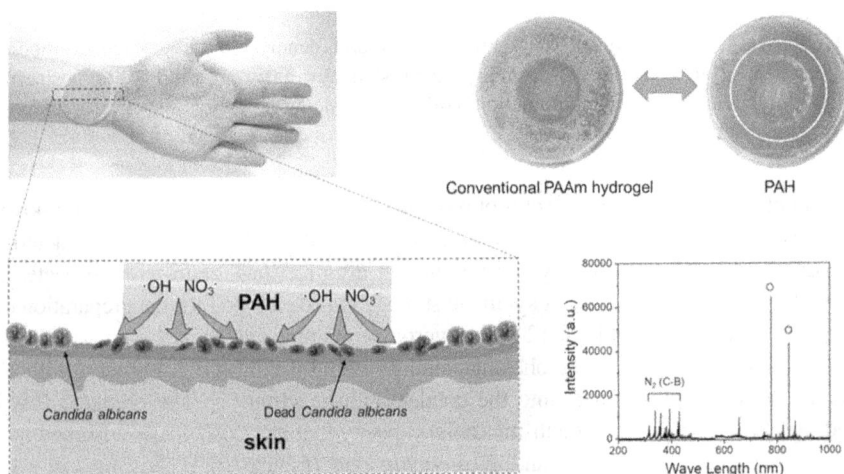

FIGURE 16.5 Polycyclic aromatic hydrogels obtained by directly dissolving acrylamide in plasma activated water and adding cross-linkers, initiators, and polymerization inhibitors for long-term anti-fungal therapy. Adapted with permission from Reference [8]. Copyright 2019, American Chemical Society.

16.4 HYDROGELS FOR ANTI-VIRAL THERAPY

16.4.1 THERAPY AGAINST VIRAL INFECTION

Virus infection causes many infectious diseases. Importantly, a continuous infection that leads to persistent inflammation can cause chronic inflammatory diseases or even cancer. Herein, we summarize the current progress on anti-viral hydrogels and provide some key information for their future development. Based on current research progress, hydrogels could be used as carriers for sustained or controlled release of anti-viral drugs or materials, or also could be directly used as anti-viral agents. To date, hydrogels were shown to obtain anti-viral activities against many kinds of viruses, including human immunodeficiency virus (HIV), hepatitis B virus (HBV), herpes simplex virus (HSV), cytomegalovirus, and human papillomavirus (HPV).

In terms of drug delivery for the treatment of HIV, hydrogels can provide sustained release of anti-HIV drugs because of the cross-linked network. A supramolecular nanofibrous hydrogel for multifunctional anti-HIV therapy was developed, which was synthesized by covalent conjugation of anti-HIV agents extracted from Epivir tablets as reverse transcriptase inhibitors and non-steroid anti-inflammatory drugs (NSAIDs) [9]. To be noted, an addition of prostatic acid phosphatase (PAP) enhanced the mechanical and elastic properties of this hydrogel.

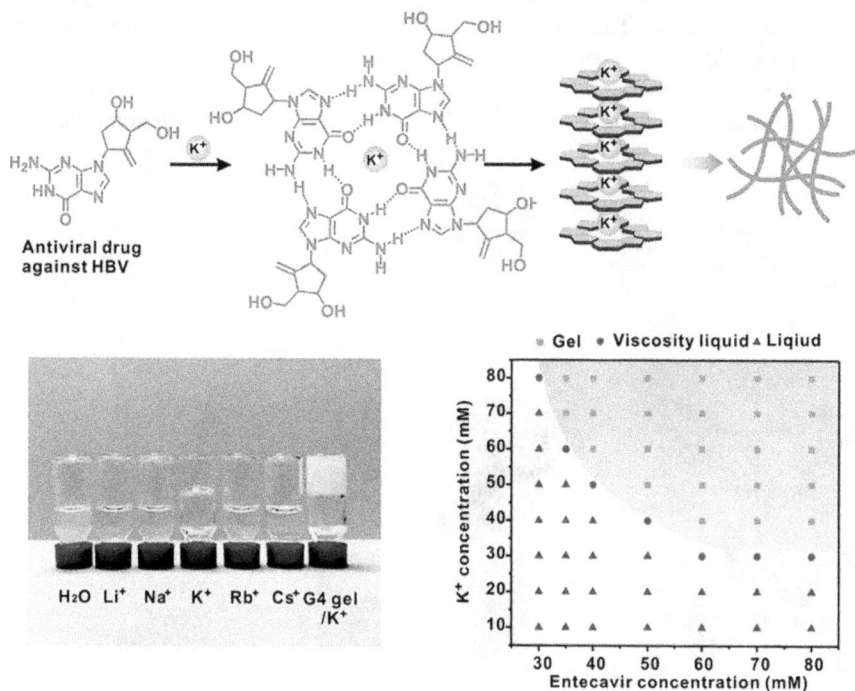

FIGURE 16.6 Guanosine tetrahydrate gel prepared by direct gelation of clinically available anti-viral drugs such as entecavir, penciclovir, and ganciclovir in the presence of K^+ for anti-viral therapy. Adapted with permission from Reference [10]. Copyright 2019, Royal Society of Chemistry.

This study, for the first time, proved that a hydrogel could be fabricated by self-assembled nanofibers for anti-HIV drug delivery. Importantly, this method could be considered as a potential approach to fight against other viruses.

In addition to HIV, the possibility of delivering clinically available anti-HBV drugs using hydrogels has been investigated. For instance, hydrogels could be made with guanine derivatives. Based on this, researchers developed a series of innovative hydrogels to load and release clinical anti-viral drugs, as shown in Figure 16.6 [10]. In detail, considering that many clinically anti-viral drugs, such as entecavir, penciclovir, and ganciclovir, all have the guanosine group. Thus, the hydrogel was prepared via the interaction between guanine in anti-viral drugs and K^+ by a hydrogen bond. Results indicated that the entecavir-loaded anti-viral hydrogel exhibited a therapeutic effect in HBV-infected HepG2.2.15 cells. Ganciclovir and penciclovir, which work as anti-HSV drugs, could also be loaded in hydrogels for anti-viral therapy. Importantly, the sustained release of the anti-viral drugs could result in an enhanced and long-lasting anti-viral therapeutic effect. In addition, a hydrogel with the property of pH-responsive release had also been invented to treat HSV. The cross-linked hydrogel was fabricated by the ionic interaction between the amino groups of chitosan and the carboxyl groups of xanthan gum (XG), of which chitosan was the main carrier and the XG enhanced its solubility [11]. In addition, the monomer 2-acrylamido-2-methylpropane sulfonic acid was used to endow the hydrogel with pH-responsive properties for the controlled release of acyclovir. Thus, considering the pH-dependent release behavior, the hydrogels could be developed with smart properties for controlled drugs release.

For the treatment of cytomegalovirus, a thermosensitive hydrogel was developed to *in situ* release ganciclovir (GCV), a nucleoside viral replication inhibition drug [12].

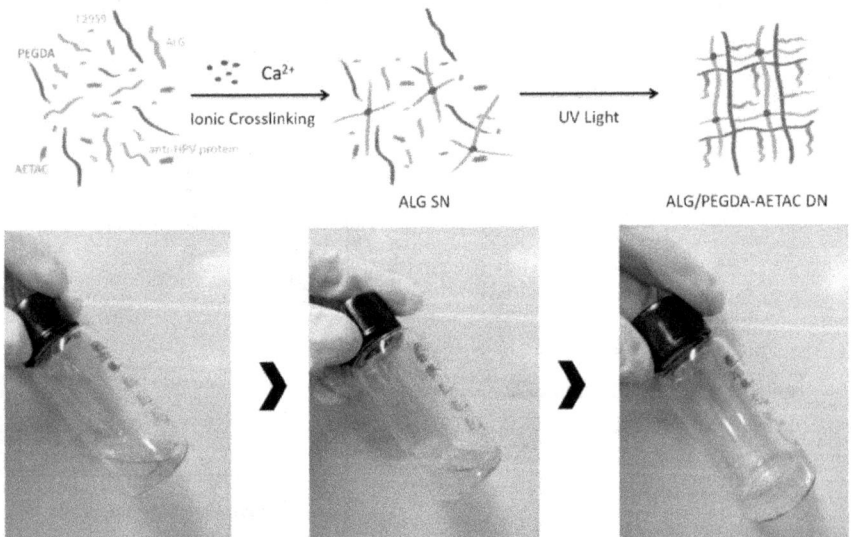

FIGURE 16.7 A double network hydrogel prepared by calcium cross-linking and photoinitiator induction for releasing anti-human papillomavirus protein. Adapted with permission from Reference [13]. Copyright 2021, MDPI.

It was noted that this hydrogel was liquid at room temperature and could become solid and provide sustained release at 33–38°C. In summary, hydrogels are popular delivery vehicles in the field of anti-viral treatment.

In terms of anti-HPV therapy based on hydrogels, the traditional single network (SN) hydrogels are limited by poor mechanical properties and rapid drug release. To solve this problem, a recent study reported a novel dual-network (DN) hydrogel; see Figure 16.7, which used an alginate (ALG)/polyethylene glycol diacrylate-diacrylate (PEGDA) network to improve the mechanical properties and network density of the hydrogel [13]. In addition, to achieve a slower release of the integrated drugs, a negatively charged anti-HPV protein (bovine-lactoglobulin modified with 3-hydroxyphthalic anhydride) was loaded into the hydrogel by electrostatic interactions with a positively charged monomer [2-(acryloyloxy) ethyl] trimethylammonium chloride (AETAC). Compared to PEGDA SN hydrogels and ALG SN hydrogels, ALG/PEGDA-AETAC DN hydrogel showed a much stronger mechanical property and a longer time for anti-viral protein release, achieving a continuous release over 14 days, which may be due to the dense network structure. In conclusion, the smarter DN or multi-functional hydrogel should be designed with controlled drug release capacity, good mechanical properties, and excellent biocompatibility for future anti-viral applications (Figure 16.7).

In addition to the delivery of drugs or small molecules, hydrogels can also be applied for the controlled delivery of nanomaterials that possess anti-viral functions. Tannic acid-modified silver nanoparticles (TA-AgNPs) were reported to prevent HSV attachment and penetration, as well as reduce inflammation simultaneously [14]. To enhance the therapeutic effect, a hydrogel made by Carbopol 974 P cross-linking with polyacrylic acid derivative was synthesized as the delivery platform for sustained release of the TA-AgNPs, which showed excellent performance for anti-HSV with both *in vitro* and *in vivo* experiments.

16.4.2 Perspective for Hydrogels against COVID-19 and Other Coronaviruses

As the pandemic caused by SARS-CoV-2 is still ongoing, it seems that coronaviruses may continue to affect us in the future. Thus, it is important to continuously develop novel methods to constrain coronaviruses. Indeed, many studies and biomaterials have been developed for anti-SARS-CoV-2, including biomimetic cytomembrane nanoparticles [15], AIE luminogens [16], and 2D-nanosheets [17]. Although many researchers emphasize developing hydrogels for vaccines of SARS-CoV-2, recent works have started to pay attention to fabricating hydrogel-based biomaterials for anti-SARS-CoV-2. We believe that hydrogels can act as a promising candidate for SARS-CoV-2 restriction, mostly because of their excellent bioactivity and biocompatibility.

For vaccine development, the construction of hydrogels to deliver a receptor binding region (RBD) of SARS-CoV-2 as an antigen has been tried in order to overcome the short life of the RBD antigen for efficient antibody production. For instance, a biodegradable and injectable hydrogel prepared by the ionic interaction of poly(N-(3-aminopropyl)methacrylamide)-co-(N-[Tris(hydroxymethyl)methyl] acrylamide)p(APMA-THMA) and hyaluronic acid was reported for sustained release of the poly (I:C)-adjuvanted SARS-CoV-2 RBD, which is because of a strong interaction of RBD with the other amino group and the hydroxyl group of *p* (APMA-THMA) and together with a high loading efficiency of the RBD antigen [18]. In another study, a polymer-nanoparticle (PNP) hydrogel made of hydrophobically modified hydroxypropylmethylcellulose derivates (HPMC-C_{12}) and degradable poly(ethyleneglycol)-b-poly(lactic acid) nanoparticles were developed to deliver the RBD antigen and adjuvant complex, including class B CpG ODN1826 (CpG) and Alhydrogel (Alum) [19]. Using this approach, a high antibody titer was generated after the injection of a hydrogel-based vaccine against different variants, *e.g.,* Alpha, Beta, and Delta.

To be noted, few studies focus on developing hydrogels to inhibit SARS-CoV-2 infection currently. However, some valuable researches could still provide valuable suggestions for future anti-SARS-CoV-2 hydrogel development. Methacrylate hyaluronic acid hydrogel microspheres that were modified with engineered ACE2-expressing cell membranes and macrophages membranes were designed to specifically suppress SARS-CoV-2 cellular infection and to inhibit the section of inflammatory cytokines, such as TNF-α, IL-1β, and IL-6 [20]. In the future work, based on the plasticity of hydrogel, we hope more efficient anti-SARS-CoV-2 hydrogels could be developed to prevent the virus infection or transmission. For example, sprayable hydrogels could be designed to stop airborne virus transmission. In addition, we believe that a combination of other anti-SARS-CoV-2 nanomaterials with a hydrogel may also show an enhanced therapeutic effect.

16.5 HYDROGELS FOR ANTI-PARASITIC THERAPY

The morbidity of parasitic diseases depends mainly on the number and virulence of the parasites invading the human bodies. Amoebiasis, malaria, trichomoniasis of the vagina, and leishmaniasis are all closely associated with our life as severe parasitic infections. In terms of anti-parasites, the porous structure and physicochemical properties, such as thermosensitive, endow the hydrogels with the ability of limiting parasite motility to prevent their further infectivity. A thermosensitive hydrogel made by pluronic F127 could be used as a physical barrier to prevent vaginal trichomoniasis, because it can restrict *Trichomonas vaginalis* motility completely [21]. A cross-linked network can be formed to immobilize the parasites once the hydrogels are placed at the condition with a higher temperature than room temperature. A similar hydrogel has also been used in capturing human pathogenic *Acanthamoeba castellanii*. The hydrogel was invented by a unique interconnected microchannel structure, which was formed by removing zinc oxide (ZnO) templates from the inner part of the hydrogel [22]. The interconnected microchannels of hydrogels showed great potential for reducing parasite infection by capturing and limiting the parasites in the interconnected network.

In addition to physical immobilization, hydrogels have also been explored in drug delivery to treat parasitic infections. Leishmaniasis is a parasitic disease of the skin, mucous membranes, or internal organs caused by the *Leishmania* protozoa, which can be transmitted to humans by the bite of the phlebotomine sandfly. Although the Buparvaquone (BPQ) exhibits effective anti-parasite activity, the poor solubility limits its clinical application. A BPQ-loaded nanoscale hydrogel was synthesized for transdermal drug delivery, which indicated an excellent permeability and release behavior *in vivo* pharmacokinetic assay [23].

16.6 CONCLUSIONS AND OUTLOOK

Hydrogels have been established as promising candidates for anti-pathogen treatment in the past few years, with yearly accelerating clinical applications. At present, the structures and functions of hydrogels have changed from SN displaying one prominent function to DN or multi-network with more complicated structures and multiple functions. Therefore, we summarized the current progress by highlighting functional hydrogels for anti-bacterial, anti-fungal, anti-viral, and anti-parasitic therapy. For anti-bacterial hydrogels, studies mainly focus on the delivery of anti-bacterial drugs, molecules, peptides, or nanomaterials for efficient therapy. In these systems, hydrogels could provide a much longer and sustained release of anti-pathogen agents. Another smarter hydrogel platform with a controlled release property, *e.g.,* pH-responsive hydrogel, was also designed to release the integrated anti-pathogen agents in specific wounds or infectious sites of different pH regions. Furthermore, innovative treatment strategies, for instance, PTT/PDT combinational anti-bacterial therapy, were also developed with hydrogels. However, extra data is still needed to ensure their potential for practical applications. Compared to anti-bacterial hydrogels, fewer studies have been reported for anti-fungal hydrogels, of which the effects mainly derive from the delivered anti-fungal agents. However, we should recognize that the resistance to widely used anti-fungal drugs is increasing; thus, we also need to explore drug-free therapeutic strategies based on novel biomaterials. On the other hand, fungal infections can also cause diseases like enteritis, pneumonia, or even death. Thus, prevention or therapy toward fungal infections in internal organs should deserve more attention; for instance, the injectable hydrogels that are used to prevent fungal infection in deep tissues. With regard to anti-viral therapeutic approaches with hydrogels, delivery hydrogels as well as directly inhibiting hydrogels have been developed. However, many studies concentrated on improving the physicochemical and structural properties of hydrogels, but their clinical application was not well studied. Thus, to improve the efficacy and safety of hydrogels in clinics, more investigations toward biosafety are highly desired in the future to ensure their potential applications. Hydrogels were also used to fight against the current pandemic COVID-19, which was mainly about delivering antigens for vaccine development against SARS-CoV-2. Thus, we think hydrogels with a controllable structure may also be used to constrain SARS-CoV-2 or other coronaviruses; for instance, sprayable hydrogels or injectable hydrogels to stop the spread of coronaviruses. In such cases, hydrogels may become a more promising candidate for future virus treatment to stop pandemics caused by coronaviruses.

Hydrogels with various modes that are used for anti-pathogens have been designed and proven recently, showing improved therapeutic effects compared to traditional treatment strategies used for pathogen prevention and cure. However, the specific mechanisms of these developed hydrogels often lack further exploration, which greatly limits their future clinical applications. Importantly, although the safety issue of hydrogels is often claimed, a more detailed biosafety evaluation of these designed materials should be provided before their future applications. We, nevertheless, anticipate that hydrogels will be further developed and investigated in more detail as time goes by, and thus be easily utilized for practical applications and for better fulfilling the current urgent needs in the field of anti-pathogen applications.

ACKNOWLEDGMENTS

This study was supported by the International Partnership Program (IPP) of CAS (172644KYSB20210011), the Austrian Research Promotion Agency (FFG) (890610), the National Natural Science Foundation of China (32171390, 82261138630, 32201154), the Natural Science Foundation of Guangdong Province (2023B1515020104 and 2023A0505050123), the Key Collaborative Research Program of the Alliance of International Science Organizations (ANSO-CR-KP-2022-01).

REFERENCES

[1] K. Zhang, C. Yang, C. Cheng, C. Shi, M. Sun, H. Hu, T. Shi, X. Chen, X. He, X. Zheng, M. Li, D. Shao. Bioactive injectable hydrogel dressings for bacteria-infected diabetic wound healing: a "pull-push" approach. *ACS Appl Mater Interfaces*. 2022, **14**, 26404–26417.

[2] S. Zeng, Z. Wang, C. Chen, X. Liu, Y. Wang, Q. Chen, J. Wang, H. Li, X. Peng, J. Yoon. Construction of rhodamine-based AIE photosensitizer hydrogel with clinical potential for selective ablation of drug-resistant gram-positive bacteria in vivo. *Adv Healthc Mater*. 2022, **11**, e2200837.

[3] Z. Ming, L. Han, M. Bao, H. Zhu, S. Qiang, S. Xue, W. Liu. Living bacterial hydrogels for accelerated infected wound healing. *Adv Sci (Weinh)*. 2021, **8**, e2102545.

[4] H. Wang, S. Zhou, L. Guo, Y. Wang, L. Feng. Intelligent hybrid hydrogels for rapid in situ detection and photothermal therapy of bacterial infection. *ACS Appl Mater Interfaces*. 2020, **12**, 39685–39694.

[5] F. Cui, J. Sun, J. Ji, X. Yang, K. Wei, H. Xu, Q. Gu, Y. Zhang, X. Sun. Carbon dots-releasing hydrogels with antibacterial activity, high biocompatibility, and fluorescence performance as candidate materials for wound healing. *J Hazard Mater*. 2021, **406**, 124330.

[6] G. Fichman, C. Andrews, N. L. Patel, J. P. Schneider. Antibacterial gel coatings inspired by the cryptic function of a mussel byssal peptide. *Adv Mater*. 2021, **33**, e2103677.

[7] S. Cheng, H. Wang, X. Pan, C. Zhang, K. Zhang, Z. Chen, W. Dong, A. Xie, X. Qi. Dendritic hydrogels with robust inherent antibacterial properties for promoting bacteria-infected wound healing. *ACS Appl Mater Interfaces*. 2022, **14**, 11144–11155.

[8] Z. Liu, Y. Zheng, J. Dang, J. Zhang, F. Dong, K. Wang, J. Zhang. A novel antifungal plasma-activated hydrogel. *ACS Appl Mater Interfaces*. 2019, **11**, 22941–22949.

[9] J. Li, X. Li, Y. Kuang, Y. Gao, X. Du, J. Shi, B. Xu. Self-delivery multifunctional anti-HIV hydrogels for sustained release. *Adv Healthc Mater*. 2013, **2**, 1586–1590.

[10] J. Hu, H. Wang, Q. Hu, Y. Cheng. G-quadruplex-based antiviral hydrogels by direct gelation of clinical drugs. *Mater Chem Front.* 2019, **3**, 1323–1327.

[11] N. S. Malik, M. Ahmad, M. U. Minhas, R. Tulain, K. Barkat, I. Khalid, Q. Khalid. Chitosan/xanthan gum based hydrogels as potential carrier for an antiviral drug: fabrication, characterization, and safety evaluation. *Front Chem.* 2020, **8**, 50.

[12] Q. Wang, C. Sun, B. Xu, J. Tu, Y. Shen. Synthesis, physicochemical properties and ocular pharmacokinetics of thermosensitive in situ hydrogels for ganciclovir in cytomegalovirus retinitis treatment. *Drug Deliv.* 2018, **25**, 59–69.

[13] C. Zhao, J. Ji, T. Yin, J. Yang, Y. Pang, W. Sun. Affinity-controlled double-network hydrogel facilitates long-term release of anti-human papillomavirus protein. *Biomedicines.* 2021, **9**, 1298.

[14] E. Szymańska, P. Orłowski, K. Winnicka, E. Tomaszewska, P. Bąska, G. Celichowski, J. Grobelny, A. Basa, M. Krzyżowska. Multifunctional tannic acid/silver nanoparticle-based mucoadhesive hydrogel for improved local treatment of HSV infection: in vitro and in vivo studies. *Int J Mol Sci.* 2018, **19**, 387.

[15] Z. Li, Z. Wang, P. C. Dinh, D. Zhu, K. D. Popowski, H. Lutz, S. Hu, M. G. Lewis, A. Cook, H. Andersen, J. Greenhouse, L. Pessaint, L. J. Lobo, K. Cheng. Cell-mimicking nanodecoys neutralize SARS-CoV-2 and mitigate lung injury in a non-human primate model of COVID-19. *Nat Nanotechnol.* 2021, **16**, 942–951.

[16] M. Y. Wu, M. Gu, J. K. Leung, X. Li, Y. Yuan, C. Shen, L. Wang, E. Zhao, S. Chen. A membrane-targeting photosensitizer with aggregation-induced emission characteristics for highly efficient photodynamic combat of human coronaviruses. *Small.* 2021, **17**, e2101770.

[17] G. Zhang, Y. Cong, F. L. Liu, J. Sun, J. Zhang, G. Cao, L. Zhou, W. Yang, Q. Song, F. Wang, K. Liu, J. Qu, J. Wang, M. He, S. Feng, D. Baimanov, W. Xu, R. H. Luo, X. Y. Long, S. Liao, Y. Fan, Y. F. Li, B. Li, X. Shao, G. Wang, L. Fang, H. Wang, X. F. Yu, Y. Z. Chang, Y. Zhao, L. Li, P. Yu, Y. T. Zheng, D. Boraschi, H. Li, C. Chen, L. Wang, Y. Li. A nanomaterial targeting the spike protein captures SARS-CoV-2 variants and promotes viral elimination. *Nat Nanotechnol.* 2022, **17**, 993–1003.

[18] J. Chen, B. Wang, J. S. Caserto, K. Shariati, P. Cao, Y. Pan, Q. Xu, M. Ma. Sustained delivery of SARS-CoV-2 RBD subunit vaccine using a high affinity injectable hydrogel scaffold. *Adv Healthc Mater.* 2022, **11**, e2101714.

[19] E. C. Gale, A. E. Powell, G. A. Roth, E. L. Meany, J. Yan, B. S. Ou, A. K. Grosskopf, J. Adamska, V. Picece, A. I. d'Aquino, B. Pulendran, P. S. Kim, E. A. Appel. Hydrogel-based slow release of a receptor-binding domain subunit vaccine elicits neutralizing antibody responses against SARS-CoV-2. *Adv Mater.* 2021, **33**, e2104362.

[20] Z. Wang, L. Xiang, F. Lin, Z. Cai, H. Ruan, J. Wang, J. Liang, F. Wang, M. Lu, W. Cui. Inhaled ACE2-engineered microfluidic microsphere for intratracheal neutralization of COVID-19 and calming of the cytokine storm. *Matter.* 2022, **5**, 336–362.

[21] S. Malli, P. M. Loiseau, K. Bouchemal. Trichomonas vaginalis motility is blocked by drug-free thermosensitive hydrogel. *ACS Infect Dis.* 2020, **6**, 114–123.

[22] S. B. Gutekunst, K. Siemsen, S. Huth, A. Möhring, B. Hesseler, M. Timmermann, I. Paulowicz, Y. K. Mishra, L. Siebert, R. Adelung, C. Selhuber-Unkel. 3D hydrogels containing interconnected microchannels of subcellular size for capturing human pathogenic acanthamoeba castellanii. *ACS Biomater Sci Eng.* 2019, **5**, 1784–1792.

[23] A. Lalatsa, L. Statts, J. Adriana de Jesus, O. Adewusi, M. Auxiliadora Dea-Ayuela, F. Bolas-Fernandez, M. Dalastra Laurenti, L. Felipe Domingues Passero, D. R. Serrano. Topical buparvaquone nano-enabled hydrogels for cutaneous leishmaniasis. *Int J Pharm.* 2020, **588**, 119734.

17 Hydrogels for Environmental Applications

Ashok Bora, Dimpee Sarmah, and Niranjan Karak
Tezpur University, Tezpur, India

17.1 INTRODUCTION

In the last few decades, the rapid development of industries and worldwide growth of the population led to more and more serious environmental problems, including air, water, and soil pollutions. In order to solve these problems, lots of research works have been performed to develop greener technologies throughout the world. Recently, researchers have found hydrogels as a good alternative in solving some industrial, biological, and ecological problems. Hydrogels were first reported in 1960 by Wichterle and Lim, and since then, this incredible material has advanced from simply inert to complex active systems. They are a three-dimensionally (3D) cross-linked network of hydrophilic polymeric chains, which can absorb and retain huge amounts of water or biological fluids within their network structures, even under certain pressure. They can be prepared in different forms such as films, beads, and nanocomposites, and are synthesized by cross-linking reactions of bio-based polymers (such as starch, guar gum, chitosan, etc.) or synthetic monomers (like acrylamide (AM), methacrylamide (MAM), acrylic acid (AA), etc.) with cross-linkers such as N, Ń-methylene bisacrylamide (MBA), epichlorohydrin (ECH), etc. However, petroleum-based hydrogels have environmental adverse impacts and are non-biodegradable, which limits their extensive usage in different fields. In contrast, bio-based hydrogels have obtained huge attention as favorable materials for different kinds of applications due to their inherent advantages, including high hydrophilic, abundant, biocompatible, non-toxic, and biodegradable features [1].

Due to the availability of several hydrophilic groups (like carboxyl, amino, amide, hydroxyl group, etc.) on polymeric chains, hydrogels become extremely hydrophilic and can absorb a significant amount of water. They are receiving great attention as soil conditioners, controlled release systems, soilless media, etc. in the field of agriculture. As a soil conditioner, they are used to enhance the water retention capacity of the soil in water-scarce areas. Further, they absorb large amounts of water and subsequently are accessible to the plants for extended periods of time. These bio-based hydrogels not only provide biodegradability but also offer a slow-release property of the system. In addition, the use of hydrogels as slow-release systems improves fertilizer

DOI: 10.1201/9781003340485-17

efficiency, which in turn reduces the cost as well as environmental contamination by the leached-out fertilizer [2].

Besides water absorbency, hydrogels are responsive to external stimuli such as electrical field, salt solution, light, pH, etc. These features make them an ideal candidate for the detection of pollutants, such as metal ions, CN^-, F^-, carbon monoxide, etc., present in the wastewater or in the atmosphere. In addition to these, their biodegradability, easy availability, and low-cost features make them suitable for food packaging material. Further, hydrogels are emerging as promising adsorbents for the removal of heavy metal ions from polluted water. The porous 3D structure and polar functional groups of hydrogels promote excellent performances for removing heavy metal ions from polluted water. In particular, the porous structure of the hydrogel provides large specific surface areas for adsorption of metal ions that are capable of binding with the polar functional groups present in the polymeric chains. Moreover, hydrogels are also applied as adsorbents for the removal of organic dyes (like methylene blue (MB), methyl violet (MV), malachite green (MG), etc.), pesticides, organic micro-pollutants, etc. from wastewater. Dye degradation is another technique i.e., used for removal of dyes from wastewater. The porous hydrogel combined with a catalyst is more efficiently used for this purpose [3].

Moreover, oil/water separation has also emerged as a promising application of hydrogels for the reduction of pollution. Additionally, hydrogels have received significant interest in different areas, such as hygiene products (particularly in disposable diapers), cosmetics, microbial fuel cells, enhanced oil recovery, etc. Thus, hydrogels have found extensive applicability as an advanced biomaterial in the environmental sector [4].

In this chapter, the current status of hydrogels and their uses for environmental applications are highlighted. Moreover, in the first part of the chapter, we have presented the preparative methods and properties essential for the environmental applications of hydrogels. Further, the applicability of hydrogels in each section of the environmental sector is discussed.

17.2 PREPARATIVE METHODS OF HYDROGELS TO BE USED FOR ENVIRONMENTAL APPLICATIONS

The preparative methods of hydrogels include both a physical and chemical cross-linking strategy. The physically cross-linking methods include preparation of hydrogel without any chemical cross-linker and, thus, these methods are comparatively safer than the chemically cross-linking methods. Physically cross-linking methods include formation of hydrogen bond, electrostatic interactions, ion exchange, etc. However, the mechanical weakness, structural imperfections, reversible nature, etc. are some serious drawbacks, due to which the preparation of hydrogels by using chemical cross-linkers is gaining much more attention [5].

The chemically cross-linking methods can be divided into various sub-categories, including direct cross-linking by using small molecules, free-radical polymerization, and grafting. Various mono- and bi-functional molecules, such as formaldehyde, glutaraldehyde, ECH, genipin, ethylene glycol diglycidyl ether, diethyl squarate, etc., are extensively used for the direct chemical cross-linking method [6].

The free-radical mechanism of hydrogel preparation involves a three-step process: (a) initiation, (b) propagation, and (c) termination of the polymer chains. The initiators, such as unimolecular initiator, light, redox initiator, heat, microwave, etc., activate the chains, which further react with the double bond of monomers to propagate the active chains. The presence of cross-linkers forms a network with the polymeric chains. Further, the chains are terminated by combination, or disproportionation, or other chain transfer processes. The initiators used in this process are potassium persulfate, ceric ammonium nitrate, ferrous ammonium sulfate, ammonium persulfate (APS), azobisisobutyronitrile, and benzoyl peroxide (BPO), etc. MBA, ethylene glycol dimethacrylate, etc. are most widely used cross-linkers, which are generally used to cross-link various vinyl monomers, such as AA, AM, vinyl chloride, 2-hydroxyethyl methacrylate, etc. [6].

In grafting methods, polymerization of other monomers is done on the backbone of preformed polymers, usually polysaccharides. To activate the polymer chain, various types of initiators are used, followed by grafting of the polymerizable monomer in the presence of cross-linking agent. Various hydrophilic monomers, such as AA, AM, etc., are widely grafted on polysaccharides. Depending on the source of activation, grafting can be divided into radiation grafting and chemical grafting. Chemical grafting methods involve the utilization of chemical initiators, such as APS, BPO, etc. [7]. Graft copolymerization of AA and AM was done on starch by Al-Aidy *et al.*, for a wastewater purification process [8].

17.3   PROPERTIES OF HYDROGELS TO BE USED FOR ENVIRONMENTAL APPLICATIONS

The sky-high utilization of hydrogels in numerous fields is due to their attractive properties. As hydrogels are hydrophilic polymers, swelling is their most important property. Swelling allows penetration of a large volume of water molecules inside the 3D network of the hydrogel without disintegrating its structure. Actually, swelling is the most intrinsic property that defines the existence of hydrogels. During this phenomenon, various other charged or neutral molecules can penetrate inside the hydrogel network, which help in the wastewater purification process. Moreover, along with the high-water absorption ability, the entrapment efficiency for various chemicals and their slow release with time are some of the most important properties of hydrogels for agricultural applications. In addition to this, there are numerous other attractive properties, such as biocompatibility, biodegradability, self-healing ability, mechanical toughness, ionic conductivity, shape memory property, stretchability, electrolyte permeability, etc. In Figure 17.1, a general representation of the desired attributes of hydrogels for environmental applications is shown. The existence of numerous properties makes them suitable for various applications, such as health and safety, food packaging, alternative energy, etc. Most importantly, properties of hydrogels can be easily tuned according to a specific application by changing the monomer ratio, cross-linking density, and polymer-monomer interactions, etc. [6]. Various preparative methods and advantages of some commonly used hydrogels for environmental applications are represented in Table 17.1.

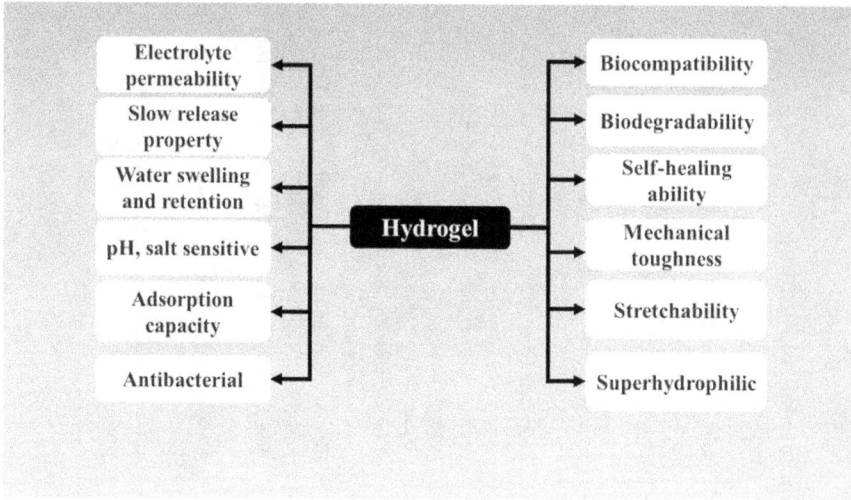

FIGURE 17.1 Desired general attributes of hydrogels for environmental applications.

17.4 VARIOUS ENVIRONMENTAL APPLICATIONS

Hydrogels have been used in different fields related to environmental applications and are represented in Figure 17.2 and discussed below.

17.4.1 AGRICULTURAL APPLICATIONS

In arid and semi-arid areas, water scarcity is a major environmental issue as a consequence of low rainfalls, which severely restrict the sustainability of agricultural products in those areas. Further, exponential growth of global population forces the agricultural sector to use a tremendous amount of mineral fertilizer to meet the global food demands. However, a significant part of the applied fertilizer is lost to the environment via leaching, surface run-off, and volatilization. Thus, the low effective utilization of fertilizer in conventional agriculture methods causes financial losses for farmers, generates serious environmental problems, and represents a waste of energy. Thus, along with inappropriate irrigation, the overuse of fertilizers has severe consequences on agriculture and the environment [19].

In order to solve problems like water scarcity and fertilizer loss, researchers have developed hydrogels for agricultural usage, which shows their versatile roles in outstanding water absorption property and water-retaining ability. The use of hydrogels in agriculture as soil conditioners can improve simultaneously the water retention capacity together with biological, chemical, and physical property of soil. Further, utilization of hydrogels as a controlled/slow-release system enhances the fertilizer/agrochemicals "utilization efficiency". In agricultural applications, poly (acrylamide) (PAM) and poly(acrylic acid) (PAA) are the most commonly used petroleum-based non-biodegradable hydrogels. As a safe alternative, bio-based

TABLE 17.1

Preparative Methods and Advantages of Some Commonly used Hydrogels for Environmental Applications

Hydrogel	Method of Preparation	Advantage of Hydrogel	Application	Reference
Cellulose-whey	Cross-linking method	Renewable, biodegradable, slow-release fertilizer, high water retention in soil	Soil conditioner, Slow-release system	Durpekova et al. [9]
AA	Free-radical polymerization	High swelling, mechanically strong, slow-release fertilizer	Slow-release fertilizer	Lipowczan et al. [10]
AC-agarose	Solvent cast technique	High mechanical, thermal, porosity, improved water retention property, enhanced seed germination rate	Soilless medium	Cao et al. [11]
Dialdehyde guar gum-chitosan	Cross-linking method	High mechanical strength, antimicrobial activity	Food packaging	Maroufi et. el. [12]
Carrageenan	Blending method	Strong antibacterial activity, high mechanical, thermal stability, potential for food packaging area	Food packaging, skin care product, biomedical	Oun et al. [13]
Chitosan-calcium alginate-bentonite	Sol-gel transition method	Excellent metal ion adsorption capacity, good reusability, improved mechanical properties	Removal of heavy metal ions from water	Lin et al. [14]
Chitosan-PAM	Chemical cross-linking method	Cost effective, biosorbent, reusability	Elimination of heavy metal ions from water	Pavithra et al. [15]
SA-AA	Free-radical polymerization	High adsorption of dye, reusability	Wastewater treatment	Mozaffari et al. [16]
SA-AA-AM-Montmorillonite	Polymerization followed by amidation	Dynamic dye adsorption property, reusable	Wastewater treatment	Wang et al. [17]
Carboxymethyl cellulose-PAM	Free-radical polymerization	High adsorption capacity, faster adsorption rate, biocompatibility, recyclability	Removal of heavy metal ions from water	Zhao et al. [18]

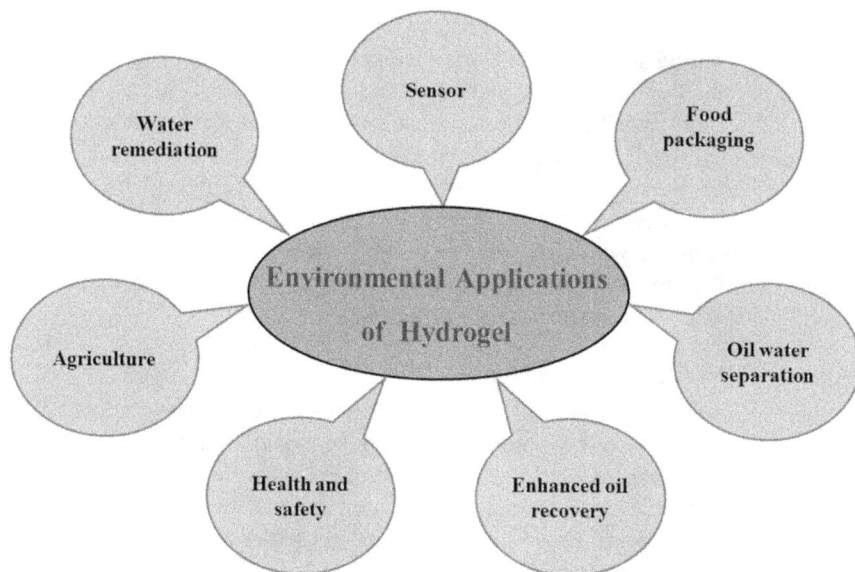

FIGURE 17.2 Different applications of hydrogels related to the environment.

hydrogels have been considered as a substitute for petroleum-based hydrogels and to produce biodegradable and eco-friendly hydrogels for agricultural applications. In this view, polysaccharide-based hydrogels are used as potential substitutes to petroleum-based hydrogels, which are eco-friendly, biodegradable, abundant, low cost, and favorable attributes. Hence, bio-based hydrogels are extensively utilized in the agriculture sector [20].

Besides using hydrogels as soil conditioners and slow-release fertilizer (SRF) system, they are also used as the soilless medium for cultivation, especially ornamental gardening. A lot of herbicides and pesticides other than fertilizers are used in agriculture to increase crop yield. However, most of the applied contents are lost to the soil, which leads to environmental pollution. Therefore, researchers have started to focus more on soilless cultivation these days by using hydrogels.

17.4.1.1 Soil Conditioner

An irrigation process is one of the most important parts of agriculture that assists in the growth of crops and plants by supplying water and fertilizers/agrochemicals to the soil. However, an irrigation process is costlier in drought areas. Therefore, to resolve this issue, superabsorbent hydrogels (SAHs) are considered an attractive replacement. SAHs provide a continuous irrigation feature and reduce the cost of this process. Moreover, due to the excellent water absorption capacity (WAC) and water-retaining ability, SAHs can help in maintaining the moisture content of the soil, which in turn aids in enhancing the plant growth rate. Further, incorporation of SAHs increases porosity of the soil, which leads to the enhancement of root growth, seeds germination rate, reduction in soil erosion, etc. Many recent research works have outlined the significance of using hydrogels as soil conditioners to assist in absorption and retention of water. For instance, Sarmah and Karak developed a hydrogel by grafting starch with AA via a radical polymerization method [21]. The prepared hydrogel exhibits biodegradability and high WAC that in turn increases the water holding capacity (WHC) of the soil by 120% on incorporation of only 0.25% of the hydrogel. Further, enhanced soil porosity and bulk density due to the addition of the hydrogel, help in growth rate of the chick-pea plant. In 2021, Durpekova *et al.* developed a biodegradable hydrogel from the derivatives of cellulose, cross-linked with citric acid, in order to utilize it as a soil conditioner in the agricultural sector [9]. It was reported that the hydrogel showed excellent WAC in soil, and it was used repetitively up to five times.

17.4.1.2   Slow-Release Fertilizer

Fertilizers are used to provide nutrients to plants to enhance the crop productivity in agriculture sectors. Therefore, utilization of conventional fertilizer such as urea, super phosphate potash, etc. have gradually increased. However, plants cannot take up all the applied fertilizers, which is later released into the environment and causes secondary pollution. Recently, hydrogels have been used as a SRF system in order to improve the fertilizers' "utilization efficiency" and to reduce their effects on the environment, which in turn make the process cost effective, too [19]. In this view, bio-based hydrogels have been used as an excellent SRF system, due to their biocompatible, non-toxic, widely available, low cost, and biodegradable nature. In 2018, Cheng *et al.* synthesized a cost-effective AA and urea-based SAH for slow release of nitrogen (N) in agriculture [22]. The results showed that the synthesized SAH exhibited a slow N release property and enhanced the maize seed germination rate, which is shown in Figure 17.3. Thombare *et al.* prepared a boron-loaded guar gum-based hydrogel to use as a SRF system in the agricultural sector [23]. The hydrogel exhibited WAC of 356 g/g and release of 38% of boron fertilizer within 30 days of incubation, which indicates a controlled release property of the hydrogel, along with WHC. Thus, the synthesized hydrogel can be considered as a potential SRF system in agriculture. In 2021, Lipowczan and Trochimczuk used a

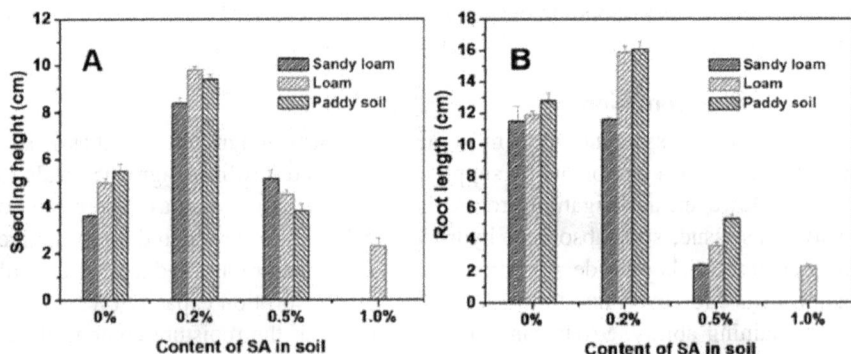

FIGURE 17.3   Effect of superabsorbent on (A) seedling height and (B) root length of maize seeds in different types of soil. Adapted with permission from Reference [22], Copyright (2018), ACS.

phosphorus-loaded hydrogel as a SRF system [10]. The hydrogel showed good WAC and high mechanical properties, which assist in the slow release of fertilizer from the hydrogel matrix.

17.4.1.3 Soilless Cultivation

Soilless cultivation is a method used to grow plants without soil or a medium. Hydroponics is considered the most commonly used method in soilless cultivation. The use of a soilless cultivation method has advantages compared to the traditional soil-based cultivation method, where soil- and water-related problems create difficulties in crop production. Moreover, the soilless method does not need soil sterilization and reduces labor requirements. Further, this method has accurate control over the supply of nutrients, water, pH, etc., which increases the productivity of the crops. Recently, hydrogels have been used as effective soilless media for cultivation because of their potential to absorb and store a huge volume of water. In addition to these, a 3D network structure of hydrogels can provide suitable conditions for plant growth and seed germination. Therefore, hydrogels are widely used as cultivation media, which can supply water as well as nutrients to the plant. Currently, there are several reports available on the utilization of hydrogels as a soilless medium. For instance, Coa and Li developed an activated carbon (AC) and agarose-based hydrogel for soilless cultivation of rapeseed seed germination [11]. The prepared hydrogel was tested as biodegradable and safe. Further, they observed that the addition of AC stimulated the germination and plant growth rate. Zhang *et al.* prepared a cellulose-based hydrogel in the presence of modified-cellulose nanofiber (CNF) for soilless cultivation [24]. They observed that the use of hydrogels with suitable carboxylate contents could promote the seed germination rate of the black sesame cultivar.

17.4.2 ENHANCED OIL RECOVERY

In the fossil fuel production process, approximately one-third of oil can be recovered from the reservoirs. As oil production is a very expensive process, this partial recovery leads to the undesirable consumption of various resources and also causes serious environmental degradation. Moreover, the discovery of a new oil field is extremely difficult and is very expensive. To meet the world energy demand, more advanced methods compared to conventional ones are urgent to enhance the oil production from wells. In this vein, enhanced oil recovery by polymer flooding has been proven as the most effective technology. An ultra-stable hydrogel for enhanced oil recovery in a highly saline condition and high temperature was formulated by Chen *et al.* [25]. Sodium tripolyphosphate was used as a syneresis inhibitor, which is a common problem faced by the conventional hydrogel, where the solvent phase splits from the hydrogel. AM/acryloyloxyethyl trimethylammonium chloride cross-linked with phenol-formaldehyde and a sodium tripolyphosphate hydrogel provided a double network structure that increased the interaction force among the polymers, resulting in a stronger grid structure to overcome the syneresis problem [25]. A cyclodextrin-based hydrophobically associating hydrogel was synthesized by reacting allyl-β-cyclodextrin and octadecyl dimethyl allyl ammonium chloride were reacted with AM for enhanced oil recovery [26]. In addition to this, there are a huge number of works that exist in literature to enhance the oil production from the reservoir. Thus, hydrogel-based systems have a pronounced effect on the oil recovery system, which indirectly reduces the environmental pollution caused by the oil production process.

17.4.3 Food Packaging

For more than 50 years, traditional packaging materials, that is synthetic plastics, have been used for packaging applications due to their mechanical, abundant, cost effective, optical, and resistance against water and grease. However, synthetic plastic materials are neither totally recyclable nor biodegradable, which causes serious environmental and waste disposal problems. Rising to these challenges, many research groups as well as industrial R&D have focused on the enhancement of the performance of eco-friendly packaging materials through the improvement of their biodegradable, mechanical, and thermal properties. Recently, to reduce the utilization of synthetic plastic-based materials, bio-based hydrogels have been applied for food packaging due to their low-cost, hydrophilic, renewable, biodegradable, antibacterial, and good moisture barrier properties. Several reports suggest that the use of hydrogels in food packaging can decrease the growth rate of mold, yeast, and spoilage bacteria on foods, and decrease the softening of dry, crispy products. In 2021, Maroufi *et al.* developed chitosan and dialdehyde guar gum–based hydrogels via the Schiff base reaction [12]. The hydrogels showed high WAC (i.e., 12000% of dry weight) and antimicrobial activity for a long time. Further, a high mechanical property makes the hydrogel suitable for food packaging. Oun and Rhim, in 2017, developed a carrageenan-based hydrogel film with the incorporation of zinc-oxide (ZnO) and copper oxide (CuO) nanoparticles (NPs) [13]. The film showed a swelling ratio of 3535% in the presence of ZnO NPs. Further, the film showed strong antibacterial activity, higher mechanical properties, and high potential for food packaging applications.

However, there are still limitations for the development of efficient food packaging material from bio-based hydrogels due to their low strength, poor water resistance, etc. Therefore, research on the development of bio-based food packaging material as a substitute for petroleum-based materials have been going on to provide new opportunities for designing productive biopolymer packaging materials with advantageous properties.

17.4.4 Health and Safety

Absorbent hygiene products (AHPs) (such as diapers, sanitary pads, female napkins, etc.) have provided a significant contribution to the life and health of millions of people. SAHs are predominantly used in hygiene products for the absorption of water and biological fluids. They have the capacity to absorb large amounts of water, i.e., up to 10,000 times their dry weight. PAAs are the primary polymer used for the preparation of SAHs. However, petroleum-based PAA hydrogels are non-biodegradable, so disposed diapers are not easily degraded in the soil, which in turn generate adverse effects on the environment in the form of solid toxic waste.

Apart from this problem, disposed solid waste releases some toxic chemicals into the environment, which in turn contaminates the soil and water bodies. Further, the use of AA causes severe skin irritation in babies. Therefore, research works have been performed in the development of sustainable alternative materials for personal care products. The best alternatives contain polysaccharides (like guar gum, starch, cellulose, etc.) and protein-based (like collagen, gelatine, etc.) SAHs. They exhibit properties like safety, biocompatibility, biodegradability, relatively low cost, etc. Therefore, they are widely used as an environment friendly and sustainable replacement for various non-renewable petroleum-based materials. Enawgaw *et al.* synthesized a bio-based personal hygiene hydrogel using cellulose as the base material via the radical polymerization method [27]. It was observed that the hydrogel exhibits WAC of 279.6 g/g and 83.3 g/g in urine solution. Further, Maso *et al.* fabricated a bio-based superabsorbent using CNF for potential use in baby diapers [28].

17.4.5 SENSOR

Due to the rapid development of industries, metal ion contamination in the groundwater has become one of the serious environmental problems, which needs to be addressed urgently. Heavy metal ions, such as Cu(II), Ni(II), Pb(II), Hg(II), etc., released from industrial activities remain contaminants in water bodies, which directly or indirectly have toxic effects on humans and aquatic life. Therefore, detection of metal ions present in water bodies is of great importance. In the past decades, various equipment (such as atomic emission spectrometry (AES), atomic absorption spectrometry (AAS), inductively coupled plasma mass spectrometry (ICP-OES), and so on) have been used to detect and analyze metal ions. These techniques are widely used and can give accurate results, but they have the disadvantage of being costly and time consuming. In this vein, a chemosensor has been developed as a powerful sensing tool for applications in environmental and biological fields [29]. They have the advantage of high selectivity towards specific metal ions. Recently, chemosensors fabricated from hydrogels are attracting increased attention because of their hydrophilic property and 3D polymer network that can allow significant entry of pollutant molecules inside the hydrogel network structure. Moreover, their properties can be modified in order to further improve their affinity for targeted metal ions. Liu *et al.* reported a rhodamine functionalized PAM-based hydrogel for the detection of Fe(III) ion [30]. They reported that the synthesized hydrogel possessed great selectivity and sensitivity for Fe(III) ion detection, which can be used as an alternative of sophisticated instruments. In 2022, an ion-imprinted hydrogel chemosensor was developed for the detection of trace strontium ions (Sr(II)) [31]. The synthesized hydrogel-based chemosensor showed a detection limit for the Sr(II) ion as low as 10^{-11} M.

17.4.6 WATER REMEDIATION

Over the last decades, a significant amount of textile dyes, heavy metal ions, hazardous organic micropollutants, pharmaceutical products, etc. have been introduced from petroleum, mining, metallic, medicinal, paper, printing, energy-storage, coating, fabric, chemicals, and cosmetics industries into the environment, causing water pollution, a serious environmental problem. Currently, different techniques, such as electro-dialysis, chemical precipitation, coagulation, ion exchange, membrane filtration, and adsorption, have been applied to remove these pollutants from wastewater, among which adsorption is considered the most advantageous due to its high removal efficiencies, ease of operation, wide applicability, and economic and flexibility in the design of adsorbents. Using this technique, various adsorbents have been applied to date for water remediation. Recently, hydrogels emerged as the potential adsorbent for wastewater treatment, due to their distinguished features, like eco-friendly, cost-effective, high adsorption capacity, etc. Further, their excellent potentiality to absorb a large volume of water within their 3D porous structures, along with the presence of chemically responsive functional groups, make them capture the pollutants from wastewater and release them under suitable conditions. Their properties, like high mechanical strength, high surface area, and porosity, are essential characteristics for hydrogels to capture contaminants from wastewater sources. Therefore, hydrogels are considered to be suitable for wastewater treatment [3].

17.4.6.1 Metal Ion Removal

Among the pollutants, heavy metal ions such as Cd(II), Cr(VI), Cu(II), Pb(II), Hg (II), and Ni(II) are considered to be the most discharged pollutants in water bodies. They can remain in those systems for long periods and accumulate in food chains. Further, they are poisonous, even at very low concentrations. To solve these issues, hydrogels have gained particular interest as an efficient adsorbent for metal ion removal, due to their high adsorption capacities, recycling abilities, non-toxicity, cost effectiveness, and eco-friendly nature. Moreover, functional groups present on the hydrogel were found to have an important role in removing heavy metal ions via electrostatic attraction [3]. For instance, Mittal et al. synthesized a biodegradable hydrogel polymer via graft polymerization of AA and AM on the surface of gum ghatti and used it for adsorption of Pb(II) and Cu(II) from an aqueous solution [32]. It was observed that the experimental data fit well with the Langmuir isotherm model with a maximum adsorption capacity (q_{max}) of 203 and 384 mg/g for Cu(II) and Pb(II), respectively. A possible mechanism for adsorption of Cu(II) and Pb(II) ions onto an Am and AN grafted gum ghatti–based hydrogel is shown in Figure 17.4. Lin et al. have prepared a cost-effective chitosan and calcium alginate-based hydrogel with the incorporation of bentonite for simultaneous adsorption of Pb(II), Cd(II), and Cu(II) ions [14]. They investigated the metal ion removal performance of the hydrogel under different conditions and observed q_{max} for Pb(II), Cd(II), and Cu(II) ions up to 434, 102, and 115 mg/g, respectively. The adsorbent also possesses recyclability. Therefore, it is considered a promising adsorbent for heavy metal ion removal from water. Pavithra et al. synthesized a biodegradable and low-cost chitosan-based hydrogel via chemical cross-linking with PAM in the presence of an orange peel [15]. The hydrogel was used for the removal of Cr(VI) and Cu(II) ions from wastewater. It was observed that it took only 240 min to remove 82% and 80% of Cu(II) and Cr(VI) ions, respectively.

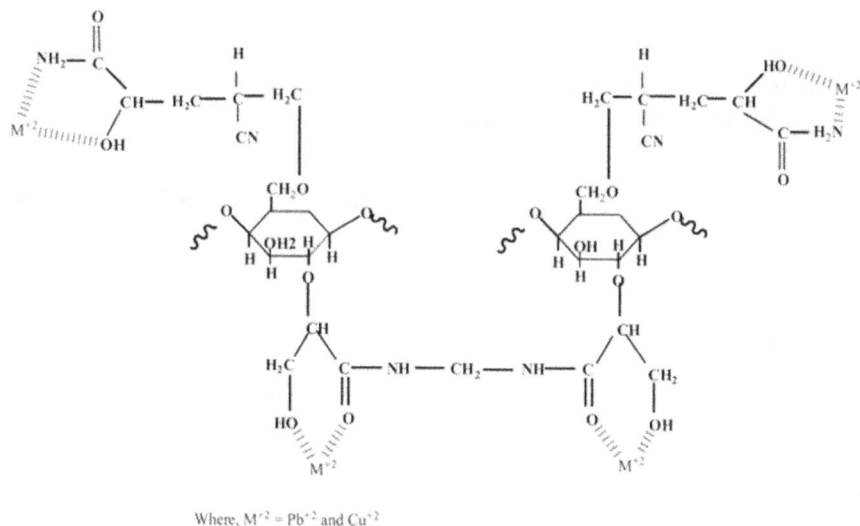

Where, $M^{+2} = Pb^{+2}$ and Cu^{+2}

FIGURE 17.4 A possible mechanism of interaction between the metal ions and functional groups present in the hydrogel polymer. Adapted with permission from Reference [32], Copyright (2014), ACS.

17.4.6.2 Dye Removal

Organic synthetic dyes are the most hazardous color-imparting agents applied, mostly in textile industries and in food, paper, cosmetics, etc. industries. Due to the presence of aromatic and azo groups, dyes are highly toxic and non-degradable in nature, which make them responsible for enormous environmental pollution. To reduce the effects of contaminated dyes, hydrogels have been widely applied for their removal from polluted water bodies using the adsorption technique. To date, various reports regarding removal of dyes using hydrogels have been reported. For example, Mozaffari et al. prepared a novel nanocomposite hydrogel via in situ cross-linking of SA in the presence of AA with incorporation of Cu(II) ions for removal of MG and crystal violet (CV) dyes [16]. They observed the dye removal capacity for both MG and CV, which was more than 96%. The synthesized hydrogel was found to possess a reusable property up to eight adsorption cycles. Wang et al. developed hydrogel beads by grafting SA with AA and AM with incorporation of montmorillonite nanoclay [17]. The prepared beads were used for the adsorption of MB, which was found to be equal to 530 mg/g. Therefore, it can be used as a potential adsorbent in actual applications. In 2021, Du et al. prepared an AA grafted xanthan gum–based hydrogel using electron beam irradiation [33]. They studied the dye adsorption capacity of the prepared hydrogel and compared the results with previously reported data, which are shown in Table 17.2.

Besides using the adsorption technique, photocatalytic degradation has attracted considerable attention to deal with toxic dyes in wastewater. In this method, a photocatalyst such as TiO_2, CdS, Cu_2O, etc. is used to treat dye pollutants in

TABLE 17.2

Comparison of Dye Adsorption Results of Xanthan Gum/AA-based Hydrogels. Adapted with Permission from Reference [33], Copyright (2021), ACS

Adsorbent	Q_{max} (mg/g)	Equilibrium Time (min)	pH
Salecan	71.6	160	7
Covalent organic polymer	131.23	4	6.5
Graphene hydrogel	341	–	5.5
Poly(AM/sodium acrylate)	469.48	–	5.5
Gum karaya	497.51	60	7
Chitosan	556.9	210	3
Surfactant impregnation of chitosan hydrogel	1,000	–	–
Boron organic polymer	1,388	200	3
Sodium alginate/kaolin	1,516	800	7
Xanthangum/AA	2,716.6	90	5

wastewater. However, the use of a photocatalyst suffers some limitations like difficulty in the separation of a catalyst after treatment, low adsorption capacity, aggregation of nano-sized catalyst, etc. Recently, hydrogels combined with a photocatalyst have been used to overcome these limitations. Further, the reports indicate that hydrogels are promising material for dye degradation due to their large surface area as well as the presence of abundant functional groups that assist the photocatalyst to react with dye molecules inside the network [18].

17.4.6.3 Pharmaceutical Removal

In the last decades, liquid waste generated from active pharmaceutical ingredients (API) have created a huge environmental pollution. Paracetamol, acetylsalicylic acid, carbamazepine, ibuprofen, etc. are the most prescribed APIs that are non-biodegradable. Furthermore, largely expired and unwanted drugs are thrown away in the environment that diffuse into the groundwater resources, which in turn leads to environmental pollution. Recently, adsorptive removal of drugs from wastewater has emerged as one of the most favorable approaches for wastewater treatment. For instance, Mahmoodi *et al.* developed a bio-based adsorbent using graphene oxide and chitosan via a mechanical mixing method [34]. It was used as a novel biodegradable adsorbent to remove diclofenac from water. The q_{max} of the adsorbent was found equal to 130 mg/g. Further, it has the advantage of recyclability and is used up to three consecutive cycles. Mota *et al.* studied diazepan (DZP) removal from water by using a hydrogel prepared by gum arabic grafted with PAA [35]. It was observed that the hydrogel could remove more than 80% DZP within 300 min and possessed a q_{max} of 15 mg/g. Further, the hydrogel showed excellent recyclability of up to five consecutive cycles.

17.4.6.4   Pesticide Removal

In the last decade, environmental contamination of pesticides has increased due to their excessive uses in the agricultural sector, which have dangerous effects on human health and the environment. Pesticides are hazardous and toxic in nature, which can remain in the environment for several years, which in turn decreases the biodiversity of the soil. In recent years, pesticide removal from wastewater using the adsorption technique of hydrogels has evolved as an effective physical method. For example, Aouada *et al.* synthesized a methyl cellulose and PAM-based hydrogel as a novel adsorbent to remove the pesticide paraquat dichloride from an aqueous solution [36]. The results suggest that the q_{max} for the hydrogel was 14.3 mg/g and it can be used for the removal of other pesticides, too. Further, to remove the highly toxic pesticide paraquat from water, Thakur *et al.* synthesized SA, dextrin, and AA-based hydrogels with bentonite clay [37]. The hydrogel possesses reusable and biocompatible properties. The report suggested that the incorporation of bentonite improved the q_{max} from 76 to 90 mg/g. Thus, the hydrogel matrix is considered a potential candidate for the removal of pesticides from wastewater.

17.4.7   OIL-WATER SEPARATION

Oily wastewater release from oil reserves and domestic and industrial activities is a serious environmental issue that threatens our health and environmental balance. Oil/water separation is a novel and very effective method to separate an oil/water mixture. So far, numerous materials have been investigated to address this issue. Among them, hydrogels have emerged as a potential option because of their surface wettability, superhydrophilic, underwater oleophobicity, and good antifouling properties. Different studies suggest that hydrogel-coated materials have excellent efficiency in separating different oil/water mixtures under different conditions. For example, Zhu *et al.* reported a novel negatively charged hydrogel membrane (HM) with high mechanical and antifouling properties and used it for the separation of oil from wastewater [38]. PVA, glutaraldehyde, and PA were used to prepare the HM, which possessed a negative surface charge density of 35.51 mmol/g. Zhang *et al.* developed a pH-responsive nanofibrous HM for the separation of oil and water from their emulsion [39]. Due to the underwater superoleophobic and superhydrophilic nature, the prepared HMs can be effectively used for the separation of oil/water emulsion in acid, alkali, and neutral environments. In 2020, Huang et al. prepared a porous polyurethane hydrogel and used it as an oil-water separation material for various oil-water mixtures bearing 99.9% oil-water separation efficiencies [40]. In Figure 17.5, the separation of an n-hexane-water mixture by a polyurethane hydrogel is shown. Further, the report suggested that high separation efficiency and recyclability of the HM make them a potential for use in oil/water separation.

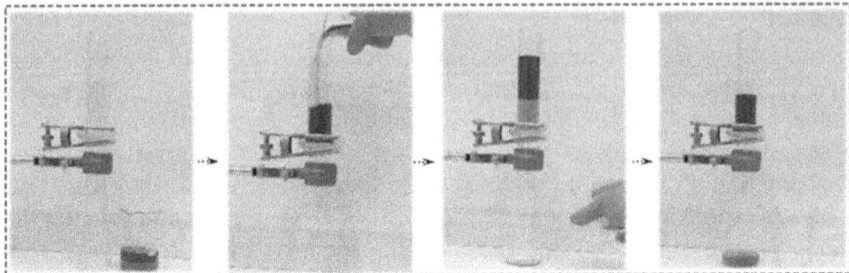

FIGURE 17.5 Separation of n-hexane-water mixture by using porous polyurethane hydrogel. Adapted with permission from Reference [40], Copyright (2020), ACS.

17.5 CONCLUSION

Environmental pollution, including air, water and soil, global warming, etc., is a serious issue across the world. Hence, it is necessary to address such issues urgently. In this vein, hydrogels have emerged as excellent material for different fields of applications. Therefore, this chapter mainly concentrated on the extensive use of hydrogels for environment-related applications. Further, a brief description of properties and preparative methods of hydrogels is provided. The reports showed that numerous hydrogel-based adsorbents have been developed for the removal of pollutants from wastewater. Bio-based hydrogels with high WAC and biodegradability are extensively used in agriculture to improve the fertilizer utilization efficiency and to solve irrigation problems. Further, this chapter also concentrated on the use of bio-based hydrogels in order to address the biodegradable issue. The use of bio-based hydrogels for chemosensors, food packaging, oil/water separation, and hygienic products not only improves the biodegradability but also reduces the cost and toxicity, which in turn leads to no/less environment pollution. Therefore, hydrogels are of great importance for different environmental applications.

REFERENCES

[1] Spagnol, C., Rodrigues, F. H., Pereira, A. G., Fajardo, A. R., Rubira, A. F., and Muniz, E. C. Superabsorbent hydrogel nanocomposites based on starch-g-poly (sodium acrylate) matrix filled with cellulose nanowhiskers. *Cellulose*, 2012, *19*(4), 1225–1237.

[2] Bortolin, A., Aouada, F. A., Mattoso, L. H., and Ribeiro, C. Nanocomposite PAAm/ methyl cellulose/montmorillonite hydrogel: Evidence of synergistic effects for the slow release of fertilizers. *Journal of Agricultural and Food Chemistry*, 2013, *61*(31), 7431–7439.

[3] Tang, S., Yang, J., Lin, L., Peng, K., Chen, Y., Jin, S., and Yao, W. Construction of physically crosslinked chitosan/sodium alginate/calcium ion double-network hydrogel and its application to heavy metal ions removal. *Chemical Engineering Journal*, 2020, *393*, 124728.

[4] Kabiri, K., Omidian, H., Zohuriaan-Mehr, M. J., and Doroudiani, S. Superabsorbent hydrogel composites and nanocomposites: A review. *Polymer Composites*, 2011, *32*(2), 277–289.

[5] Sinha, V., and Chakma, S. Advances in the preparation of hydrogel for wastewater treatment: A concise review. *Journal of Environmental Chemical Engineering*, 2019, *7*(5), 103295.

[6] Bashir, S., Hina, M., Iqbal, J., Rajpar, A. H., Mujtaba, M. A., Alghamdi, N. A., and Ramesh, S. Fundamental concepts of hydrogels: Synthesis, properties, and their applications. *Polymers*, 2020, *12*(11), 2702.

[7] Ali, A., and Ahmed, S. Recent advances in edible polymer-based hydrogels as a sustainable alternative to conventional polymers. *Journal of Agricultural and Food Chemistry*, 2018, *66*(27), 6940–6967.

[8] Al-Aidy, H., and Amdeha, E. Green adsorbents based on polyacrylic acid-acrylamide grafted starch hydrogels: The new approach for enhanced adsorption of malachite green dye from aqueous solution. *International Journal of Environmental Analytical Chemistry*, 2021, *101*(15), 2796–2816.

[9] Durpekova, S., Di Martino, A., Dusankova, M., Drohsler, P., and Sedlarik, V. Biopolymer hydrogel based on acid whey and cellulose derivatives for enhancement water retention capacity of soil and slow release of fertilizers. *Polymers*, 2021, *13*(19), 3274.

[10] Lipowczan, A., and Trochimczuk, A. W. Phosphates-containing interpenetrating polymer networks (ipns) acting as slow-release fertilizer hydrogels (srfhs) suitable for agricultural applications. *Materials*, 2021, *14*(11), 2893.

[11] Cao, L., and Li, N. Activated-carbon-filled agarose hydrogel as a natural medium for seed germination and seedling growth. *International Journal of Biological Macromolecules*, 2021, *177*, 383–391.

[12] Maroufi, L. Y., Tabibiazar, M., Ghorbani, M., and Jahanban-Esfahlan, A. Fabrication and characterization of novel antibacterial chitosan/dialdehyde guar gum hydrogels containing pomegranate peel extract for active food packaging application. *International Journal of Biological Macromolecules*, 2021, *187*, 179–188.

[13] Oun, A. A., and Rhim, J. W. Carrageenan-based hydrogels and films: Effect of ZnO and CuO nanoparticles on the physical, mechanical, and antimicrobial properties. *Food Hydrocolloids*, 2017, *67*, 45–53.

[14] Lin, Z., Yang, Y., Liang, Z., Zeng, L., Zhang, A. Preparation of chitosan/calcium alginate/bentonite composite hydrogel and its heavy metal ions adsorption properties. *Polymers*, 2021, *13*(11), 1891.

[15] Pavithra, S., Thandapani, G., Sugashini, S., Sudha, P. N., Alkhamis, H. H., Alrefaei, A. F., and Almutairi, M. H. Batch adsorption studies on surface tailored chitosan/orange peel hydrogel composite for the removal of Cr(VI) and Cu(II) ions from synthetic wastewater. *Chemosphere*, 2021, *271*, 129415.

[16] Mozaffari, T., Vanashi, A. K., and Ghasemzadeh, H. Nanocomposite hydrogel based on sodium alginate, poly (acrylic acid), and tetraamminecopper (II) sulfate as an efficient dye adsorbent. *Carbohydrate Polymers*, 2021, *267*, 118182.

[17] Wang, W., Fan, M., Ni, J., Peng, W., Cao, Y., Li, H., and Song, S. Efficient dye removal using fixed-bed process based on porous montmorillonite nanosheet/poly (acrylamide-co-acrylic acid)/sodium alginate hydrogel beads. *Applied Clay Science*, 2022, *219*, 106443.

[18] Zhao, H., and Li, Y. Eco-friendly floatable foam hydrogel for the adsorption of heavy metal ions and use of the generated waste for the catalytic reduction of organic dyes. *Soft Matter*, 2022, *16*(29), 6914–6923.

[19] Noppakundilograt, S., Pheatcharat, N., and Kiatkamjornwong, S. Multilayer-coated NPK compound fertilizer hydrogel with controlled nutrient release and water absorbency. *Journal of Applied Polymer Science*, 2015, *132*(2).

[20] Patra, S. K., Poddar, R., Brestic, M., Acharjee, P. U., Bhattacharya, P., Sengupta, S., and Hossain, A. Prospects of hydrogels in agriculture for enhancing crop and water

productivity under water deficit condition. *International Journal of Polymer Science*, 2022, *2022*, 15.

[21] Sarmah, D., and Karak, N. Biodegradable superabsorbent hydrogel for water holding in soil and controlled-release fertilizer. *Journal of Applied Polymer Science*, 2020, *137*(13), 48495.

[22] Cheng, D., Liu, Y., Yang, G., and Zhang, A. Water-and fertilizer-integrated hydrogel derived from the polymerization of acrylic acid and urea as a slow-release N fertilizer and water retention in agriculture. *Journal of Agricultural and Food Chemistry*, 2018, *66*(23), 5762–5769.

[23] Thombare, N., Mishra, S., Shinde, R., Siddiqui, M. Z., and Jha, U. Guar gum based hydrogel as controlled micronutrient delivery system: Mechanism and kinetics of boron release for agricultural applications. *Biopolymers*, 2021, *112*(3), e23418.

[24] Zhang, H., Yang, M., Luan, Q., Tang, H., Huang, F., Xiang, X., and Bao, Y. Cellulose anionic hydrogels based on cellulose nanofibers as natural stimulants for seed germination and seedling growth. *Journal of Agricultural and Food Chemistry*, 2017, *65*(19), 3785–3791.

[25] Chen, L., Zhang, G., Ge, J., Jiang, P., Zhu, X., Ran, Y., and Han, S. Ultrastable hydrogel for enhanced oil recovery based on double-groups cross-linking. *Energy & Fuels*, 2015, *29*(11), 7196–7203.

[26] Zou, C., Zhao, P., Hu, X., Yan, X., Zhang, Y., Wang, X., and Luo, P. β-Cyclodextrin-functionalized hydrophobically associating acrylamide copolymer for enhanced oil recovery. *Energy & Fuels*, 2013, *27*(5), 2827–2834.

[27] Enawgaw, H., Tesfaye, T., Yilma, K. T., and Limeneh, D. Y. Synthesis of a cellulose-Co-AMPS hydrogel for personal hygiene applications using cellulose extracted from corncobs. *Gels*, 2021, *7*(4), 236.

[28] Patino-Maso, J., Serra-Parareda, F., Tarres, Q., Mutje, P., Espinach, F. X., and Delgado-Aguilar, M. TEMPO-oxidized cellulose nanofibers: A potential bio-based superabsorbent for diaper production. *Nanomaterials*, 2019, *9*(9), 1271.

[29] Nishiyabu, R., Kobayashi, H., and Kubo, Y. Dansyl-containing boronate hydrogel film as fluorescent chemosensor of copper ions in water. *RSC Advances*, 2012, *2*(16), 6555–6561.

[30] Liu, X., Chen, Z., Gao, R., Kan, C., and Xu, J. Portable quantitative detection of Fe^{3+} by integrating a smartphone with colorimetric responses of a rhodamine-functionalized polyacrylamide hydrogel chemosensor. *Sensors and Actuators B: Chemical*, 2021, *340*, 129958.

[31] Liu, Y. Q., Ju, X. J., Zhou, X. L., Mu, X. T., Tian, X. Y., Zhang, L., and Chu, L. Y. A novel chemosensor for sensitive and facile detection of strontium ions based on ion-imprinted hydrogels modified with guanosine derivatives. *Journal of Hazardous Materials*, 2022, *421*, 126801.

[32] Mittal, H., Maity, A., and Sinha Ray, S. The adsorption of Pb2+ and Cu2+ onto gum ghatti-grafted poly (acrylamide-co-acrylonitrile) biodegradable hydrogel: Isotherms and kinetic models. *The Journal of Physical Chemistry B*, 2015, *119*(5), 2026–2039.

[33] Du, J., Yang, X., Xiong, H., Dong, Z., Wang, Z., Chen, Z., and Zhao, L. Ultrahigh adsorption capacity of acrylic acid-grafted xanthan gum hydrogels for rhodamine B from aqueous solution. *Journal of Chemical and Engineering Data*, 2021, *66*(3), 1264–1272.

[34] Mahmoodi, H., Fattahi, M., and Motevassel, M. Graphene oxide–chitosan hydrogel for adsorptive removal of diclofenac from aqueous solution: Preparation, characterization, kinetic and thermodynamic modelling. *RSC Advances*, 2021, *11*(57), 36289–36304.

[35] Mota, H. P., and Fajardo, A. R. Development of superabsorbent hydrogel based on Gum Arabic for enhanced removal of anxiolytic drug from water. *Journal of Environmental Management*, 2021, *288*, 112455.

[36] Aouada, F. A., Pan, Z., Orts, W. J., and Mattoso, L. H. Removal of paraquat pesticide from aqueous solutions using a novel adsorbent material based on polyacrylamide and methylcellulose hydrogels. *Journal of Applied Polymer Science*, 2009, *114*(4), 2139–2148.

[37] Thakur, S., Verma, A., Raizada, P., Gunduz, O., Janas, D., Alsanie, W. F., and Thakur, V. K. Bentonite-based sodium alginate/dextrin cross-linked poly (acrylic acid) hydrogel nanohybrids for facile removal of paraquat herbicide from aqueous solutions. *Chemosphere*, 2022, *291*, 133002.

[38] Zhu, X., Zhu, L., Li, H., Xue, J., Ma, C., Yin, Y., and Xue, Q. Multifunctional charged hydrogel nanofibrous membranes for metal ions contained emulsified oily wastewater purification. *Journal of Membrane Science*, 2021, *621*, 118950.

[39] Zang, L., Ma, J., Lv, D., Liu, Q., Jiao, W., and Wang, P. A core–shell fiber-constructed pH-responsive nanofibrous hydrogel membrane for efficient oil/water separation. *Journal of Materials Chemistry A*, 2017, *5*(36), 19398–19405.

[40] Huang, J., Zhang, Z., Weng, J., Yu, D., Liang, Y., Xu, X., and Wu, X. Molecular understanding and design of porous polyurethane hydrogels with ultralow-oil-adhesion for oil–water separation. *ACS Applied Materials and Interfaces*, 2020, *12*(50), 56530–56540.

18 Hydrogels for Wastewater Cleaning and Water Recovery

Sofia Paulo-Mirasol, David Naranjo,
Sonia Lanzalaco, Elaine Armelin,
José García-Torres, and Juan Torras
Universitat Politècnica de Catalunya-Barcelona Tech,
Barcelona, Spain

18.1 INTRODUCTION

Fresh water is essential for the life and health of human beings. Water use has been increasing about 1% per year since the 1980s, and a similar rate of increase is expected through 2050. Current demand comes mainly from developed countries and emerging economies, although there is a clear difference in the per capita water use of the latter, which is far below the water use of developed countries (UNESCO World Water Assessment Program 2019). Although 71% of the earth's surface is water, which could be assumed to not be a scarce resource, most of this water is found in the seas and only 2.5% is freshwater. However, most of that freshwater (70%) is located in glaciers and snow in a solid state. The freshwater needed by humans must be taken from the remaining 30% of liquid freshwater located in groundwater and surface water, which represents only 0.79% of the earth's water (Salehi et al. 2020).

Unfortunately, anthropogenic pressures on freshwater sources typically act simultaneously at the global level. The increase in the human population leads to a non-stop consumption increase of goods and services, which in turn, is promoting an excess of freshwater consumption, wastewater production, and contamination. On the whole, the excessive pressure by humankind has caused greater losses of biodiversity and altered the ecosystem functioning, leading to losses in ecosystem services such as food and water provisioning (Feio et al. 2023). Moreover, climate change will be ever more critical for water availability. In the last decades, drought events have increased in frequency and duration along several areas, such as Africa, eastern Asia, southern Australia, and southern Europe. Indeed, access to water for people and natural ecosystems is an issue of concern, especially in those areas where water scarcity is incremented by an overexploited and mismanaged use of water resources.

In recent decades, concern for the environment and the sustainability of natural resources has increased a lot. This has led to continuous research to identify new methods of water purification at a lower cost and lower energy consumption, as well

DOI: 10.1201/9781003340485-18

as less environmental impact than that being employed nowadays. The production of drinking water to cope with water scarcity is a subject of active research. Among them, the new wastewater treatments stand out (Shannon et al. 2008), although lately more emphasis has been placed on direct water production processes, such as desalination of brackish water and sea water (Elimelech and Phillip 2011) and harvesting water from humid air (Zhou et al. 2020). However, each process has its corresponding scope of use. Wastewater treatment is an important and present problem in industrialized areas that must necessarily be addressed. Water desalination is an industrial process that can only be used in coastal areas but is not suitable for dry and desert areas. Finally, collecting water from humid air, which could be appropriate for more desert areas, has a lower production capacity compared to the desalination process and, to be more efficient, requires a relative humidity level greater than 30%. However, a high yield of water harvested from air at a very low relative humidity was reported by Kim et al. using a superabsorbent metal-organic framework (MOF) made from zirconium-based materials (MOF-801) (Kim et al. 2017). A production of 2.8 L of water per kg of adsorbent per day was achieved at 20% relative humidity without any additional energy source, except sunlight. This suggests that the production of water in arid and desert regions with very low relative humidity content could be achieved through the design and use of appropriate superabsorbent materials.

The development of the most innovative techniques in this field go hand in hand with obtaining and development of new materials with more adaptable properties. In recent years, the research and development of new hydrogels and their application in the field of environmental remediation and pollution control has increased dramatically (Khan and Lo 2016). Hydrogels consist of three-dimensional flexible polymeric networks that can absorb a large amount of water without dissolving. This characteristic is due to the presence of a large number of hydrophilic groups in the polymeric network (Figure 18.1). These polar groups also allow the elimination of potential aqueous contaminants through electrostatic interactions, just like the classic ion exchange resins. Furthermore, hydrogels present intrinsic flexibility, a great uptake of water, and a larger surface both outside and inside the swollen three-dimensional hydrogel network, unlike polymeric membranes that are much more rigid and compact.

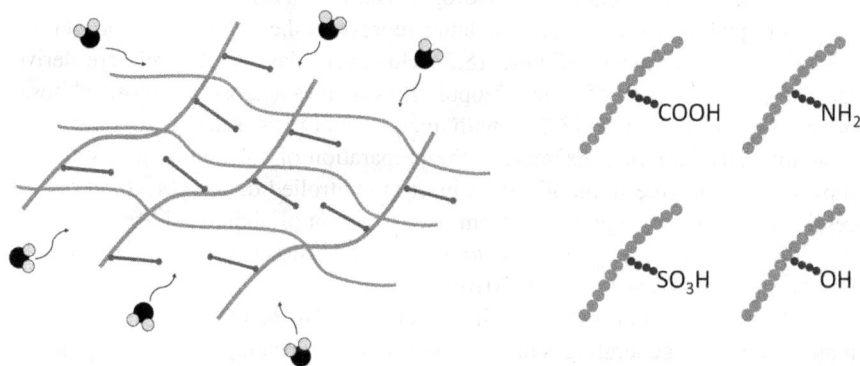

FIGURE 18.1 Scheme of the functional groups present in the structure of hydrogels.

Moreover, hydrogels allow to reduce the enthalpy of water evaporation within its structural matrix, which facilitates and reduces the energy needed to dehydrate the system once it is swollen, with respect to other hygroscopic materials already used. This makes the transport of water through the hydrogel to its surface and its subsequent evaporation quite fast and allows it to be used in advanced systems for collecting atmospheric water or solar desalination (Z. Zhang et al. 2022). These characteristics make hydrogels a very interesting and expanding material for wastewater treatment, purification processes, and freshwater production. In this chapter, we present the main applications for wastewater treatment and water purification, where hydrogels play an important role due to their unique properties that make them indispensable in order to increase energy efficiency and reduce the environmental impact of future generations' technologies for water remediation.

18.2 HYDROGEL MATERIALS APPLIED FOR WASTEWATER TREATMENT AND WATER PURIFICATION

Hydrogels can be chemically and physically cross-linked and assembled as double network systems, which consist of two networks, composed of a material with rigid properties and a compound with soft properties as the main components. In chemically cross-linked hydrogels, covalent bonds are promoted by radical polymerization, and usually cross-linker agents and/or co-monomers are necessary for network formation. Physically cross-linked hydrogels have non-covalent bonds, and the interactions during gelation come from van der Waals forces, electrostatic forces, hydrogen bonds, hydrophilic-hydrophobic associations, and/or chain en-tanglements. Double network hydrogels have the advantages of better mechanical strength, flexibility, and elasticity than the formers. They are able to absorb external stress and dissipate the energy absorbed from the external forces by using the duality of a soft-rigid nature. For the preparation of such structures, the two components should be chemically or physically cross-linked to be stable (Guo et al. 2020). Also, the control over the morphology during gelation is a key factor to ensure good swelling and strength properties, among others, for any hydrogel class.

Regarding the chemical nature, hydrogels can be classified as natural, synthetic, composite, and hybrid materials; the latter represents the joining of a soft gel with non-polymeric compounds (Figure 18.2). However, classic hydrogels are derived from the polymerization of natural biopolymers such as agarose, cellulose, chitosan, and alginic acid (Figure 18.2a), with mainly inorganic salts. One typical and technologically important example is the preparation of calcium alginates, usually employed for enzyme immobilization in drug-controlled delivery systems, wound dressings, etc. It can be produced from the extraction of alginic acid from seaweed with calcium carbonate, giving rise to a three-dimensional structure known as the "egg-box model" (Armelin et al. 2016).

In addition to natural polymers, hydrogels can also be obtained from synthetic small monomers, generating some of the polymer-repeating units exemplified in Figure 18.2b. Moreover, the biopolymers mentioned above can be chemically cross-linked or physically entangled with synthetic polymers to offer water insoluble

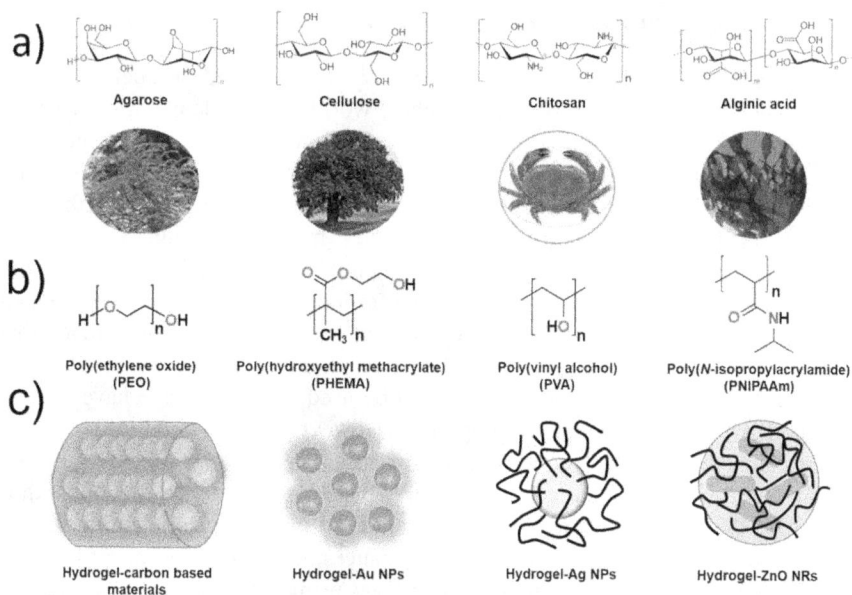

FIGURE 18.2 Examples of the chemical structures of polymers employed to prepare hydrogels and illustration of hydrogel combinations with other compounds being used in wastewater treatment technologies: a) biopolymers such as natural polysaccharides; b) synthetic polymers; and c) hydrogel composites and hybrid materials. Adapted with permission (Lanzalaco et al. 2022). Copyright 2022, Wiley.

systems for wastewater treatment and water recovery. Kyzas and co-workers have achieved 99.99% of Cu^{+2} adsorption on chitosan/poly(vinyl alcohol (PVA) beads functionalized with poly(ethylene glycol) (PEG) (Trikkaliotis et al. 2020). They employed glutaraldehyde as a cross-linker agent and sodium tri-polyphosphate as a counter-ion to the chemical and physical gelation of a chitosan matrix. This work represents one of the multiple options being reported in literature for different contaminants in wastewater (Sinha and Chakma 2019). On the other hand, thermoresponsive polymers, such as poly(N-isopropylamide) (PNIPAAm) (Figure 18.2b), combined with several co-monomers and cross-linker molecules, have proved to be a powerful alternative to the development of solar water evaporators to produce freshwater (Lanzalaco et al. 2022).

Nowadays, one strategy to improve the water hydration or dehydration (depending on the application envisaged) of polymer hydrogels is based on the addition of solar absorbers, by the combination of them with metallic nanoparticles (NPs), nanorods (NRs), carbon nanotubes (CNTs), graphene oxide (GO), conducting polymers (CPs), and others (Figure 18.2c). Such modifications have revealed an enormous ability to reach ultra-high efficiency in processes involving heavy metal ions and organic pollutants treatments, seawater and brackish desalination, and moisture harvesting. Some examples of hydrogels being investigated will be exemplified in the next sections.

18.3   HYDROGELS IN OIL/WATER SEPARATION

The industrial development has boosted the use of oil-derived products and, as a consequence, oily wastewater has turned into a worldwide challenge. Moreover, oil accidents that occurred in the last decade caused serious ecology and environmental damages to nature and human life. New technologies have been engineered in order to treat and separate water-oil waste. Currently, the use of traditional methods to separate oil from water is based on filtration membranes, e.g., coagulation, flocculation, and absorbing materials, still present some limitations such as high energy cost, easy fouling, and low separation efficiency. As a consequence, much effort has been invested in the development of new separation membranes that allow only oil to pass through while water is repelled, or vice versa.

In order to achieve it, all the materials to be used as a membrane have to hold hydrophilicity or hydrophobicity properties. However, this method still presents some weaknesses since hydrophobic membranes can be fouled by oil and other non-polar contaminants (due to strong hydrophobic-hydrophobic interactions), thus decreasing the flux and dropping the separation efficiency. In order to avoid oil-fouling and improve novel membrane performances, many different materials have been studied and employed. Here, some examples using a hydrogel as an active part of the membrane to separate both water/oil phases are presented. A hydrogel is a well-known material used in many applications because of its capacity of water retention and easy fabrication, which can be easily tuned to its hydrophilic/hydrophobic nature by incorporating inorganic or organic compounds by chemical or physical absorption (Figure 18.3a).

FIGURE 18.3   a) Simplified scheme of hydrogel material for oil/water separation showing the effect of roughness (up) and a material with hydrophilic/hydrophobic properties (down). Adapted with permission (Wan Ikhsan et al. 2021). Copyright 2021, Elsevier; b) Nickel and PAM hydrogel coated nickel foams showing the visual change of wettability towards a water droplet (up) and oil/water separation process (down) using PAM hydrogel coated nickel foam. Oily phase (blue solution) was blocked on top of the hydrogel composite while water (colorless solution) passes through the modified foam. Reproduced with permission (Chen et al. 2015). Copyright 2015, Royal Society of Chemistry.

Jiang's and co-workers reported a hydrogel-coated filter paper (by cross-linking PVA with cellulose using glutaraldehyde as a cross-linker) that exhibited high oil/water separation ability in saline environments and with different pH media. The filter showed high resistance and robustness, as well as high efficiency of up to 99.9% (Fan et al. 2015). Elkamel and co-workers fabricated polyacrylamide (PAM)- and poly (acrylamide/sodium acrylate) copolymer (PAM/Na-Ac)-based hydrogels to be coated on a metallic mesh (Madhuranthakam, Alsubaei, and Elkamel 2018). A PAM hydrogel reinforced the metallic mesh support and improved the anti-oil fouling. Both tested hydrogels presented water recovery values of 93–98%. Nevertheless, the water efficiency decreased when the oil concentration increased. A PAM/Na-Ac copolymer showed lower separation time due to the ionic effect of the sodium acrylate (Na-Ac), which improved the swelling ratio and water fraction that can pass easily through the membrane. Chen et al. reported a filter made of a PAM hydrogel deposited on top of a nickel foam mesh with a pore size of 300 μm (Chen et al. 2015). They used N-N-methylenebisacrylamide (MBA) as a cross-linker. After the device's fabrication, the superhydrophilicity and underwater superoleophobicity were tested (Figure 18.3b). The water efficiency was higher than 99.5% and the membrane could be reused six times, with high separation efficiency as well.

Despite the good results published using hydrogels as active material in the membranes, there are certain problems that should be taken into account, such as low efficiency when working with complex oil/water mixtures. In order to solve this problem, many research groups have been working on hydrogel surface modification to improve hydrophilicity, antibacterial properties, mechanical stability, etc. Recently, Lui et al. presented a cellulose hydrogel coated onto a biodegradable cotton fabric by the sol-gel method (H. Liu et al. 2022). During the gelation process, Ag|AgCl nanoparticles (NPs), which presented anti-oil-fouling and antibacterial behavior, were added to reinforce and improve the mechanical properties of the membrane. By adding Ag|AgCl NPs to the hydrogel network, it was observed to also help in the membrane photoinduced self-cleaning.

18.4 HYDROGELS IN WASTEWATER TREATMENTS

Anthropization has dramatically increased the variety of substances in nature with the potential and proven harmful effects on the environment and humans. Over the last century, a plethora of pollutants, such as heavy metals, dyes, nitrosamines, halogenated aromatic hydrocarbons, phosphates, and nitrates, have been found in water bodies. Moreover, the so-called contaminants of emerging concern, which include micro- and nano-pollutants, pesticides, pharmaceuticals, and personal care products, have been identified in wastewater (Van Tran, Park, and Lee 2018). Consequently, wastewater treatments (WWTs) have undergone constant adaptation and innovation, not only to deal with such a wide variety of pollutants effectively but also to be sustainable alternatives that guarantee water security. In this regard, the use of hydrogels in WWTs—typically as a tertiary treatment—has yielded promising results, enabling them as capable agents for sustained water security (Van Tran, Park, and Lee 2018).

This section provides a general overview of hydrogels in WWTs as novel materials that are able to significantly outperform other well-known WWTs processes when used during them, as well as serve as enabling agents for innovative alternative approaches to existing WWTs. We will focus on the use of hydrogels in the adsorptive removal of pollutants, advanced oxidation processes, and biological processes.

18.4.1 Hydrogels in the Adsorptive Removal of Pollutants

Adsorption is one of the most appropriate processes for the remediation of wastewater contaminated with some of the pollutants described above (Loo et al. 2021). The adsorption mechanism is conceptually simple: it is a process by which atoms, ions, or molecules (the adsorbates) adhere to the surface of a material (the adsorbent). The type of adsorbent employed determines the effectiveness of adsorbing specific pollutants (Van Tran, Park, and Lee 2018).

Adsorbent materials employed in WWTs are typically based on activated carbon, sand, and zeolites. However, the usage of hydrogels in the remediation process has demonstrated higher performance due to their advantageous properties compared to the materials mentioned above. One of these essential properties is the presence of abundant functional groups, as schematized in Figure 18.1, which have a high chemical affinity with a wide diversity of pollutants, enhancing enormously the sorption capacity of the hydrogel (Loo et al. 2021).

The broad variety of functional groups present in the structure of hydrogels can interact with charged species, and the adsorption mechanism usually is based on electrostatic interactions, physical sorption, or complexation, making hydrogels a versatile and wide-spectrum sorbent. In this regard, hydrogels that bear anionic groups have been used to adsorb a wide variety of cations, such as Pb^{2+}, Zn^{2+}, Fe^{3+}, Cu^{2+}, Cd^{2+}, and Al^{3+}; and also including rare earth elements such as La^{3+}, Ce^{3+}, Nd^{3+}, and Eu^{3+}; and radionuclides such as U^{6+} (Van Tran, Park, and Lee 2018). However, hydrogels bearing cationic groups have been used to adsorb inorganic anions such as nitrites, nitrates, sulfates, bisulfides, bisulfites, and oxyanions. Hydrogels that bear both negative and positive charges present an amphoteric behavior, i.e., they can be adjusted to absorb either anions or cations depending on the pH value of the solution (Loo et al. 2021). In addition, this capacity of hydrogels to interact with charged species makes them excellent agents for removing a wide variety of organic dyes that are soluble in water. As an example, hydrogels based on chitosan—which is rich in hydroxyl and primary amine groups—have yielded superior adsorption capacity than a typical activated carbon when removing dyes and oxyanions from water (Wittmar, Klug, and Ulbricht 2020).

The functional groups present in the structure of the hydrogel usually show a strong affinity to water molecules, which endow them with high hydrophilicity and even a great moisture-capturing capacity (Loo et al. 2021). This high hydrophilic behavior allows for the efficient uptake of water-soluble sorbates and their subsequent transportation to the hydrogel active sites. While the adsorption process occurs on the surface of the adsorbate when typical adsorbents are employed in WWTs, in the case of hydrogels, this process may occur in their flexible tridimensional structure since hydrogels are able to mobilize water into their network due to their hydrophilicity. As a consequence, pollutants rapidly diffuse into the hydrogel network within the water molecules (Loo et al. 2021).

On the other hand, when the functional groups are weak acids or bases, hydrogels present a buffering capacity, which provides a broad operational pH range. It has been observed, for example, that the carboxylate groups in poly(sodium acrylate) (PSA) form a R-COOH/R-COO⁻ buffering system capable of maintaining the pH of a solution between 8.8 and 11.7. In addition, this buffering effect enhanced the removal efficiency of Cr^{6+} due to the fact that the reduction to Cr^{3+} is favored between a pH range of 8.5 and 9, followed by a precipitation process (Jia et al. 2018). Also, the removal of Cr^{6+} strongly depended on the initial pH of the solution when a cellulosic hydrogel was employed, which has no weak acid or base functional groups and consequently did not show such buffering effect (Wittmar, Klug, and Ulbricht 2020).

In addition to the versatility that a variety of functional groups provide to hydrogels, one of the most appealing aspects currently lies in the fact that they constitute a greener alternative to conventional adsorbents. Many polymeric gels have good reusability characteristics, since they may be regenerated for numerous operation cycles without substantial loss in adsorption capabilities, using appropriate solutions such as organic solvents or acidic/basic treatments (Loo et al. 2021). Nevertheless, alternative regenerative methods have been investigated to reduce and even eliminate the need to use either hazardous chemicals or caustic solutions. Interesting advances in this field were introduced by smart polymers; for example, hydrogels based on PNIPAAm combined with chelating agents, the latter being the compounds that form complexes with the metal ions in water. Then, these complexes are adsorbed onto the PNIPAAm hydrogel through stable hydrophobic interactions above the lower critical solution temperature (LCST) (Figure 18.4a). Finally, the ion

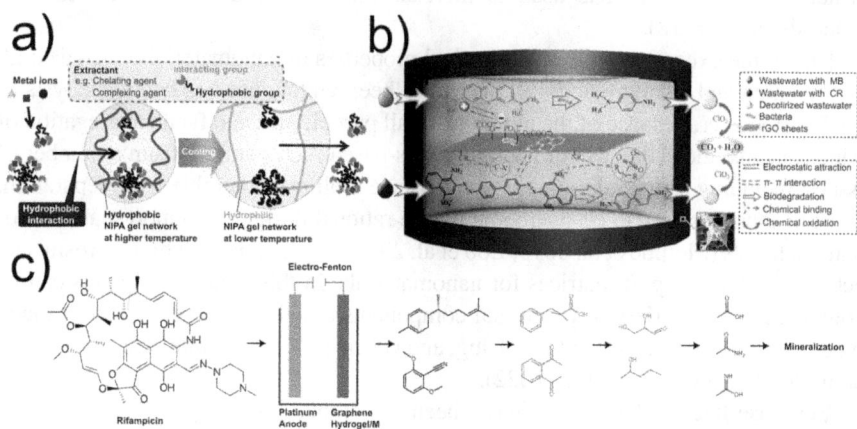

FIGURE 18.4 a) Schematic diagram of temperature-swing solid-phase extraction of a metal ion complexed with an extractant onto the poly(N-isopropyl acrylamide) hydrogel. Reproduced with permission (Tokuyama and Iwama 2007). Copyright 2007, American Chemical Society. b) Schematic diagram of a dye removal process employing a hydrogel containing live *Shewanella xiamenensis* and reduced graphene oxide (rGO) sheets. Reproduced with permission (Shen et al. 2019). Copyright 2019, American Chemical Society. c) Schematic rifampicin removal mechanism during electro-Fenton process employing graphene hydrogel/M nanocomposites (where M is a coupling metal: Cu, Co, Ni) as cathodes. Adapted with permission (Ebratkhahan et al. 2022).Copyright 2022, Elsevier.

complexes extracted from the water are desorbed from the hydrogel by reducing the temperature below the LCST, which causes a phase change in the hydrogel towards a more hydrophilic behavior (Tokuyama and Iwama 2007; Loo et al. 2021).

18.4.2 Hydrogels for Advanced Oxidation Process (AOPs) in Wastewater Treatments

Advanced oxidation processes (AOPs) are appealing water treatment solutions. During AOPs, highly reactive oxidant species are produced, such as hydroxyl radical or persulfate (Meijide et al. 2018). These species react non-selectively during the treatment, breaking down the molecules of water pollutants and transforming them into biodegradable compounds, which is of the utmost importance when considering the so-called refractory organic contaminants (refractory term is coined to refer to the organic matter which is apparently non-accessible or resistant to rapid microbial degradation). AOPs are a group of a diverse set of oxidative treatments, including Fenton, photo-Fenton, electrochemical oxidation, supercritical water oxidation processes, and photocatalysis (Figure 18.4b and 18.4c). The Fenton process and its variants differ from photocatalysis in that the latter is based on the generation of hydroxyl radicals through oxidation-reduction reactions that occur on the surface of a photocatalyst due to the action of UV light and the presence of an oxidizing agent, such as hydrogen peroxide, while Fenton processes require the presence of Fe^{2+} as a catalyst to generate hydroxyl radicals from H_2O_2, even in photo-Fenton processes in which UV-Vis radiation is used to increase the photoreduction of Fe^{3+} to Fe^{2+} (Meijide et al. 2018).

Due to the exceptional physicochemical properties of nanomaterials, such as high surface area and catalytic activity, they have been widely used as heterocatalysts in AOPs. However, because of the nature of small particle size and facile aggregation of nanocatalysts, the actual use of nanomaterials-based AOPs suffers from a number of issues, including restricted mass transfer, surface fouling, and problematic separation. Most of these issues can be overcome by the rational design of support matrices for nanomaterials (Meijide et al. 2018; Loo et al. 2021). Hydrogels with high porosity can act as excellent support matrices for nanomaterials. In this regard, hydrogels can be combined with a variety of functional components to aid in the breakdown of many types of water pollutants (Wittmar, Klug, and Ulbricht 2020; Meijide et al. 2018; Shen et al. 2019; Ebratkhahan et al. 2022).

Iron-base-hydrogel catalysts have been developed for carrying out heterogeneous catalysis to remove the pesticide acetamiprid via the electro-Fenton (EF) process. The hydrogel was based on PVA-alginate and goethite as a source of iron; by combining it with an adsorption process, the pesticide was completely mineralized, and the inorganic ions were removed (Meijide et al. 2018). The number and variety of organic contaminants in urban wastewater and seawater in ports are very high. In this sense, EF processes have demonstrated to be more selective than conventional methods. A series of mesoporous carbonaceous materials, based on natural polymer hydrogels (chitosan and agarose), were synthesized and implemented by Sirés and co-workers in air-diffusion cathodes employed in photoelectro-Fenton (PEF) techniques. The systems were used to electrocatalyze the oxygen reduction reaction (ORR) for the in-situ production of H_2O_2 and for the complete mineralization and elimination of pharmaceutical drugs from urban water (Daniel et al. 2020; Y. Zhang et al. 2022). In addition, graphene hydrogel-metal nanocomposites were prepared as cathodes by Ebratkhahan *et al.* within the EF process to effectively remove the antibiotic rifampicin from polluted water (Ebratkhahan et al. 2022) (Figure 18.4c).

Regarding the field of photocatalysis, when using a ZnO and PHEMA composite hydrogel as an adsorbent, the additional photocatalytic function of the ceramic component allows the in-situ degradation of adsorbed dyes after light irradiation and the consequent regeneration of the hydrogel adsorbent (Ussia et al. 2018). Moreover, the composited PSA cryogels functionalized with nano zero-valent iron (nZVI) are both efficient adsorbents as well as efficient agents for reducing Cr^{6+} to Cr^{3+}, leading to total Cr removal. Indeed, implementing the PSA-nZVI composites shows a reaction rate of approximately 2.5 higher than that of free nZVI (Jia et al. 2018).

It is important to mention that the majority of photoactive hydrogel systems have an adsorption and degradation synergistic effect. In this sense, the hydrogel system can act not only as a carrier to ensure a higher utilization rate as well as the recyclability of nano photocatalytic materials but also to promote the reaction efficiency of the hydrogel composites. However, in order for the efficiency of the composite hydrogel to reach higher performance, the choice of nano photocatalyst and hydrogel is critical, and correcting the influencing variables within a suitable range is a priority (Meijide et al. 2018; Loo et al. 2021).

Aside from catalysis, the insertion of specific units and biomolecules into the hydrogel structure might improve its selectivity toward the removal of a target pollutant, which may be impacted in the presence of interfering species. Careful choice of such units may allow this interference caused by other species in the water to be overcome (Meijide et al. 2018; Loo et al. 2021).

18.4.3 Hydrogels in Biological Processes of Wastewater Treatments

Hydrogels' high water content and fluidity provide an appropriate environment for entrapping and maintaining the activity of microorganisms endowed with a natural capability for degrading recalcitrant pollutants as well as aromatic compounds such as phthalic acids and phenols (Jia et al. 2018; Loo et al. 2021).

Microbes such as ammonia-oxidizing bacteria, anaerobic ammonium oxidation bacteria, or activated sludge can be confined onto a hydrogel structure, which leads to outperforming effects compared to the case where the treatment tank has microorganisms suspended as free cells. Thus, when hydrogels are employed within WWTs, they improve the retention of biomass during the remediation of wastewater while at the same time boosting the microorganism concentration within the hydrogel. As a consequence, even starting from a lower requirement of seeding biomass, it is possible to improve the degradation rate, showing, at the same time, enhanced tolerance to high-strength wastewater (Bae et al. 2017; Xu et al. 2017).

Since mass transport properties are critical for the efficiency of biodegradation, high porosity hydrogels also show superior performance during bioremediation processes; however, it is of utmost importance that the polymeric matrix is compatible with microbes. In this sense, biocompatible polymers such as alginate, PVA, poly(N-vinyl pyrrolidone), and polyethylene glycol are typically employed for synthesizing the biomass entrapment based on hydrogel matrices (Bae et al. 2017; Xu et al. 2017).

It has been observed that not only the biocompatibility but also the specific surface area of PVA hydrogels were significantly improved by compositing them with a graphene oxide-glutamic acid nano-composite. This fact also led to a higher microorganism loading capacity and a consequent increase in the chemical oxygen demand removal rate when the hydrogels are employed in WWTs (Yu, Shu, and Ye 2018). Moreover, the study of the growth of microorganisms on porous polyurethane hydrogels showed that ammonium-oxidizing bacteria performed a surface-attached growth, whereas anammox microcolonies had embedded-like growth, i.e., the biodegradation can occur both on the surface and inside the network of the hydrogel (W. Liu et al. 2022).

Analogous to the case of AOPs, a synergistic effect consisting of adsorption and biodegradation has been observed in the treatment of cationic and anionic dyes using graphene hydrogels as a matrix for live bacteria encapsulation (Shen et al. 2019).

18.5 HYDROGELS IN WATER PURIFICATION

To address the current growing water shortage that nowadays affects more than 3 billion people, the development of technologies that can directly utilize and treat unconventional water resources (seawater, waste-water, rainfalls, etc.) for our daily water supply is needed. Classical thermal- and membrane-based technologies for water purification are largely employed, but drawbacks such as fouling, salt rejection, and high-energy consumption have stimulated the study and utilization of innovative materials, such as the hydrogels-based materials. Because of their highly tunable physicochemical properties and their cross-linked architecture is easily engineered, the hydrogels represent new polymer platforms with a high potential to be implemented in novel water treatment systems.

18.5.1 SOLAR WATER PURIFICATION

In the solar water purification field, the usage of solar energy in many different ways, as heat and light as well as chemical energy, is largely studied; since Chile, way back in the 1880s, built the first commercial-scale desalination plant using solar energy, various technologies have been developed for water treatment. All these techniques are based on the most positive feature about solar water purification: no requirement of fuel. It's precisely due to the lack of fuel that makes solar applications relatively more interesting than conventional sources of energy, as it does not cause pollution or health hazards associated with it. Among the more employed solar water treatments, we can find solar water disinfection (SODIS), solar water pasteurization, and solar vapor generation (SVG) (Goel, Verma, and Tripathi 2022). These technologies are quite simple and easy to understand, usually require a low financial investment, and have proven their effectiveness. Additionally, they can remove contaminants and

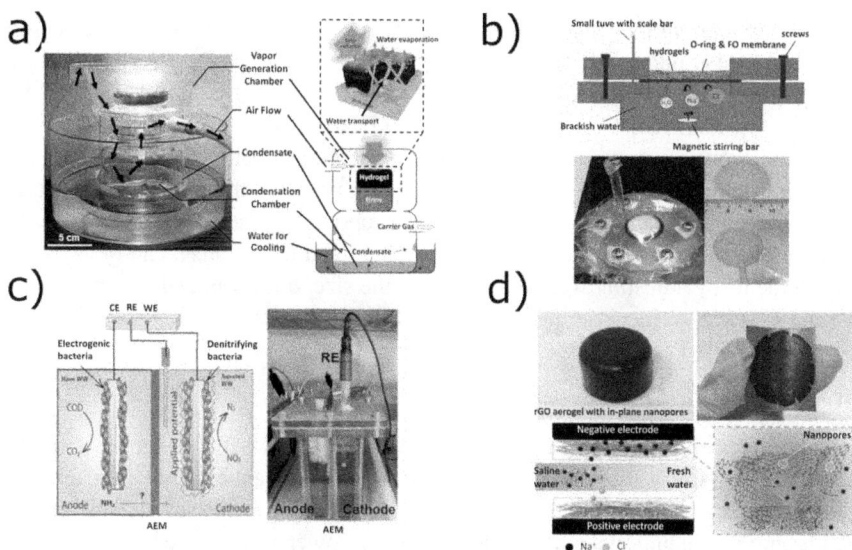

FIGURE 18.5 a) Photograph and schematic illustration of a solar water distillation based on a light-absorbing sponge-like hydrogel (LASH). Adapted with permission from (Guo et al. 2019). Copyright 2019 American Chemical Society. b) Schematic image of forward osmosis (FO) devices under water treatment process. Photograph illustration of hydrogel disc with polymer particles. The hydrogel disc can be handled and it is robust for performed water cleaning treatment. Reprinted from (Cai et al. 2013). Copyright 2013 Elsevier. c) Schematic illustration and photograph of a two-chambered electrochemical denitrification system. Both chambers are separated by an anion exchange membrane (AEM). Reproduced with permission from (Babiak et al. 2022); published by MDPI, 2022. d) Image of rGo hydrogel and scheme of CDI process. Reprinted from (W. Shi et al. 2016). Copyright 2016 Springer Nature.

microbes through repeated boiling and condensation processes. One of the most interesting technologies is the SVG, a simple system where a membrane covered with an absorbent is capable of separating the liquid water from the water vapor generated on its surface after irradiation with sunlight. The material employed must be capable of promoting the transport of water to its surface, as well as the specific interactions of the water molecules, in order to improve its condensation and recovery (Figure 18.5a). However, it remains a challenge to maximize the evaporation rate and the energy conversion efficiency concurrently, which requires the coordination of chemical properties of materials used and process engineering, among other aspects.

The last technologies are based on the design of new materials to be inserted in the solar water purification plants, able to improve the two key aspects of heat and water management is of great interest. A team of researchers, led by Guihua Yu at The University of Texas at Austin, recently developed an inexpensive hydrogel able to significantly increase the volume of water evaporated compared to conventional methods, only needing naturally occurring levels of sunlight (Zhou et al. 2018; Guo et al. 2019). The hydrogel is based on polyvinyl alcohol (PVA) and polypyrrole (PPy), where the PPy solar absorbers in the PVA matrix allow for the solar energy to be directly delivered to the water in the molecular meshes, with an energy efficiency of the gel of ~94% under the equivalent of radiation of 1 kW·m^{-2}, whereas, in other systems, an efficiency greater than 90% could only be achieved with highly concentrated solar radiation above 4 kW·m^{-2} (Zhao et al. 2018). The team had and continues carrying out an intense scientific investigation about different materials that can optimize the process. This innovative approach proceeds through the optimization of (i) heat management (by tailoring surface topography, selecting adequate solar absorbers to minimize heat loss to the atmosphere and bulk water), and of (ii) water management (by adjusting wettability; tuning the size, direction, and structure of the internal channel; to increase the water content in hydrogels).

18.5.2 REVERSE OSMOSIS

Reverse osmosis (RO) is a separation technology where a semi-permeable membrane separates two different solutions. Applying an external pressure, the water flow through the membranes and the molecules and ions stack on the membrane, producing clean water. RO requires power to increase the pressure of the saline solution. The rate of diffusion permeation of water and the external pressure needed depends on the pressure, permeability, concentration of contaminants, and temperature. RO membrane must be a non-porous and must be excludes particles and low molar mass species. In addition, these membranes have to present mechanical durability, high salt rejection, low cost, chlorine resistance, and fouling resistance (Li and Wang 2010). Taking these properties into account, the capabilities of hydrogel membranes make them a very interesting material to be used in this type of application.

One of the main inconveniences in working with a RO membrane is the formation of a layer on the membrane surface which influence negatively in its efficiency. There are different types of membrane fouling depending on the type of "particle" that adheres; for example salt precipitation, organic fouling, colloidal fouling, or biofouling. The one that shows the highest challenge is biofouling. In water clean treatment, biofouling is a phenomenon where bacteria adhere to the membrane surface, forming biofilms. This biofilm influences the properties of the semi-membrane obstructing the pores. In order to prevent this phenomenon, many groups have been working on different strategies, such as the addition of nanoparticles that confer antibacterial behavior to the network, modify the hydrogel surface, or the covalent addition of an antibacterial polymer to the hydrogel surface.

Yang et al. published a composite of polyamide with different polymers as poly (2-hydroxyethyl methacrylate) (PHEMA), poly[poly(ethylene glycol)methacrylate] (pPEG), and poly[[(2-methacryloyloxy)ethyl]dimethyl[3-sulfopropyl]ammonium hydroxide (pMEDSAH), which present good stability and biocompatibility (Yang et al. 2019). The objective was to study the structural influence of these composites against biofouling. It was shown that parameters such as hydration and chain length play an important role in order to fabricate anti-biofouling membranes.

Nikolaeva et al. reported a hydrophilic poly(amidoamine)(PAMAM) bonded to a poly(amide) (PA) layer (Nikolaeva et al. 2015). PA is a well-known hydrogel that has been used for RO for quite some time. There are many published articles showing the functionalization of the PA layer with different multifunctional groups, such as amines, carboxylic groups, and carbonyl chloride groups, to improve membrane properties. PAMAM was coated using either methanol (TFC-PAMAM2) or water (TFC-PAMA1) by spraying, creating a thin film composite (TFC). The use of PAMAM allowed an increase in permeability of the membrane by 25%. The salt rejection ration did not present any difference comparing both the unmodified and the modified membrane (98.2% and 98.0%, respectively) in an aqueous solution, while in a methanolic solution the salt retention decreased (95.4%). Regarding the protein absorption tested using bovine serum albumin in a static mode presented two different trends compared with modified membrane. In the case of aqueous PAMAM, the protein ratio adsorption is higher than in an unmodified membrane due to its higher electrostatic interaction. However, methanolic PAMAM presents lower protein absorption because it is more homogeneous and covers the entire membrane.

18.5.3 FORWARD OSMOSIS

Forward osmosis (FO) is a technique that gained attention to be employed in many fields, such as water treatment, desalination, power generation, etc. Forward osmosis employs a cell separated by a semi-permeable membrane. On one side of the cell, a salt solution is added, creating an osmotic pressure difference. The osmotic pressure difference drives the water that passes through the semi-permeable membrane, thus obtaining clean water free of salts and contaminants (Qin et al. 2018). The water is recovered from a membrane or draw agent by different techniques, such as distillation, membrane separation, extraction, precipitation, etc. FO membranes have a huge potential due to their reduced cost, ease of cleaning, fast regeneration of draw agent, and the great advantage of using osmotic pressure instead of hydraulic pressure.

Hydrogels can play an important role in forward osmosis since they have the ability to absorb water through a membrane by its swelling capability, and its insoluble cross-linked structure allows dewatering under mechanical and thermal stimuli. However, a hydrogel presents poor liquid water recovery rate. Although the hydrogel has proved to be a potential draw agent, there are still many issues to be solved, such as low flux and low yield of clean water. The high energy costs for water recovery are still an issue for FO desalination technologies.

In order to improve water recovery, Cai et al. synthesized a thermally responsive semi-interpenetrated network (semi-IPN) (Cai et al. 2013). Semi-IPN was synthesized by polymerizing the thermosensitive polymer N-isopropylacrylamide (NIPAm) in the presence of linear hydrophilic chains of polysodium acrylate (PSA) or polyvinyl alcohol (PVA) (Figure 18.5b). The semi-IPN hydrogel showed a better balance of swelling and dehydration compared to a pure PNIPAAm hydrogel copolymerized with sodium acrylate (Zeng et al. 2013). After swelling, the semi-IPN hydrogel released 90% of the water by heating the hydrogel to 40°C and subsequently cooling it to room temperature.

Gorbani's group published a biodegradable and biocompatible polymer hydrogel based on carboxymethylcellulose and acrylic acid to be used as a draw agent in forward osmosis (Shakeri et al. 2019). In order to improve the swelling ratio and water flux, quaternary ammonium graphene oxide (GCO) was added to the hydrogel matrix. The positive charges of quaternary ammonium groups from GCO interact with the carboxylate groups (negative charges) from hydrogels influencing the physicochemical properties of the membrane. The swelling ratio and water recovery were the parameters where some improvement was observed. The swelling ratio between pure hydrogel and GCO-hydrogel improved from 379 to 537 g of water/g of hydrogel. In addition, the water recovery of GCO-hydrogel was 41.32% less than 6 hours, while the water recovery of the pure hydrogel was 51.76% in 6 hours. Thus, the GCO hydrogel has better properties in water recovery and in the swelling ratio than the pure hydrogel.

18.5.4 ELECTRODIALYSIS

Electrodialysis (ED) is a technology used for brackish water cleaning when an external electric field is applied to remove salts from the water by being forced through an ion exchange membrane (IEM) (Arana Juve et al. 2022). These membranes are based on cation exchange and anion exchange pairs. When an electric field is applied, cations and anions migrate to the cathode and anode through a semipermeable cation and anion exchange membrane, respectively, which act as a selective ion filter. IEMs must hold some properties that facilitate this selectivity, i.e., they have to be permeable for specific ions, low electrical resistance, mechanical properties, and high cost effectiveness. Initially, the materials used in wastewater treatment by ED were mainly metal oxide, carbon nanotubes, and carbon-based materials. However, according to the literature, these materials present some inconvenience, such as its functional groups are isolated, thus decreasing the permselectivity of the mobile ions. In addition, the nanomaterials contained in the membrane tend to aggregate, which affects the current intensity and the uniformity of the membrane. Also, the concentration of the nanoparticles also affects electrical resistance and the membrane structure. Due to their interesting properties, hydrogels have become an interesting material to replace IEMs, thanks to the ease of adjusting their exchange properties by creating new positive or negative groups, thus improving the selectivity of the membrane.

Nemati et al. reported a cation exchange membrane using a composite made from 2-acrylamido-2-methylpropanesulfonic acid (AMAH) particles deposited on a PVC (poly vinyl chloride) membrane (Nemati, Hosseini, and Shabanian 2017). This new composite had high water content thanks to its higher porosity and the hydrophilic nature of the hydrogel. Various tests to determine the permeability against specific ions, the ability to remove Pb and Ni in solution, and the life cycles of the membrane, were performed. In fact, the authors managed to eliminate K (99.9%), Pb (99.9%), and Ni (96.9%) from the sample. The composite membrane presents a higher permeability and flux compared to Na and Ba in solution. Increased AMAH in the membrane decreased transport capacity and its permeability (PQ). Furthermore, the mechanical stability and the swelling rate were not interrupted by adding AMAH NP to PVC.

Babiak et al. developed an anion exchange membrane based on a hydrogel deposited on a ceramic support for wastewater treatment (Babiak et al. 2022). All tests performed were compared to a commercially available AMI-7001 hydrogel membrane (Figure 18.5c). Four different and environmentally friendly hydrogels were fabricated (although diallyldimethylammonium was chosen as the best material to use as anion exchange membrane, hereafter AEM). Compared with the commercial membrane, the new hydrogels presented similar properties. However, they did show some improvements when working with their AEM. More specifically, the ceramic composite presented a strong physical resistance and was self-supporting, thanks to the ceramic platform. Furthermore, the hydrogel could be deposited on different porous ceramic supports. It was also observed how the permselectivity of EAM can change by modifying charged functional groups of the monomers.

18.5.5 Capacitive Deionization

Capacitive deionization (CDI) technology is based on the removal of charged species in a solution through the formation of an electrical double layer by applying a voltage (Kalfa et al. 2020). CDI technology is mainly used for desalination of brackish water; however, in the last years, it has attracted attention to be used for water treatment processes. A CDI cell is formed by two electrodes (cathode and anode) separated by a channel that contains the solution to be desalted. When a voltage is applied, an electric double layer is created in both electrodes; thus, the charged substances from the solution diffuse to the cathode (positive species) and to the anode (negative species) by means of electrosorbing on both electrodes (Figure 18.5d). When the ionic storage of electrodes reaches their maximum capacity, electrodes must be regenerated. The entire process is divided into two parts: i) electroabsorbance of the ions contained in the water and ii) regeneration of electrodes by discharge (Xing et al. 2020). CDI presents some advantages compared with other conventional wastewater treatment technologies, such as its high energy efficiency, low cost effectiveness, and low salt rejection. In recent decades, CDI has focused on the development of new electrode materials, searching for the best combination of a higher electrical conductivity, larger surface area, and increased electrosorption capacity. Up to date, most published papers are based on carbonaceous materials, such as activated carbon (AC), porous carbon, carbon nanotubes, graphene, etc.

One of the proposals to improve the CDI electrodes is to add nanoparticles or hydrogels to AC. Yasin et al. reported a ZnO nanoparticle-incorporated activated carbon/graphene hydrogel composite (Yasin et al. 2021). This composite shows better conductivity and wettability, electrosorptive capacity, and larger ion transport compared with bare graphene hydrogel (GH) and activate carbon. Such an improvement is due to the presence of ZnO NPs, which avoid the agglomeration of graphene sheets in AC, reducing the internal resistance. In addition, the composite presented the highest salt removal efficiency (83.65%) compared with AC and GH (11.4% and 18.18%, respectively).

Liu et al. present a new CDI electrode with the purpose to be used as a desalination, disinfection (*E. coli*, *P. aeruginosa*), and heavy metal removal (Pb^{2+} and Cd^{2+}) (N. Liu et al. 2021). Polyhexamethylene guanidine (PHMG) and polydopamine (PDA) were deposited on the surface of an AC electrode. PHMG is a well-known cationic polymer with antimicrobial activity, while PDA is generally used to modify surface properties, where its hydrophilic and hydroxyl groups on the molecule help improve the wettability of AC. The effectiveness of the PDA/PHMG electrode against *E. coli* and *P. aeruginosa* bacterial elimination rate, which was calculated by the count of cells after and before the treatment, was checked. Pristine AC electrode showed a 30% of *E. coli* elimination while the new composite showed *E. coli* antibacterial activity of 99.11% after 60 min and 98.67% of *P. aeruginosa* after 60 min. In addition, the heavy metal test was performed to investigate the adsorption capacity of the electrode using Pb^{2+} and Cd^{2+}. Consequently, the adsorption capacity of Pb^{+2} using AC was 16.51 mg·g^{-1}, while using the AC-PDA/PHMG2 composite electrode it was 20.30 mg·g^{-1} after 100 min. The increase in the absorption rate of lead was due to the amino groups of PDA and PHMG that can easily coordinate with metal cations. No significant improvement was observed when salt adsorption tests were measured. The salt adsorption capacity obtained at the equilibrium was 10.23, 11.54, 10.44, 10.24, and 10.09 mg g^{-1} for AC, AC-PDA, AC-PDA/PHMG0.5, AC-PDA/PHMG1, and AC-PDA/PHMG2, respectively. Furthermore, the authors showed that both electrodes can be regenerated after five cycles without the loss of desalination capacity.

18.6   FUTURE OF HYDROGELS IN FRESHWATER PRODUCTION AND WASTEWATER PURIFICATION

The shortage of freshwater resources has prompted researchers to conduct detailed investigations on advanced water treatment and water harvesting technologies based on the use of hydrogel networks. From the perspective of the increased use of such materials for wastewater cleaning and potable water supply in the next decades, the most promising materials seem to be that derived from double-network systems, due to their superior mechanical strength, thermal stability, and swelling properties. In this way, cross-linked hydrogels modified with conducting polymers, and hybrid systems with active carbon compounds and graphene sheets seem to be the most promising materials to come into force in the next years. They have proved to have a high efficiency reinforcing the physicochemical processes of pollutants adsorption, the electrochemical capacity of the new electrodes involved in the purification processes, as well as facilitating the energy transfer in solar water harvesting processes. Unfortunately, the full-scale implementation of hydrogel networks in wastewater treatment plants should overcome some important limitations, such as improvement of their reusability; mechanical durable life, self-healing, and antifouling properties after swollen state; and the efficiency and capability of large freshwater production comparable to the classical methods.

Among the technologies discussed in the present chapter, solar desalination from brackish water and seawater by using hydrogel-based solar evaporators should represent a powerful alternative to reduce the cost of implementation of desalination plants based on RO, FO, and related membrane processes for water purification. If researchers are able to reproduce the same efficiency and capability of such plants to generate water with composite hydrogels or hybrid hydrogels, it will represent an advanced challenge to overcome the scarcity of freshwater in coastal areas.

Similarly, to the encouraging implementation of electrical batteries in cars to replace fuel-powered vehicles or the development and installation of solar cells for self-energy generation to decrease the dependence of non-environmentally friendly processes that came into force in the last two decades, small, scaled-up prototypes for house self-freshwater generation would be also possible in the next years. The wastewater from showers and dish- and clothes-washing machines, among others, are discarded after use, and could be converted into inputs to be purified at home with the appropriate hydrogel-derived material. It would be possible if engineers are able to control the limitations mentioned above.

ACKNOWLEDGMENTS

Grant PID2021-125257OB-I00 funded by MCIN/AEI/10.13039/501100011033 and, by ERDF "A way of making Europe", by the European Union.

REFERENCES

Arana Juve, Jan-Max, Frederick Munk S Christensen, Yong Wang, and Zongsu Wei. 2022. "Electrodialysis for Metal Removal and Recovery: A Review." *Chemical Engineering Journal* 435: 134857. 10.1016/j.cej.2022.134857.

Armelin, Elaine, Maria M Pérez-Madrigal, Carlos Alemán, and David Díaz Díaz. 2016. "Current Status and Challenges of Biohydrogels for Applications as Supercapacitors and Secondary Batteries." *Journal of Materials Chemistry A* 4 (23): 8952–8968. 10.1039/C6TA01846G.

Babiak, Peter, Geoff Schaffer-Harris, Mami Kainuma, Viacheslav Fedorovich, and Igor Goryanin. 2022. "Development of a New Hydrogel Anion Exchange Membrane for Swine Wastewater Treatment." *Membranes* 12 (10): 984. 10.3390/membranes121 00984.

Bae, Hyokwan, Minkyu Choi, Yun Chul Chung, Seockkheon Lee, and Young Je Yoo. 2017. "Core-Shell Structured Poly(Vinyl Alcohol)/Sodium Alginate Bead for Single-Stage Autotrophic Nitrogen Removal." *Chemical Engineering Journal* 322: 408–416. 10.101 6/j.cej.2017.03.119.

Cai, Yufeng, Wenming Shen, Siew Leng Loo, William B. Krantz, Rong Wang, Anthony G. Fane, and Xiao Hu. 2013. "Towards Temperature Driven Forward Osmosis Desalination Using Semi-IPN Hydrogels as Reversible Draw Agents." *Water Research* 47 (11): 3773–3781. 10.1016/J.WATRES.2013.04.034.

Chen, Baiyi, Guannan Ju, Eiichi Sakai, and Jianhui Qiu. 2015. "Underwater Low Adhesive Hydrogel-Coated Functionally Integrated Device by a One-Step Solution-Immersion Method for Oil–Water Separation." *RSC Advances* 5 (106): 87055–87060. 10.1039/C5 RA13657A.

Daniel, Giorgia, Yanyu Zhang, Sonia Lanzalaco, Federico Brombin, Tomasz Kosmala, Gaetano Granozzi, Aimin Wang, Enric Brillas, Ignasi Sirés, and Christian Durante. 2020. "Chitosan-Derived Nitrogen-Doped Carbon Electrocatalyst for a Sustainable Upgrade of Oxygen Reduction to Hydrogen Peroxide in UV-Assisted Electro-Fenton Water Treatment." *ACS Sustainable Chemistry & Engineering* 8 (38): 14425–14440. 10.1021/acssuschemeng.0c04294.

Ebratkhahan, Masoud, Mahmoud Zarei, Ibtihel Zaier Akpinar, and Önder Metin. 2022. "One-Pot Synthesis of Graphene Hydrogel/M (M: Cu, Co, Ni) Nanocomposites as Cathodes for Electrochemical Removal of Rifampicin from Polluted Water." *Environmental Research* 214 (November): 113789. 10.1016/j.envres.2022.113789.

Elimelech, Menachem, and William A Phillip. 2011. "The Future of Seawater Desalination: Energy, Technology, and the Environment." *Science* 333 (6043): 712–717. www.sciencemag.org.

Fan, Jun-Bing, Yongyang Song, Shutao Wang, Jingxin Meng, Gao Yang, Xinglin Guo, Lin Feng, and Lei Jiang. 2015. "Directly Coating Hydrogel on Filter Paper for Effective Oil–Water Separation in Highly Acidic, Alkaline, and Salty Environment." *Advanced Functional Materials* 25 (33): 5368–5375. 10.1002/adfm.201501066.

Feio, Maria João, Robert M Hughes, Sónia R Q Serra, Susan J Nichols, Ben J Kefford, Mark Lintermans, Wayne Robinson, et al. 2023. "Fish and Macroinvertebrate Assemblages Reveal Extensive Degradation of the World's Rivers." *Global Change Biology* 29 (2): 355–374. 10.1111/gcb.16439.

Goel, Malti, V.S. Verma, and Neha Goel Tripathi. 2022. *Solar Energy. Made Simple for a Sustainable Future.* Singapore: Springer Nature. https://link.springer.com/bookseries/8059.

Guo, Youhong, Jiwoong Bae, Zhiwei Fang, Panpan Li, Fei Zhao, and Guihua Yu. 2020. "Hydrogels and Hydrogel-Derived Materials for Energy and Water Sustainability." *Chemical Reviews* 120 (15): 7642–7707. 10.1021/acs.chemrev.0c00345.

Guo, Youhong, Xingyi Zhou, Fei Zhao, Jiwoong Bae, Brian Rosenberger, and Guihua Yu. 2019. "Synergistic Energy Nanoconfinement and Water Activation in Hydrogels for Efficient Solar Water Desalination." *ACS Nano* 13 (7): 7913–7919. 10.1021/acsnano.9b02301.

Jia, Zhenzhen, Yuehong Shu, Renlong Huang, Junguang Liu, and Lingling Liu. 2018. "Enhanced Reactivity of NZVI Embedded into Supermacroporous Cryogels for Highly Efficient Cr(VI) and Total Cr Removal from Aqueous Solution." *Chemosphere* 199 (May): 232–242. 10.1016/j.chemosphere.2018.02.021.

Kalfa, Ayelet, Barak Shapira, Alexey Shopin, Izaak Cohen, Eran Avraham, and Doron Aurbach. 2020. "Capacitive Deionization for Wastewater Treatment: Opportunities and Challenges." *Chemosphere* 241 (February): 125003. 10.1016/j.chemosphere.2019.125003.

Khan, Musharib, and Irene M.C. Lo. 2016. "A Holistic Review of Hydrogel Applications in the Adsorptive Removal of Aqueous Pollutants: Recent Progress, Challenges, and Perspectives." *Water Research* 106: 259–271. 10.1016/j.watres.2016.10.008.

Kim, Hyunho, Sungwoo Yang, Sameer R Rao, Shankar Narayanan, Eugene A Kapustin, Hiroyasu Furukawa, Ari S Umans, Omar M Yaghi, and Evelyn N Wang. 2017. "Renewable Resources Water Harvesting from Air with Metal-Organic Frameworks Powered by Natural Sunlight." *Science* 356 (6336): 430–434. https://www.science.org.

Lanzalaco, Sonia, Júlia Mingot, Juan Torras, Carlos Alemán, and Elaine Armelin. 2023. "Recent Advances in Poly(N-Isopropylacrylamide) Hydrogels and Derivatives as Promising Materials for Biomedical and Engineering Emerging Applications." *Advanced Engineering Materials* 25 (4): 2201303. 10.1002/adem.202201303.

Li, Dan, and Huanting Wang. 2010. "Recent Developments in Reverse Osmosis Desalination Membranes." *Journal of Materials Chemistry* 20 (22): 4551–4566. 10.1039/B924553G.

Liu, Hongyu, Jiaojiao Shang, Yafang Wang, Yazhou Wang, Jianwu Lan, Baojie Dou, Lin Yang, and Shaojian Lin. 2022. "Ag/AgCl Nanoparticles Reinforced Cellulose-Based Hydrogel Coated Cotton Fabric with Self-Healing and Photo-Induced Self-Cleaning Properties for Durable Oil/Water Separation." *Polymer* 255: 125146. 10.1016/j.polymer.2022.125146.

Liu, Wenru, Han Zhou, Wei Zhao, Caixia Wang, Qian Wang, Jianfang Wang, Peng Wu, Yaoliang Shen, Xiaoming Ji, and Dianhai Yang. 2022. "Rapid Initiation of a Single-Stage Partial Nitritation-Anammox Process Treating Low-Strength Ammonia Wastewater: Novel Insights into Biofilm Development on Porous Polyurethane Hydrogel Carrier." *Bioresource Technology* 357 (August): 127344. 10.1016/j.biortech.2022.127344.

Liu, Nian, Panyu Ren, Atif Saleem, Wei Feng, Jingjing Huo, Huifang Ma, Sheng Li, Peng Li, and Wei Huang. 2021. "Simultaneous Efficient Decontamination of Bacteria and Heavy Metals via Capacitive Deionization Using Polydopamine/Polyhexamethylene Guanidine Co-Deposited Activated Carbon Electrodes." *ACS Applied Materials & Interfaces* 13 (51): 61669–61680. 10.1021/acsami.1c20145.

Loo, Siew Leng, Lía Vásquez, Athanassia Athanassiou, and Despina Fragouli. 2021. "Polymeric Hydrogels—A Promising Platform in Enhancing Water Security for a Sustainable Future." *Advanced Materials Interfaces.* John Wiley and Sons Inc 8 (24): 2100580. 10.1002/admi.202100580.

Madhuranthakam, Chandra Mouli R., Amal Alsubaei, and Ali Elkamel. 2018. "Performance of Polyacrylamide and Poly(Acrylamide/Sodium Acrylate) Hydrogel-Coated Mesh for Separation of Oil/Water Mixtures." *Journal of Water Process Engineering* 26 (October): 62–71. 10.1016/j.jwpe.2018.09.009.

Meijide, Jessica, Sandra Rodríguez, M. Angeles Sanromán, and Marta Pazos. 2018. "Comprehensive Solution for Acetamiprid Degradation: Combined Electro-Fenton and Adsorption Process." *Journal of Electroanalytical Chemistry* 808 (January): 446–454. 10.1016/j.jelechem.2017.05.012.

Nemati, M, S M Hosseini, and M Shabanian. 2017. "Novel Electrodialysis Cation Exchange Membrane Prepared by 2-Acrylamido-2-Methylpropane Sulfonic Acid; Heavy Metal Ions Removal." *Journal of Hazardous Materials* 337: 90–104. 10.1016/j.jhazmat.2017.04.074.

Nikolaeva, Daria, Christian Langner, Ahmad Ghanem, Mona Abdel Rehim, Brigitte Voit, and Jochen Meier-Haack. 2015. "Hydrogel Surface Modification of Reverse Osmosis Membranes." *Journal of Membrane Science* 476 (February): 264–276. 10.1016/J.MEMSCI.2014.11.051.

Qin, Detao, Zhaoyang Liu, Zhi Liu, Hongwei Bai, and Darren Delai Sun. 2018. "Superior Antifouling Capability of Hydrogel Forward Osmosis Membrane for Treating Wastewaters with High Concentration of Organic Foulants." *Environmental Science and Technology* 52 (3): 1421–1428. 10.1021/acs.est.7b04838.

Salehi, Ali Akbar, Mohammad Ghannadi-Maragheh, Meisam Torab-Mostaedi, Rezvan Torkaman, and Mehdi Asadollahzadeh. 2020. "Hydrogel Materials as an Emerging Platform for Desalination and the Production of Purified Water." *Separation and Purification Reviews* 50 (4): 1–20. 10.1080/15422119.2020.1789659.

Shakeri, Alireza, Hasan Salehi, Mahdi Taghvay Nakhjiri, Ehsan Shakeri, Neda Khankeshipour, and Farnaz Ghorbani. 2019. "Carboxymethylcellulose-Quaternary Graphene Oxide Nanocomposite Polymer Hydrogel as a Biodegradable Draw Agent for Osmotic Water Treatment Process." *Cellulose* 26 (3): 1841–1853. 10.1007/S10570-018-2153-0.

Shannon, Mark A, Paul W Bohn, Menachem Elimelech, John G Georgiadis, Benito J Mariñas, and Anne M Mayes. 2008. "Science and Technology for Water Purification in the Coming Decades." *Nature* 452 (7185): 301–310. 10.1038/nature06599.

Shen, Liang, Ziheng Jin, Wenhao Xu, Xia Jiang, Yue Xiao Shen, Yuanpeng Wang, and Yinghua Lu. 2019. "Enhanced Treatment of Anionic and Cationic Dyes in Wastewater through Live Bacteria Encapsulation Using Graphene Hydrogel." *Industrial and Engineering Chemistry Research* 58 (19): 7817–7824. 10.1021/acs.iecr.9b01950.

Shi, Wenhui, Haibo Li, Xiehong Cao, Zhi Yi Leong, Jun Zhang, Tupei Chen, Hua Zhang, and Hui Ying Yang. 2016. "Ultrahigh Performance of Novel Capacitive Deionization Electrodes Based on A Three-Dimensional Graphene Architecture with Nanopores." *Scientific Reports* 6 (January): 18966. 10.1038/srep18966.

Sinha, Vibha, and Sumedha Chakma. 2019. "Advances in the Preparation of Hydrogel for Wastewater Treatment: A Concise Review." *Journal of Environmental Chemical Engineering* 7 (5): 103295. 10.1016/j.jece.2019.103295.

Tokuyama, Hideaki, and Takahiko Iwama. 2007. "Temperature-Swing Solid-Phase Extraction of Heavy Metals on a Poly(N -Isopropylacrylamide) Hydrogel." *Langmuir* 23 (26): 13104–13108. 10.1021/la701728n.

Trikkaliotis, Dimitrios G, Achilleas K Christoforidis, Athanasios C Mitropoulos, and George Z Kyzas. 2020. "Adsorption of Copper Ions onto Chitosan/Poly(Vinyl Alcohol) Beads Functionalized with Poly(Ethylene Glycol)." *Carbohydrate Polymers* 234: 115890. 10.1016/j.carbpol.2020.115890.

UNESCO World Water Assessment Program, 2019. 2019. "WWAP (UNESCO World Water Assessment Program). 2019. The United Nations World Water Development Report 2019: Leaving No One Behind. Paris, UNESCO." *UNESCO* 186.

Ussia, Martina, Alessandro Di Mauro, Tommaso Mecca, Francesca Cunsolo, Giuseppe Nicotra, Corrado Spinella, Pierfrancesco Cerruti, Giuliana Impellizzeri, Vittorio Privitera, and Sabrina C. Carroccio. 2018. "ZnO-PHEMA Nanocomposites: An Ecofriendly and Reusable Material for Water Remediation." *ACS Applied Materials and Interfaces* 10 (46): 40100–40110. 10.1021/acsami.8b13029.

Van Tran, Vinh, Duckshin Park, and Young Chul Lee. 2018. "Hydrogel Applications for Adsorption of Contaminants in Water and Wastewater Treatment." *Environmental Science and Pollution Research.* Springer Verlag. 10.1007/s11356-018-2605-y.

Wan Ikhsan, Syarifah Nazirah, Norhaniza Yusof, Farhana Aziz, Ahmad Fauzi Ismail, Juhana Jaafar, Wan Norharyati Wan Salleh, and Nurasyikin Misdan. 2021. "Superwetting Materials for Hydrophilic-Oleophobic Membrane in Oily Wastewater Treatment." *Journal of Environmental Management* 290 (July): 112565. 10.1016/J.JENVMAN.2021.112565.

Wittmar, Alexandra S.M., Jonathan Klug, and Mathias Ulbricht. 2020. "Cellulose/Chitosan Porous Spheres Prepared from 1-Butyl-3-Methylimidazolium Acetate/Dimethylformamide Solutions for Cu2+ Adsorption." *Carbohydrate Polymers* 237 (June): 116135. 10.1016/j.carbpol.2020.116135.

Xing, Wenle, Jie Liang, Wangwang Tang, Di He, Ming Yan, Xiangxi Wang, Yuan Luo, Ning Tang, and Mei Huang. 2020. "Versatile Applications of Capacitive Deionization (CDI)-Based Technologies." *Desalination* 482: 114390. 10.1016/j.desal.2020.114390.

Xu, Xiaoyi, Zhaoxia Jin, Bin Wang, Chenpei Lv, Bibo Hu, and Dezhi Shi. 2017. "Treatment of High-Strength Ammonium Wastewater by Polyvinyl Alcohol–Sodium Alginate Immobilization of Activated Sludge." *Process Biochemistry* 63 (December): 214–220. 10.1016/j.procbio.2017.08.016.

Yang, Zhe, Daisuke Saeki, Hao Chen Wu, Tomohisa Yoshioka, and Hideto Matsuyama. 2019. "Effect of Polymer Structure Modified on RO Membrane Surfaces via Surface-Initiated ATRP on Dynamic Biofouling Behavior." *Journal of Membrane Science* 582 (July): 111–119. 10.1016/J.MEMSCI.2019.03.094.

Yasin, Ahmed S, Ahmed Yousef Mohamed, Dong Hyun Kim, Thi Luu Luyen Doan, S.S. Chougule, Namgee Jung, Sungchan Nam, and Kyubock Lee. 2021. "Design of Zinc Oxide Nanoparticles and Graphene Hydrogel Co-Incorporated Activated Carbon for Efficient Capacitive Deionization." *Separation and Purification Technology* 277 (December): 119428. 10.1016/j.seppur.2021.119428.

Yu, Yaru, Ying Shu, and Lin Ye. 2018. "In Situ Crosslinking of Poly (Vinyl Alcohol)/ Graphene Oxide-Glutamic Acid Nano-Composite Hydrogel as Microbial Carrier:

Intercalation Structure and Its Wastewater Treatment Performance." *Chemical Engineering Journal* 336 (March): 306–314. 10.1016/j.cej.2017.12.038.

Zeng, Yao, Ling Qiu, Kun Wang, Jianfeng Yao, Dan Li, George P. Simon, Rong Wang, and Huanting Wang. 2013. "Significantly Enhanced Water Flux in Forward Osmosis Desalination with Polymer-Graphene Composite Hydrogels as a Draw Agent." *RSC Advances* 3 (3): 887–894. 10.1039/C2RA22173J.

Zhang, Yanyu, Giorgia Daniel, Sonia Lanzalaco, Abdirisak Ahmed Isse, Alessandro Facchin, Aimin Wang, Enric Brillas, Christian Durante, and Ignasi Sirés. 2022. "H2O2 Production at Gas-Diffusion Cathodes Made from Agarose-Derived Carbons with Different Textural Properties for Acebutolol Degradation in Chloride Media." *Journal of Hazardous Materials* 423: 127005. 10.1016/j.jhazmat.2021.127005.

Zhang, Zhibin, Hiroshi Fu, Zheng Li, Jianying Huang, Zhiwei Xu, Yuekun Lai, Xiaoming Qian, and Songnan Zhang. 2022. "Hydrogel Materials for Sustainable Water Resources Harvesting & Treatment: Synthesis, Mechanism and Applications." *Chemical Engineering Journal* 439: 135756. 10.1016/j.cej.2022.135756.

Zhao, Fei, Xingyi Zhou, Ye Shi, Xin Qian, Megan Alexander, Xinpeng Zhao, Samantha Mendez, Ronggui Yang, Liangti Qu, and Guihua Yu. 2018. "Highly Efficient Solar Vapour Generation via Hierarchically Nanostructured Gels." *Nature Nanotechnology* 13 (6): 489–495. 10.1038/s41565-018-0097-z.

Zhou, Xingyi, Hengyi Lu, Fei Zhao, and Guihua Yu. 2020. "Atmospheric Water Harvesting: A Review of Material and Structural Designs." *ACS Materials Letters* 2 (7): 671–684. 10.1021/acsmaterialslett.0c00130.

Zhou, Xingyi, Fei Zhao, Youhong Guo, Yi Zhang, and Guihua Yu. 2018. "A Hydrogel-Based Antifouling Solar Evaporator for Highly Efficient Water Desalination." *Energy and Environmental Science* 11 (8): 1985–1992. 10.1039/c8ee00567b.

19 Hydrogels for Soft Robotics

Akhiri Zannat
University of South Asia, Dhaka, Bangladesh

Zinnat Morsada
Clarkson University, Potsdam, New York, USA

Md. Milon Hossain
Cornell University, Ithaca, New York, USA

19.1 INTRODUCTION

Robots are widely used in the medical field, industrial assistance, and even in daily life to mimic the usual or repetitive motion mostly inspired by biological organisms. However, the conventional mechanisms and components of robots are rigid and not feasible for human-robot interfaces/assistive technologies. Soft robots, unlike their traditional counterpart, are compliant structures constructed from soft materials and can be designed on different scales. In the past decade, soft robotic research progressed exponentially due to the increasing demand for seamless human-machine interactions. Additionally, soft-bodied robots have a high degree of freedom and reduced mechanical mismatch between robotic components and living tissues resulting in safe and comfortable interaction with humans [1]. Different soft and compliant materials have been explored for the fabrication of soft robots. Hydrogels made of water and hydrophilic polymers are promising materials because of their high mechanical compliance as a biological system, good ionic conductivity, biocompatibility, and accessibility to various molecules. Hydrogels respond to different environmental stimuli through changes in hydrogen bonding, electrostatic interactions, and van der Waals forces [2]. For example, actuation of soft robots based on hydrogels use polymer-solvent interaction, polymer-ion interactions, and polymer elasticity. As a result, shape transformation, mechanical movements, and even actuation at the micro level are possible by the hydrogel-based actuators. A soft robot usually consists of three components: an actuator for providing motion, a sensor to create and analyze signals, and a control unit to supply energy or maintain the whole system [3]. Though the integration of hydrogels in all of these units is still challenging, hydrogels have expanded from actuators to the control unit of the soft machines. Indeed, there should be some rigid parts of any robotic system; the mechanical and electrical wiring and coupling between the hard and soft parts of any device can perform complementary functions. As an integrated part, the hydrogel can show movements or mechanical motion responding

DOI: 10.1201/9781003340485-19

to various stimuli, sense different changes, perform as semiconductors, and convert energy [4,5]. This chapter briefly discusses the fundamentals of hydrogels, including different types of hydrogels. Various components of soft robots fabricated by hydrogels, such as actuators, sensors, and control units, are highlighted. Patterning and manipulation of hydrogels in different scales are discussed before concluding the chapter.

19.2   FUNDAMENTALS OF HYDROGELS

A hydrogel typically indicates a three-dimensional interconnected network of polymers with hydrophilic groups in its chemical structure. The unique character of this polymer network is to hold a very good amount of water and simultaneously retain its structure because of the cross-linking of polymer chains when water must be at least 10% of its weight. Hydrogels may consist of cross-linked micro-sized water-soluble monomers, nano-sized nano-fibrils/nanotubes, or polymers as the building blocks, as shown in Figure 19.1.

Cross-linking, the most important parameter that prevents the hydrogels from being dissolved into water or any aqueous medium, may be either physical or chemical. In chemical cross-linking, the structure uses covalent bonds, e.g., disulfide linkage, whereas, in physical cross-linking, intermolecular forces, e.g., hydrogen bonding, or physical entanglement is used. The degree of cross-linking can be determined by the distance between cross-links where water is absorbed. Water absorption is another essential parameter of hydrogels that indicates the ability to absorb a wide range of water, i.e., 10–95% by hydrogel structures. The hydrophilic groups that make it water absorbent are usually $-NH2$, $-COOH$, $-OH$, $-CONH2$, $-CONH$, and $-SO3H$ [6]. The water absorption property of hydrogels can be influenced by factors such as temperature, pH, ionic strength, and the nature of the solvent. The degree of water intake or the polymeric volume of hydrogels in the swollen state can be calculated by the following formula:

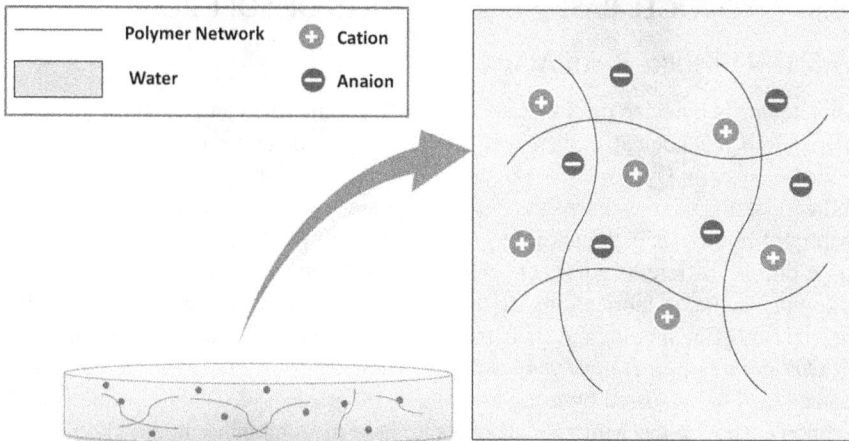

FIGURE 19.1   Simplified illustration of the polymer network of the hydrogel.

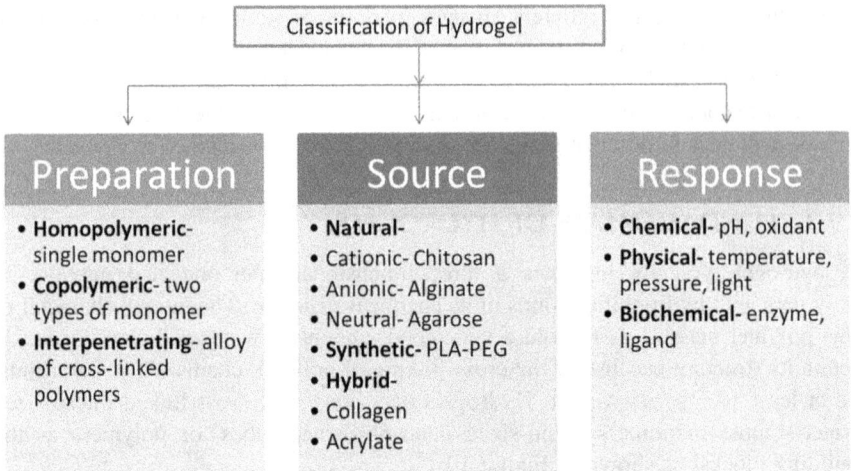

FIGURE 19.2   Classification of hydrogels.

$$S = \frac{m_t - m_0}{m_0} \times 100$$

where S indicates the percentage of swelling, m_t indicates net weight, and m_0 indicates the dry weight of the hydrogel. The water content in a hydrogel can be explained in three states, i.e., i) strongly bound water, ii) weekly bound water, and iii) free (non-bound) water. The physical structure of a hydrogel is heterogeneous and depends on some factors such as types of monomer, synthesis method, solvent conditions, and mechanical loading. Hydrogels can be classified based on preparation, source, and response [6]. The classification is illustrated in Figure 19.2.

19.3   HYDROGEL-BASED COMPONENTS OF SOFT ROBOTICS

19.3.1   HYDROGEL-BASED ACTUATORS

An actuator is a device or component of any machine that can convert any form of energy into mechanical movements or the creation of any motion to the specific device or system. Hydrogel-based actuators are very reliable for their fast responses and good deformations due to the flexible polymer network. The most advantageous feature of hydrogels is the ability to respond to both physical and chemical stimuli. In addition, different types of engineering techniques are mentioned in the following actuator fabrication. Different types of hydrogel-based actuators, their mechanism of actuation, and their response rate are compared in Table 19.1.

Thermally responsive hydrogel actuators show deformations against a temperature gradient and are prepared by using two phases by gradient [2,10], bi-layering [11,12], or interpenetrating mechanisms, where each phase may fall either in the lower critical solution temperature (LCST) or upper critical solution temperature (UCST) [1,18]. Poly (N-isopropyl acrylamide) (pNIPAM) is a good example that describes LCST by

TABLE 19.1

Details of the Different Hydrogel-Based Actuators

Mechanism of Actuation	Hydrogel	Structure	Rate of Actuation	Reference
pH Value	PAAc	Bilayer	1.6/s	[2]
	PDMA-PAAc	Bilayer	0.16/s	[7]
	PAAm-PAAc	Bilayer	0.56/min	[8]
	PNIPAm-PAAm	Bilayer	0.25/s	[9]
Temperature	PNIPAm-TCNCs	Gradient	4.8/s	[2]
	GO-PDMAEMA	Gradient	2.7/s	[10]
	PNIPAm-PAAc	Bilayer	6.7/s	[11]
	Alg-PDMAEMA	Bilayer	5.0/min	[12]
Electrical Field	P(AMPS-co-AA)	Interpenetrating network	0.80/s	[13]
	PAMPS-PAAm	Uniform	0.73/s	[14]
	PAMPS-PAAm-GO	Uniform	0.044/(mm*min)	[15]
Light	PEG-aCD-Sti	Uniform	1.2/s	[2]
	PNIPAm-GO	Gradient	7.5/s	[12]
Magnetic Field	Fe_3O_4-MWCNT-PNIPAm	Bilayer	0.60/s	[16]
	Fe^{3+}-PNIPAm	Bilayer	0.12/(mm*min)	[17]

NP = nanoparticles; MWCNT = multiwalled carbon nanotube; PDMA = poly(N,N-dimethylacrylamide); GO = graphene oxide; aCD = a-cyclodextrin; Sti = stilbene; TCNC = tunicate cellulose nanocrystals; Alg = alginate; PMAAc = poly(methacrylic acid); POEGMA = poly(oligo(ethylene glycol) methacrylate)

displaying shrinkage at a higher temperature than the critical temperature and returning to a normal state by swelling at lower temperatures [18]. On the other hand, poly (acrylic acid-co-acrylamide) shows the opposite reactions of swelling at higher temperatures and shrinkage at lower temperatures (Figure 19.3a) [18]. Traditional mechanisms required the use of aqueous media for the swelling activity of thermally responsive hydrogels to actuate any device. However, this limitation can be decreased to a large extent by using such bilayer hydrogel-based structures that facilitate the transformation of the water molecules from the LCST to the UCST layer. Consequently, the actuation process can be performed even in a non-aqueous medium.

Chemically responsive hydrogel actuators, when exposed to the chemical environment, such as various solvents, pH, and biomolecules, can affect the actuation of the polymeric structure of a hydrogel, providing actuation. For example, a polymer of intrinsic micro-porosity demonstrates a good attraction for the solvents, such as acetone, ethanol, and dimethyl sulfoxide (DMSO), while maintaining a hydrophobic nature to a large extent that eventually provides actuation in the solvents rather than the water. The mechanism of the good interaction of a polymer network and solvents (Figure 19.3b) [19] is strengthened by using a modulus gradient, combining other material's layers, and altering the geometric design to change or modify the actuation

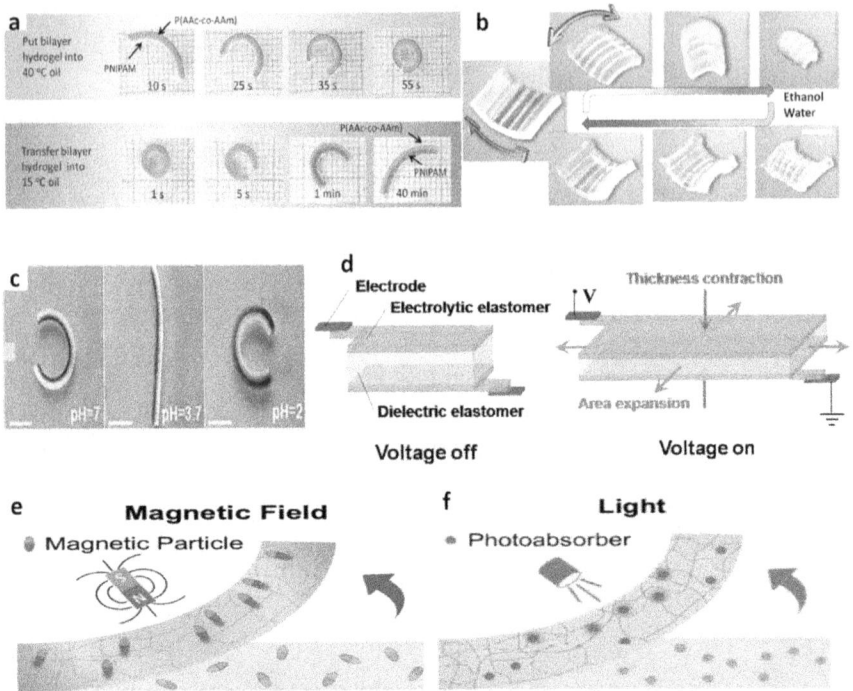

FIGURE 19.3 Working mechanisms of various stimuli responsive actuators. a) Actuation mechanism of the temperature-responsive hydrogel at different time considerations [18]. b) Representation of the hydrogel-based actuators that respond to the solvent [19]. c) Illustration of the working technique of the pH-responsive hydrogels [1]. d) Schematic representation of the actuation responding to the electricity [1]. e) Working mechanism of a magnetic field for actuators. Adapted with permission from [2]. Copyright (2018) John Wiley and Sons. f) Working procedure and gripping technique by an actuator under the effect of light [1,2].

direction [19,20]. Volumetric changes toward bio-molecular components can be used in achieving actuation by the cross-linking of biomolecular complexes in the polymer network of the hydrogel. Once the biomolecular complexes are grafted into the polymer network, the dynamic cross-links are used for the dissociation and association of polymer chains when the structure is in contact with any target biomolecules. The target molecules can be an antigen, glucose, DNA, bisphenol A (BPA), etc. depending on the biomolecular complexes used. Cross-linking length/density are the main factors for the volumetric phase transition of the hydrogel. A pH-responsive polymer, due to the existence of acidic or basic groups in its chemical structure, can respond to pH-based changes by attaining or losing protons. Polyelectrolytes can be considered as such examples of polymers that contain a large number of ionizable groups. In a basic environment, anionic polyelectrolytes lose protons, causing an increase in electrostatic repulsions between the chains that eventually allow water molecules to enter and swell the hydrogel [2]. However, the acidic media helps an acidic polymer to achieve protons, causing a decrease in charge density and a collapse in the volume of the hydrogel. Because of containing both hydrophilic and hydrophobic parts, amphiphilic hydrogels can show a two-phase transition, regardless of the acidic and basic environmental conditions (Figure 19.3c) [1]. The actuation motion of the hydrogel between the extreme expansion and contraction takes place within a very small apparent dissociation constant, pKa, of the polymer that is almost similar to the pKa of the ionizable groups [21].

Electrically responsive hydrogel actuators can achieve quicker and more accurate responses than those requiring stimuli conversions, due to the application of Maxwell stress. In electrically responsive hydrogel actuators, a dielectric elastomeric layer is sandwiched between conductive polymeric layers or electrodes (Figure 19.3d) [1,22]. When voltage is applied to the electrodes, ions of opposite charges gather alternately on both sides, i.e. positive charge on one side and negative on the other. It causes a local contraction due to the attraction of the opposite charges and expansion on the other side. This type of actuator provides greater design flexibility, fast response, and greater deformations. Osmotic pressure is also used for tuning the actuation of polyelectrolyte hydrogel where ionic charge plays the role of actuation instead of electric charge. The polymer networks such as poly(2-Acrylamido-2-methyl-1-propane-sulfonic acid-acrylic acid P(AMPS-co-AA) [13], and poly(2-acryloyl amino-2-methyl-1-propane sulfonic acid-acrylamide) (PAMPS-PAAm) [14] are preferred due to the availability of charged functional groups and mobile counter ions in their chemical structure. When an electric field is applied to the polymer structure in an aqueous medium, the charges opposite to the counter ions of the polymer networks migrate to the electrode. To satisfy charge neutrality, ions in an aqueous medium with the opposite charge as the counter ions of the polyelectrolyte hydrogel simultaneously migrate toward the same electrode. As a result, the creation of ionic gradients leads to the swelling of polyelectrolyte hydrogels.

Magnetically responsive hydrogel actuators can respond to the external magnetic field when they are filled with some magnetic particles, such as ferrous oxide (Fe_2O_3), ferric oxide (Fe_3O_4), and cobalt ferrite ($CoFe_2O_4$). The nano/micro-sized magnetic particles are dispersed in the hydrogel and when the hydrogel-based actuators come in contact with the magnetic field, the magnetic particles receive the magnetic force that eventually is transmitted to the hydrogel network, providing volumetric changes (Figure 19.3e) [2,16,17].

Optically responsive hydrogel actuators follow the mechanism of shape transformation by light irradiation. The mechanism does not require any energy transmission, but rather responds to lights of different wavelengths, as shown in (Figure 19.3f) [2]. In chemistry, moiety means a part of a molecule provided with a particular name because that attached molecule can be a part of other molecules as well. Some such moieties like spiropyan and azobenzene can respond to light and reversible isomerization, respectively. Spiropyan can dissociate when it is under an ultraviolet ray and go back into its ring structure while in visible light. Copolymerization of the spiropyan group and the acrylic groups in the hydrogel structure can yield the fastest photo-responsiveness [23]. On the other hand, the presence of azobenzene ensures the fastest bending toward the light sources under UV rays. The main mechanism of this bending is the isomerizations of the azobenzene that can influence the length of the polymer chain and the cross-linking density in the polymeric structure that eventually transform the trans structure in visible light into the cis structure in the UV light.

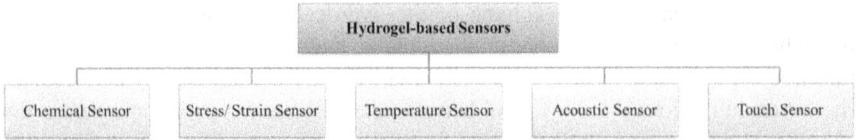

FIGURE 19.4   Classification of different sensors.

Responsiveness of the hydrogels toward the optical and the thermal stimuli simultaneously have been studied with the hydrogel of pNIPAM. However, the integration of extra particles such as graphene oxide (GO) [12] into the hydrogel structure can increase the responsiveness of the actuators. Carbon nanotubes and gold nanoparticles are used as the additive particles inside the hydrogel structure because of their surface plasmon resonance. When the frequency of light matches the resonance frequency of the electrons, the nanoparticles can have strong absorption and scattering. Because of this plasmonic effect, the hydrogel can absorb and respond to the rays of selective wavelengths, causing fast actuation [1]. Light-controlled skeletal muscle-powered actuators can possibly be fabricated by using the optical responsive hydrogel that can contract under the light of a particular wavelength.

Different types of sensors are used with soft robots to perform various functions. These sensors can monitor or record different changes to the stimuli and can be classified as shown (Figure 19.4). The following section will highlight different sensors and their mechanisms based on the signals received.

Hydrogel-based chemical sensors have the capability to distinguish different analytes and convert them into electrical and optical signals. To get the optical signals, the hydrogels are modified at the nanoscale by changing thickness while layering or using different spaces in the periodic structure of hydrogels. The formation of the structure should be in a way that can fulfill the requirements of Bragg's law that can diffract imposed visible light rays to demonstrate the changes in the color of the hydrogels when they undergo swelling and deswelling because of the chemical stimuli absorption [24]. In Figure 19.5a,b, [24] a chemical sensor is illustrated to show the color changes of the hydrogel upon undergoing the chemical stimuli. When the additive particles come in contact with the chemical environment, the particles show changes accordingly [1]. The changes are induced in the transmission of the visible lights through the transparent hydrogels, creating optical signals. Metal nanoparticles, organic dyes, and bacterial cells are good examples of chemically responsive additives for wound monitoring, wearable patches, and in vivo cellular cytotoxicity sensors (Figure 19.5c,d) [1]. Following Bragg's law, optical signals can also be found by temperature-responsive hydrogel-based actuators. The application of hydrogel-based chemical sensors is used in different soft machines, due to their portability, ease of operation, and cost. Various biomolecular components of the human body, such as ions, pH, glucose, lactate, and uric acid, are monitored and measured by chemical sensors for proper health monitoring. Sweat in the human body during various physical activities or rest time shows respective analyte profiles. The hydrogel-based sensor can detect these profiles by electrochemical detection method or colorimetric detection method [25].

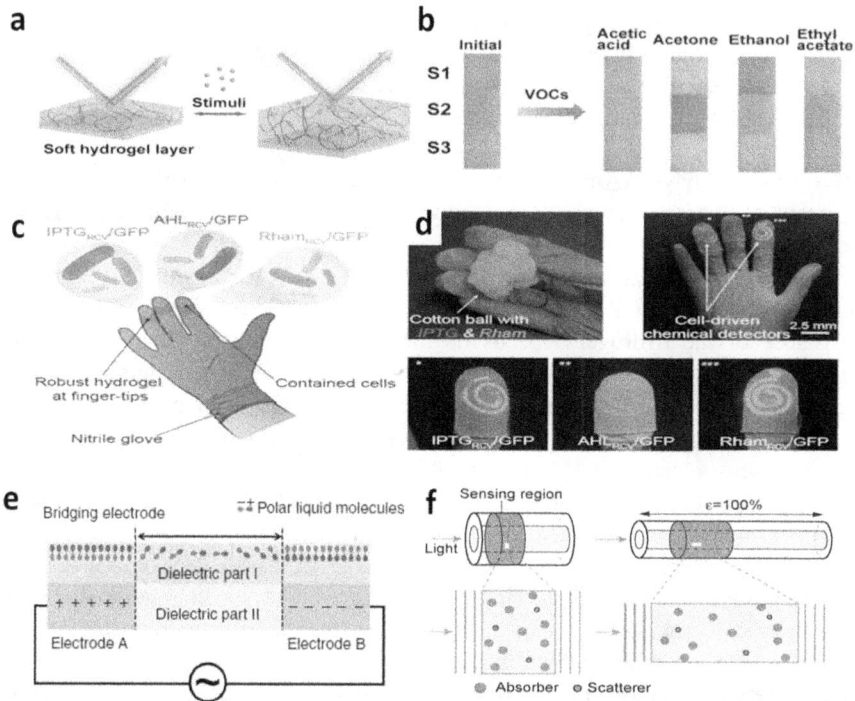

FIGURE 19.5 Different sensors creating optical and electrical signals. a, b) Light diffraction by hydrogel-based chemical sensor due to the responsiveness to various solvents. Adapted with permission from [24] Copyright (2018) John Wiley and Sons. c) Illustration of bacterial-cell integrated hydrogel that works as a chemical sensor and creates green fluorescence when comes to chemicals [27]. d) Wearing a chemical sensor based patched on the finger [27]. e) Hydrogel-based temperature sensor showing a decrease in resistance when it is heated and returned to the original state after cooling [1]. f) Illustration of the optical loss/reflected light due to the strain energy when the hydrogel is stretched and contracted [1].

In the electrochemical method, current or potential is measured by functionalized electrodes using sweat as an electrolyte. In colorimetric methods, reagents are used so that they can show color changes when coming in contact with the analytes. Electric signals can be achieved by chemical stimuli conversion. When two ion-based hydrogels are used as electrodes separated by dielectric materials and connected in an AC source, the structure can work similarly to a capacitor (Figure 19.5e) [1]. The capacitance power of the structure can verify the polarity of the hydrogels, creating electric signals.

Hydrogel-based strain sensors exhibit high stretchability and sensitivity due to their favorable structures. Alginate-polyacrylamide hydrogels are good examples of these characteristics when acrylamide (AAm) is used as a doping molecule. The sample is prepared by polymerizing and cross-linking alginate-polyacrylamide as a precursor solution. Depending on the concentration of acrylamide in the hydrogel, changes in the optical loss can be observed when the sample is strained (Figure 19.5f) [1]. Further, the dye molecules can be incorporated into different parts of the hydrogel. Each dye is able to sense the stretching applied to each region based on its wavelength when stretched. Different types of materials, such as polyelectrolytes, nanoparticles, inorganic electrolytes, and biomolecules, are incorporated into the hydrogels to fabricate flexible strain sensors. PEDOT: PSS–polyAAm array, polyaniline (PANI)–poly(AAm-co-HEMA), and polyNIPAAm/PANI are some conductive hydrogels that are used in strain sensors [26]. These sensors are suitable for long-term sensing, are biocompatible, and possess self-healing properties.

Hydrogel-based temperature sensors follow an easy mechanism of only attaching a hydrogel to the LCR meter, as shown in Figure 19.6a,b [28]. The electrical resistance of the hydrogel depends on the electron migration in it. As electron migration is directly dependent on the temperature, the changes in the temperature can be measured by the measurement of the electron migration with an LCR meter.

Hydrogel-based touch sensors can convert touch sensation into electric signals. A touch sensor can be either surface capacitive or triboelectric. Hydrogels and electrodes are the main components of the surface capacitive touch sensor. The electrodes have been integrated with the hydrogel so that they are evenly distributed throughout the structure [29]. Hydrogels and electrodes can be made from polyacrylamide and lithium chloride salts, respectively. The whole fabricated panel is subjected to the same voltage. Because the human body is connected to the ground, when a conductor like a finger touches the panel, a potential difference is created between the touch point and the hydrogel panel [29]. Due to this potential difference, current flows from electrodes to the finger. Shorter distances between

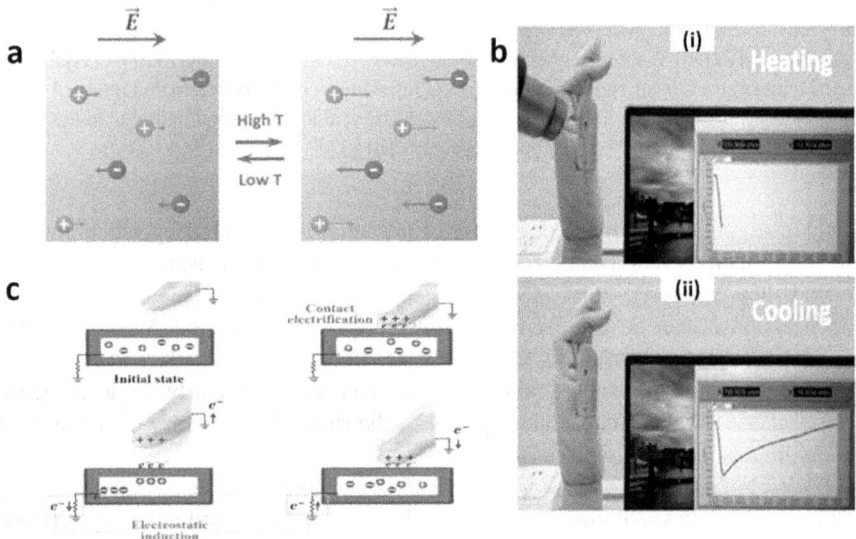

FIGURE 19.6 Different sensors creating electrical signals. a,b) Schematic representation of the capacitive stress-strain sensor [28]. c) Schematic representation and image of triboelectric touch sensor [1].

electrodes and touchpoints lead to more current flowing. It is possible to determine the location of the touched place by comparing the magnitudes of the current flow.

Hydrogels surrounded by dielectric elastomeric material can be used as triboelectric touch sensors that generate and electrify charges in response to touch [1]. When two surfaces encounter friction, opposite static charges are produced on both surfaces according to their positions in the triboelectric series. A voltage is produced when conductive material connects the surfaces and charges can pass (Figure 19.6c) [1].

19.4 MORPHING OF HYDROGEL-BASED STRUCTURE FOR SOFT ROBOTICS

The fundamental form of the hydrogels are based on chemical synthesis that must follow other subsequent manipulations to perform the shape-changing activities. Hydrogels are either organized to assist in some actuation mechanism such as bending, twisting, folding, and extension or incorporated with other components. Some examples of structuring include composites of soft and rigid materials, spatially related microstructures, gradients, and geometrical shapes. The details of different hydrogel patterning with the responsiveness to stimuli to provide shape changes are described in the following section.

19.4.1 FOLDING AND BENDING

When several layers of hydrogels of different features are combined following a particular sequence and orientation, internal stress is created that results in bending or structural deformation (Figure 19.7a,b) [30,31]. For example, two layers of soft and rigid hydrogels possess the characteristics of swelling and non-swelling, respectively, in active conditions. When the structure undergoes the active situation, the soft portion of the hydrogel tends to expand and contract while the rigid portion creates resistance to the deformation of the soft hydrogel. These two opposite actions create the stress gradient that eventually yields deformation, even in the creation of capsules (Figure 19.7c) [32]. The deformation depends on some factors, such as the mechanical and swelling characteristics of the hydrogel layers, the shape of the structure, and geometric factors. When there is a large difference in the mechanical properties between the layers of the hydrogel, the soft portion (active layer) might not be able to deform the whole rigid portion (passive layer) due to less stress creation [32]. Consequently, a crease or wrinkle is produced in the structure, instead of any significant deformation. In the same way, when two layers of hydrogels with different swelling properties are combined, actuation in the form of deformation can be achieved due to the action of the active and passive layers. Some examples of deformation due to various factors are described in Table 19.2.

19.4.2 MICRO AND MESO PATTERNING

Hydrogel-based structures can also be patterned microscopically and mesoscopically to convey shape transformations. It is possible to combine hydrogels or

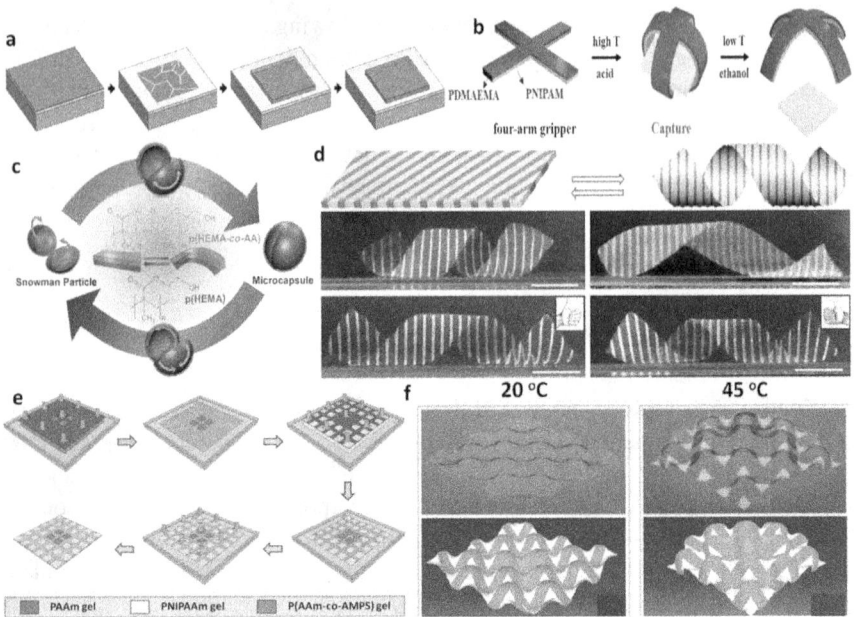

FIGURE 19.7 Illustration of different structures of hydrogels that show shape transformation. a) Fabrication of a trilayer self-folding origami. Adapted with permission from [30] Copyright (2014) John Wiley and Sons. b) Schematic representation of a bilayer hydrogel gripper. Adapted with permission from [31] Copyright (2018) Elsevier. c) Bilayer hydrogel that can perform reversible actuation. Adapted with permission from [32] Copyright (2018) John Wiley and Sons. d) Patterning of two hydrogels following side-by-side placement at a specific angle and helical shape transformation of the fabricated structure. Adapted with permission from [33] Copyright (2011) John Wiley and Sons. e) Illustration of the multi-step polymerization to fabricate a patterned hydrogel actuator. Adapted with permission from [35] Copyright (2018) John Wiley and Sons. f) Shape changes of periodically patterned hydrogels at different temperatures. Adapted with permission from [35] Copyright (2018) John Wiley and Sons.

additives on a micro or meso scale by placing them next to each other [33], or by photolithography with a variety of photo masks to create different patterns using different sections of a single hydrogel cross-linked at different levels [35]. The main similarity between layering and micro and meso patterning is their differences in mechanical and swelling properties. Depending on the characteristics of the hydrogels, stress is developed across the structure of the hydrogel and its shape changes. Different lengths, shapes, thicknesses, and placement angles of the side-by-side hydrogels (Figure 19.7d) [33] influence shape transformation such as bending, twisting, rolling, and folding. In addition, 3D shapes can be created by combining two fiber-like regions of hydrogels with different elastic modulus, shrinkage, and chemical properties [36]. The design follows the photopatterning method, where an alternate alignment of two hydrogels such as pNIPAM and pNIPAM/2-acrylamide-2-methyl propane sulfonic acid (pAMPS) gel is ensured at 30°, 60°, and ±45° angles. Two types of findings can be explained from this study:

TABLE 19.2

Details of the Factors of Layering Techniques

Active Layer	Passive Layer	Responsiveness Factor	Actuation	Reference
p(HEMA-co-AA)	p(HEMA)	pH	Folding into capsulation	[33]
P(NIPAM-AA-BA) or P(NIPAM-BA)	PCL or P (MMABA)	Temperature	Folding	[34]
pNIPAM	poly(p-methylstyrene)	Temperature	self-folding origami	[30]
PNIPAM-PVA, PDMAEMA-PSS	–	Temperature, and pH	Reversible bending	[31]

NIPAm-AAc = N-isopropyl-acrylamide- acrylic acid; PEODA = poly-ethylene oxide diacrylate; p(HEMA-co-AA) = poly(2-hydroxyethyl methacrylate-co-acrylic acid; p(HEMA) = poly(2-hydroxyethyl methacrylate); (PCL) = polycaprolactone; P(MMABA) = poly(methylmethacrylate-co-benzophenone acrylate); PVA = poly(vinyl alcohol); PDMAEMA-PSS = poly(2-(dimethylamino) ethyl methacrylate) poly(sodium-p-styrene sulfonate)

transforming into helices shapes while in contact with the NaCl and changes in the width can provide more complex shapes. The decrease in the change of hydrogel strips in comparison to their thickness follows a ratio or pattern. If the width is decreased below a critical size compared to the thickness of the films, the helical shape transformation is superseded by the expansion of the hydrogels along the direction of interfaces. Based on the requirements of the more complex shapes, the investigations regarding combining different stimuli-responsive hydrogels and layering of already patterned sheets play a significant role. For example, the composite of two polymers, pAAc, p(P-co-AAm), and pNIPAM, can be programmed to show various deformations. Another way to get 3D shape actuation is using pAAm, pNIPAM, and p(AAm-co-AMPS), placed side by side following a different pattern by lithography technique (Figure 19.7e,f) [35].

19.4.3 ANISOTROPY USING ADDITIVES AND ALIGNMENT

The microstructure of the internal structure of many living organisms in nature reveals complex shape transformations. It is mostly the swelling and deswelling properties of fibrils orientated in different directions that cause these anisotropic movements. Programming approaches for anisotropic behavior are of growing interest to researchers in addition to layering and patterning techniques. A variety of polymer chains can be aligned within the hydrogel matrix to achieve anisotropy, such as hydrogel-fiber composites, nanoparticles, or sheet-based hydrogel composites [32]. A composite ink that consists of a hydrogel (Figure 19.8a) [37], fiber, and clay can be used for 3D printed structures that can respond to different stimuli. Different concentrations of the components, such as 0.73% NFC, 9.7% Laponite

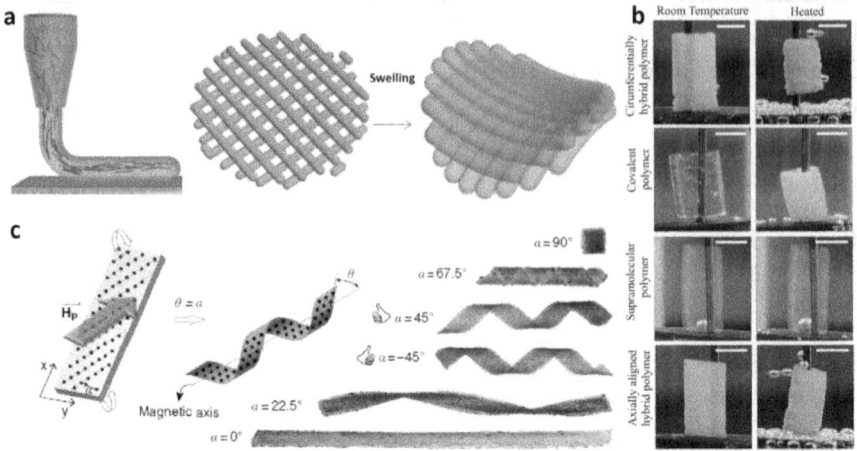

FIGURE 19.8 Illustration of different structures of hydrogels that show shape transformation. a) Schematic representation of direct ink writing by hydrogel-fibril composite and before-after swelling condition [37]. b) Image of anisotropic supramolecular hydrogels during actuation. Adapted with permission from [32] Copyright (2019) John Wiley and Sons. c) Magnetic nanocomposite hydrogel gradient structure and actuation technique. Adapted with permission from [32] Copyright (2019) John Wiley and Sons.

XLG clay, and 7.8% monomer, made the ink feasible for maintaining proper viscosity for good printing [37]. The effectiveness of the printing by the produced ink can be determined by the anisotropic elastic and swelling properties changes according to the direction of the fibrils in the hydrogel composites. To achieve the anisotropic characteristics in the hydrogel structure, nanoparticles, nanosheets, and nanorods can be considered effective additives [32]. In a study, titanate nanosheets with a thickness of 0.75 nm and length of several μm were incorporated into the NIPAM hydrogel structure [38]. A magnetic field that ensured the perpendicular alignment of the nanosheets with respect to the field was applied during polymerization. By using the anisotropic responsiveness of the produced structure to the optical stimuli, earthworm-like crawling can be achieved.

19.4.4 Gradients

Monomers and cross-linkers that have varying concentrations can be used to shape the gradient form by mechanical forces, magnetic forces, and electric forces in order to mimic nature. Two monomers with low and high concentrations are mixed in a programmable mixer and injected into a Hele-Shaw cell to produce the hydrogel structure [39]. The produced lateral radial gradients of varying cross-linked monomers have the capability to show 3D shape configuration when it is activated in the hot bath, i.e., responding toward the temperature. Copolymerization of two polymer solutions between two similar substrates can provide a hydrogel with uniform distribution of the monomers with no shape-changing capability in the acidic solution. This type of hydrogel can display the actuation or motion changes in the acidic solution. Gradients are also fabricated by the combination of stimuli-responsive hydrogels and magnetic nanoparticles (MNPs) [33]. A pre-gel solution of N-isopropylacrylamidepolyacrylamide-poly(ethylene glycol)diacrylate (NIPAM-AAm-pEGDA) and MNPs are patterned, sandwiching them between the glass mask and silicon wafer. A gradient structure is produced (Figure 19.8b,c) [33] by the gravitational sedimentation, followed by the uniform alignment and the polymerization of the hydrogel by the application of UV curing.

19.5 CONCLUSION

Hydrogels are an important class of materials with a wide range of chemical, mechanical, and electrical properties. For smooth and uninterrupted performance, soft robotics requires a flexible, soft, and stretchable integrated system. Therefore, hydrogel-based components are continuing to be incorporated into soft robotics. Hydrogels play an important role in this engineering process, which focuses on simulating various natural entities through human-robot interfaces. So many studies have been conducted in recent decades to improve the stretchability, mechanical toughness, and conductivity of hydrogels. An extensive study has been conducted on hydrogel-based actuators, sensors, and interconnects. Polymer networks of hydrogels exhibit a wide range of stimuli-responsive properties. Despite this, there are some limitations regarding manufacturability, cost, reproducibility, viability, and scalability. The focus of any hydrogel-based component's synthesis and characterization is given to a specific stimulus. However, in reality, different stimuli may be present at the same time, creating challenges. Moreover, in many cases, the viability of the studied structure remains uncertain outside such a highly restrictive environment.

REFERENCES

[1] Y. Lee, W. J. Song, and J. Y. Sun, "Hydrogel soft robotics," *Mater. Today Phys.*, vol. 15, p. 100258, 2020.

[2] M. Ding *et al.*, "Multifunctional soft machines based on stimuli-responsive hydrogels: from freestanding hydrogels to smart integrated systems," *Mater. Today Adv.*, vol. 8, 2020.

[3] J. Xiong, J. Chen, and P. S. Lee, "Functional fibers and fabrics for soft robotics, wearables, and human–robot interface," *Adv. Mater.*, vol. 33, no. 19, pp. 1–43, 2021.

[4] K. B. Kim, J. H. Han, H. C. Kim, and T. D. Chung, "Polyelectrolyte junction field effect transistor based on microfluidic chip," *Appl. Phys. Lett.*, vol. 96, no. 14, pp. 14–17, 2010.

[5] C. Yang and Z. Suo, "Hydrogel ionotronics," *Nat. Rev. Mater.*, vol. 3, no. 6, pp. 125–142, 2018.

[6] B. Shahid *et al.*, "Fundamental concepts of hydrogels: Synthesis, properties and their applications," Polymers (Basel), vol. 12, no. 11, pp. 1–60, 2020.

[7] C. Yang *et al.*, "Fabrication of a biomimetic hydrogel actuator with rhythmic deformation driven by a pH oscillator," *Soft Matter*, vol. 16, no. 12, pp. 2928–2932, 2020.

[8] S. Wu, F. Yu, H. Dong, and X. Cao, "A hydrogel actuator with flexible folding deformation and shape programming via using sodium carboxymethyl cellulose and acrylic acid," *Carbohydr. Polym.*, vol. 173, pp. 526–534, 2017.

[9] P. Sun *et al.*, "Super tough bilayer actuators based on multi-responsive hydrogels crosslinked by functional triblock copolymer micelle macro-crosslinkers," *J. Mater. Chem. B*, vol. 7, no. 16, pp. 2619–2625, 2019.

[10] W. Fan *et al.*, "Dual-gradient enabled ultrafast biomimetic snapping of hydrogel materials," *Sci. Adv.*, vol. 5, no. 4, pp. 1–7, 2019.

[11] H. Lin *et al.*, "Fabrication of asymmetric tubular hydrogels through polymerization-assisted welding for thermal flow actuated artificial muscles," *Chem. Mater.*, vol. 31, no. 12, pp. 4469–4478, 2019.

[12] Y. Yang *et al.*, "Photothermal nanocomposite hydrogel actuator with electric-field-induced gradient and oriented structure," *ACS Appl. Mater. Interfaces*, vol. 10, no. 9, pp. 7688–7692, 2018.

[13] Z. Ying, Q. Wang, J. Xie, B. Li, X. Lin, and S. Hui, "Novel electrically-conductive electro-responsive hydrogels for smart actuators with a carbon-nanotube-enriched three-dimensional conductive network and a physical-phase-type three-dimensional interpenetrating network," *J. Mater. Chem. C*, vol. 8, no. 12, pp. 4192–4205, 2020.

[14] Y. Li *et al.*, "Electric field actuation of tough electroactive hydrogels cross-linked by functional triblock copolymer micelles," *ACS Appl. Mater. Interfaces*, vol. 8, no. 39, pp. 26326–26331, 2016.

[15] C. Yang *et al.*, "Reduced graphene oxide-containing smart hydrogels with excellent electro-response and mechanical properties for soft actuators," *ACS Appl. Mater. Interfaces*, vol. 9, no. 18, pp. 15758–15767, 2017.

[16] J. C. Kuo, H. W. Huang, S. W. Tung, and Y. J. Yang, "A hydrogel-based intravascular microgripper manipulated using magnetic fields," *Sensors Actuators, A Phys.*, vol. 211, pp. 121–130, 2014.

[17] J. Tang, Q. Yin, Y. Qiao, and T. Wang, "Shape morphing of hydrogels in alternating magnetic field," *ACS Appl. Mater. Interfaces*, vol. 11, no. 23, pp. 21194–21200, 2019.

[18] J. Zheng *et al.*, "Mimosa inspired bilayer hydrogel actuator functioning in multi-environments," *J. Mater. Chem. C*, vol. 6, no. 6, pp. 1320–1327, 2018.

[19] E. Palleau, D. Morales, M. D. Dickey, and O. D. Velev, "Reversible patterning and actuation of hydrogels by electrically assisted ionoprinting," *Nat. Commun.*, vol. 4, pp. 1–7, 2013.

[20] K. Polak-Kraśna *et al.*, "Solvent sorption-induced actuation of composites based on a polymer of intrinsic microporosity," *ACS Appl. Polym. Mater.*, vol. 3, no. 2, pp. 920–928, 2021.

[21] M. Bahram, N. Mohseni, and M. Moghtader, "An introduction to hydrogels and some recent applications," *Emerg. Concepts Anal. Appl. Hydrogels*, no. April 2018, 2016.

[22] C. Keplinger, J. Y. Sun, C. C. Foo, P. Rothemund, G. M. Whitesides, and Z. Suo, "Stretchable, transparent, ionic conductors," *Science (80-.).*, vol. 341, no. 6149, pp. 984–987, 2013.

[23] K. Iwaso, Y. Takashima, and A. Harada, "Fast response dry-type artificial molecular muscles with [c2]daisy chains," *Nat. Chem.*, vol. 8, no. 6, pp. 625–632, 2016.

[24] M. Qin *et al.*, "Bioinspired hydrogel interferometer for adaptive coloration and chemical sensing," *Adv. Mater.*, vol. 30, no. 21, pp. 1–7, 2018.

[25] M. Bariya, H. Y. Y. Nyein, and A. Javey, "Wearable sweat sensors," *Nat. Electron.*, vol. 1, no. 3, pp. 160–171, 2018.

[26] D. Zhang *et al.*, "From design to applications of stimuli-responsive hydrogel strain sensors," *J. Mater. Chem. B*, vol. 8, no. 16, pp. 3171–3191, 2020.

[27] X. Liu *et al.*, "Stretchable living materials and devices with hydrogel-elastomer hybrids hosting programmed cells," *Proc. Natl. Acad. Sci. U. S. A.*, vol. 114, no. 9, pp. 2200–2205, 2017.

[28] Z. Lei and P. Wu, "large deformation," *Nat. Commun.*, no. 2019, pp. 1–9.

[29] C. C. Kim, H. H. Lee, K. H. Oh, and J. Y. Sun, "Highly stretchable, transparent ionic touch panel," *Science (80-.)*, vol. 353, no. 6300, pp. 682–687, 2016.

[30] J. H. Na *et al.*, "Programming reversibly self-folding origami with micropatterned photo-crosslinkable polymer trilayers," *Adv. Mater.*, vol. 27, no. 1, pp. 79–85, 2015.

[31] Y. Cheng, K. Ren, D. Yang, and J. Wei, "Bilayer-type fluorescence hydrogels with intelligent response serve as temperature/pH driven soft actuators," *Sensors Actuators, B Chem.*, vol. 255, pp. 3117–3126, 2018.

[32] O. Erol, A. Pantula, W. Liu, and D. H. Gracias, "Transformer hydrogels: A review," *Adv. Mater. Technol.*, vol. 4, no. 4, pp. 1–27, 2019.

[33] T. S. Shim, S. H. Kim, C. J. Heo, H. C. Jeon, and S. M. Yang, "Controlled origami folding of hydrogel bilayers with sustained reversibility for robust microcarriers," *Angew. Chemie - Int. Ed.*, vol. 51, no. 6, pp. 1420–1423, 2012.

[34] G. Stoychev, S. Zakharchenko, S. Turcaud, J. W. C. Dunlop, and L. Ionov, "Shape-programmed folding of stimuli-responsive polymer bilayers," *ACS Nano*, vol. 6, no. 5, pp. 3925–3934, 2012.

[35] P. Ma *et al.*, "Sequentially controlled deformations of patterned hydrogels into 3D configurations with multilevel structures," *Macromol. Rapid Commun.*, vol. 40, no. 3, pp. 1–5, 2019.

[36] Z. L. Wu *et al.*, "Three-dimensional shape transformations of hydrogel sheets induced by small-scale modulation of internal stresses," *Nat. Commun.*, vol. 4, pp. 1586–1587, 2013.

[37] A. Sydney Gladman, E. A. Matsumoto, R. G. Nuzzo, L. Mahadevan, and J. A. Lewis, "Biomimetic 4D printing," *Nat. Mater.*, vol. 15, no. 4, pp. 413–418, 2016.

[38] Z. Sun *et al.*, "An anisotropic hydrogel actuator enabling earthworm-like directed peristaltic crawling," *Angew. Chemie.*, vol. 130, no. 48, pp. 15998–16002, 2018.

[39] Y. Klein, E. Efrato, and E. Sharon, "Shaping of elastic sheets by prescription of non-euclidean metrics," *Science (80-.)*, vol. 315, no. 5815, pp. 1116–1120, 2007.

Index

For Product Safety Concerns and Information please contact our EU
representative GPSR@taylorandfrancis.com
Taylor & Francis Verlag GmbH, Kaufingerstraße 24, 80331 München, Germany

9 781032 375038